T0304675

An Integrated Approach to Modeling and Optimization in Engineering and Science

An Integrated Approach to Modeling and Optimization in Engineering and Science examines the effects of experimental design, mathematical modeling, and optimization processes for solving many different problems. The Experimental Design Method, Central Composite, Full Factorial, Taguchi, Box-Behnken, and D-Optimal methods are used, and the effects of the datasets obtained by these methods on mathematical modeling are investigated.

This book will help graduates and senior undergraduates in courses on experimental design, modeling, optimization, and interdisciplinary engineering studies. It will also be of interest to research and development engineers and professionals working in scientific institutions based on design, modeling, and optimization.

An Integrated Approach to Modeling and Optimization in Engineering and Science

Melih Savran and Levent Aydin

CRC Press

Taylor & Francis Group

Boca Raton London New York

CRC Press is an imprint of the
Taylor & Francis Group, an **informa** business

First edition published 2025
by CRC Press
2385 NW Executive Center Drive, Suite 320, Boca Raton FL 33431

and by CRC Press
4 Park Square, Milton Park, Abingdon, Oxon, OX14 4RN

CRC Press is an imprint of Taylor & Francis Group, LLC

ISBN: 978-1-032-78279-9 (hbk)
ISBN: 978-1-032-79978-0 (pbk)
ISBN: 978-1-003-49484-3 (ebk)

DOI: 10.1201/9781003494843

Typeset in Times
by KnowledgeWorks Global Ltd.

Contents

Preface..ix
Authors..xiii

Chapter 1   Introduction 1
      1.1   Literature Review 1
      1.2   Motivation... 6
      References 9

Chapter 2   Design of Experiment, Mathematical Modeling, and
      Optimization... 11
      2.1   Design of Experiment.. 11
      2.2   Design of Experiment Methods... 13
           2.2.1   One Variable at a Time Design 15
           2.2.2   Factorial Design (FD)... 15
           2.2.3   Central Composite Design (CCD)........................ 17
           2.2.4   Box-Behnken Design (BBD) 17
           2.2.5   Taguchi Design... 18
           2.2.6   D-Optimal Design.. 21
      2.3   Mathematical Modeling 22
           2.3.1   Neuro Regression Method (NRM)....................... 23
           2.3.2   Stochastic Neuro Regression Method (SNRM)...... 23
           2.3.3   Mathematical Models.. 24
      2.4   Mathematica and Optimization... 24
           2.4.1   NMinimize and NMaximize Functions............... 28
           2.4.2   Random Search Algorithm.................................... 30
           2.4.3   Simulated Annealing Algorithm.......................... 33
           2.4.4   Differential Evolution Algorithm 38
           2.4.5   Nelder Mead Algorithm 41
           2.4.6   FindMinimum 44
      References 46

Chapter 3   Comparison of ANN and Neuro Regression Methods in
      Mathematical Modeling 48
      Case Study 3.1.. 48
      Case Study 3.2.. 53
      References 59

Chapter 4   Evaluation of R^2 as a Model Assessment Criterion 60

Case Study 4.1 60
Case Study 4.2 66
References 70
Appendix 71

Chapter 5   Questioning the Adequacy of Using
Polynomial Structures 74

Case Study 5.1 74
Case Study 5.2 77
References 85
Appendix 86

Chapter 6   The Effect of Using the Taguchi Method in Experimental
Design on Mathematical Modeling 93

Case Study 6.1 96
References 105
Appendix 106

Chapter 7   Comparison of Different Test and Validation Methods
Used in Mathematical Modeling 142

Case Study 7.1 143
Reference 153
Appendix 153

Chapter 8   Comparison of Different Model Assessment Criteria
Used in Mathematical Modeling 212

Case Study 8.1 212
References 223
Appendix 224

Chapter 9   Comparison of the Effects of Experimental Design
Methods on Mathematical Modeling 227

Case Study 9.1 231
References 247
Appendix 248

Chapter 10 Special Functions in Mathematical Modeling 292

10.1   Bessel Function of the First Kind-$J_n(x)$ 292

10.2   Chebyshev Function-$T_n(x)$.. 293

10.3   Error Function-$erf(x)$.. 294

10.4   The Exponential Integral Function-$E_i(x)$......................... 295

10.5   Fresnel Integrals-$S(z)$ and $C(z)$ 296

10.6   Hermite Polynomial-$H_n(x)$.. 299

10.7   Hypergeometric Function... 299

10.8   Legendre Function.. 301

10.9   RamanujanTauTheta Function.. 302

10.10 Riemann-Siegel Theta Function 303

Case Study 10.1.. 304

References 311

Appendix 312

Chapter 11   Conclusion 321

Index.. 327

Preface

The fields of data science, additive manufacturing, and artificial intelligence have recently garnered heightened recognition for their significance. At present, pivotal processes such as data analysis, modeling, and the utilization of outcomes for design and optimization stand as imperative undertakings. However, reliance solely on statistics and mathematics during the modeling phase, without holistic consideration of the scientific problem within the realms of engineering or fundamental sciences such as physics, chemistry, and biology, often yields model structures that diverge from realistic outcomes. Consequently, these models lack practical applicability or are erroneously perceived as correct.

Furthermore, a pressing concern lies in the potential for misguiding practitioners when employing artificial intelligence approaches during the modeling stages of engineering problems. Succinctly put, solutions proffered by individuals possessing rudimentary knowledge in modeling, design optimization, and data science frequently yield incongruous and unviable outcomes. In addition to this, many of the available books in the field tend to be either too theoretical or present computational algorithms.

The primary objective of this book is three-fold: (1) to illuminate the pitfalls of such misapplications, (2) to present a methodological framework aimed at circumventing these issues, and (3) to underscore the practical significance of this subject matter through elucidation via engineering exemplars.

It is aimed to carry out research for the most effective and efficient use of experimental design, mathematical modeling, and optimization methods in solving engineering problems. In this context, the effects of experimental design, mathematical modeling, and optimization processes on problem-solving were examined in solving many different problems, both original problems selected from the literature and defined in the text. As the experimental design method, Central Composite, Full Factorial, Taguchi, Box-Behnken, and D-Optimal methods, which are the most common in the literature, are used, and the effects of the datasets obtained by these methods on mathematical modeling are investigated.

In the mathematical modeling process, the evaluation of methods such as Regression, Surface Response Method, and Artificial Neural Networks, which are frequently used in the literature, are included, and the applications of **Neuro Regression** and **Stochastic Neuro Regression** methods, which are introduced as new modeling methods within the scope of this book, on different problems are shown. Of the two original modeling methods introduced within the scope of the study, while Neuro Regression aims to create a mathematical model by taking advantage of artificial intelligence and regression; Stochastic Neuro Regression uses stochastic optimization techniques in addition to artificial intelligence and regression to minimize the difference between actual and predicted values and to determine the most appropriate model coefficients. Effects of different methods followed in the stage of separating the data set as training and testing on modeling and model success are investigated using k-fold cross-validation and bootstrap data separation techniques. Different model evaluation criteria are used to measure the success of mathematical models. Within the scope of the study, model success was measured by using 22 different model evaluation criteria determined as a result of the literature research, and the success criteria were compared with each other. The boundedness check criterion, introduced within this book's scope, is the only criterion that could give direct information about the usability of the model, unlike all other model success evaluation criteria.

In the optimization process, Differential Evolution, Nelder Mead, Random Search, and Simulated Annealing methods have been preferred optimization algorithms for problem solving throughout the book. In the book study, Mathematica, Matlab, DesignExpert, and Minitab programs are used for experimental design, modeling, and optimization. Comparisons of Full Factorial, D-Optimal, Central Composite, Taguchi, and Box-Benhken are made.

Another research topic in the book is whether special functions can be used as an alternative model type in mathematical modeling. Special function types including Bessel, ChebyShevT, Erf, Exp-IntegralIE, Fresnel, Hermite, HyperGeometric, LegendreP, RamanujanTauTheta, and Riemann-Siegel Theta are used for modeling, and comparisons are made with the basic mathematical functions used in the study.

The material has been divided into an introduction followed by 10 chapters.

Chapter 1 aims to define the main concepts such as Data Mining, Design of Experiment, Mathematical Modeling, Optimization, Stochastic Approaches, and Artificial Intelligence. It emphasizes that the experimental design, modeling, and optimization processes should be considered as a whole, and any negativity in one of these three stages will affect the other processes. Chapter 2 provides an introduction to experimental design, mathematical modeling, and stochastic optimization methods, discussing the necessary conditions that must be met for the correct and effective use of the most commonly known experimental design methods. Chapter 3 compares the hybrid Neuro Regression, which is proposed as a new modeling method and combines the advantages of Artificial Neural Networks (ANN) and regression approach, with ANN for original case study problems. Chapter 4 highlights that relying solely on R-squared (R^2) as a model assessment criterion is inadequate to confirm a model's success. Chapter 5 examines the limitations of polynomial structures in mathematical modeling and explores alternative models that may be required. Using engineering problems, different mathematical model types are compared and analyzed. Chapter 6 investigates the effectiveness of the Taguchi method, which is a widely used experimental design and modeling method. Chapter 7 compares the effects of methods such as cross-validation, and bootstrap used in model testing and validation on the mathematical model creation process utilizing the original modeling techniques of Neuro Regression and Stochastic Neuro Regression. In Chapter 8, the most frequently used model assessment criteria in the literature were determined, and the performance of the proposed mathematical models for a specific problem was measured using the model assessment criteria. Chapter 9 emphasizes the effect of experimental design methods on modelling and optimization. Chapter 10 delves into the exploration of special functions as potential alternative model types in mathematical modeling. Chapter 11 shared the results obtained from each chapter in the book of the intricate relationship between experimental design, mathematical modelling, and optimization, which are crucial components in solving engineering problems.

It is thought to produce answers to the following questions about experimental design, modeling, and optimization:

1. What are the constraints and limitations of ANN, which is frequently preferred in the literature for solving different types of problems as a mathematical modeling method?
2. How meaningful is it to use R^2, one of the most preferred criteria in measuring model success, as an evaluation criterion alone and to make a decision as successful or unsuccessful for the model?
3. Are the polynomial structures used in the regression and Response Surface Modeling methods sufficient to explain the relationship between input and output parameters? Or is there a need to use different mathematical functions?
4. What should be considered when choosing the experimental design method? Is it appropriate to use Taguchi as an experimental design method on modeling? Which experimental design method is more advantageous to use?
5. In mathematical modeling, what are the effects of the different methods followed during the separation of the dataset into training and testing, on modeling and model success?
6. Which criteria are meaningful to use to measure model success?
7. How meaningful is the use of special mathematical functions in modeling?

Melih Savran and Levent Aydin

Authors

Melih Savran earned a BS degree in mechanical engineering at Manisa Celal Bayar University in 2013. He earned MS and PhD degrees in mechanical engineering at İzmir Katip Çelebi University in 2017 and 2023, respectively. He continues to work as a researcher at the same university. His research areas include mechanics of solids, design and mathematical modeling, machine learning, stochastic optimization, and hybrid natural/synthetic composites. He has international publications on stochastic optimization and modeling in engineering, including book chapters, journal articles, and conference papers.

Levent Aydin is an Associate Professor of Mechanical Engineering at İzmir Katip Çelebi University. He earned a PhD degree in mechanical engineering at İzmir Institute of Technology in 2011. His main research interests are stochastic optimization, mechanics of solids, biocomposites, biosensors, advanced engineering mathematics, hybrid neuro regression, and artificial intelligence modeling. Dr. Aydin has written more than 100 international publications on stochastic optimization and modeling in engineering, including book chapters, journal articles, and conference papers. He is also a consultant for many industrial research and development projects of international engineering firms. Dr. Aydin is the founder of the Optimization, Modeling and Applied Math Research Group (OMA-RG). He is the editor or author of *Designing Engineering Structures Using Stochastic Optimization Methods*, *Bioelectrochemical Interface Engineering*, *Hybrid Natural Fiber Composites*, *Vegetable Fiber Composites and Their Technological Applications*, and *Fiber Technology for Fiber-Reinforced Composites*.

1 Introduction

1.1 LITERATURE REVIEW

Experimental design, mathematical modeling, and optimization represent fundamental concepts frequently employed in addressing engineering challenges. Experimental design, a statistical procedure, aims to accurately identify the design parameters (factors or inputs) impacting the output parameter in production or analytical studies. This systematic approach facilitates the observation of interactions between design parameters and output, thereby streamlining time and cost in engineering problem-solving. In the academic discourse, various experimental design methodologies such as Full Factorial design (FFD), Box-Behnken design (BBD), Central Composite design (CCD), Latin Hypercube sampling (LHS), Taguchi, and D-Optimal designs are prevalent. The overarching goal of these methodologies is to minimize the requisite number of experiments. Notably, the Taguchi method stands out as a widely embraced experimental design approach, offering streamlined design opportunities with minimal experimental iterations through the use of orthogonal arrays, even in scenarios involving numerous design parameters. Consequently, the Taguchi method has garnered significant attention across diverse engineering disciplines and problem domains. However, given the focus of this text on mechanical engineering problems and applications, the examples and assessments drawn from the literature are primarily limited to this field.

In this context, Raju et al. [1] conducted experiments to examine the impact of 3D additive manufacturing process parameters on surface roughness, hardness, and mechanical strength output parameters. Employing the Taguchi experimental design method, they executed 18 experiments, a significant reduction from the 54 experiments that would have been necessary with the FFD approach. Similarly, Jahari and Akyüz [2] utilized the Taguchi experimental design method to optimize the cooling time in vehicle brake discs. By implementing this methodology, they conducted 16 experiments, a drastic reduction from the 1728 experiments that would have been mandated by the FFD method. Additionally, Pahange and Abolbashari [3] employed the Taguchi method to minimize displacement in aircraft wing design, conducting 18 experiments instead of the 486 that would have been requisite with conventional methods.

BBD and CCD are experimental design methodologies commonly employed within the Response Surface Method (RSM). BBD is applicable when all design parameters possess three levels, whereas the CCD method necessitates all design parameters to have five levels. Nonetheless, a specialized version known as Face Central Composite design (FCCD) can be utilized when all design parameters consist of three levels. Despite their limitations compared to the Taguchi method,

DOI: 10.1201/9781003494843-1

BBD and CCD are still frequently favored. Previous research has often integrated BBD and CCD methods with Analysis of Variance (ANOVA) and regression modeling.

Kechagias and Vidakis [4] investigated the impacts of building orientation, layer thickness, and building temperature on the maximum tensile strength of polyamide material in 3D additive manufacturing. They conducted a study employing the BBD and FFD experimental design methods, assessing their influences on the outcomes. With 81 experiments conducted using FFD and 15 using BBD, they established mathematical models through regression modeling. The results indicated that the model generated utilizing the BBD experimental setup performed nearly on par with the FFD approach. Thus, the successful utilization of the BBD method instead of FFD in experimental design and modeling is plausible.

Veljković et al. [5] undertook a study comparing the efficacy of experimental design methods, including BBD, CCD, and FFD, in modeling the biodiesel fuel production process. The findings revealed that BBD and CCD, proposing simpler model structures, could be proficiently employed for statistical modeling of biodiesel production processes, substituting the more intricate, labor-intensive, and costly FFD methodology. Patel et al. [6] applied BBD and CCD experimental design methodologies to establish the correlation between input and output parameters in the casting process. They investigated the influence of four design parameters—applied pressure duration, compression pressure, mold temperature, and casting temperature, each with three levels—on output parameters including density, hardness, and secondary dendrite arm spacing. Comparative analysis of models generated using BBD and CCD methods revealed that while CCD exhibited superior performance in modeling secondary dendrite arm spacing and stiffness outputs, both BBD and CCD demonstrated similar efficacy in modeling density output.

D-Optimal is a contemporary and widely used experimental design methodology that constructs an optimized experiment set. Sharma et al. [7] explored the performance and exhaust characteristics of a dual-fuel compression ignition engine utilizing biogas as the primary fuel, diesel as the pilot injection fuel, and oxyhydrogen as the booster. Design parameters encompassed oxyhydrogen flow rate, pilot injection timing, and engine load, while thermal performance of the engine and variations in exhaust emissions served as output parameters. Employing D-Optimal as the experimental design method, the study utilized ANOVA, RSM, and Desirability approaches to scrutinize the modeling effects and parameter impacts. Integration of oxyhydrogen improved combustion in biogas-diesel-fueled engines, leading to enhanced fuel efficiency and reduced carbon-based emissions, albeit with a marginal increase in nitrogen oxide (NOx) levels. Models proposed using RSM and ANOVA to establish the relationship between input and output parameters demonstrated high success ($R^2 = 0.99$). Kuram et al. [8] conducted a study to assess the viability of vegetable-based cutting fluids as alternatives to synthetic cutting fluids for end milling. Design parameters included cutting speed, depth of cut, and feed rate, with experiments determined using the D-Optimal experimental design method. This approach reduced the number of experiments from 320 to 27, as BBD, CCD, and Taguchi methods were deemed inappropriate due to differing levels of the design parameters. Results indicated that for minimizing specific energy and surface roughness, canola cutting fluid with lubrication additives was more suitable, whereas maximizing tool life favored commercial semi-synthetic cutting fluid.

Rajmohan et al. [9] delved into the tribological behavior of polyether ether ketone (PEEK) and glass fiber-reinforced PEEK using a wear tester under varying applied loads, speeds, and sliding distances. Employing an RSM-based D-optimal design, they modeled and optimized tribological parameters. Evaluation of the wear performances of PEEK matrix composites utilized specific wear rate and friction coefficient as performance metrics. Employing an RSM-based multi-objective desirability approach, they optimized both numerical and categorical factors impacting the sliding wear of PEEK matrix composites. Optimal conditions were determined to be a sliding speed of 2.92 m/s and a load of 33 N to minimize specific wear rate and friction coefficient in the sliding wear of the PEEK matrix. Results suggested a sliding distance of 1053 m, advocating for the selection of PEEK/30 GF composite.

Experimental design stands as the foundational and pivotal initial step in problem-solving, as it entails determining the design parameters crucial in defining the problem and output parameters. The outcomes derived from experiments employing experimental design are indispensable for establishing the relationship between input and output parameters in subsequent stages of mathematical modeling. A flaw in the experimental design process directly impacts the modeling process, underscoring the necessity of comprehensive examination encompassing experimental design rather than solely evaluating the modeling process.

Mathematical modeling encompasses various techniques, including regression, Artificial Neural Networks (ANN), Adaptive Neural Fuzzy Inference System (ANFIS), Support Vector Machines (SVM), Decision Trees Method (DTM), Fuzzy Logic (FL), RSM, and ANOVA. Researchers predominantly opt for package programs such as Minitab, Design Expert, Matlab, SPSS, SAS, and WEKA to develop models, eschewing manual creation of mathematical models through coding.

Among the modeling methods outlined above, RSM, ANN, and ANOVA stand out as the most prominent. RSM utilizes polynomial models to articulate the relationship between input and output parameters. In RSM modeling, ANOVA plays a pivotal role in assessing the effectiveness of model terms on output parameters and determining their inclusion in the model.

For instance, Abbasi et al. [10] embarked on mathematically modeling the behavior of vehicle bodies concerning bending and torsional frequency modes using RSM and ANOVA. They selected the thickness of plates forming the vehicle body as the design parameter and employed the CCD experimental design method to create experimental sets, each featuring five levels of design parameters. Second-order polynomial models yielded results closely aligned with actual values for both bending and torsional frequency modes. Likewise, Ghenai et al. [11] focused on modeling and optimizing double-sided usable photovoltaic solar energy systems. They utilized the FFD method in the experimental design process to generate the experimental set, and RSM and ANOVA methods in the modeling and optimization stages. Design parameters impacting the annual energy production capacity of the photovoltaic solar energy system were determined to be the placement angle, height, and light reflecting capacity (Albedo) of the solar panels. ANOVA analysis concluded that each design parameter individually, as well as their interactions, significantly affect annual energy production and warrant inclusion in the mathematical model. Researchers recommended employing a second-order polynomial model for modeling annual energy production. The proposed model demonstrated a high level of fit, with an R^2 value of 0.99, indicative of strong consistency between actual and predicted values.

Vidyarthy et al. [12] investigated the influence of welding current, welding speed, and flux coating density on output parameters of TIG welding, including penetration depth, weld zone width, depth-width ratio, and weld fusion zone area in weld bead geometry. Employing CCD, RSM, and ANOVA, they established a relationship between input and output parameters, presenting a successful second-order polynomial model. Notably, welding current emerged as the most critical input parameter impacting welding bead geometry.

ANN represent another popular modeling approach. They facilitate information transfer across layers and neurons within these layers, capable of delivering rapid results even with complex and large-volume datasets. Gajic et al. [13] employed ANN to model the impact of steel's chemical composition, melted in electric arc furnaces, on electrical energy consumption within the iron and steel industry. Design parameters included the percentage weight ratios of additives in the chemical content of stainless steel. Utilizing an experimental set comprising 46 data points, they determined the optimal ANN structure that best elucidates electrical energy consumption. Model assessment criteria encompassed R-squared (R^2), Mean Squared Error (MSE), Mean Absolute Error (MAE), and Root Mean Squared Error (RMSE). Results indicated that the hyperbolic tangent model successfully met all assessment criteria, with the proposed 5–10–1 hyperbolic tangent network structure proficiently modeling the effect of steel's chemical composition on electrical energy consumption.

Sagbas et al. [14] employed RSM approaches, including the ANN-based backpropagation algorithm and CCD, to model and predict surface roughness in the electroerosion process. They assessed the performance of the developed ANN and RSM prediction models using metrics such as the coefficient of determination and RSME. Results indicated that the ANN model offered more accurate predictions compared to the RSM model.

In a separate study, Dey et al. [15] utilized RSM and ANN modeling methods to estimate output parameters of a single-cylinder diesel engine operating with biodiesel blends. They estimated brake thermal efficiency, fuel consumption, and NOx emissions. The ANN model was trained on motor experimental data using the Levenberg-Marquardt backpropagation training algorithm with a logistic-sigmoid activation function. Various statistical measures including MSE, R^2, Mean Absolute Percentage Error (MAPE), and Kling Gupta Efficiency (KGE) were employed to assess the errors and correlations of the estimated models. The

ANN model demonstrated relatively lower prediction error and higher correlation compared to RSM.

Optimization stands as a widely applied technique across diverse fields, aiming to identify the optimal solution to a problem by maximizing or minimizing a specific objective function. Numerous optimization methods have proven successful in literature, with stochastic methods among the most commonly utilized. Stochastic methods such as Genetic Algorithm, Non-Dominated Sorting Genetic Algorithm, Ant Colony Optimization, Particle Swarm Optimization, Simulated Annealing, Differential Evolution, and Artificial Bee Colony are non-traditional approaches that do not necessitate derivative information in the solution phase.

The objective function plays a pivotal role in optimization, representing the parameters to be maximized or minimized to attain the optimal solution. Its formulation varies depending on the problem at hand. If an analytical formula is available, it serves as the objective function. However, in the absence of such a formula, a mathematical model describing the relationship between input and output parameters can be utilized as the objective function.

The precision of experimental design and modeling methods directly impacts the optimization process. Thus, conducting these processes meticulously is crucial. Inaccuracies in experimental design and mathematical modeling, despite employing successful optimization methods, may hinder the attainment of a usable solution meeting desired conditions, potentially leading to misleading results.

Stochastic methods are frequently favored in solving diverse engineering problems. Genç et al. [16] proposed a hybrid methodology combining RSM and Simulated Annealing to explore the viability of elastomer-based shock absorber springs in automobile clutch systems under dynamic conditions. This hybrid approach optimized the shock absorber spring system's design, enhancing its performance and reliability. Gnanavelbabu and Saravanan [17] investigated the abrasive water jet machining of titanium alloy (Ti-6Aa-4V), analyzing the effect of machining parameters on kerf taper angle and surface roughness. Employing BBD experimental design and RSM modeling methods, they utilized evolutionary optimization techniques such as Particle Swarm Optimization, Cuckoo Search Algorithm, and Simulated Annealing to determine the optimal configuration. The Cuckoo Search Algorithm outperformed other algorithms, delivering superior designs. Raju et al. [1] explored the effect of 3D additive manufacturing process parameters on surface roughness, hardness, and mechanical strength to identify optimal process parameters. Employing Particle Swarm Optimization, Bacterial Foraging Optimization, and their hybrid version, they found that the hybrid algorithm offered the most significant improvement, followed by Particle Swarm and Bacterial Foraging Optimization. Shui et al. [18] aimed to design a housing package for electric vehicle battery modules, selecting the thickness of specific points of the battery housing package as the design parameter and displacement, natural frequency, and weight as output parameters. Utilizing CCD-ANN methods, they employed mathematical models as the objective function for the optimization stage. The Non-Dominated Sorting Genetic Algorithm yielded significant improvements in displacement, natural frequency, and weight compared to initial designs.

Apart from stochastic optimization methods, the desirability approach, gray relational analysis, and Taguchi are frequently used to obtain optimum designs. These methods are preferred because they are generally easy to use and provide quick solutions. Technique for Order of Preference by Similarity to Ideal Solution (TOPSIS), Desirability Approach, and Gray Relational Analysis methods are usually preferred in multi-objective optimization problems.

Pahange and Abolbashari [3] utilized Taguchi and Gray Relational Analysis methods to minimize deformation caused by bird strikes on an aircraft wing while simultaneously minimizing wing weight. The Taguchi method was employed to minimize deformation and wing weight independently, while Gray Relational Analysis determined the appropriate design when both displacement and weight were equally important and required minimization. Results indicated that Gray Relational Analysis significantly improved upon the initial design. However, these methods are limited to selecting the best designs generated within the experimental set. Tabet et al. [19] employed the RSM-based Desirability Approach and Genetic Algorithm methods to model and optimize drilling parameters' effect on delamination damage in a bio-sandwich structure. Findings revealed that all three methods yielded similar results.

In conclusion, selecting an appropriate optimization method proven effective in diverse problems is crucial. However, ensuring precision in experimental design and mathematical modeling steps during the optimization process is equally vital for obtaining a reliable and usable solution meeting desired conditions.

1.2 MOTIVATION

Experimental design, mathematical modeling, and optimization constitute essential components in the realm of engineering problem-solving, necessitating both analytical and numerical approaches for effective resolution. While analytical solutions are favored for situations requiring precision, many engineering challenges surpass the capabilities of analytical methods, prompting the utilization of numerical techniques. Within this framework, experimental design, mathematical modeling, and optimization play pivotal roles, serving as critical stages in data analysis and offering significant convenience in problem-solving. The extensive utilization of these processes across various studies underscores their importance in engineering practice.

Several reasons underscore the crucial role of experimental design, mathematical modeling, and optimization in engineering:

1. **Efficient Parameter Examination**: In experimental studies, exploring the effects of multiple parameters is often essential. However, conducting numerous experiments can be impractical in terms of time and cost. Experimental design facilitates parameter examination with fewer experiments, thereby optimizing resources and enhancing efficiency.

2. **Time and Cost Savings**: Experimental design reduces the number of required experiments, leading to time and cost savings. By strategically selecting experiments, experimental design ensures comprehensive coverage of the problem boundaries, enhancing the reliability of results.
3. **Consideration of Parameter Interactions**: Experimental design accounts for the interactions among design parameters and output parameters. Properly selected experiments provide insight into parameter interactions, yielding more accurate results compared to randomly conducted experiments.
4. **Behavior Determination through Mathematical Modeling**: Mathematical modeling elucidates the behavior of output parameters within a defined problem space. This obviates the need for exhaustive experimentation or simulation to understand how output parameters vary with changes in design parameters.
5. **Simplicity and Efficiency**: While analytical solutions demand intensive mathematical operations, mathematical modeling methods offer simpler yet effective alternatives. These models provide valuable insights into system behavior with reduced computational complexity.
6. **Optimization for Best Design**: Optimization is indispensable in engineering, aiming to identify the optimal design while considering specific constraints. It enables engineers to navigate complex design spaces and identify solutions that fulfill performance requirements.
7. **Generation of Alternative Designs**: Optimization often yields multiple alternative designs, known as elite designs, closely approaching the optimal solution. These alternatives offer flexibility and enable engineers to explore diverse design options while maintaining quality and usability.

In conclusion, experimental design, mathematical modeling, and optimization are indispensable facets of engineering problem-solving, offering efficient and effective strategies for navigating complex design spaces, optimizing resources, and achieving desired performance outcomes. Their integration into engineering practice underscores their significance in addressing diverse challenges across various disciplines.

It is crucial to emphasize that experimental design, mathematical modeling, and optimization are indispensable stages in defining and analyzing engineering problems. These processes are fundamental in various scientific fields, particularly in primary disciplines such as physics, chemistry, mathematics, engineering, and medicine.

The main reasons for writing a book on this subject are as follows:

1. **Importance Across Scientific Fields**: Experimental design, mathematical modeling, and optimization are vital processes applicable across numerous scientific domains. Understanding and mastering these techniques are essential for advancing research and problem-solving in diverse fields.

2. **Popularity of Experimental Design Methods**: Experimental design methods, notably Taguchi, are widely utilized due to their ability to save time and costs in problem-solving endeavors. However, there is a lack of comprehensive literature on the efficacy of these methods and their comparison with other approaches.

3. **Need for Evaluation of the Taguchi Method**: The Taguchi method is extensively employed in various industries, yet there is a dearth of comprehensive studies evaluating its advantages and disadvantages. A thorough examination of this method is essential for enhancing understanding and optimizing its application.

4. **Impact of Artificial Intelligence**: Recent advancements in artificial intelligence have revolutionized mathematical modeling and optimization techniques. These developments have increased the relevance and importance of these methods across diverse sectors, necessitating comprehensive exploration and analysis.

5. **Continuous Evolution of Modeling Tools**: With the continuous development of new programs and methods for mathematical modeling, there is a need to simplify and enhance the comprehensibility of these processes. The high demand for popular modeling methods such as ANN, ANFIS, FL, and RSM underscores the necessity for comprehensive resources in this field.

6. **Demand for Quicker Results**: The evolution of technology and the need for rapid results have heightened the significance of mathematical modeling. These techniques facilitate the efficient measurement of parameter effects without the need for extensive experimental observation, simulation, or analysis, thereby expediting problem-solving processes.

In summary, a book focusing on experimental design, mathematical modeling, and optimization is warranted due to their critical importance across scientific disciplines, the popularity and need for evaluation of experimental design methods like Taguchi, the impact of artificial intelligence on modeling and optimization, the continuous evolution of modeling tools, and the demand for quicker results in problem-solving endeavors. Such a resource would provide valuable insights and guidance for researchers, practitioners, and students navigating these essential processes in their respective fields.

The aim of the book is to provide a comprehensive methodology for problem-solving by examining experimental design, mathematical modeling, and optimization processes. To achieve this objective, recent studies in the literature utilizing these methods were analyzed, particularly focusing on those published in high-impact international journals. The book aims to offer insights into the use of Taguchi, ANN, and RSM methods in experimental design, modeling, and optimization.

Additionally, the book introduces a novel modeling method termed Neuro Regression (NRM), which is a hybrid approach combining ANN and regression modeling methods. Stochastic optimization techniques were incorporated to enhance the effectiveness and usability of NRM. Furthermore, Stochastic Neuro Regression (SNRM) is proposed as a modeling method tailored to address complex problems.

The book also introduces new assessment criteria beyond the commonly used R^2-based metrics, including boundedness checking and stability testing. These criteria provide a more comprehensive evaluation of model performance.

Key original contributions of the book include the following:

1. Introduction of novel hybrid modeling methods (NRM and SNRM) combining ANN, regression, and Stochastic Optimization techniques.
2. Exploration of alternative modeling approaches beyond polynomial models, such as trigonometric, logarithmic, rational, and hybrid models, as well as special functions like Bessel, ChebyShev, Erf, Fresnel, Hermite, HyperGeometric, and LegendreP.
3. Comparison of different test and validation methods, such as cross-validation, holdout, and bootstrap, to assess model performance.
4. Evaluation of the performance of the Taguchi experimental design method and comparison with original laminated composite problems, highlighting its weaknesses.
5. Comparison of the performance of experimental design methods (BBD, D-Optimal, Taguchi, FFD, and CCD) in the same problem, discussing their robustness and weaknesses against each other.

Overall, the book emphasizes the holistic consideration of experimental design, mathematical modeling, and optimization processes and introduces innovative methodologies and assessment criteria to enhance problem-solving effectiveness in engineering and scientific domains.

REFERENCES

[1] Raju M, Gupta MK, Bhanot N, Sharma VS. A hybrid PSO–BFO evolutionary algorithm for optimization of fused deposition modelling process parameters. Journal of Intelligent Manufacturing 2019; 3: 2743–2758.
[2] Jafari R, Akyüz R. Optimization and thermal analysis of radial ventilated brake disc to enhance the cooling performance. Case Studies in Thermal Engineering 2022; 3: 101731.
[3] Pahange H, Abolbashari MH. Mass and performance optimization of an airplane wing leading edge structure against bird strike using Taguchi-based grey relational analysis. Chinese Journal of Aeronautics 2016; 29(4): 934–944.
[4] Kechagias JD, Vidakis N. Parametric optimization of material extrusion 3D printing process: an assessment of Box-Behnken vs. full-factorial experimental approach. The International Journal of Advanced Manufacturing Technology 2022; 121(5–6): 3163–3172.
[5] Veljković VB, Veličković AV, Avramović JM, Stamenković OS. Modeling of biodiesel production: Performance comparison of Box–Behnken, face central composite and full factorial design. Chinese Journal of Chemical Engineering 2019; 27(7): 1690–1698.

[6] G.C Manjunath Patel, Krishna P, Parappagoudar MB. Squeeze casting process modeling by a conventional statistical regression analysis approach. Applied Mathematical Modelling 2016; 40(15–16): 6869–6888.

[7] Sharma P, Balasubramanian D, Khai CT, Venugopal IP, Alruqi M, Josephin F, Wae-Hayee M, et al. Enhancing the performance of renewable biogas powered engine employing oxyhydrogen: Optimization with desirability and D-optimal design. Fuel 2023; 341: 127575.

[8] Kuram E, Ozcelik B, Bayramoglu M, Demirbas E, Simsek BT. Optimization of cutting fluids and cutting parameters during end milling by using D-optimal design of experiments. Journal of Cleaner Production 2013; 42: 159–166.

[9] Rajmohan T, Palanikumar K, Ranganathan S. Evaluation of mechanical and wear properties of hybrid aluminium matrix composites. Transactions of Nonferrous Metals Society of China 2013; 23(9): 2509–2517.

[10] Abbasi M, Fard M, Khalkhali A. Dynamic Stiffness Investigation of an Automotive Body-in-White by Utilizing Response Surface Methodology 2018; (No. 2018-01-1479): SAE Technical Paper.

[11] Ghenai C, Ahmad FF, Rejeb O, Hamid AK. Sensitivity analysis of design parameters and power gain correlations of bi-facial solar PV system using response surface methodology. Solar Energy 2021; 223: 44–53.

[12] Vidyarthy RS, Dwivedi DK, Muthukumaran V. Optimization of A-TIG process parameters using response surface methodology. Materials and Manufacturing Processes 2018; 33(7): 709–717.

[13] Gajic D, Savic-Gajic I, Savic I, Georgieva O. Di Gennaro S. Modelling of electrical energy consumption in an electric arc furnace using artificial neural networks. Energy 2016; 108: 132–139.

[14] Sagbas A, Gürtuna F, Polat U. Comparison of ANN and RSM modeling approaches for WEDM process optimization. Materials Testing 2021; 63(4): 386–392.

[15] Dey S, Reang NM, Das PK, Deb M. Comparative study using RSM and ANN modelling for performance-emission prediction of CI engine fuelled with bio-diesohol blends: A fuzzy optimization approach. Fuel 2021; 292: 120356.

[16] Genç MO, Konakçı S, Kaya N, Kartal S, Serbest AK. Development of a novel testing procedure and optimisation of a rubber spring using constrained simulated annealing algorithm for automobile clutch system. International Journal of Vehicle Design 2022; 88(1): 33–55.

[17] Gnanavelbabu A, Saravanan P. Experimental investigations of abrasive waterjet machining parameters on titanium alloy Ti-6Al-4V using RSM and evolutionary computational techniques. In Advances in Unconventional Machining and Composites: Proceedings of AIMTDR 2018; 413–425, Singapore: Springer Singapore.

[18] Shui L, Chen F, Garg A, Peng X, Bao N, Zhang J. Design optimization of battery pack enclosure for electric vehicle. Structural and Multidisciplinary Optimization 2018; 58: 331–347.

[19] Tabet Z, Belaadi A, Boumaaza M, Bourchak M. Drilling of a bidirectional jute fibre and cork-reinforced polymer biosandwich structure: ANN and RSM approaches for modelling and optimization. The International Journal of Advanced Manufacturing Technology 2021; 117(11–12): 3819–3839.

2 Design of Experiment, Mathematical Modeling, and Optimization

2.1 DESIGN OF EXPERIMENT

Design of experiment (DOE) is the observation and interpretation of variations in the output of a process by making changes to the factors affecting the process in order to improve the performance of a process. A process and its components are generally shown in Figure 2.1 [1].

A process can be defined as a sequence of activities that utilize various resources such as machinery, materials, methods, and human labor, which interact with each other to produce a specific output, whether it be a product or a service. Within this framework, factors, also known as design variables, play a crucial role. These factors can be categorized as controllable or uncontrollable, depending on their susceptibility to manipulation during the process.

Controllable factors are those for which values can be intentionally assigned and maintained constant throughout the operation of the process. Examples of controllable factors include material type, machine settings, and production methods.

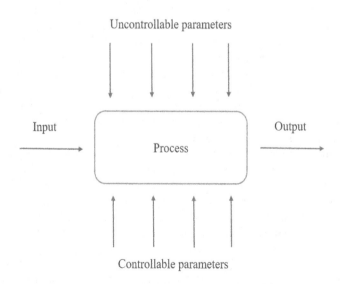

FIGURE 2.1 Flow chart of a process [2]

DOI: 10.1201/9781003494843-2

Uncontrollable factors, such as environmental conditions like humidity and temperature, are more challenging to regulate and may fluctuate unpredictably during the process [1].

The DOE emerges as a powerful statistical and mathematical methodology for enhancing the performance of systems or processes by systematically manipulating inputs to optimize outputs. In the context of DOE, factors or design variables represent the controllable and uncontrollable elements that influence the process [3]. Levels denote the various settings or adjustments in the values that each factor can assume, while outputs, or responses, represent the performance indicators of the process, reflecting how factors affect the outcome [4].

Utilizing the principles of DOE enables the development of robust designs, which are characterized by their ability to maintain desired performance levels despite variations in uncontrollable external factors. By systematically designing experiments to investigate the effects of factors on outputs, DOE facilitates process improvement and optimization, ultimately leading to enhanced performance and efficiency.

The DOE is useful to determine the cause–effect relationships of the process of interest, which helps in optimizing the response variable. Experimental design helps maximize the accuracy of the result by reducing the number of experiments to improve system performance. In addition to identifying the most important factors affecting the output, experimental design also reveals the interactions between factors [5].

In experimental design, the primary objective is to assess the variability in the outcomes of a process by systematically manipulating the factors influencing it. The success of statistical experimental design hinges on the accuracy of the collected data. Therefore, decisions regarding data collection methods and the number of observations per experiment must be made at the design stage. The collected data should be independent and adequate for statistical interpretation. Three fundamental principles are employed in statistical experimental design to ensure these conditions: replication, randomness, and blocking.

Replication involves conducting multiple experiments for a given set of conditions. It serves two critical purposes: first, it enables the estimation of experimental error and, second, it ensures a complete and accurate estimation of the effect of a factor when using sample means. Increased replication leads to higher accuracy in the results obtained from experimental design.

Randomness is a fundamental aspect of statistical methods utilized in experimental design. It involves random assignment of operators, machines, materials, and the sequence of experiments. The aim is to minimize the impact of extraneous factors on the collected data, ensuring independence among observations. For example, if variations in collected data arise from machine heating during experiments, randomization ensures that this effect is consistent across all trial combinations.

Blocking is a technique employed to enhance the accuracy and precision of experiments. It involves categorizing collected data into homogeneous subsets, known as blocks. Each block is evaluated independently, allowing for the

elimination of the influence of extraneous factors, such as materials and operators, on the process. By utilizing blocking, only the effects of the factors of interest on the process are determined.

Overall, replication, randomness, and blocking are essential principles in statistical experimental design, contributing to the reliability and validity of the collected data and facilitating accurate interpretations of the effects of factors on the
process under investigation.

For instance, in an experiment evaluating the performance of three different machines, if variations in outcomes are observed due to differences between
operators using these machines, considering operators as blocking variables during experimental design helps mitigate the effects of these differences on the
experiment's outcome.

The implementation process of experimental design is depicted schematically
in Figure 2.2 [2]. The initial step in the implementation phase is defining the problem. During this stage, comprehensive information regarding the experiment's
objective is gathered, and the purpose is clearly delineated. It is crucial to collect
necessary information from all relevant units regarding the process under examination. A clear problem statement contributes to a thorough understanding and
effective resolution of the issue.

Once the problem is articulated, the factors influencing the process of interest
and their respective levels need to be determined. This involves specifying how
the factors will be controlled and measured at different levels. Additionally, the
output variable, which provides insights into the process, should be identified. It
is imperative to ensure that the chosen output variable yields relevant information
about the process of interest.

Next, the appropriate experimental design technique is selected based on the
problem's objectives. This step encompasses determining the sample size (number of repetitions), the sequence of experiment execution, whether blocking will
be implemented, and other constraints related to randomness. The primary criterion in selecting the experimental design is aligning it with the experiment's
purpose. The choice of experimental design technique is tailored to the specific
objectives of the experiment.

Following the selection of the experimental design technique, experiments
are conducted according to the designated experimental design. Vigilance
during the experimental phase is critical to ensure adherence to the planned
framework. Any errors occurring during this phase can compromise the experiment's validity. Thus, meticulous planning is imperative for the experiment's
success.

2.2   DESIGN OF EXPERIMENT METHODS

This chapter aims to introduce the subject by showing the basic techniques used
in practice; therefore, instead of explaining in detail all the experimental design
techniques in the literature, the focus is on the most frequently encountered
techniques and in what situations they can and cannot be used. It is possible to

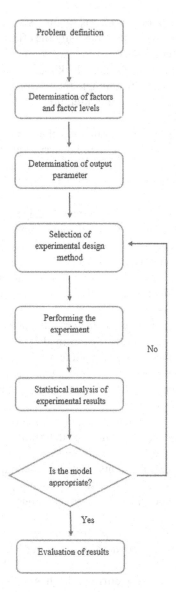

FIGURE 2.2 Flow chart for implementing the design of experiment

access many articles and books in the literature containing detailed explanations of experimental design techniques. Another issue that wants to be emphasized within the scope of this chapter is to create awareness that the experimental design, modeling, and optimization processes should be considered as a whole and that any negativity that occurs in one of these three stages will be reflected in the other processes. Experimental design, which is the first stage of this triple process, includes the steps of understanding the problem, defining it correctly, determining factors, levels, constraints, and outputs, and seeking answers to the question of "What experiments I should perform to observe the effects of factors on the output accurately?"

The easiest and most frequently applied method in experimental design approaches is to observe the change in product or process performance by changing the level of a single factor in each experiment. The most challenging thing is to evaluate their effects and interactions by keeping them all under control in an environment where many factors affect product performance.

2.2.1   One Variable at a Time Design

Indeed, the one variable at a time (OVAT) approach serves as a straightforward experimental design method. In this method, the impact of individual factors on the output parameter is assessed by varying one factor at a time while keeping others constant. However, OVAT does not enable simultaneous examination of multiple factors' effects on the output parameter, thereby limiting its ability to reveal interactions between factors.

To illustrate, consider an experiment investigating the tensile strength of a composite plate subjected to bending load. Let the factors influencing tensile strength be the fiber angles (A), thickness (B), and number (C) of layers. When employing the OVAT approach, factors B and C are held constant while varying factor A to discern its effect on tensile strength. Consequently, insights into the impact of factor A on the output parameter can be gained. However, OVAT does not facilitate the exploration of interactions between factors.

The absence of factor interactions is a prerequisite for obtaining reliable analysis results in OVAT. This means that the effect of any factor on the output parameter at its current level remains unaffected by the level of another factor. Since OVAT does not account for interactions between factors, it may not accurately depict the system's behavior when interactions exist. Consequently, achieving the desired optimum point becomes challenging in experimental setups with factor interactions.

In summary, while OVAT offers simplicity in experimental design and analysis, its inability to capture factor interactions may limit its effectiveness in identifying optimal conditions in systems where such interactions play a significant role. Therefore, more sophisticated experimental design methods, capable of capturing interactions between factors, are often preferred for comprehensive analysis and optimization [6].

2.2.2   Factorial Design (FD)

The experimental design method that seems to be the most "optimal" in evaluating and interpreting the effects of factors is Full Factorial design (FFD) because all possible combinations of different levels of factors are evaluated in this method. In this type of design, equal numbers of experimental results are taken from each level of factor and compared with each other. It is a balanced design, but FFD can only be used when a few factors are involved because the number of required experiments increases rapidly with the number of factors and their levels [7].

A different approach that can be used instead is the Fractional Factorial Design Method (FFDM). In FFDM, only a consciously chosen part is tried and evaluated out of all options. This way, all resources, especially workforce, time, and money, are saved in conducting experiments. However, there is also a disadvantage. Depending on the fractional factorial level chosen, and thus the reduction in the number of experiments, one or more main effects and interactions will be mixed, and their effects cannot be estimated separately. Table 2.1 shows the number of experiments that need to be performed when using FFD and FFDM for cases with different 2-level factor numbers.

The number of experiments to be performed in FFDM is determined according to the 2^{k-p} formula. Here, "k" indicates the number of factors, while "p" takes value according to the fraction ratio in partial factorial design. For example, suppose the number of experiments is to be reduced by half (1/2), one-fourth (1/4), or one-eighth (1/8) compared to the FFD. In that case, the "p" parameter takes the values 1, 2, and 3, respectively.

Another factorial design method is Multi-Level Factorial Design (MFD). This method can determine the number of experiments to be performed to observe the effect of factors with different level values on the output parameter. MFD helps us understand all factors' impact on the output parameter alone and when interacting with other factors. It contains all combinations of experiments that need to be performed. We can say that the MFD method includes 2- and 3-level factorial design methods, which are other factorial design methods.

If time, opportunity, and financial resources allow, FD should be preferred to decide on the experiments that need to be performed. Otherwise, FFDM or other experimental design methods can be used as an alternative method.

TABLE 2.1

Number of Experiments Depending on the Number of Factors When Using FFD and FFDM [2]

Factor Number	Level Number	FFD	FFDM (1/2)	FFDM (1/4)	FFDM (1/8)	FFDM (1/16)	FFDM (1/32)
2	2	4	2	–	–	–	–
3	2	8	4	–	–	–	–
4	2	16	8	–	–	–	–
5	2	32	16	8	–	–	–
6	2	64	32	16	8	–	–
7	2	128	64	32	16	8	–
8	2	256	128	64	32	16	–
9	2	512	256	128	64	32	16

2.2.3 CENTRAL COMPOSITE DESIGN (CCD)

CCD, one of the experimental design methods that can reduce the time and cost impact of using FD, is used to create an experiment set in problems consisting of 5-level factors. In a problem composed of "k" factors, the CCD method includes up to 2k factorial experiments, 2k axis experiments, and at least one central point experiment. In other words, when the CCD method is used, the number of experiments to be performed is determined by the formula $N = 2^k + 2k + 1$ [8]. In addition to the factors' lower, upper, and center point level values, the CCD method determines the level values for the points at a distance of "α" from the center point and includes them in the design. Different formulas are used to determine the "α" values depending on the number of factors and whether the resulting design is rotatable, spherical, or orthogonal [9]. Calculations of the "α" parameter for different design types are given in Table 2.2.

The CCD method is sometimes preferred to create an experiment set for problems of 3-level factors. Here, the factors take only the lower, upper, and center point level values, and the level values of the points at a distance of α are not included in the design. These designs are defined as Face Central Composite Design (FCCD). Figure 2.3 shows the different design types obtained using the CCD method.

2.2.4 BOX-BEHNKEN DESIGN (BBD)

BBD can only create an experimental set when all factors are at the 3-level. Unlike CCD, the center points of the factors are more common in the experimental sets created using the BBD experimental design method. The BBD method creates a spherical design with a diameter of $\sqrt{2}$ units. In addition, there are no lower and upper-level values for any factor on the horizontal and vertical axis in the cubic representation. While this provides an advantage in cases where experiments at corner points on the cubic element are difficult and costly to perform, it is a disadvantage where these points may significantly impact the output [10].

TABLE 2.2
α Values for Different CCD Types [2]

CCD Types	α Value
Spherical	$\alpha = \sqrt{k}$
Rotatable	$\alpha = \sqrt[4]{2^k}$
Orthogonal	$\alpha = \sqrt[4]{\dfrac{2^k}{4}\left(\sqrt{2^k + 2k + 1} - \sqrt{2^k}\right)^2}$
Face central	$\alpha = 1$

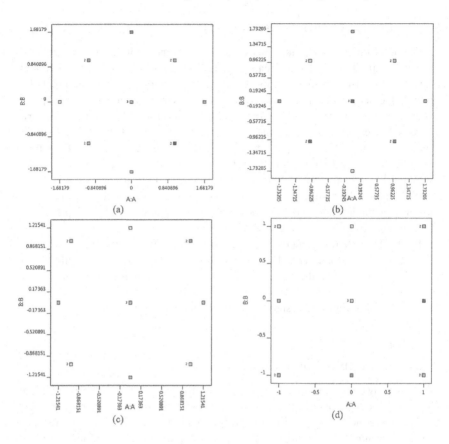

FIGURE 2.3 Types of Central Composite Design: (a) Rotational, (b) Spherical, (c) Orthogonal, and (d) Central [2]

2.2.5 TAGUCHI DESIGN

Indeed, Taguchi's robust design concept, consisting of system design, parameter design, and tolerance design stages, aims to develop systems resilient to unwanted noise effects. This is achieved by strategically selecting controllable factors and minimizing the influence of uncontrollable factors. As the number of factors impacting the problem's output increases, the number of required experiments escalates rapidly. The Taguchi method, rooted in the principle of ensuring quality at the design stage, effectively addresses this challenge by employing orthogonal arrays to streamline experiments while mitigating the effects of uncontrollable factors.

However, it's essential to acknowledge a potential limitation: the reduction in the number of experiments facilitated by Taguchi may overlook some interactions between factors. In other words, while Taguchi efficiently minimizes the experiment count, it may neglect certain factor interactions crucial for comprehending output parameter behavior. Consequently, it cannot be universally deemed advantageous to solely rely on the Taguchi experimental design method [11].

In summary, while Taguchi's approach offers significant advantages in terms of efficiency and resource optimization, particularly in scenarios with numerous factors, its efficacy may be compromised when intricate interactions between factors are disregarded. Therefore, a balanced consideration of experimental design methods is necessary, accounting for both efficiency and the need for comprehensive understanding of system behavior.

In a typical experimental design with "k" factors, 2^k experiments should be conducted as per the FFD. In the Taguchi method, this number of trials is reduced by simultaneously changing the levels of several factors. For example, when FD design is used for 8 factors with 2 levels, 256 experiments must be performed, while Taguchi says 12 experiments are sufficient under the same conditions. Reducing the number of experiments to this level provides significant time and cost savings, but as previously mentioned, this may cause problems in analyzing the problem correctly. Compared to other experimental design methods, Taguchi offers many options regarding the number of factors and the level values of these factors. While CCD and BBD can only produce experimental sets for specific level values, Taguchi offers the opportunity to create a wide range of experimental design sets using orthogonal arrays for different factors and level values.

Table 2.3 shows the experimental sets recommended by the Taguchi method when factors with the same level values are used. It is seen that the Taguchi method offers the opportunity to carry out the design with a significantly lower number of experiments compared to the FD design when all factors consist of 2 or 3 levels. Additionally, the number of factors that can be used is relatively high in both 2-level and 3-level cases. The Taguchi method offers the opportunity to study using 31 different factors when all factors consist of 2 levels and 23 different factors when all factors are comprised of 3. When you want to perform experiments using the FFD method for these two cases, the number of experiments is 231 and 223, respectively. In contrast, the Taguchi method says it is sufficient to perform 32 and 36 experiments for the same situation, respectively. It is possible to make similar evaluations when the factors consist of 4 and 5 levels [12].

Table 2.4 shows the experiments that can be performed with the Taguchi experimental design method when factors with different level values are used together. Here, too, it can be seen that the Taguchi method offers a wide range of design possibilities. For instance, while the number of 2-level factors varies between 1 and 11 and 3-level factors varies between 2 and 12, the Taguchi method says it is sufficient to perform 36 experiments to express the relationship between output and input accurately. This provides the opportunity to reach results by using different numbers and levels of factors and performing far fewer experiments compared to other methods.

TABLE 2.3

Taguchi Experiment Sets When Factors with the Same Level Value Are Used [2]

	Number of Levels										
2			**3**			**4**			**5**		
P=2	S=2	L4	P=2	S=3		P=2	S=4		P=2	S=5	
P=3	S=2		P=3	S=3	L9	P=3	S=4		P=3	S=5	
P=4	S=2		P=4	S=3		P=4	S=4	L16	P=4	S=5	L25
P=5	S=2	L8	P=5	S=3		P=5	S=4		P=5	S=5	
P=6	S=2		P=6	S=3	L18	P=6	S=4		P=6	S=5	
P=7	S=2		P=7	S=3		P=7	S=4		P=7	S=5	
P=8	S=2		P=8	S=3		P=8	S=4	L32	P=8	S=5	
P=9	S=2	L11	P=9	S=3		P=9	S=4		P=9	S=5	L50
P=10	S=2		P=10	S=3		P=10	S=4		P=10	S=5	
P=11	S=2		P=11	S=3	L27				P=11	S=5	
P=12	S=2		P=12	S=3					P=12	S=5	
P=13	S=2	L16	P=13	S=3							
P=14	S=2		P=14	S=3							
P=15	S=2		P=15	S=3							
P=16	S=2		P=16	S=3							
P=17	S=2		P=17	S=3							
P=18	S=2		P=18	S=3							
P=19	S=2		P=19	S=3	L36						
P=20	S=2		P=20	S=3							
P=21	S=2		P=21	S=3							
P=22	S=2		P=22	S=3							
P=23	S=2		P=23	S=3							
P=24	S=2	L32									
P=25	S=2										
P=26	S=2										
P=27	S=2										
P=28	S=2										
P=29	S=2										
P=30	S=2										
P=31	S=2										

Number of Parameters (row axis label)

Considering the time and cost constraints, the Taguchi method offers the opportunity to design with fewer experiments and is attractive to many researchers. However, it should not be forgotten that with the Taguchi method, it will not always be possible to easily define the input and output relationship with such few experiments, and the possibility of encountering unsuccessful designs should be taken into consideration when trying to save time and cost.

TABLE 2.4

Taguchi Experiment Sets When Using Factors with Different Level Values [2]

Trial	Level 2	3	4	6	8
L8	1–4		1		
L16	2–12		1		
L16	1–9		2		
L16	1–6		3		
L16	1–3		4		
L16	1–8				1
L18	1		1–7		
L18		1–6		1	
L32	1		2–9		
L36	1–11	2–12			
L36	1–3	13			
L54	1	3–25			

2.2.6 D-Optimal Design

The D-Optimal method stands out as an optimization-based experimental design technique that offers several advantages over traditional methods like BBD, CCD, and Taguchi. Unlike these methods, which generate consistent experimental sets with fixed factor and level values, D-Optimal allows for the creation of different experimental sets in each run. This variability provides greater flexibility and adaptability to the experimental design process.

One key advantage of the D-Optimal method is its ability to handle both numerical and categorical factors. This versatility enables researchers to explore a wider range of variables and factor types in their experiments. Moreover, D-Optimal allows for the study of factors with different level values, providing a more comprehensive understanding of the system under investigation.

Another significant feature of D-Optimal is its capability to incorporate constraints into the experimental design process. These constraints can represent mathematical relationships or interactions between factors, allowing researchers to model complex systems more accurately. By including critical factor interactions as constraints, researchers can ensure that the experimental set captures the most relevant aspects of the problem.

Overall, the D-Optimal method offers researchers greater flexibility, versatility, and precision in experimental design compared to traditional methods. Its ability to handle various types of factors, explore different factor levels, and incorporate constraints makes it a powerful tool for optimizing experimental designs and improving the efficiency and effectiveness of research efforts [9, 13].

2.3   MATHEMATICAL MODELING

Mathematical modeling, which constitutes the second stage of the DOE-Modeling-Optimization process, aims to define the relationship between the input and output parameters in the data set obtained by the experimental design using mathematical models. To define this relationship correctly, the most crucial stage in the experimental design process is to determine correctly the factors and level values and the experiments that need to be carried out using the appropriate experimental design method.

Many approaches and methods regarding mathematical modeling are used in the literature. Artificial Neural Networks (ANN), Response Surface Methods (RSM), Support Vector Regression, Decision Trees, Regression, Fuzzy Logic, and Adaptive Neural Fuzzy Inference Systems (ANFIS) are the most well-known. These methods are frequently preferred in modeling the input and output relationship in all kinds of data sets, regardless of data type distinction and preliminary study. However, as mentioned previously, modeling is not an independent process on its own. The success of the mathematical modeling process does not mean creating a model that can predict the output parameters exactly. Deficiencies and inadequate approaches frequently encountered in the mathematical modeling process can be listed as follows:

1. In modeling engineering systems, it is necessary to consider the interactions of all design factors affecting the process. Changing the value of a factor and keeping the values of other factors constant to observe the effect of a factor (OVAT) may lead to ignoring the nonlinear effects. Therefore, to model the system accurately, nonlinear interactions between variables must be taken into account. This approach will allow more accurate modeling of the system.

2. Most studies on mathematical modeling choose only one or two traditional polynomial-based regression models as the objective function of their problem. However, as the nonlinearity level of the relationship between input and output parameters increases, there may be situations where polynomial models are inadequate.

3. R^2 is an essential criterion in evaluating the model's prediction performance created in modeling the results obtained in experimental and simulation studies. However, a high R^2 value does not describe all physical phenomena of the engineering process. The R^2 value expresses how close the prediction results of the created model are to experimental or simulation results. In other words, even if the R^2 value is very high for real systems, this does not always mean successive modeling. The model only fits the experimental data well but does not describe the underlying behavior of the output parameter. Therefore, it is necessary to try new modeling approaches that include different regression forms and techniques.

4. An essential feature of the model function proposed in engineering problems is that it must be bounded. Being finite relates to the realistic

modeling of engineering systems, and all engineering parameters are known to be finite. Therefore, before the optimization step, whether the selected models are bounded under the engineering parameter ranges should be checked.

For these reasons, the need to present a new approach that evaluates the experimental design-modeling-optimization processes has arisen. This approach is a hybrid method in which ANN, Regression, and Stochastic Optimization are used together in modeling. Thus, it aims to introduce a more effective method that combines the advantages of these methods used in modeling and optimization. Detailed information about the recommended modeling methods is given next.

2.3.1 NEURO REGRESSION METHOD (NRM)

NRM is a hybrid method that combines the strengths of Regression analysis and ANN to increase the accuracy of predictions at the modeling stage. The NRM approach divides the dataset into training, testing, and validation. During the training phase, data is used to create mathematical models. The aim here is to enable the model to learn the physical and mathematical nature of the problem by using the dataset consisting of inputs and outputs included in the system. Although the data used in the training phase is usually set to be 80% of the entire dataset, this is not a rule. If successful models can be created by keeping the ratio of the training set at lower levels, this is a more desirable situation. The trained model is tested in the second stage of the modeling process. The aim here is to measure the performance of the trained model in predicting data that it has not encountered before. Thus, it can be understood whether this success in the training phase is due to real learning. At this stage, it is essential to distinguish between the model that learns the physical nature of the problem and the model that memorizes it. The third stage of the modeling process is validation. At this stage, where a procedure similar to the testing process is followed, the model is subjected to a second test process with less data than the one used in the testing process. Thus, the robustness of the model is tried to be guaranteed. However, as mentioned under the modeling heading, the model that successfully passes these stages may still produce meaningless results inconsistent with the problem's physical nature. In this case, examining the maximum and minimum output parameter values produced by the model, defined as boundedness control, can enable us to evaluate the model more accurately. Models that do not meet the boundedness control criterion produce results incompatible with the problem's nature and cannot be achieved. These models are not usable no matter how high their predictive performance is.

2.3.2 STOCHASTIC NEURO REGRESSION METHOD (SNRM)

This modeling method is similar to NRM in terms of operation. While the least-squares method determines the model coefficients in NRM, stochastic optimization methods are utilized to find the model coefficients in SNRM. Similar to NRM, data is divided into three groups: training, testing, and validation, and it is required to meet the necessary success criteria. Trying to determine model coefficients using stochastic optimization methods allows the production of many alternative models and offers the opportunity to treat the model creation process as an optimization problem. In this respect, compared to NRM, it has a much greater variety of models and a more advanced methodology that can help in cases where successful models cannot be produced using NRM.

In SNRM, the error function showing the difference between actual and predicted values is defined as the objective function, and model coefficients that will minimize its value are tried to be determined.

NRM and SNRM, two original mathematical modeling methods introduced within the scope of this study, were created through the coding process using the Mathematica program. Differential Evolution (DE), Nelder Mead (NM), Simulated Annealing (SA), and Random Search (RS) methods in Mathematica were used as optimization methods to determine the models.

2.3.3 MATHEMATICAL MODELS

In the mathematical modeling phase, linear, polynomial, trigonometric, logarithmic, rational, exponential, and hybrid forms were used as mathematical models to define the relationship between input and output parameters. Examples of the mentioned mathematical models are given in Table 2.5.

2.4 MATHEMATICA AND OPTIMIZATION

The Mathematica software has a collection of commands that enable analytical–numerical optimization to solve linear–nonlinear and unconstrained–constrained problems. In this context, while the NMinimize and NMaximize commands are used in numerical global optimization methods, Minimize and Maximize commands are suitable for analytical global optimization. Numerical local optimization is performed using the FindMinimum command. All the previously mentioned commands can be used for linear–nonlinear and constrained–unconstrained optimization problems [13, 14]. Detailed explanations of the commands, algorithms, and the types of problems they are used to solve are given in Table 2.6 and Figure 2.4. Numerical global optimization algorithms for constrained nonlinear problems can be classified as gradient-based and direct search methods. Gradient-based methods use first or second derivatives of the objective function and constraints for calculation, while Direct search methods have a probabilistic process and do not need derivative information. This section explains the Mathematica commands (FindMinimum, NMaximize, and Nminimize) and optimization algorithms (DE, MN, SA, RS) used within the book's scope.

TABLE 2.5
Mathematical Model Forms

Model Name	Nomenclature	Formula
Multiple linear	L	$Y = \sum_{i=1}^{4}(a_i Q_i) + c$
Multiple linear rational	LR	$Y = \dfrac{\sum_{i=1}^{4}(a_i Q_i) + c_1}{\sum_{j=1}^{4}(\beta_j Q_j) + c_2}$
Second order multiple nonlinear	SON	$Y = \sum_{k=1}^{4}\sum_{j=1}^{4}(a_j Q_j Q_k) + \sum_{i=1}^{4}(\beta_i Q_i) + c$
Second order multiple nonlinear rational	SONR	$Y = \dfrac{\sum_{i=1}^{4}\sum_{j=1}^{4}(a_j Q_j Q_i) + \sum_{i=1}^{4}(\gamma_i Q_i) + c_1}{\sum_{i=1}^{4}\sum_{m=1}^{4}(\beta_m Q_m Q_i) + \sum_{n=1}^{4}(\theta_n Q_n) + c_2}$
Third order multiple nonlinear	TON	$Y = \sum_{l=1}^{4}\sum_{m=1}^{4}\sum_{p=1}^{4}(a_l Q_l Q_m Q_p) + \sum_{k=1}^{4}\sum_{j=1}^{4}(\sigma_j Q_j Q_k) + \sum_{i=1}^{4}(\beta_i Q_i) + c$
Third order multiple nonlinear rational	TONR	$Y = \dfrac{\sum_{l=1}^{4}\sum_{m=1}^{4}\sum_{p=1}^{4}(a_l Q_l Q_m Q_p) + \sum_{k=1}^{4}\sum_{j=1}^{4}(\delta_j Q_j Q_k) + \sum_{i=1}^{4}(\sigma_i Q_i) + c_1}{\sum_{j=1}^{4}\sum_{r=1}^{4}\sum_{v=1}^{4}(\alpha_r Q_r Q_j Q_v) + \sum_{s=1}^{4}\sum_{t=1}^{4}(\beta_t Q_s Q_t) + \sum_{n=1}^{4}(\gamma_n Q_n) + c_2}$
Fourth order multiple nonlinear	FON	$Y = \sum_{r=1}^{4}\sum_{s=1}^{4}\sum_{t=1}^{4}\sum_{v=1}^{4}(a_s Q_r Q_s Q_t Q_v) + \sum_{l=1}^{4}\sum_{m=1}^{4}\sum_{p=1}^{4}(\beta_l Q_l Q_m Q_p) + \sum_{k=1}^{4}\sum_{j=1}^{4}(\gamma_j Q_j Q_k) + \sum_{i=1}^{4}(\delta_i Q_i) + c$
Fourth order multiple nonlinear rational	FONR	$Y = \dfrac{\sum_{g=1}^{4}\sum_{h=1}^{4}\sum_{w=1}^{4}\sum_{y=1}^{4}(a_g Q_g Q_h Q_w Q_y) + \sum_{l=1}^{4}\sum_{m=1}^{4}\sum_{p=1}^{4}(\beta_l Q_l Q_m Q_p) + \sum_{k=1}^{4}\sum_{j=1}^{4}(\alpha_j Q_j Q_k) + \sum_{i=1}^{4}(\gamma_i Q_i) + c_1}{\sum_{c=1}^{4}\sum_{u=1}^{4}\sum_{d=1}^{4}(\delta_c Q_c Q_u Q_d) + \sum_{j=1}^{4}\sum_{r=1}^{4}\sum_{v=1}^{4}(\varepsilon_f Q_f Q_j Q_v) + \sum_{s=1}^{4}(\theta_s Q_s Q_i) + \sum_{m=1}^{4}(\sigma_n Q_n) + c_2}$
First order trigonometric multiple nonlinear	FOTN	$Y = \sum_{i=1}^{4}(a_i Sin[Q_i] + \beta_i Cos[Q_i]) + c$

(Continued)

TABLE 2.5 *(Continued)*
Mathematical Model Forms

Model Name	Nomenclature	Formula
First order trigonometric multiple nonlinear rational	FOTNR	$Y = \dfrac{\sum_{i=1}^{4}(\beta_i Sin[Q_i] + \theta_i Cos[Q_i]) + c_1}{\sum_{m=1}^{4}(\alpha_m Sin[Q_m] + \gamma_m Cos[Q_m]) + c_2}$
Second order trigonometric multiple nonlinear	SOTN	$Y = \sum_{i=1}^{4}(a_i Sin[Q_i] + \beta_i Cos[Q_i]) + \sum_{j=1}^{4}(\theta_j Sin[Q_j] + \gamma_j Cos[Q_j])^2 + c$
Second order trigonometric multiple nonlinear rational	SOTNR	$Y = \dfrac{\sum_{i=1}^{4}(a_i Sin[Q_i] + \beta_i Cos[Q_i]) + \sum_{m=1}^{4}(\theta_m Sin^2[Q_m] + \gamma_m Cos^2[Q_m]) + c_1}{\sum_{k=1}^{4}(\sigma_k Sin[Q_k] + \alpha_k Cos[Q_k]) + \sum_{n=1}^{4}(\delta_n Sin^2[Q_n] + \tau_n Cos^2[Q_n]) + c_2}$
First order logarithmic multiple nonlinear	FOLN	$Y = \sum_{i=1}^{4}(a_i Log[Q_i]) + c$
First order logarithmic multiple nonlinear rational	FOLNR	$Y = \dfrac{\sum_{i=1}^{4}(a_i Log[Q_i]) + c_1}{\sum_{j=1}^{4}(\beta_j Log[Q_j]) + c_2}$
Second order logarithmic multiple nonlinear	SOLN	$Y = \sum_{k=1}^{4}\sum_{j=1}^{4}(a_j Log[Q_j Q_k]) + \sum_{i=1}^{4}(\beta_i Log[Q_i]) + c$
Second order logarithmic multiple nonlinear rational	SOLNR	$Y = \dfrac{\sum_{k=1}^{4}\sum_{i=1}^{4}(a_j Log[Q_j Q_k]) + \sum_{i=1}^{4}(\beta_i Log[Q_i]) + c_1}{\sum_{m=1}^{4}\sum_{i=1}^{4}(\gamma_l Log[Q_l Q_m]) + \sum_{n=1}^{4}(\delta_n Log[Q_n]) + c_2}$
Hybrid	H	$Y = \sum_{i=1}^{4}(a_i Q_i) + \sum_{i=1}^{4}(\gamma_i Sin[Q_i] + \beta_i Cos[Q_i]) + \sum_{i=1}^{4}(\delta_i Log[Q_i]) + c$
Hybrid Rational	HR	$Y = \dfrac{\sum_{i=1}^{4}(a_i Q_i) + \sum_{i=1}^{4}(\gamma_i Sin[Q_i] + \beta_i Cos[Q_i]) + \sum_{i=1}^{4}(\delta_i Log[Q_i]) + c_1}{\sum_{i=1}^{4}(\rho_i Q_i) + \sum_{i=1}^{4}(\theta_i Sin[Q_i] + \epsilon_i Cos[Q_i]) + \sum_{i=1}^{4}(\mu_i Log[Q_i]) + c_2}$

TABLE 2.6

Optimization Methods and Related Commands [2]

Optimization	Optimization Algorithm	Mathematica Command
Numerical Local Optimization	Linear Programming Methods Nonlinear Interior Point Algorithms	FindMinimum FindMaximum
Numerical Global Optimization	Linear Programming Methods Differential Evolution Nelder-Mead Simulated Annealing Random Search	NMinimize NMaximize
Exact Global Optimization	Linear Programming Methods Cylindrical Algebraic Decomposition Lagrange Multipliers Integer Linear Programming	Minimize Maximize
Linear Optimization	Linear Programming Methods (simplex, revised simplex, interior point)	LinearProgramming

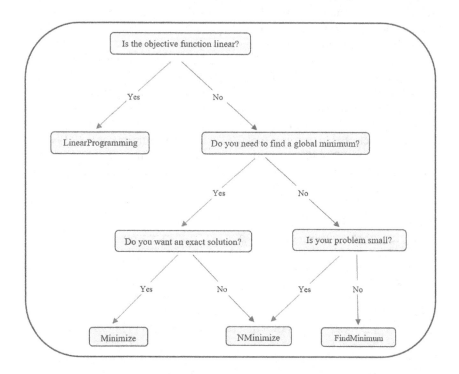

FIGURE 2.4   The Mathematica optimization process [2]

2.4.1 NMINIMIZE AND NMAXIMIZE FUNCTIONS

These functions in Mathematica allow complex problems in science and engineering to be optimized using search algorithms with specific characteristics. Although these methods effectively find global optima, it may be difficult to find optimum results when constraints and boundary conditions are not defined. The best way to deal with this situation would be to optimize the given function under different initial conditions.

The constraints can be either a list or a rational combination of domain options, equalities, and inequalities. For instance, if you want to specify results in integer form, you should include the unknown parameter "z" as "z ∈ Integers" in the line. This constraint will restrict the possible solutions to only integers. Additionally, the NMinimize command requires a rectangular starting region to begin optimization. This means each variable in the given function should have a finite upper and lower bound.

Method options can be applied to other search algorithms to provide non-automatic set solutions. If the objective function is minimized or maximized (called the objective function) and constraints are linear, the LinearProgramming method is the default setting in the solving process. If the primary part of the objective function is not numerical, and the variables are in integer form, DE is the algorithm as default. In other situations, NM is the search algorithm to be used. If NM does not provide desirable solutions, it switches to DE to obtain optimal values [15].

Examples of maximizing and minimizing a selected objective function using the NMinimize and NMaximize commands are given next.

Test functions whose values are desired to be maximized and minimized having more than one or many local minima are used. This kind of global optimization problem is quite difficult when an algorithm is not designed appropriately, and it can be inserted into the local minima without finding the global minimums or not all global minimums. In this regard, the Bukin function N6 and Eggholder function are selected as two test functions with global minima at f(-10,1)=0 and g(512,404.2319)=-959.6407, respectively [16]. The following commands give the Mathematica syntax for the definition of test functions and its 3D plot in an interval in Figures 2.5 and 2.6.

```
Clear["Global`*"]
f[x1_,x2_]:=100 Sqrt[Abs[x2-0.01x1^2]]+0.01 Abs[x1+10];
Plot3D[f[x1, x2], {x1, -15, -5}, {x2, -3, 3}, PlotRange ->
{0, 250}, AxesLabel -> {x1, x2, y}, ColorFunction ->
"CMYKColors"]
```

```
Clear["Global`*"]
g[x1_,x2_]:=-(x2+47) Sin[Sqrt[Abs[x2+x1/2+47]]]-x1 Sin[Sqrt
[Abs[x1-(x2+47)]]];
Plot3D[g[x1,x2],{x1,-512,512},{x2,-512,512},PlotRange->{-1000,
1500},AxesLabel->{x1,x2,y}, ColorFunction->Function[{x,y,z},
Hue[z]]]
```

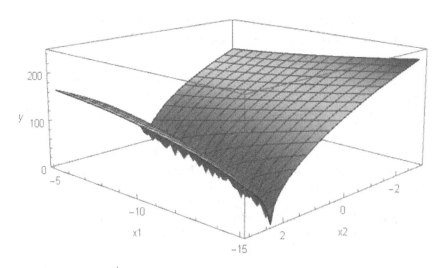

FIGURE 2.5   3D plot of the Bukin N6 test function

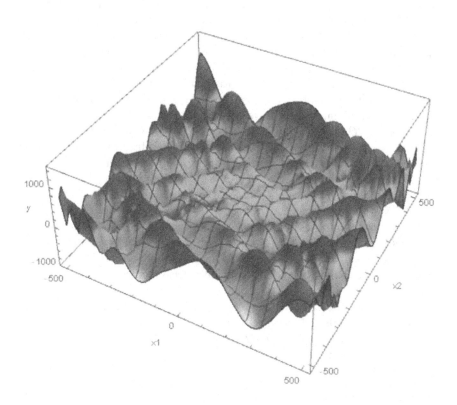

FIGURE 2.6   3D plot of the Eggholder test function

The results obtained using the NMinimize and NMaximize commands for the test functions Bukin N6 (f [x1,x2]) and Eggholder (g[x1,x2]) are given next.

```
Clear["Global`*"]
NMinimize[{f[x1,x2],-15<= x1<=- 5,-3<= x2<= 3},{x1,x2}]
Out[1]= {0.0370489,{x1->-6.29511,x2->0.396285}}
NMinimize[{g[x1,x2],-512<= x1<=512,-512<= x2<= 512},{x1,x2}]
Out[2]= {-507.874,{x1->-408.72,x2->-156.101}}
```

According to initial results, global minimum values of the Bukin N6 and Eggholder functions could not be achieved. Here, the NMinimize command is run with default values. Adjusting the parameters or changing the restriction region might be effective in obtaining global results. In the following sections, explanations of the optimization methods in Mathematica and the results of each optimization method for the two specified test functions will be given, and the effectiveness of the methods will be discussed.

2.4.2 RANDOM SEARCH ALGORITHM

This algorithm starts by creating a population of random starting points. Then, it uses the FindMinimum local search method to evaluate the convergence behavior of the starting points to the local minimum. The RS algorithm has several options to customize the optimization process. The SearchPoints option determines the number of starting points, which is calculated as "min(10 f,100)", where f is the number of variables. The RandomSeed option adjusts the starting value for the random number producer. The Method option selects which method to use for minimizing the objective function. For unconstrained optimization problems, the *"Quasi-Newton"* search method is used, which doesn't need the second derivatives (Hessians matrix) to be computed. Instead, the Hessian is updated by analyzing successive gradient vectors. For constrained optimization problems, the nonlinear interior point is used as the search method. Finally, the *"PostProcess"* option can be chosen as either Karush-Kuhn-Tucker (KKT) conditions or FindMinimum.

The RS algorithm uses several options by default, such as *"InitialPoints"*, *"Method"*, *"PenaltyFunction"*, *"PostProcess"*, and *"SearchPoints"*, and selects appropriate option values according to the optimization problem. The algorithm follows the procedure in Figure 2.7, and the best local minimum is chosen as the solution [15].

To assess the effectiveness of the RandomSearch algorithm in finding the global minimum, selected test functions Bukin N6 and Eggholder, which have global minima at f(-10,1)=0 and g(512,404.2319)=-959.6407, are utilized.

```
Clear["Global`*"]
NMinimize[{f[x1,x2],-15<= x1<=- 5,-3<= x2<= 3},{x1,x2},Method->
"RandomSearch"]
Out[3]= {0.010488,{x1->-10.35,x2->1.07123}}
```

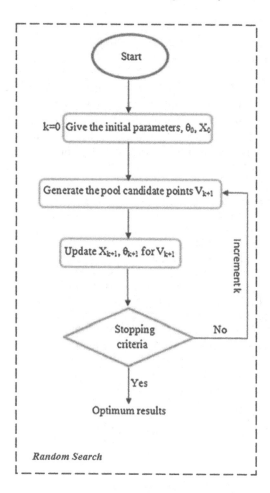

FIGURE 2.7   Flowchart of the Random Search algorithm [17]

Without altering its options, the RS command may not be able to find a global minimum. Sometimes changing the *"SearchPoints"* option that specifies the number of points to start searches can be effective in finding a global minimum.

```
Do[Print[NMinimize[{bukin6[x1,x2],-15≤x1≤-5,-3≤x2≤3},{x1,x2}
Method→{"RandomSearch", "SearchPoints" →i}]],{i,500,5000,500}]
{0.000483073,{x1->-9.96414,x2->0.99284}}
{0.000483073,{x1->-9.96414,x2->0.99284}}
{0.000483073,{x1->-9.96414,x2->0.99284}}
{0.000340469,{x1->-10.029,x2->1.0058}}
{0.000340469,{x1->-10.029,x2->1.0058}}
{0.000340469,{x1->-10.029,x2->1.0058}}
{0.000340469,{x1->-10.029,x2->1.0058}}
```

```
{0.000340469,{x1->-10.029,x2->1.0058}}
{0.000340469,{x1->-10.029,x2->1.0058}}
{0.000340469,{x1->-10.029,x2->1.0058}}
```

The effect of the *"RandomSeed"* option, which constitutes the starting value for the random number generator, can be investigated below. In the previous example, while *"SearchPoints"* is insufficient to reach the global minimum, it improved compared to the default settings. In the following example, a global minimum tried to be obtained by setting the values of the *"SearchPoints"* and the *"RandomSeed"*.

```
Do[Print[NMinimize[{f[x1,x2],-15≤x1≤-5,-3≤x2≤3}, {x1,x2},
Method→ {"RandomSearch", "SearchPoints"→500,"RandomSeed"→i}]],
{i,20}]
{0.00068578,{x1->-9.94584,x2->0.989197}}
{0.000878471,{x1->-9.93398,x2->0.986839}}
{0.000860442,{x1->-9.92384,x2->0.984827}}
{0.000563407,{x1->-9.96189,x2->0.992393}}
{0.00060851,{x1->-10.0508,x2->1.01019}}
{0.000989473,{x1->-10.0848,x2->1.01703}}
{0.000657367,{x1->-9.97222,x2->0.994452}}
{0.000542692,{x1->-10.045,x2->1.00903}}
{0.0006332,{x1->-10.0323,x2->1.00646}}
{0.000598592,{x1->-10.0176,x2->1.00353}}
{0.000559985,{x1->-10.042,x2->1.00842}}
{0.00027939,{x1->-10.0018,x2->1.00035}}
{0.000321799,{x1->-9.98477,x2->0.996957}}
{0.000519004,{x1->-9.95615,x2->0.991248}}
{0.000598531,{x1->-9.96535,x2->0.993083}}
{0.00122135,{x1->-9.90017,x2->0.980134}}
{0.000536666,{x1->-9.97038,x2->0.994084}}
{0.000389224,{x1->-10.0246,x2->1.00492}}
{0.000657191,{x1->-9.94941,x2->0.989908}}
{0.000929304,{x1->-9.91117,x2->0.982312}}
```

The *"RandomSeed"* option was also insufficient to achieve the global optimum result (f[x1,x2]=0). However, it has been shown that this option can be used to further improve the result. By utilizing this option, a value of f[-10.0018, 1.00035]=0.00027939 was obtained, which is very close to the global optimum.

```
Print[NMinimize[{f[x1,x2],-15≤x1≤-5,-3≤x2≤3},{x1,x2},Method→
{"RandomSearch","InitialPoints"→Flatten[Table[{i,j},{i,-10,0,5},
{j,-2,2,1}],1]}]]
{0.,{x1->-10.,x2->1.}}
```

"Initialpoints" can be produced on a grid to facilitate problem-solving. Assigning the starting point assists in finding the solution, particularly if the approximate range of the solution is known.

The *"PostProcess"* option is not of primary importance for this problem. PostProcess methods KKT and FindMinimum give the same results.

```
Print[NMinimize[{f[x1,x2],-15≤x1≤-5,-3≤x2≤3},{x1,x2},Method→
{"RandomSearch","SearchPoints"→3000,"PostProcess"→KKT}]]
{0.000340469,{x1->-10.029,x2->1.0058}}
```

```
Print[NMinimize[{f[x1,x2],-15≤x1≤-5,-3≤x2≤3},{x1,x2},Method→
{"RandomSearch","SearchPoints"→3000,"PostProcess"→FindMinimum}]]
{0.000340469,{x1->-10.029,x2->1.0058}}
```

```
Clear["Global`*"]
NMinimize[{Eggholher[x1,x2],-512≤x1≤512,-512≤x2≤512},{x1,x2},
Method→ "RandomSearch"]
{-959.641,{x1->512.,x2->404.232}}
```

The RS algorithm finds the global minimum point for the test function Eggholder without working any alteration of its options.

2.4.3 SIMULATED ANNEALING ALGORITHM

The SA algorithm is a stochastic method that imitates the physical annealing process of solids. Its primary purpose is to find the maximum or minimum values of functions with multiple variables, and to locate the minimum of nonlinear functions with several local minimums. The algorithm derives its name from the simulation of the perfect arrangement of atoms of solid bodies and minimizing potential energy during the cooling process. SA algorithm assists the structure to move away from the local minimum and explore and discover a better global minimum. This method is particularly useful in solving optimization problems that require a global optimum solution [18].

In each iteration, the algorithm generates a startup solution "Z". Then, a new solution "Znew" is produced in the neighborhood of the current point, "Z". Finally, the best solution "Zbest" is defined.

If $f(Z_{new}) \leq f(Z_{best})$, Z_{new} replaces Z_{best} and Z. Otherwise, Z_{new} replaces with Z. In this loop, the options InitialPoints, SearchPoints, and RandomSeed determine the initial guess, the number of guesses, and the starting value, respectively. The SA algorithm performs random motions in the search space based on the Boltzmann probability distribution $e^{D(k,\Delta f, f0)}$. The function D is defined by the option BoltzmannExponent, k is the current iteration, and Δf is the change in the objective function.

When not manually selected, B is defined as $\frac{-\Delta f \log(k+1)}{10}$ by Boltzmann Exponent within Mathematica.

The process outlined previously is used as the starting point for the algorithm. The algorithm continues until it either converges to a point or reaches the maximum number of iterations set by the Level Iterations option [19]. The SA algorithm follows the steps given in Figure 2.8.

Test functions Bukin N6 and Eggholder are utilized to evaluate the performance of the SA algorithm for finding the global minimum.

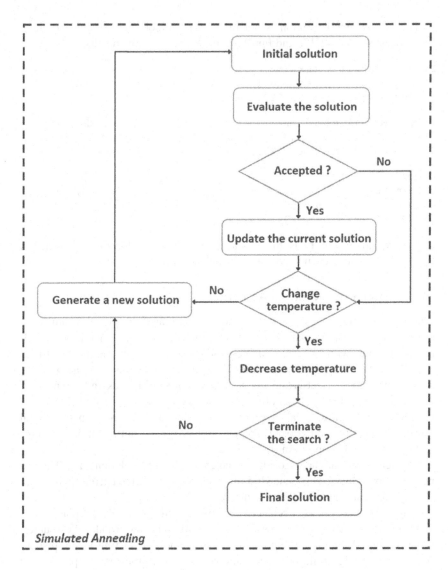

FIGURE 2.8   Flowchart of the Simulated Annealing algorithm [17]

The default options of the SA algorithm may not find a global minimum.

```
Clear["Global`*"]
f[x1_,x2_]:=100 Sqrt[Abs[x2-0.01x1^2]]+0.01 Abs[x1+10];
(*bukin n.6*)
NMinimize[{f[x1,x2],-15<= x1<=- 5,-3<= x2<= 3},{x1,x2},
Method-> "SimulatedAnnealing"]
Out[68]= {0.0420169,{x1->-13.3977,x2->1.79498}}
```

```
NMinimize[{f[x1,x2],-15<= x1<=- 5,-3<= x2<= 3},{x1,x2},
Method-> {"SimulatedAnnealing","BoltzmannExponent"->
Function[{i,df,f0},-df/(Exp[i/5])]}]
Out[101]= {0.00519003,{x1->-9.55843,x2->0.913636}}
Do[Print[NMinimize[{f[x1,x2],-15<= x1<=- 5,-3<= x2<= 3},
{x1,x2},Method-> {"SimulatedAnnealing","PerturbationScale"->
i}]],{i,15}]
{0.0420169,{x1->-13.3977,x2->1.79498}}
{0.00693808,{x1->-10.1023,x2->1.02056}}
{0.00879267,{x1->-10.3251,x2->1.06609}}
{0.0383077,{x1->-6.26297,x2->0.392248}}
{0.0307205,{x1->-12.0043,x2->1.44103}}
{0.0387912,{x1->-6.39833,x2->0.409386}}
{0.0269431,{x1->-12.2134,x2->1.49167}}
{0.013208,{x1->-8.75729,x2->0.766901}}
{0.0416711,{x1->-6.21507,x2->0.386271}}
{0.0490345,{x1->-14.8641,x2->2.20941}}
{0.0414215,{x1->-6.10944,x2->0.373253}}
{0.061582,{x1->-14.9993,x2->2.2498}}
{0.0416711,{x1->-6.21507,x2->0.386271}}
{0.0416711,{x1->-6.21507,x2->0.386271}}
{0.0416711,{x1->-6.21507,x2->0.386271}}
Do[Print[NMinimize[{f[x1,x2],-15<= x1<=- 5,-3<= x2<= 3},
{x1,x2},Method-> {"SimulatedAnnealing","SearchPoints"-> i}]],
{i,500,3000,500}]
{0.0564983,{x1->-5.,x2->0.25}}
{0.00446741,{x1->-9.94404,x2->0.98884}}
{0.0190515,{x1->-11.0795,x2->1.22755}}
{0.0359381,{x1->-7.17735,x2->0.515144}}
{0.0176009,{x1->-8.33379,x2->0.69452}}
{0.0450136,{x1->-13.3313,x2->1.77724}}
Do[Print[NMinimize[{f[x1,x2],-15<= x1<=- 5,-3<= x2<= 3},
{x1,x2},Method-> {"SimulatedAnnealing","RandomSeed"->i}]],
{i,0,20}]
{0.0420169,{x1->-13.3977,x2->1.79498}}
{0.0172848,{x1->-11.5721,x2->1.33914}}
{0.0301351,{x1->-12.0219,x2->1.44525}}
{0.0375189,{x1->-6.35022,x2->0.403253}}
{0.0109509,{x1->-10.3562,x2->1.07251}}
{0.0444877,{x1->-5.61579,x2->0.315372}}
{0.0264381,{x1->-12.3348,x2->1.52147}}
{0.0158634,{x1->-11.2169,x2->1.25818}}
{0.0568719,{x1->-5.00033,x2->0.250033}}
{0.0158629,{x1->-11.3113,x2->1.27946}}
{0.0486088,{x1->-5.41572,x2->0.2933}}
{0.015118,{x1->-8.50841,x2->0.723931}}
{0.0548695,{x1->-14.2714,x2->2.03674}}
{0.00521231,{x1->-10.4862,x2->1.09961}}
{0.0335872,{x1->-7.25085,x2->0.525748}}
{0.0250491,{x1->-8.08192,x2->0.653175}}
```

```
{0.0585989,{x1->-14.8782,x2->2.2136}}
{0.0537045,{x1->-5.4578,x2->0.297876}}
{0.0553902,{x1->-14.9835,x2->2.24505}}
{0.0151499,{x1->-9.33279,x2->0.87101}}
{0.0413764,{x1->-5.91857,x2->0.350295}}
Do[Print[NMinimize[{f[x1,x2],-15<= x1<=- 5,-3<= x2<= 3},
{x1,x2},Method-> {"SimulatedAnnealing","PerturbationScale"->
2, "SearchPoints"-> 1000, "RandomSeed"->i}]],{i,0,20,1}]
{0.0319362,{x1->-13.0877,x2->1.71288}}
{0.0281585,{x1->-7.28045,x2->0.530049}}
{0.0509739,{x1->-5.,x2->0.25}}
{0.00104053,{x1->-9.94861,x2->0.989748}}
{0.0332591,{x1->-7.53578,x2->0.567879}}
{0.0383163,{x1->-12.7204,x2->1.61808}}
{0.0257383,{x1->-7.56489,x2->0.572275}}
{0.0172252,{x1->-8.36268,x2->0.699344}}
{0.0473644,{x1->-5.30414,x2->0.281339}}
{0.0257528,{x1->-11.5737,x2->1.33949}}
{0.0212155,{x1->-8.67546,x2->0.752636}}
{0.0161807,{x1->-8.71827,x2->0.760083}}
{0.0555851,{x1->-5.04672,x2->0.254694}}
{0.0534486,{x1->-5.,x2->0.25}}
{0.0548202,{x1->-5.,x2->0.25}}
{0.056041,{x1->-5.,x2->0.25}}
{0.062665,{x1->-15.,x2->2.25}}
{0.0166399,{x1->-8.35081,x2->0.69736}}
{0.0169809,{x1->-8.93984,x2->0.799207}}
{0.00909523,{x1->-9.81725,x2->0.963784}}
{0.00640446,{x1->-10.053,x2->1.01062}}
Do[Print[NMinimize[{f[x1,x2],-15<= x1<=- 5,-3<= x2<= 3},
{x1,x2},Method-> {"SimulatedAnnealing","BoltzmannExponent"->
Function[{i,df,f0},-df/ (Exp[i/5])], "PerturbationScale"-> 2,
"SearchPoints"-> 1000,"RandomSeed"-> i}]],{i,0,20,1}]
{0.051775,{x1->-5.00651,x2->0.250651}}
{0.0590955,{x1->-14.997,x2->2.24911}}
{0.0307994,{x1->-7.57637,x2->0.574013}}
{0.0410629,{x1->-5.92817,x2->0.351432}}
{0.000753764,{x1->-9.95206,x2->0.990434}}
{0.020579,{x1->-12.0081,x2->1.44195}}
{0.050011,{x1->-5.08875,x2->0.258954}}
{0.0517649,{x1->-5.00184,x2->0.250184}}
{0.0160514,{x1->-8.41241,x2->0.707686}}
{0.0439246,{x1->-13.5256,x2->1.82941}}
{0.0252636,{x1->-12.5168,x2->1.5667}}
{0.0418112,{x1->-6.67966,x2->0.446179}}
{0.0468911,{x1->-5.39816,x2->0.291401}}
{0.00183645,{x1->-9.8211,x2->0.96454}}
{0.0276218,{x1->-7.6636,x2->0.587308}}
{0.0217276,{x1->-8.02402,x2->0.643849}}
{0.0503916,{x1->-5.00033,x2->0.250033}}
```

```
{0.0525499,{x1->-14.9991,x2->2.24972}}
{0.050598,{x1->-5.00077,x2->0.250077}}
{0.0141484,{x1->-8.62651,x2->0.744167}}
{0.0508397,{x1->-5.00653,x2->0.250654}}
Clear["Global`*"]
g[x1_,x2_]:=-(x2+47) Sin[Sqrt[Abs[x2+x1/2+47]]]-x1 Sin[Sqrt[Abs[x1-
(x2+47)]]];
NMinimize[{g[x1,x2],-512<= x1<=512,-512<= x2<= 512},{x1,x2},
Method-> "SimulatedAnnealing"]
Out[85]= {-935.338,{x1->439.481,x2->453.977}}
NMinimize[{g[x1,x2],-512<= x1<=512,-512<= x2<= 512},{x1,x2},
Method-> {"SimulatedAnnealing","BoltzmannExponent"-> Function
[{i,df,f0},    -df/(Exp[i/100])]}]
Out[113]= {-959.641,{x1->512.,x2->404.232}}
Do[Print[NMinimize[{g[x1,x2],-512<= x1<=512,-512<= x2<= 512},
{x1,x2},Method-> {"SimulatedAnnealing","PerturbationScale"->
i}]], {i,15}]
{-935.338,{x1->439.481,x2->453.977}}
{-955.255,{x1->479.045,x2->434.505}}
{-894.579,{x1->-465.694,x2->385.717}}
{-959.641,{x1->512.,x2->404.232}}
{-959.641,{x1->512.,x2->404.232}}
{-959.641,{x1->512.,x2->404.232}}
{-888.949,{x1->347.327,x2->499.415}}
{-888.949,{x1->347.327,x2->499.415}}
{-959.641,{x1->512.,x2->404.232}}
{-959.641,{x1->512.,x2->404.232}}
{-959.641,{x1->512.,x2->404.232}}
{-821.196,{x1->-313.972,x2->512.}}
{-888.949,{x1->347.327,x2->499.415}}
{-821.196,{x1->-313.972,x2->512.}}
{-959.641,{x1->512.,x2->404.232}}
Do[Print[NMinimize[{g[x1,x2],-512<= x1<=512,-512<= x2<= 512},
{x1,x2},Method-> {"SimulatedAnnealing","SearchPoints"-> i}]],
{i,50,300,50}]
{-959.641,{x1->512.,x2->404.232}}
{-959.641,{x1->512.,x2->404.232}}
{-959.641,{x1->512.,x2->404.232}}
{-959.641,{x1->512.,x2->404.232}}
{-959.641,{x1->512.,x2->404.232}}
{-959.641,{x1->512.,x2->404.232}}
Do[Print[NMinimize[{g[x1,x2],-512<= x1<=512,-512<= x2<= 512},
{x1,x2},Method-> {"SimulatedAnnealing","RandomSeed"->i}]],
{i,0,20}]
{-935.338,{x1->439.481,x2->453.977}}
{-959.641,{x1->512.,x2->404.232}}
{-894.579,{x1->-465.694,x2->385.717}}
{-821.196,{x1->-313.972,x2->512.}}
{-959.641,{x1->512.,x2->404.232}}
{-821.196,{x1->-313.972,x2->512.}}
```

```
{-894.579,{x1->-465.694,x2->385.717}}
{-959.641,{x1->512.,x2->404.232}}
{-821.196,{x1->-313.972,x2->512.}}
{-894.579,{x1->-465.694,x2->385.717}}
{-894.579,{x1->-465.694,x2->385.717}}
{-959.641,{x1->512.,x2->404.232}}
{-894.579,{x1->-465.694,x2->385.717}}
{-959.641,{x1->512.,x2->404.232}}
{-888.949,{x1->347.327,x2->499.415}}
{-959.641,{x1->512.,x2->404.232}}
{-821.196,{x1->-313.972,x2->512.}}
{-821.196,{x1->-313.972,x2->512.}}
{-894.579,{x1->-465.694,x2->385.717}}
{-888.949,{x1->347.327,x2->499.415}}
{-959.641,{x1->512.,x2->404.232}}
```

2.4.4 DIFFERENTIAL EVOLUTION ALGORITHM

DE is a popular stochastic search algorithm for complex design problems. DE consists of four main steps: *"initialization"*, *"mutation"*, *"crossover"*, and *"selection"*. While DE efficiently searches for solutions by exploring a population of solutions in iterations rather than a single solution, it is computationally expensive. Nevertheless, DE is a reliable method to obtain global optimum and is known for its robustness. However, like other search methods, there is still some uncertainty in finding the global optimum points [2].

In the DE algorithm, a new population of k points is produced in iterations. To generate the jth new point, three random points, z_1, z_2, and z_3, are taken from the previously generated population. Then, a new formation is built by calculating $z_s = z_3 + s(z_1 - z_2)$, where s is the real scaling factor.

A new point, z_{new}, is generated from z_j and zs by randomly choosing either the ith coordinate of z_j or another coordinate of z_s with a probability of ρ. If the function of $h(z_{new})$ is smaller than the function of $h(z_j)$, then znew replaces z_j [15].

The DifferentialEvolution command includes several adjustment options such as *"CrossProbability"*, *"InitialPoints"*, *"PenaltyFunction"*, *"PostProcess"*, *"RandomSeed"*, *"ScalingFactor"*, *"SearchPoints"*, and *"Tolerance"*. However, none of these options guarantees finding the global optima. The process flow chart of the algorithm is illustrated in Figure 2.9 [17].

As with previous search algorithms, we evaluate the DE algorithm's ability to find the global minimum using the same test functions of Bukin N6 and Eggholder.

The minimum values obtained from the test functions when using the DE method, with consideration to the impact of parameters such as CrossProbability, InitialPoints, RandomSeed, ScalingFactor, and SearchPoints are given here.

```
Clear["Global`*"]
f[x1_,x2_]:=100 Sqrt[Abs[x2-0.01x1^2]]+0.01 Abs[x1+10];
(*bukin n.6*)
```

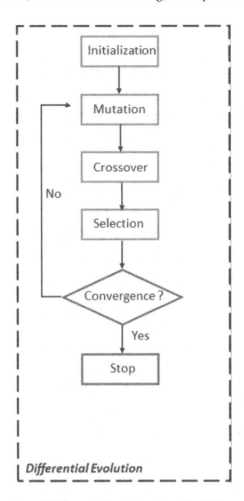

FIGURE 2.9   Flowchart of Differential Evolution algorithm [17]

```
NMinimize[{f[x1,x2],-15<= x1<=- 5,-3<= x2<= 3},{x1,x2},
Method-> "DifferentialEvolution"]
Out[116]= {0.0315634,{x1->-7.20753,x2->0.519485}}
NMinimize[{f[x1,x2],-15<= x1<=- 5,-3<= x2<= 3},{x1,x2},
Method-> {"DifferentialEvolution","ScalingFactor"->0.2}]
Out[130]= {0.0117477,{x1->-9.54497,x2->0.911064}}
NMinimize[{f[x1,x2],-15<= x1<=- 5,-3<= x2<= 3},{x1,x2},Method->
{"DifferentialEvolution", "CrossProbability"->0.5,
"ScalingFactor"-> 0.2,"SearchPoints"->1000}]
Out[146]= {0.00407254,{x1->-10.383,x2->1.07807}}
Do[Print[NMinimize[{f[x1,x2],-15<= x1<=- 5,-3<= x2<= 3},
{x1,x2}, Method-> {"DifferentialEvolution","CrossProbability"->
0.5, "ScalingFactor"->0.2,"SearchPoints"-> i}]],
{i,1000,3000,500}]
```

```
{0.00407254,{x1->-10.383,x2->1.07807}}
{0.0173933,{x1->-10.8352,x2->1.17401}}
{0.00923781,{x1->-10.2566,x2->1.05199}}
{0.00969754,{x1->-9.75805,x2->0.952194}}
{0.0123601,{x1->-9.6189,x2->0.925232}}
Do[Print[NMinimize[{f[x1,x2],-15<= x1<=- 5,-3<= x2<= 3},
{x1,x2},Method-> {"DifferentialEvolution","CrossProbability"->
0.5, "ScalingFactor"->0.2,"SearchPoints"-> 1000,"RandomSeed"->
i}]], {i,0,20}]
{0.00407254,{x1->-10.383,x2->1.07807}}
{0.0144943,{x1->-9.16508,x2->0.839987}}
{0.0185387,{x1->-8.89292,x2->0.790841}}
{0.0124773,{x1->-9.55109,x2->0.912232}}
{0.0217938,{x1->-8.2655,x2->0.683185}}
{0.0194102,{x1->-8.23301,x2->0.677824}}
{0.0104907,{x1->-10.3203,x2->1.06509}}
{0.0197497,{x1->-8.88077,x2->0.788681}}
{0.00554957,{x1->-9.49807,x2->0.902133}}
{0.0313111,{x1->-7.73043,x2->0.597595}}
{0.0194827,{x1->-8.90377,x2->0.792771}}
{0.00721605,{x1->-10.0705,x2->1.01414}}
{0.0217444,{x1->-8.68049,x2->0.753509}}
{0.00851923,{x1->-9.25129,x2->0.855864}}
{0.0217104,{x1->-8.15944,x2->0.665764}}
{0.0178786,{x1->-10.9831,x2->1.20629}}
{0.00502506,{x1->-9.57927,x2->0.917624}}
{0.02943,{x1->-7.34657,x2->0.53972}}
{0.0242952,{x1->-8.36707,x2->0.700078}}
{0.00781364,{x1->-9.28104,x2->0.861377}}
{0.00415195,{x1->-9.6147,x2->0.924424}}
NMinimize[{f[x1,x2],-15<= x1<=- 5,-3<= x2<= 3},{x1,x2},
Method-> {"DifferentialEvolution", "CrossProbability"->0.5,
"ScalingFactor"-> 0.2,"SearchPoints"-> 1000,"InitialPoints"->
Flatten[Table[{i,j}, {i,-10,0,5},{j,-2,2,1}],1]}]
{0.,{x1->-10.,x2->1.}}
Clear["Global`*"]
g[x1_,x2_]:=-(x2+47) Sin[Sqrt[Abs[x2+x1/2+47]]]-x1Sin[Sqrt[Abs[x1-
(x2+47)]]];
NMinimize[{g[x1,x2],-512<= x1<=512,-512<= x2<= 512},{x1,x2},
Method-> "DifferentialEvolution"]
{-888.949,{x1->347.327,x2->499.415}}
NMinimize[{g[x1,x2],-512<= x1<=512,-512<= x2<= 512},{x1,x2},
Method-> {"DifferentialEvolution", "ScalingFactor"->0.5}]
{-959.641,{x1->512.,x2->404.232}}
NMinimize[{g[x1,x2],-512<= x1<=512,-512<= x2<= 512},{x1,x2},
Method-> {"DifferentialEvolution", "SearchPoints"->500}]
{-959.641,{x1->512.,x2->404.232}}
NMinimize[{g[x1,x2],-512<= x1<=512,-512<= x2<= 512},{x1,x2},
Method-> {"DifferentialEvolution", "CrossProbability"->0.1}]
{-959.641,{x1->512.,x2->404.232}}
```

To find the global minimum of the Bukin N6 test function, the InitialPoints option needs to be set. However, for the Eggholder test function, the global minimum point can be found easily by making simple changes to the default settings.

2.4.5 NELDER MEAD ALGORITHM

The NM algorithm or simplex is a derivative-free optimization method used for local search problems. It was initially designed for unconstrained optimization problems [2]. In an m-dimensional space, given a function of m variables, this method keeps a set of m+1 points that generate the vertices of a polytope. It should be noted that this method is not the same as the simplex method used for linear programming. The process involves forming m+1 points, numbered y_1, y_2, y_3,..., y_{m+1}, which are arranged in the order of their function values as $h(y_1) \le h(y_2) \le h(y_3) \le ...h(y_{m+1})$. After a new point is produced, it replaces the previous worst point y_{m+1}. A polytope can be defined in terms of its centroid, which is the average position of all the points of an object, represented as $c = \Sigma_{i=1}^{m} y_i$. At this point, a trial point is defined as y_t. It is produced by reflecting the worst point until the centroid, $y_t = c + \propto (c - y_{m+1})$, where $\propto$ is a parameter that is larger than 0. The new point need not be a new worst point or a new best point. Hence, $h(y_1) \le h(y_t) \le h(y_m)$, and y_t replaces y_{m+1}. If a new point is obtained that is better than the previous best point, it means that the reflection is successful. Further, it can be continued with $y_e = c + \beta(y_t - r)$, where β is a parameter that is larger than 1 and is used to enlarge the polytope. If $h(y_e)$ is obtained as smaller than $h(y_t)$, it means that the expansion process is achieved, and y_e replaces y_{m+1}. Alternatively, if y_t performs worse than the second-worst point, $h(y_t) \le h(y_m)$, the polytope is considered too large and needs to be contracted. A new trial point is obtained using the following expressions [2].

$$y_c = \left(\begin{array}{l} c + \gamma \left(y_{m+1} - c \right), \quad \text{if } h\left(y_t \right) \ge h\left(y_{m+1} \right) \\ c + \gamma \left(y_t - c \right), \quad \text{if } h\left(y_t \right) < h\left(y_{m+1} \right) \end{array} \right)$$

The parameter γ can range between 0 to 1. If contraction is achieved, it means that the value of $h(y_c)$ is smaller than the minimum value between $h(y_{m+1})$ and h(yt). On the other hand, if successful contraction is not achieved, more processing is required.

NM is an algorithm that comes with adjustable options such as *"Contract Ratio", "Expand Ratio", "Initial Points", "Penalty Function", "Post Process", "Random Seed", "Reflect Ratio", "Shrink Ratio", and "Tolerance"*. This algorithm may not provide all the specifications required for true global optimization, but it tends to work well for problems with fewer local minima. Like previous algorithms, Nelder Mead can be used to obtain global minimum for Bukin N6 and Eggholder test functions [15]. The process flow chart of the algorithm is illustrated in Figure 2.10 [17].

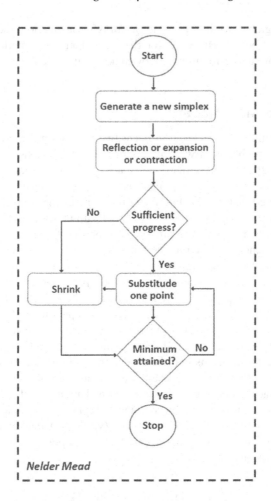

FIGURE 2.10 Flowchart of the Nelder Mead algorithm [17]

The global minimum results for the test functions Bukin N6 and Eggholder and the effect of changes made in the NM algorithm options on the results can be seen in the following trials performed.

```
Clear["Global`*"]
f[x1_,x2_]:=100 Sqrt[Abs[x2-0.01x1^2]]+0.01 Abs[x1+10];
(*bukin n.6*)
NMinimize[{f[x1,x2],-15<= x1<=- 5,-3<= x2<= 3},{x1,x2},
Method-> "NelderMead"]
{0.0370489,{x1->-6.29511,x2->0.396285}}
Do[Print[NMinimize[{f[x1,x2],-15<= x1<=- 5,-3<= x2<= 3},
{x1,x2}, Method-> {"NelderMead","RandomSeed"->i}]],{i,0,20}]
{0.0370489,{x1->-6.29511,x2->0.396285}}
{0.0332237,{x1->-13.3224,x2->1.77486}}
```

```
{0.0273448,{x1->-12.7345,x2->1.62167}}
{0.0275884,{x1->-12.7588,x2->1.62788}}
{0.0395644,{x1->-6.04356,x2->0.365246}}
{0.0461063,{x1->-14.6104,x2->2.13464}}
{0.0184811,{x1->-8.15189,x2->0.664534}}
{0.0498296,{x1->-14.983,x2->2.24489}}
{0.00997675,{x1->-9.00232,x2->0.810419}}
{0.0487399,{x1->-14.874,x2->2.21236}}
{0.0417074,{x1->-14.1707,x2->2.0081}}
{0.00280245,{x1->-10.2802,x2->1.05683}}
{0.0489411,{x1->-5.10602,x2->0.260714}}
{0.00694663,{x1->-9.30534,x2->0.865893}}
{0.0493259,{x1->-14.9326,x2->2.22982}}
{0.017279,{x1->-8.2721,x2->0.684276}}
{0.043348,{x1->-14.3348,x2->2.05486}}
{0.0276831,{x1->-12.7683,x2->1.6303}}
{0.0257374,{x1->-7.42626,x2->0.551493}}
{0.00119802,{x1->-9.8802,x2->0.976183}}
{0.00671256,{x1->-9.32874,x2->0.870255}}
Do[Print[NMinimize[{f[x1,x2],-15<= x1<=- 5,-3<= x2<= 3},
{x1,x2}, Method-> {"NelderMead", "ShrinkRatio"->0.5,
"ContractRatio"->0.8, "ReflectRatio"->2, "RandomSeed"->i}]],
{i,5}]
{0.0316293,{x1->-12.8629,x2->1.65453}}
{0.0230235,{x1->-11.5121,x2->1.32528}}
{0.0593528,{x1->-14.6578,x2->2.14852}}
{0.00627351,{x1->-10.3029,x2->1.06151}}
{0.0598293,{x1->-14.9914,x2->2.24742}}
Print[NMinimize[{f[x1,x2],-15<= x1<=-5,-3<= x2<= 3},{x1,x2},
Method-> {"NelderMead", "ShrinkRatio"-> 0.5, "ContractRatio"->
0.8,"ReflectRatio"->2, "InitialPoints"->Flatten[Table[{i,j},
{i,-10,0,5}, {j,-2,2,1}],1]}]]
{0.,{x1->-10.,x2->1.}}

Clear["Global`*"]
g[x1_,x2_]:=-(x2+47) Sin[Sqrt[Abs[x2+x1/2+47]]]-x1 Sin[Sqrt
[Abs[x1-(x2+47)]]];
NMinimize[{g[x1,x2],-512<= x1<=512,-512<= x2<= 512}, {x1,x2},
Method-> "NelderMead"]
{-507.874,{x1->-408.72,x2->-156.101}}
Do[Print[NMinimize[{g[x1,x2],-512<= x1<=512,-512<= x2<= 512},
{x1,x2},Method-> {"NelderMead", "RandomSeed"->i}]],{i,0,20}]
{-507.874,{x1->-408.72,x2->-156.101}}
{-565.998,{x1->-105.877,x2->423.153}}
{-633.842,{x1->-174.914,x2->-512.}}
{-633.842,{x1->-174.914,x2->-512.}}
{-718.167,{x1->283.076,x2->-487.126}}
{-306.064,{x1->-15.6579,x2->-342.777}}
{-202.539,{x1->151.596,x2->-100.611}}
{-633.842,{x1->-174.914,x2->-512.}}
```

```
{-888.949,{x1->347.327,x2->499.415}}
{-633.842,{x1->-174.914,x2->-512.}}
{-821.196,{x1->-313.972,x2->512.}}
{-306.72,{x1->-76.1459,x2->191.155}}
{-718.167,{x1->283.076,x2->-487.126}}
{-894.579,{x1->-465.694,x2->385.717}}
{-718.167,{x1->283.076,x2->-487.126}}
{-716.672,{x1->399.559,x2->-367.691}}
{-821.196,{x1->-313.972,x2->512.}}
{-821.196,{x1->-313.972,x2->512.}}
{-821.196,{x1->-313.972,x2->512.}}
{-211.18,{x1->-171.139,x2->-96.1191}}
{-206.572,{x1->13.2258,x2->151.328}}
Do[Print[NMinimize[{g[x1,x2],-512<= x1<=512,-512<= x2<= 512},
{x1,x2},Method-> {"NelderMead", "ShrinkRatio"->0.5,
"ContractRatio"-> 0.5,"ReflectRatio"->7,"RandomSeed"->i}]],
{i,5}]
{-935.338,{x1->439.481,x2->453.977}}
{-718.167,{x1->283.076,x2->-487.126}}
{-786.526,{x1->-456.886,x2->-382.623}}
{-955.255,{x1->479.045,x2->434.505}}
{-493.961,{x1->150.232,x2->301.314}}
Print[NMinimize[{g[x1,x2],-512<= x1<=512,-512<= x2<= 512},
{x1,x2}, Method-> {"NelderMead", "ShrinkRatio"->0.5,
"ContractRatio"->0.8,  "ReflectRatio"->2, "InitialPoints"->
Flatten[Table[{i,j},{i,-512,512,10},{j,512,512,10}],1]}]]
{-956.918,{x1->482.353,x2->432.878}}
```

2.4.6 FINDMINIMUM

The FindMinimum command is a useful tool for finding the local minimum of a function in both unconstrained and constrained optimization problems [14]. There are several options available when using the FindMinimum command, including *"Method", "MaxIterations", "WorkingPrecision", "PrecisionGoal", and "AccuracyGoal"* [14].

The Method option allows you to choose which method to use to solve the problem. For unconstrained optimization problems, there are several methods available, such as *"Newton", "QuasiNewton", "LevenbergMarquardt", "ConjugateGradient", and "PrincipalAxis"*. Each method has its own strengths and weaknesses, so it's important to choose the one that is best suited for your problem. For constrained optimization problems, the only available method is InteriorPoint.

The MaxIterations option sets the maximum number of iterations that the algorithm will perform. The default value for constrained optimization problems is 500.

The WorkingPrecision, PrecisionGoal, and AccuracyGoal options specify the number of digits of precision that you want. WorkingPrecision controls the internal computations, while PrecisionGoal and AccuracyGoal check the final result. By default, WorkingPrecision is set to MachinePrecision, but you can set it to a specific value if needed. If you choose Automatic for PrecisionGoal and AccuracyGoal, the default values are set to WorkingPrecision/3 and Infinity, respectively [14].

To test the effectiveness of the FindMinimum command and its options in finding local minima, we can use the Schwefel function, which has global minima at h (420.9687,420.9687)=0 [16].

The following commands give the Mathematica syntax for the definition of Schwefel test function and its 3D plot in an interval in Figure 2.11.

```
Clear["Global`*"]
h[x1_,x2_]:=418.9829 2-(x1 Sin[Sqrt[Abs[x1]]]+x2
Sin[Sqrt[Abs[x2]]]);
Plot3D[h[x1,x2],{x1,-500,500},{x2,-500,500},PlotRange->
{0,1800},  AxesLabel-> {x1,x2,y}, ColorFunction->"DarkBands"]

FindMinimum[{h[x1,x2],-500<= x1<= 500,-500<= x2<=
500},{x1,x2}],{830.075, {x1->5.2392,x2->5.2392}}
FindMinimum[{h[x1,x2],-500<= x1<= 500,-500<= x2<= 500},{x1,x2},
Method-> "InteriorPoint"]
{830.075,{x1->5.2392,x2->5.2392}}
```

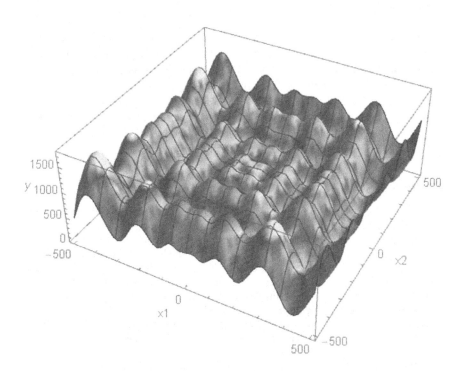

FIGURE 2.11   3D plot of the Schwefel test function

```
Do[Print[FindMinimum[{h[x1,x2],-500<= x1<= 500,-500<= x2<=
500},{x1,x2}, Method-> "InteriorPoint","MaxIterations"-> i]],
{i,{1,10,100,500,1000,2000,4000,8000}}]
{833.772,{x1->2.11162,x2->2.11162}}
{830.075,{x1->5.2392,x2->5.2392}}
{830.075,{x1->5.2392,x2->5.2392}}
{830.075,{x1->5.2392,x2->5.2392}}
{830.075,{x1->5.2392,x2->5.2392}}
{830.075,{x1->5.2392,x2->5.2392}}
{830.075,{x1->5.2392,x2->5.2392}}
{830.075,{x1->5.2392,x2->5.2392}}
Table[Print[FindMinimum[{h[x1,x2],-500<= x1<= 500,-500<= x2<=
500}, {{x1,RandomReal[{-500, 500}]},{x2,RandomReal[{-500,
500}]}},Method-> "InteriorPoint"]],{10}]
{118.438,{x1->420.969,x2->-302.525}}
{238.394,{x1->420.969,x2->-500.}}
{750.248,{x1->65.5479,x2->-25.8774}}
{455.533,{x1->-500.,x2->203.814}}
{335.578,{x1->-302.525,x2->203.814}}
{335.578,{x1->-302.525,x2->203.814}}
{0.0000254551,{x1->420.969,x2->420.969}}
{394.9,{x1->420.969,x2->-25.8774}}
{335.578,{x1->-302.525,x2->203.814}}
{434.279,{x1->203.814,x2->203.814}}
```

After conducting optimization studies with test functions, it was found that the algorithms could easily find the global minimum in the Eggholder test function using default options. However, special adjustments were required in the algorithm options to achieve the same for the Bukin N6 test function. It was observed that DE, NM, SA, and RS algorithms can effortlessly solve nonlinear problems with constraints. On the other hand, the FindMinimum algorithm can tackle problems with constraints using the InteriorPoint method, and the other methods included cannot be used in problems involving constraints.

REFERENCES

[1] Montgomery DC. Design and Analysis of Experiment. 7th Edition. John Wiley Sons. 649 p, 2008.
[2] Savran M. A new systematic approach for design, modeling and optimization of the engineering problems based on stochastic multiple-nonlinear neuro-regression analysis and non traditional search algorithms (PhD thesis). İzmir Katip Çelebi University, 2023. https://tez.yok.gov.tr/UlusalTezMerkezi/tezSorguSonucYeni.jsp
[3] Antony J. Design of Experiments for Engineers and Scientists. In Design of Experiments for Engineers and Scientists 2014.
[4] Sundararajan K. Design of Experiments – A Primer. 2018. Retrieved from https://www.isixsigma.com/tools-templates/design-of-experiments-doe/design-experiments-ᵇᵉ₅-primer/#comments
[5] Wahdame B, Candusso D, Franc X, Harel F. Design of experiment techniques for fuel cell characterisation and development. International Journal of Hydrogen Energy 2009; 34: 967–980.

[6] Multisimplex AG. Multisimplex users guide, 1999. www.multisimplex.com

[7] Savaşkan M. Deney tasarımı yöntemlerinin karşılaştırmalı kullanımı ile ince sert seramik kaplı matkap uçlarının performans değerlendirmesi ve optimizasyonu (doktora tezi). İstanbul Teknik Üniversitesi; 2003. https://tez.yok.gov.tr/UlusalTezMerkezi/tezSorguSonucYeni.jsp

[8] Anderson, M. J. *DOE simplified: practical tools for effective experimentation.* 3rd Ed. CRC Press.

[9] Design expert. Applications of DOE in engineering and science: A collection of 26 case studies. https://cdn.statease.com/media/public/documents/Applications%20of%20DOE%20in%20Engineering%20and%20Science%202019%20Revised-LYE%20Sept%2017.pdf

[10] Ferreira SLC, Bruns RE, Ferreira HS, Matos GD, David JM, Brandão GC, dos Santos WNL, et al. Box-Behnken design: An alternative for the optimization of analytical methods. Analytica Chimica Acta 2007; 597(2): 179–186.

[11] Taguchi G, Clausing D. Robust Quality. Harvard Business Review, 1990; January-February 65–76.

[12] Gökçe B, Taşgetiren S. Kalite için Deney tasarımı. Makine Teknolojileri Elektronik Dergisi 2009; 6(1): 71–83.

[13] Aydin L, Artem HS, Oterkus S et al. Designing Engineering Structures using Stochastic Optimization Methods. 1st Ed. CRC Press.

[14] Wolfram Mathematica software 12.

[15] Maaroof OW, Dede MİC, Aydin L. A robot arm design optimization method by using a kinematic redundancy resolution technique. Robotics. 2022; 11(1): 1.

[16] Virtual library of simulation experiments: Test functions and datasets. https://www.sfu.ca/~ssurjano/bukin6.html

[17] Savran M. Development of vibration performances of hybrid laminated composite materials by using stochastic methods (Master thesis). İzmir Katip Çelebi University, 2017. https://tez.yok.gov.tr/UlusalTezMerkezi/tezSorguSonucYeni.jsp

[18] Karaboğa D. Yapay zeka optimizasyon algoritmaları. Nobel akademik yayıncılık. 2018. Ankara.

[19] Ingber L. Simulated Annealing: Practice versus Theory. Mathematical and Computer Modelling 1993; 18(11): 29–57.

3 Comparison of ANN and Neuro Regression Methods in Mathematical Modeling

Artificial neural networks (ANN) are commonly used in mathematical modeling across various scientific fields. However, while ANN has its advantages, it also has some limitations that must be considered. One such limitation is that using data for modeling purposes without any row limitation may lead to misleading results. Additionally, scaling data into a specific range for modeling leads to the need for another transformation to use the results obtained. Compared to other methods like Response Surface Methodology (RSM) and regression, ANN models are more complex and require further transformation.

In this section, we re-examine problems where ANN was used for mathematical modeling. We question the usability of these models, which met the model assessment criteria with high success, with different test methods. Also, we compare the modeling performance of Neuro Regression Method (NRM), a new modeling method, and popular ANN in an original problem. Our results and evaluations show that mathematical models obtained using ANN have some limitations, and the high prediction performance of the models may be misleading.

CASE STUDY 3.1

In a recent study [1], the impact of the chemical composition of molten steel in electric-arc furnaces on the electrical energy consumption in the iron and steel industry was modeled using ANN. The study considered the effect of additives such as carbon (C), chromium (Cr), nickel (Ni), silicon (Si), and iron (Fe) on the chemical content of stainless steel as design parameters. An attempt was made to determine the optimal ANN structure that mathematically best defines electrical energy consumption using an experimental set of 46 lines. The dataset used in the study is given in Table 3.1.

In the reference study, the ANN modeling results showed that the 5–10–1 hyperbolic tangent network structure can be used to model the effect of the chemical composition of steel on electrical energy consumption. This part of the study aims to test the stability and accuracy of the models proposed by ANN in expressing the relationship between input and output parameters. The reason for this research is the tendency in the literature to use ANN in engineering problems

DOI: 10.1201/9781003494843-3

TABLE 3.1

Dataset Related to Energy Consumption in Arc Furnaces [1]

Trial	C (%)	Cr (%)	Ni (%)	Si (%)	Fe (%)	E (kWh/t)
1	1.12	17.46	7.97	0.28	73.17	483
2	1.15	19.56	8.58	0.24	70.47	484
3	1.02	17.32	8.49	0.35	72.82	432
4	1.02	17.1	7.24	0.12	74.53	452
5	1.55	19.75	8.76	0.2	69.75	455
6	1.26	19.62	8.85	0.15	70.14	503
7	1.21	18.31	8.26	0.3	71.94	419
8	1.18	15.8	9.45	0.25	73.33	443
9	1.43	18.93	8.14	0.09	71.42	521
10	1.21	16.75	7.71	0.18	74.15	567
11	1.24	17.54	7.06	0.13	74.04	527
12	1.44	16.77	7.65	0.54	73.6	436
13	1.24	15.15	9.89	0.47	73.26	529
14	0.8	15.3	8.11	0.32	75.47	462
15	1.32	18.25	9.21	0.31	70.91	601
16	1.34	19.07	8.74	0.25	70.6	592
17	1.46	19.8	7.71	0.18	70.87	572
18	1.35	18.39	8.17	0.35	71.75	570
19	1.33	18.65	5.83	0.22	73.97	585
20	1.65	18.41	7.42	0.24	72.29	557
21	1.3	18.23	7.68	0.22	72.58	510
22	1.54	18.82	9.22	0.19	70.24	504
23	0.84	16.57	7.49	0.05	75.06	483
24	1.47	20.12	7.8	0.14	70.48	533
25	0.93	16.11	8.11	0.06	74.8	447
26	1.12	18.52	8.54	0.12	71.7	547
27	1.05	16.59	7.72	0.47	74.18	445
28	1.31	17.57	8.6	0.68	71.85	458
29	0.93	15.59	7.77	0.06	75.65	482
30	1.29	19.73	7.45	0.03	71.51	524
31	1.26	20.23	9.38	0.05	69.09	477
32	1.33	20.5	8.11	0.03	70.03	509
33	1.17	19.55	7.49	0.04	71.76	548
34	1.42	19.04	7.38	0.09	72.08	542
35	1.42	17.13	7.21	0.37	73.88	426
36	1.35	18.28	8.25	0.2	71.93	543
37	1.47	18.84	7.97	0.32	71.41	567
38	1.27	18.88	7.61	0.15	72.1	571
39	1.38	17.87	7.67	0.22	72.87	564
40	1.12	14.83	7.38	0.35	76.34	473
41	1.2	16.87	7.69	0.16	74.08	564

(Continued)

TABLE 3.1 *(Continued)*
Dataset Related to Energy Consumption in Arc Furnaces [1]

Trial	C (%)	Cr (%)	Ni (%)	Si (%)	Fe (%)	E (kWh/t)
42	1.62	19.45	7.04	0.32	71.57	675
43	1.07	16.7	7.57	0.12	74.55	521
44	1.4	18.79	8.04	0.16	71.61	482
45	1.36	19.78	7.89	0.07	70.91	534
46	1.38	19.3	8.32	0.2	70.81	475

without considering any restrictions. However, in the reference study, the 5–10–1 hyperbolic tangent network structure's mathematical model with specific coefficients was unclear. Therefore, the mathematical modeling process was repeated using the nntool ANN approach embedded in Matlab, with the help of the dataset in Table 3.1.

It is found that the energy consumption output can be accurately predicted utilizing the hyperbolic tangent activation function within a 5–10–10–5–1 network structure. The ANN structure created during the trial is detailed in Figure 3.1. The fitting performance of the mathematical model in the training, testing, and validation stages is given in Figure 3.2.

The findings in Figure 3.2 demonstrate that the proposed model by ANN satisfies the assessment criteria. Unlike the reference study and other research in the field, stability testing was also considered to test the model's usability, in addition to the assessment criteria based on R-squared (R^2).

In the stability test, the values of input parameters are modified, and the effect on the output parameter is analyzed. Suppose the changes in the input parameter

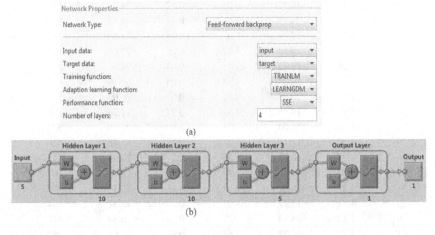

FIGURE 3.1 Artificial Neural Network proposed for modeling energy consumption: (a) network properties and (b) network structure

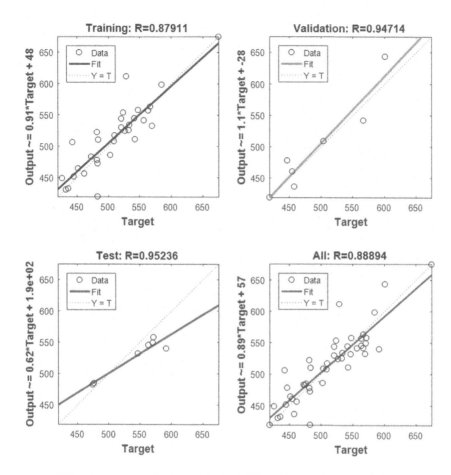

FIGURE 3.2 Assessment of proposed Artificial Neural Network model

values lead to output parameter changes that are inconsistent with reality or meaningless from an engineering perspective. In that case, the model is deemed to be unstable.

Figure 3.3 compares the output parameter values when input parameters are increased or decreased by 1%, 5%, 10%, 20%, and 50%, respectively. The actual values of the output parameter when input parameters remains constant also given for comparison.

When comparing the values of the output parameter resulting from a 1% change in the input parameters with the actual values of the output parameter, consistent results are observed. A similar evaluation can be made for the 5% decrease in input parameters. However, when input parameters are increased by 5%, we see that the output values corresponding to the input data rows in the ranges 14–18 and 20–29 remain constant or change only slightly, which is inconsistent with the actual results.

FIGURE 3.3 Comparison of the values of the output parameter and its actual values depending on the increase and decrease rates of the input parameters: (a) 1%, (b) 5%, (c) 10%, (d) 20%, and (e) 50% *(Continued)*

For a 10% increase in input parameters, the output values corresponding to input data rows in the range 15–32 remain constant, indicating that changes in input parameters do not affect the output. Again, these results are inconsistent with the actual results. The most significant effect is observed when input parameters are altered by 50% and the output parameter takes a constant value that does not change. This result implies that changing the proportions of additives in the chemical content of stainless steel does not affect electrical energy consumption. From an engineering perspective, this result is unrealistic.

Therefore, based on the results mentioned previously, it is concluded that the model cannot produce realistic results or maintain stability when input parameters are changed by 5%, 10%, 20%, and 50%. This shows that R^2 used as an assessment criterion to measure the relationship between actual and predicted

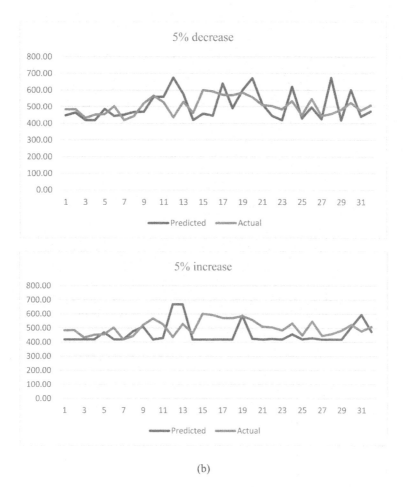

(b)

FIGURE 3.3   *(Continued)*

values, may not be sufficient. Additionally, it has been demonstrated that using ANN networks in modeling every problem and data type may not be appropriate.

CASE STUDY 3.2

This study modeled the effect of fiber angle orientation on stress in laminated composite plates using ANN [2]. The considered layered composite comprised 16 layers of symmetrical/balanced flax-epoxy material. Fiber angles were chosen as the design parameter and stress was selected as the output parameter. The laminated composite plate having a/b = 2 aspect ratio is considered to be subjected to Nxx = 100 N/m and Mxx = 10 Nm loads. To model the laminated composite stress behavior under these conditions, an experimental set was created using the Box-Behnken design (BBD) method.

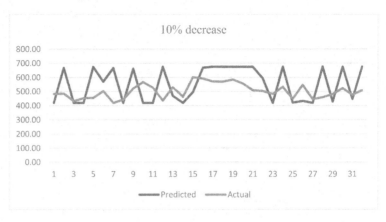

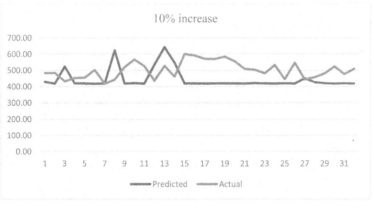

(c)

FIGURE 3.3 *(Continued)*

Table 3.2 provides the fiber angle design parameters and corresponding level values for creating the design set.

The goal is to assess the ANN's ability to model the stress behavior of laminated composite materials based on the experiment set in Table 3.3. An ANN structure was developed using the ANN toolbox in Matlab to accomplish this.

Figure 3.4 illustrates the network structure proposed for modeling the stress behavior of the laminated composite plate, based on the orientation of the fiber angle. The model's assessment result obtained using this network structure is also presented. The 4–5–1 ANN structure and the tangent hyperbolic activation function have been identified as suitable network structures for modeling the stress output. The performance of this network structure in predicting actual stress values was evaluated through training, testing, and validation model assessment processes. The results presented in Figure 3.4 demonstrate that the proposed model successfully meets all the required criteria.

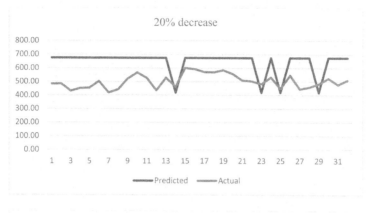

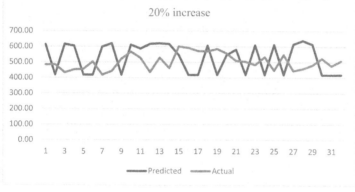

(d)

FIGURE 3.3 *(Continued)*

While ANN seems to be a successful modeling method for the example problem, the question arises whether it is realistic to use this approach as a modeling method in solving problems in different fields. To investigate this, a problem was selected from composite material mechanics, which is not very common in the literature and is a challenging problem type for modeling and optimization methods. Another reason for selecting this problem is that it is possible to obtain

TABLE 3.2

Level of Fiber Angles Selected as Design Parameters

Design Parameters	Level 1	Level 2	Level 3
x1 (fiber) angle)	0	45	90
x2 (fiber) angle)	0	45	90
x3 (fiber) angle)	0	45	90
x4 (fiber) angle)	0	45	90

(e)

FIGURE 3.3   *(Continued)*

results using an analytical solution, and the success of the proposed model can be questioned by producing different test data.

In the previous evaluation, it was observed that the model proposed by ANN was quite successful in estimating the actual values in the training, testing, and validation stages. To validate the consistency of the model's success, additional test data was produced, and the performance of the model proposed by ANN in predicting the actual values found using the analytical formula was measured. The generated random experiment set used for testing is presented in Table 3.4.

The test dataset given in Table 3.4 was simulated in ANN and compared with the actual results obtained by the analytical formula. Figure 3.5 presents the prediction results of the model proposed by ANN and the actual results obtained using the analytical formula. Upon examination of the results, it is evident that there is a significant difference between the actual values and the predicted ones. The fact that the model assessment criterion R^2 was found to be 0.36 indicates that

TABLE 3.3

Experiment Set for Fiber Angle–Stress Relation Using the Box-Behnken Design Method

x1	x2	x3	x4	Stress
45	0	0	45	54.32
45	0	90	45	63.9
90	0	45	45	66.99
0	45	45	90	65.72
0	45	0	45	58.54
90	45	45	90	38.64
0	45	90	45	65.81
90	45	0	45	47.86
45	90	0	45	46.36
45	45	90	90	46.8
45	90	90	45	52.1
45	0	45	0	62.54
90	90	45	45	40.65
45	0	45	90	64.14
90	45	45	0	36.59
45	45	0	0	44.99
45	45	0	90	46.65
45	45	45	45	45.1
90	45	90	45	41.46
45	45	90	0	44.37
0	90	45	45	66.94
45	90	45	0	47.44
45	90	45	90	50.01
0	45	45	0	64.54
0	0	45	45	49.53

the high success performance of the model in the previously mentioned training, testing, and validation stages is misleading.

These results show the confidence in modeling using ANN in the literature and that ANN as a modeling method should be questioned again.

It's important to consider that ANN is not a statistical process and requires a large dataset for modeling compared to techniques like RSM and regression. Therefore, it's not a universal modeling method that can be applied to all problems or data types, regardless of the field of study. In many studies, attempting to model using ANN without considering any data-related criteria may lead to misleading results due to limited data numbers or inappropriate data types, even if the model's prediction performance is quite good. This study highlights potential issues that may arise when modeling with ANN and the need for greater awareness of using this method.

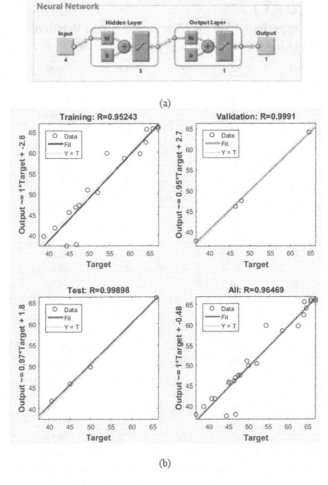

FIGURE 3.4   Artificial Neural Network model: (a) network architecture and (b) model assessment

TABLE 3.4
Randomly Generated Dataset
to Test the ANN Model

x1	x2	x3	x4
54	86	57	17
79	2	16	13
88	36	63	80
65	74	8	40
71	12	5	40
62	87	35	28

(Continued)

TABLE 3.4 *(Continued)*
**Randomly Generated Dataset
to Test the ANN Model**

46	7	80	66
57	78	16	10
55	38	43	24
80	25	52	56
82	81	59	56
8	43	35	58
64	14	7	23
45	75	76	51
65	19	73	41
13	27	67	33
88	84	45	75
49	36	17	85
73	51	12	73
69	41	84	87
13	47	47	1
37	22	74	33
27	10	36	85
0	65	76	59
82	46	18	68

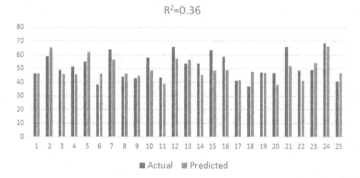

FIGURE 3.5 Comparison of the results predicted by the Artificial Neural Network model and the actual results

REFERENCES

[1] Gajic D, Savic-Gajic I, Savic I, Georgieva O. Di Gennaro S. Modelling of electrical energy consumption in an electric arc furnace using artificial neural networks. Energy 2016; 108: 132–139.

[2] Savran M, Aydin L. Stochastic optimization of graphite-flax/epoxy hybrid laminated composite for maximum fundamental frequency and minimum cost. Engineering Structures 2018; 174: 675–687.

4 Evaluation of R^2 as a Model Assessment Criterion

To evaluate the accuracy and reliability of a proposed model, it is essential to use multiple assessment criteria. Using only R-squared (R^2) as an assessment criterion can predict the results accurately, but it does not necessarily mean the model can accurately explain the problem phenomenon.

Thus, boundedness control is another essential criterion the developed model functions must meet. Since all engineering parameters are finite, the proposed models must meet the limitation criterion to achieve realistic modeling in engineering systems. It is necessary to verify whether the proposed models depend on the engineering parameter ranges in the defined problem. Testing the stability of models that successfully meet these criteria is a step that must be considered. The stability test examines the effect of increasing or decreasing the input parameter values at certain levels on the output values. If the output values produced remain within the realistic range in terms of engineering, the candidate model is stable.

In addition to model assessment criteria, the path followed in creating a model is important. Instead of creating a model using the entire dataset, it is necessary to divide it into two or three parts, such as training–test and training–test–validation. This approach allows the model created using training data to be tested and validated with data it has never encountered before. By doing so, a more reliable model can be obtained compared to the model achieved using all data.

Hence, the model that successfully passes these stages is much more reliable than the model proposed by Response Surface Method (RSM). This situation is discussed below with comparative explanations of different engineering problems.

CASE STUDY 4.1

This study [1] aimed to create a mathematical model for the relationship between input and output parameters in a free-piston Stirling engine. The researchers used the RSM and Artificial Neural Network (ANN) methods to model the output parameters, which included displacer and piston amplitudes (xd and xp), operating frequency (f), and output power (pout). The Box-Behnken design (BBD) design method was used to create the dataset, and the design parameters and their level values are given in Table 4.1.

DOI: 10.1201/9781003494843-4

TABLE 4.1
Design Parameters and Their Levels

Parameters		Level		
		L1	L2	L3
Hot-end temperature	x_1	723	823	923
Cold-end temperature	x_2	288	300.5	313
Coefficient of load damping	x_3	12	13	14
Spring stiffnesses of the displacer	x_4	12,000	15	4.4
Spring stiffnesses of the piston	x_5	10,000	13,000	16,000

The RSM results suggest that using a second-order polynomial is enough to model the relationship between the four output parameters and the inputs. In the reference study, the R^2 prediction performance of the second-order polynomial models was found to be 0.99, which is impressive. However, it's important to ask whether R^2 alone is an adequate assessment criterion for modeling. Can we assume that the data is accurately defined just because the R^2 fits the data well?

To answer these questions, the dataset given in the reference study was divided into 75% training and 25% testing. Then, the coefficients of the second-order polynomial models for each output were determined using the training data. The Analysis of Variance (ANOVA) method was used to decide which terms should be included or removed from the model, and the final forms of the second-order polynomial models were obtained.

The expectation is that if R^2 accurately fits the data and enables us to define the problem phenomenon correctly, then second-order polynomial models should be able to predict the output values of the remaining 25% of the data with high accuracy.

Table 4.2 presents the mathematical models that were obtained for each output parameter using a training dataset of 40 rows. When comparing the R^2 assessment

TABLE 4.2
Model Assessment of Proposed Models

Output Parameters	Model Assessment	
	R^2	R^2 Adjusted
Displacer amplitudes (x_d)	0.996	0.992
Piston amplitudes (x_p)	0.997	0.994
Operating frequency (f)	0.999	0.999
Output power (p_{out})	0.998	0.997

Full form of models are given in appendix (Table A4.1).

criteria of the resulting models for each output parameter with the R^2 criteria obtained from the reference study's models, which used all 55 rows of data, no differences were found. In both cases, the R^2 criterion was at 0.99 levels, and it was concluded that the second-order polynomial models could be effectively used in modeling the outputs.

It was decided by the ANOVA method whether the parameters of the mathematical models obtained for each output parameter using the training data were important enough to require their inclusion in the model. Table 4.3 shows assessment results that determine the importance levels of the terms in the models by looking at the "p" values for each output.

A p-value of less than 0.05 indicates that the model terms may significantly impact the output parameter. In this study, the following model terms were found to be important: x_1, x_2, x_3, x_4, x_5, $x_1 x_2$, $x_1 x_3$, $x_1 x_4$, $x_3 x_4$, $x_4 x_5$, x_1^2, and x_4^2. Values greater than 0.1 indicate that model terms are not significant. If there are many insignificant model terms in the model, lowering the model order may increase the success of the model.

TABLE 4.3
ANOVA Analysis of Second Order Polynomial Model

	p-value − x_d	p-value − x_p	p-value − f	p-value − p_{out}
Model	<0.0001	<0.0001	<0.0001	<0.0001
x_1	<0.0001	<0.0001	<0.0001	<0.0001
x_2	<0.0001	<0.0001	<0.0001	<0.0001
x_3	<0.0001	<0.0001	0.9290	<0.0001
x_4	<0.0001	<0.0001	<0.0001	<0.0001
x_5	<0.0001	<0.0001	<0.0001	<0.0001
$x_1 x_2$	0.0073	0.0018	0.3770	0.0031
$x_1 x_3$	0.0135	0.0067	0.2909	0.1191
$x_1 x_4$	0.6990	0.2761	<0.0001	<0.0001
$x_1 x_5$	0.0128	0.0740	0.4785	<0.0001
$x_2 x_3$	0.3354	0.2501	0.8587	0.3718
$x_2 x_4$	0.3827	0.4620	0.0001	<0.0001
$x_2 x_5$	0.5763	0.3060	0.2909	0.0014
$x_3 x_4$	0.0302	0.0531	0.7220	0.1016
$x_3 x_5$	0.1018	0.0430	0.7220	0.9439
$x_4 x_5$	<0.0001	<0.0001	<0.0001	<0.0001
x_1^2	<0.0001	<0.0001	<0.0001	0.5913
x_2^2	0.9597	0.7590	<0.0001	0.2228
x_3^2	0.7313	0.6590	<0.0001	0.9327
x_4^2	<0.0001	<0.0001	0.9290	<0.0001
x_5^2	0.1669	0.2187	<0.0001	0.8574

TABLE 4.4

Prediction Performance of Models Obtained by Removing Unnecessary Terms in ANOVA Analysis

Output Parameters	Model Assessment	
	R^2	R^2 Adjusted
Displacer amplitudes (x_d)	0.996	0.993
Piston amplitudes (x_p)	0.996	0.994
Operating frequency (f)	0.999	0.999
Output power (p_{out})	0.997	0.996

Full form of models are given in appendix (Table A4.2).

Table 4.4 displays the final models obtained after removing insignificant terms through ANOVA. The R^2 prediction performance of these models aligns with the findings of the ANOVA. Terms that were excluded due to their perceived lack of significant impact on the model did not affect the R^2 prediction performance of the models.

This study used RSM to create mathematical models for four different outcomes. The R^2 values of all proposed models are 0.99, indicating that the models fit the data very well.

As previously emphasized, the crucial question is whether the models can perform similarly during the testing phase. To address this, test data, comprising 25% of all data, was utilized as input for the models employed in modeling each output parameter. Subsequently, the models were tasked with estimating the actual values. The prediction performances of the models for the test data, which they have not encountered before, are presented in Figure 4.1.

As shown in Figure 4.1, the proposed mathematical model for output x_d failed in the testing phase, while the other three output parameters successfully completed the testing phase. However, as mentioned before, the boundedness check must also be carried out to evaluate whether the model produces realistic results within the value ranges of the defined design parameters.

Table 4.5 displays the maximum and minimum values of the proposed models for the four output parameters, which should be considered when evaluating the boundedness check criterion. According to these results, all models, except for those expressing the operating frequency output, failed to meet the boundedness check criterion. The minimum values obtained from the respective models of these three outputs are unrealistic and practically impossible. Only the proposed model for the operating frequency output successfully met all the criteria.

Regarding the results obtained in this study, it becomes evident that selecting models based solely on R^2 can lead to errors. While a high R^2 value indicates that the model's predictions align well with the actual results, in engineering problems, our expectation extends beyond merely fitting the data; we aim to generate

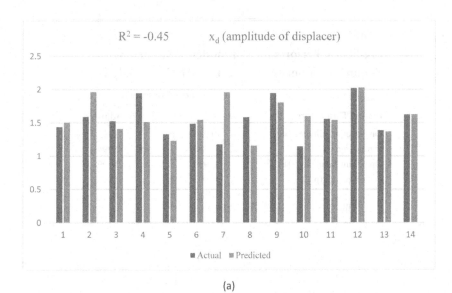

(a)

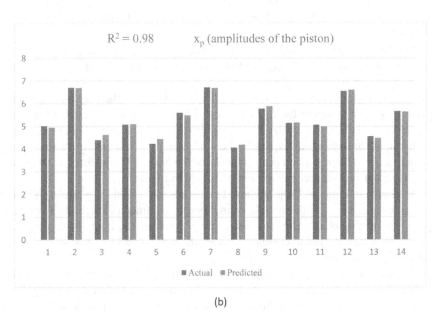

(b)

FIGURE 4.1 Model assessment in the test phase *(Continued)*

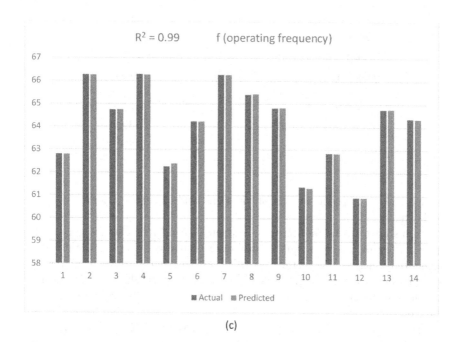

(c)

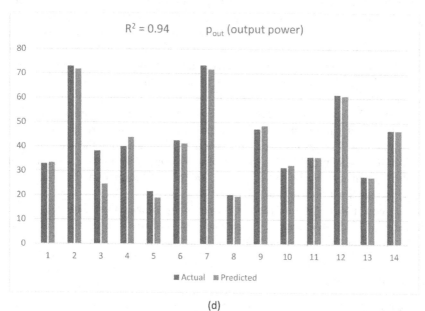

(d)

FIGURE 4.1 *(Continued)*

TABLE 4.5

Boundedness Check Control for Proposed Models

Output Parameters	Maximum	Minimum
Displacer amplitudes (x_d)	2.671	−0.476
Piston amplitudes (x_p)	8.692	−1.392
Operating frequency (f)	66.455	59.264
Output power (p_{out})	133.444	−28.901

models capable of producing realistic and practical outcomes consistent with the problem at hand.

In this context, the models were evaluated for all four outputs, considering training, testing, and boundary control criteria. It was found that the proposed model only successfully met the specified testing criteria for the operating frequency output. Consequently, it is not feasible to assert that the second-order polynomial models presented in the reference study [1] can be applied to the other three outputs.

Stability is an important factor when evaluating a model. Out of the four output parameters, only the proposed model for operating frequency was able to successfully meet the training, testing, and boundedness check criteria. Therefore, it will be sufficient to test the stability of this model. In the stability test, the input parameter values are increased or decreased at specific levels, and the effect of this situation on the output values is examined. If the obtained output values remain within a realistic range in terms of engineering, the candidate model is deemed suitable.

Figure 4.2 illustrates the change in the operating frequency output parameter when the input parameter values are increased or decreased by 1%, 5%, 10%, 20%, and 50%. The results depicted in Figure 4.2 suggest that the second-order polynomial model proposed for the operating frequency demonstrates significant stability. Even with variations of up to 50% in the input parameter values, the output remains within a reasonable range.

Consequently, unlike the models proposed for other output parameters, the polynomial model for operating frequency successfully satisfies the criteria of training, testing, boundedness control, and stability analysis assessments.

CASE STUDY 4.2

In the reference study [2], researchers investigated the impact of various manufacturing parameters, such as building orientation, layer thickness, and feed rate, on the mechanical properties of polylactic acid (PLA) samples produced by the 3D fused deposition method. The study aimed to determine the strength and modulus of the material under different loads. A dataset comprising 36 samples, generated using the Full Factorial design (FFD) method, was utilized to develop a

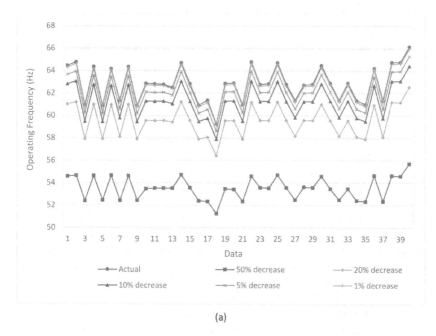

(a)

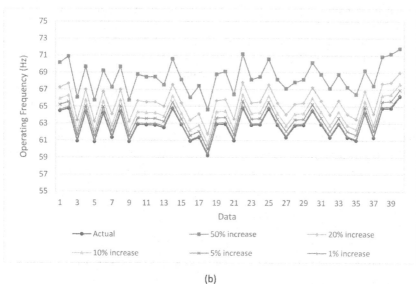

(b)

FIGURE 4.2 Stability test of the proposed model for the operating frequency output parameter (a) when the values of the input parameters are reduced at specific rates and (b) when the values of the input parameters are increased at specific rates

TABLE 4.6
3D Printing Process Design Parameters and Their Levels [2]

		Level			
Parameters	Symbol	L1	L2	L3	L4
Layer thickness	x_1	0.06	0.12	0.18	0.24
Feed rate	x_2	20	50	80	–
Build orientation	x_3	1 (Upright)	2 (Edge)	3 (Flat)	–

mathematical model. The design parameters and their corresponding values are presented in Table 4.6.

To assess the mechanical performance of the PLA material, the researchers utilized RSM and ANOVA. The success of the model obtained using the entire dataset was evaluated based on the R^2 criterion. However, as demonstrated in this study, relying solely on this evaluation criterion is insufficient. Therefore, it is necessary to recreate the mathematical modeling process according to the evaluation criteria proposed within the scope of this section.

Instead of duplicating similar steps for all output parameters, we specifically discuss the results obtained by applying the mentioned modeling process solely for tensile strength. The objective here is not to model each output parameter separately, but rather to underscore the significance of each step in the modeling process. Therefore, if desired, mathematical models for other output parameters can be created by following the same modeling steps.

Table 4.7 shows proposed mathematical models and their performance in predicting actual tensile strength. Here, the first model is created using the entire dataset. Then, the data is divided into two groups: training and testing. While the second model is created using training data, the model's prediction performance is measured with test data. After that, the dataset was divided into three parts— training, testing, and validation—and used to create the third, fourth, and fifth models and measure their performance. When the first and second models

TABLE 4.7
Proposed Mathematical Models for Tensile Strength

No.	Model*	R^2_{train}	R^2_{adj}	R^2_{test}	$R^2_{validation}$	Maximum	Minimum
1	SON	0.95	0.94	–	–	–	–
2	SON	0.94	0.92	0.95	–	–	–
3	SON	0.96	0.95	0.88	−0.39	–	–
4	TON	0.94	0.93	0.95	0.74	93.15	23.33
5	SOTN	0.97	0.96	0.98	0.88	364.54	−1846.82

* Full form of models are given in appendix (Table A4.3).

are examined, it is seen that the second-order polynomial is quite successful in predicting the actual results. However, when the third model, similar to the first and second models, is subjected to validation, a low value for prediction performance is obtained ($R^2 = -0.39$). While only the training or training and testing phases are considered, the second-order polynomial seems to be a usable model. Still, this model's prediction performance is insufficient in the validation phase. The other two models, third-order polynomial (fourth model) and second-order trigonometric (fifth model), successfully met the training, testing, and validation criteria. If the performance of these two models is compared considering R^2, the trigonometric model is more successful. However, when looking at the maximum and minimum limit values that the models can take, the trigonometric model produces unrealistic maximum and minimum results. In contrast, the third-order polynomial model produces reasonable results consistent with the experimental results given in the reference study. While the experimental maximum and minimum tensile strength were obtained as 89.1 MPa and 20.2 MPa, the proposed third-order model found the tensile strength to be 93.15 MPa and 23.33 MPa, respectively.

It can be said that only the third-order polynomial model successfully meets the training, testing, validation, and boundedness check criteria. As a final step, stability testing was applied to measure the robustness of this model. Figure 4.3 shows the change in the tensile strength results of the PLA material when the input parameter values are increased or decreased by 1%, 5%, 10%, 20%, and 50%.

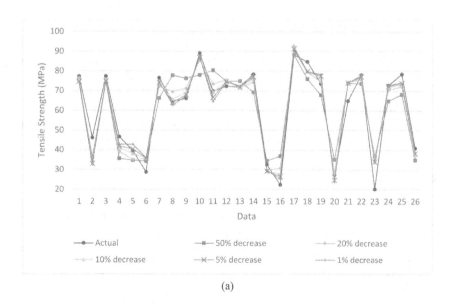

(a)

FIGURE 4.3 Stability test of the proposed model for the tensile strength output parameter (a) when the values of the input parameters are reduced at specific rates and (b) when the values of the input parameters are increased at specific rates *(Continued)*

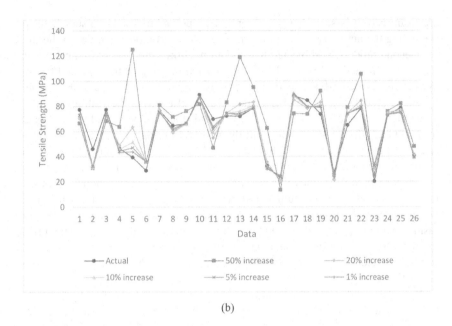

(b)

FIGURE 4.3 *(Continued)*

According to the results, the tensile strength maintains its consistency depending on the change in the values of the input parameters and takes values ranging between 20 MPa and 120 MPa, which we can call reasonable. Although there are models with better prediction performance in the training, testing, and validation stages, the third-order polynomial model has become the only suitable model for modeling the tensile strength of materials produced using the 3D printed manufacturing method.

This finding highlights that relying solely on R^2 to assess model success may be insufficient. To evaluate the model accurately, it's necessary to consider other criteria, such as testing, validation, boundedness control, and stability.

REFERENCES

[1] Ye W, Wang X, Liu Y, Chen J. Analysis and prediction of the performance of free-piston Stirling engine using response surface methodology and artificial neural network. Applied Thermal Engineering 2021; 188: 116557.
[2] Chacón JM, Caminero MA, García-Plaza E, Núñez PJ. Additive manufacturing of PLA structures using fused deposition modelling: Effect of process parameters on mechanical properties and their optimal selection. Materials & Design 2017; 124: 143–157.

APPENDIX

TABLE A4.1
Proposed Mathematical Models to Define Input-Output Parameters Relation (Related to Table 4.2)

Output Parameters	Model
Displacer amplitudes (x_d)	$xd = 7.83903000 + 0.00135428x_1 - 0.00001182x_1^2 - 0.03397270x_2 +$ $0.00004440x_1x_2 - (6.87778000 \times 10^{-6})x_2^2 - 0.04780320x_3 +$ $0.00050250x_1x_3 - 0.00146000x_2x_3 - 0.00732465x_3^2 + 0.00037970x_4 +$ $(2.41667000 \times 10^{-8})x_1x_4 + (4.40000000 \times 10^{-7})x_2x_4 + 0.00001442x_3x_4 -$ $(3.15955000 \times 10^{-8})x_4^2 - 0.00027196x_5 + (1.69167000 \times 10^{-7})x_1x_5 -$ $(2.80000000 \times 10^{-7})x_2x_5 - 0.00001058x_3x_5 + (2.21782000 \times 10^{-8})x_4x_5 -$ $(3.39178000 \times 10^{-9})x_5^2$
Piston amplitudes (x_p)	$xp = 10.24200000 - 0.00201157x_1 - 0.00003633x_1^2 - 0.03840240x_2 +$ $0.00015680x_1x_2 - 0.00012229x_2^2 + 0.23703900x_3 + 0.00164250x_1x_3 -$ $0.00512000x_2x_3 - 0.02752430x_3^2 + 0.00122400x_4 +$ $(2.01667000 \times 10^{-7})x_1x_4 + (1.08000000 \times 10^{-6})x_2x_4 + 0.00003708x_3x_4 -$ $(9.03993000 \times 10^{-8})x_4^2 - 0.00035488x_5 + (3.40000000 \times 10^{-7})x_1x_5 -$ $(1.51333000 \times 10^{-6})x_2x_5 - 0.00003900x_3x_5 + (6.06620000 \times 10^{-8})x_4x_5 -$ $(8.76968000 \times 10^{-9})x_5^2$
Operating frequency (f)	$f = 43.75330000 - 0.00377167x_1 + 0.00462000x_2 + (10^{-5})x_1x_2 +$ $0.06064170x_3 - 0.00015000x_1x_3 + 0.00020000x_2x_3 + 0.00094725x_4 +$ $(2.41667000 \times 10^{-7})x_1x_4 - (1.66667000 \times 10^{-6})x_2x_4 +$ $(1.66667000 \times 10^{-6})x_3x_4 + 0.00076406x_5 - (3.33333000 \times 10^{-8})x_1x_5 +$ $(4 \times 10^{-7})x_2x_5 - (1.66667000 \times 10^{-6})x_3x_5 - (1.28144000 \times 10^{-8})x_4x_5$
Output power (p_{out})	$pout = 193.66500000 + 0.72884800x_1 - 0.00003703x_1^2 - 3.05304000x_2 -$ $0.00161600x_1x_2 + 0.00546989x_2^2 - 10.06040000x_3 - 0.00972500x_1x_3 +$ $0.04360000x_2x_3 + 0.05800350x_3^2 + 0.03950150x_4 + 0.00002818x_1x_4 -$ $0.00007920x_2x_4 - 0.00034167x_3x_4 - (1.30239000 \times 10^{-6})x_4^2 -$ $0.01890820x_5 - 0.00001841x_1x_5 + 0.00005960x_2x_5 +$ $0.00001417x_3x_5 + (7.17616000 \times 10^{-7})x_4x_5 - (1.38715000 \times 10^{-8})x_5^2$

TABLE A4.2
Performance Assessment of the Models Obtained by Removing Unnecessary Terms with ANOVA (Related to Table 4.4)

Output Parameters	Model
Displacer amplitudes (x_d)	$xd = 14.606400 + 0.001128x_1 - 0.000011x_1^2 - 0.054126x_2 + 0.000044x_1x_2 - 0.676974x_3 + 0.000503x_1x_3 + 0.000520x_4 + 0.000014x_3x_4 - (3.120350 \times 10^{-8})x_4^2 - 0.000367x_5 + (1.691670 \times 10^{-7})x_1x_5 - 0.000011x_3x_5 + (2.219550 \times 10^{-8})x_4x_5 - (2.999810 \times 10^{-9})x_5^2$
Piston amplitudes (x_p)	$xp = 46.67240000 - 0.00225757x_1 - 0.00003434x_1^2 - 0.18193100x_2 + 0.00015680x_1x_2 - 2.01715000x_3 + 0.00164250x_1x_3 + 0.00164806x_4 + 0.00003708x_3x_4 - (8.82231000 \times 10^{-8})x_4^2 - 0.00086772x_5 + (3.40000000 \times 10^{-7})x_1x_5 - 0.00003900x_3x_5 + (6.07577000 \times 10^{-8})x_4x_5 - (6.59352000 \times 10^{-9})x_5^2$
Operating frequency (f)	$f = 40.49780000 - 0.00271667x_1 + 0.02065000x_2 + 0.00062500x_3 + 0.00096892x_4 + (2.41667000 \times 10^{-7})x_1x_4 - (1.66667000 \times 10^{-6})x_2x_4 + 0.00086259x_5 - (3.33333000 \times 10^{-8})x_1x_5 - (1.28144000 \times 10^{-8})x_4x_5$
Output power (p_{out})	$pout = 55.48000000 + 0.54146600x_1 - 2.21691000x_2 - 0.00161600x_1x_2 + 0.00631095x_2^2 - 8.39500000x_3 + 0.03461980x_4 + 0.00002818x_1x_4 - 0.00007920x_2x_4 - (1.28807000 \times 10^{-6})x_4^2 - 0.00118835x_5 - 0.00001841x_1x_5 + (7.18470000 \times 10^{-7})x_4x_5$

TABLE A4.3
Proposed Mathematical Models for Tensile Strength Output (Related to Table 4.7)

Nomenclature	Model
SON	$-69.113 + 121.06x_1 + 334.88x_1^2 + 0.38468x_2 - 1.1759x_1x_2 - 0.0037315x_2^2 + 107.29x_3 - 80.528x_1x_3 + 0.046458x_2x_3 - 19.371x_3^2$
SON	$-78.5106 + 166.952x_1 + 308.624x_1^2 + 0.486716x_2 - 1.43012x_1x_2 - 0.00394115x_2^2 + 111.447x_3 - 91.9896x_1x_3 + 0.0315031x_2x_3 - 19.9193x_3^2$
SON	$-70.204 + 165.61x_1 + 239.45x_1^2 + 0.38448x_2 - 0.71643x_1x_2 - 0.0044189x_2^2 + 105.03x_3 - 100.09x_1x_3 + 0.041421x_2x_3 - 17.789x_3^2$
TON	$-0.8808 + 10.34x_1 - 7.053x_1^2 + 0.8935x_1^3 - 0.05287x_2 - 1.665x_1x_2 + 13.02x_1^2x_2 - 0.007979x_2^2 + 0.008832x_1x_2^2 + 0.00003712x_2^3 + 21.28x_3 + 18.22x_1x_3 + 4.055x_1^2x_3 + 0.6594x_2x_3 - 2.08x_1x_2x_3 - 0.001365x_2^2x_3 + 14.38x_3^2 + 4.583x_1x_3^2 - 0.04038x_2x_3^2 - 5.342x_3^3$

(Continued)

TABLE A4.3 *(Continued)*
**Proposed Mathematical Models for Tensile Strength Output
(Related to Table 4.7)**

Nomenclature	Model
SOTN	$467.492 + 311.531\text{Cos}[x_1] + 158.603\text{Cos}[x_1]^2 - 289.772\text{Cos}[x_2] + 1098.02\text{Cos}[x_1]\text{Cos}[x_2] + 2.83281\text{Cos}[x_2]^2 - 1894.2\text{Cos}[x_3] + 3104.6\text{Cos}[x_1]\text{Cos}[x_3] + 0.563021\text{Cos}[x_2]\text{Cos}[x_3] - 252.27\text{Cos}[x_3]^2 - 4067.44\text{Sin}[x_1] + 4338.2\text{Cos}[x_1]\text{Sin}[x_1] + 149.423\text{Cos}[x_2]\text{Sin}[x_1] + 593.452\text{Cos}[x_3]\text{Sin}[x_1] + 199.34\text{Sin}[x_1]^2 - 953.246\text{Sin}[x_2] + 662.154\text{Cos}[x_1]\text{Sin}[x_2] + 697.076\text{Cos}[x_2]\text{Sin}[x_2] + 1.70547\text{Cos}[x_3]\text{Sin}[x_2] + 142.035\text{Sin}[x_1]\text{Sin}[x_2] + 451.804\text{Sin}[x_2]^2 + 1267.56\text{Sin}[x_3] - 3645.1\text{Cos}[x_1]\text{Sin}[x_3] - 3.0852\text{Cos}[x_2]\text{Sin}[x_3] - 1552.49\text{Cos}[x_3]\text{Sin}[x_3] - 589.532\text{Sin}[x_1]\text{Sin}[x_3] + 0.952449\text{Sin}[x_2]\text{Sin}[x_3] + 695.344\text{Sin}[x_3]^2$

5 Questioning the Adequacy of Using Polynomial Structures

In the modeling process, the relationship between input and output parameters is defined using mathematical models. The mathematical function that will most accurately express the input and output relationship must be introduced in this process. Therefore, the variety of mathematical models helps to define the process and input–output relationship being studied with the least error. Considering the mathematical modeling and optimization-based studies in the literature, it can be seen that polynomial-type models are used predominantly. Although polynomial-type models are generally quite successful in describing processes in problems with low levels of nonlinearity, polynomial structures alone are insufficient for producing solutions with sufficient precision in more complex problems with high levels of nonlinearity. The failure here is seen in the testing and validation stages of the proposed polynomial models. To overcome this situation, it is necessary to increase the variety of mathematical models. Within the scope of this section, mathematical model types and their comparisons are made through examples of engineering problems, and it is emphasized that in some cases, polynomial models are insufficient to describe the process accurately or that there may be other mathematical models that can describe the process better. In this context, two problems in the mechanical engineering field of interest were handled. Detailed explanations about the problems are given in the following sections.

CASE STUDY 5.1

This study aimed to design a multi-layer armored system subjected to shock loading in an underwater explosion [1]. Design parameters and their levels are presented in Table 5.1.

The dataset was created using the Central Composite design (CCD) method, taking into account the design parameters. The output parameters of stress and displacement in the target layer were determined. Table 5.2 contains the dataset with the input and output parameters mentioned.

The main purpose of examining this study is to demonstrate that the R^2 success criterion of the proposed linear model for the stress output, which was determined as 0.68, can be improved using different mathematical models. The evaluations made in the previous sections have shown that creating a model using all the data may cause misleading results. Therefore, checking the success of

  DOI: 10.1201/9781003494843-5

TABLE 5.1
Design Parameters Determined for the Armor System and Their Level Values [2]

Parameters	Symbol	L1	L2	L3	L4	L5
RHA steel thickness	x_1	6.6	10	15	20	23.4
Silicon carbide thickness	x_2	1.6	5	10	15	18.4
Scaled distance	x_3	1.33	1.4	1.5	1.6	1.66

(Level spans L1–L5)

the proposed mathematical models with test data that has not been encountered before increases the model's reliability.

Therefore, unlike the reference study, the modeling process was partitioned into two groups for this study: training and testing. Table 5.3 shows the R^2 values and maximum and minimum stress predictions for various mathematical models.

Upon examination of the results, it was found that all models had a success rate of 0.78 or higher during the training phase. SONR was the most successful model,

TABLE 5.2
Experiment Set for Armor System Using the CCD Method [2]

	x_1	x_2	x_3	Stress (MPa)	Deflection (mm)
1	15	10	1.5	36.29	13.03
2	10	5	1.4	36.83	24.2
3	15	10	1.5	36.29	13.03
4	15	10	1.5	36.29	13.03
5	20	15	1.6	34.7	12.41
6	10	15	1.6	33.28	11.78
7	15	10	1.5	36.29	13.03
8	15	10	1.66	32.62	11.09
9	15	18.4	1.5	35.48	7.49
10	20	15	1.4	37.9	7.31
11	15	10	1.5	36.29	13.03
12	10	15	1.4	37.13	14.16
13	6.6	10	1.5	36.67	21.73
14	15	10	1.5	36.29	13.03
15	15	1.6	1.5	40.07	21.83
16	10	5	1.6	35.61	22.24
17	23.4	10	1.5	36	7.44
18	15	10	1.33	42.1	15.08
19	20	5	1.6	34.8	12.41
20	20	5	1.4	38.9	14.37

TABLE 5.3

Assessment of Mathematical Models Proposed for Stress

Model*	R^2_{train}	R^2_{adj}	R^2_{test}	Maximum	Minimum
L	0.78	0.7	0.71	41.85	31.43
LR	0.79	0.71	0.66	42.67	32.23
SON	0.93	0.9	−715.92	154.85	−2.24
SONR	1	1	−55.3	9.47E+08	−1372450
TON	0.97	0.95	0.86	253.52	−180.36
TONR	0.99	0.98	0.84	1.87E+08	−102689
FOTN	0.88	0.84	−14.82	42.43	23.38
FOTNR	0.98	0.97	0.77	419299	−1.05E+14
SOTN	0.99	0.98	0.86	49.74	−5.49
SOTNR	0.98	0.97	0.77	5.76E+06	−3.29E+09
FOLN	0.83	0.77	0.88	42.87	31.83
FOLNR	0.85	0.8	0.97	59.77	30.94
SOLN	0.93	0.89	−0.03	45.07	32.95
SOLNR	0.97	0.95	0.72	60704.1	−1664.4

* Full form of models are given in appendix (Table A5.1).

while L was the least successful. All models presented in this study appear to be more successful than the linear model proposed in the reference study.

However, during the testing phase, it was observed that the SON, SONR, FOTN, and SOLN models failed to predict the actual values. Similarly, when the maximum and minimum stress values predicted by the models were considered, it was observed that the results produced by models other than L, LR, FOTN, FOLN, FOLNR, and SOLN were unacceptable stress values regarding engineering.

Based on these evaluations, the models that met all the criteria were L, LR, FOLN, and FOLNR. Among these four models, FOLNR had the highest prediction performance.

Figure 5.1 depicts the relationship between predicted and observed values for the FOLNR model, the error levels of predicted values, and the evaluation of predicted values in terms of statistical confidence bands.

When examining the results for the FOLNR model, it is evident from the first graph that the errors have a normal distribution. Additionally, the second graph displays the ratio of the error obtained from the difference between the actual and predicted values for each output value to the standard error of this difference. The statistically acceptable range of point distributions showing errors is between +3 and −3. The third graph shows that the predicted values cluster very close to the regression line, indicating that the observed and predicted values are consistent. Finally, in the fourth chart, most data falls within the 99% confidence interval, with some in the 95% confidence interval. This demonstrates that the statistical estimates are reasonable and consistent with the results in the other graphs.

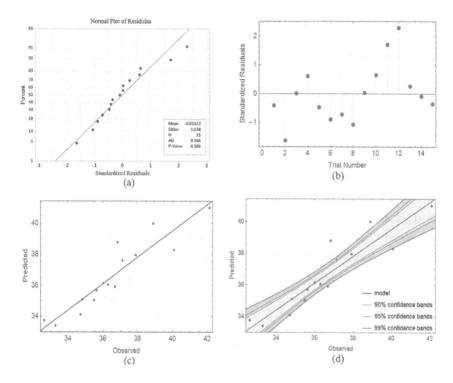

FIGURE 5.1   Statistical analysis of FOLNR model for stress output: (a) normal distribution of errors, (b) standard error distribution, (c) relationship between predicted and observed values, and (d) confidence interval bands

Based on these results, it can be concluded that the FOLNR model outperforms the L model and can serve as a viable alternative. Additionally, the literature review highlights the importance of not limiting the mathematical modeling process to polynomial models.

CASE STUDY 5.2

Due to the End of Life Vehicles regulation and emission limits, the automotive industry is now required to investigate environmentally friendly and lightweight electric vehicle components (Euro 5 & Euro 6) [3]. A lithium-ion battery pack, consisting of battery modules, is one of the most essential main components for electric vehicles. The battery pack and its enclosure are subjected to different mechanical loads that cause stresses and deformations during the electric vehicle's operation.

This reference study aimed to design electric vehicle battery pack enclosures while taking into account maximum natural frequency, minimum weight, and deformation. In the reference study, the CCD and Latin Hypercube sampling (LHS) design methods were used to create a dataset for the battery pack enclosure. Design parameters such as battery box wall thickness, battery box bottom thickness, module bottom thickness, long wall thickness of the battery module, large wall thickness, and temperature were considered. Table 5.4 contains the dataset with the above input and output parameters.

This section emphasizes the importance of using alternative mathematical model types in modeling. In some cases, polynomial models may not be enough. Therefore, this section aims to examine the effect of different model types in modeling the relationship between maximum natural frequency output and design parameters more accurately. The reference study was reviewed, and the relationship between natural frequency and design parameters was defined with polynomial, rational, trigonometric, and logarithmic models.

TABLE 5.4

Experiment Set for the Design of Electric Vehicle Battery Package Enclosure Using LHS and CCD Methods [2]

Inputs						Outputs		
						Deflection	Natural Frequency	Weight
x_1	x_2	x_3	x_4	x_5	x_6	(m)	(Hz)	(kg)
1.21	1.55	1.17	1.32	3.25	34.79	0.0019	90.12	15.81
1.62	1.71	1.39	1.32	3.31	33.65	0.0018	88.66	17.65
1.32	1.66	1.28	1.62	3.04	36.07	0.0018	89.28	16.78
1.21	1.51	1.37	1.31	2.92	38.76	0.0019	92.07	15.61
1.38	1.7	1.18	1.58	2.92	38.46	0.0018	89.3	16.73
1.29	1.53	1.31	1.26	2.52	39.26	0.0019	92.73	15.37
1.01	1.13	1.26	1.39	3.24	38.96	0.0019	93.25	14.41
1.35	1.49	1.25	1.37	3.13	42.83	0.0018	90.64	16.12
1.27	1.54	1.14	1.25	2.84	34.16	0.0019	91.15	15.43
1.24	1.79	1.2	1.6	2.88	38.56	0.0019	89.23	16.57
1.05	1.33	1.28	1.55	3.08	46.43	0.0019	92.52	15.15
1.26	1.38	1.06	1.5	2.47	40.66	0.002	90.11	14.94
1.17	1.39	1.12	1.42	2.57	45.26	0.002	91.3	14.79
1.37	1.76	1.24	1.29	2.95	37.19	0.0018	90.43	16.51
1.43	1.91	1.09	1.26	2.18	42.55	0.0019	88.97	16.1
1.22	1.69	1.22	1.14	2.79	38.06	0.0019	90.88	15.54
1.07	1.36	1.55	1.36	2.87	39.55	0.0019	93.15	15.09
.	.	.	.	.	.	.	.	.
1.14	1.61	1.12	1.41	3.18	37.85	0.0019	89.86	15.75
1.28	1.57	1.2	1.48	2.56	47.84	0.0019	91.23	15.68
1.25	1.5	1.28	1.45	2.64	39.65	0.0019	90.72	15.58
1.25	1.53	1.33	1.33	3.7	44.03	0.0018	90.25	16.48
1.34	1.64	1.16	1.36	3.22	42.69	0.0018	89.46	16.4
1.17	1.74	1.01	1.38	3.01	41.54	0.002	88.27	15.79
1.52	1.24	0.97	1.11	2.91	44.7	0.002	89.69	15.08

TABLE 5.5

Model Assessment of Natural Frequency Output

Model*	R^2_{train}	R^2_{adj}	R^2_{test}	R^2_{valid}	Maximum	Minimum
L	0.76	0.74	0.82	0.92	96.52	84.72
SON	0.87	0.86	0.82	0.67	100.16	74.75
TON	1	1	−235	−86	513.31	−371.76
FOTN	0.82	0.8	0.84	0.94	95.53	83.35
SOTN	0.89	0.88	0.89	0.84	98.8	73.2
FOLN	0.76	0.74	0.84	0.92	96.94	84.87
SOLN	0.86	0.85	0.85	0.67	100.92	75.78
LR	0.81	0.8	0.6	0.7	641169	−2.61E+07
SONR	0.98	0.98	−6.2	0.1	1.70E+11	−3.50E+07
TONR	0.83	0.82	0.78	0.91	100.72	84
FOTNR	0.87	0.86	0.7	−0.03	5.00E+07	−1.17E+06
FOLNR	0.81	0.8	0.6	0.53	5.12E+04	−1.20E+15
SOLNR	0.92	0.91	−2.11	0.81	3381	−21396

* Full form of models are given in appendix (Table A5.2).

Table 5.5 shows the assessment results of the mathematical model, taking into account the R^2 criterion and the maximum and minimum natural frequency values that provide information about the boundedness of the model.

The SONR model demonstrated the highest degree of success during the training phase, whereas the L and FOLN models exhibited comparatively lower levels of success. However, models such as TON, SONR, FOTNR, and SOLNR were ultimately excluded due to their inability to accurately forecast real-world outcomes during both testing and validation phases. Notably, despite its prominence in the training phase, the SONR model displayed suboptimal performance during testing and validation. Boundedness assessments, incorporating maximum and minimum natural frequency values derived from the models, revealed that the TON, LR, SONR, FOTNR, FOLNR, and SOLNR models yielded inconsequential outcomes.

When the successful models were ranked based on their average R^2 prediction performance, computed by averaging their performances across the training, adjusted, testing, and validation phases, the SOTN and FOTN models emerged as the most successful, as shown in Table 5.6.

The model types used in the RSM approach are limited to non-rational polynomials. This situation causes deficiencies in process modeling as well as errors in obtaining the optimum design. Therefore, proposing alternative model types that can be used in the mathematical modeling process is an issue that must be studied to define the process more accurately and reach the optimum design. In this study, the SOTN was the most successful in presenting the appropriate mathematical model for estimating the natural frequency for electric vehicle battery pack enclosure.

TABLE 5.6

Assessment of Model Performance Considering Average R^2 and Number of Model Terms

Model	R^2_{train}	R^2_{adj}	R^2_{test}	R^2_{valid}	Avg R^2	Model Term Number	Max.	Min.
SOTN	0.89	0.88	0.89	0.84	0.875	50	98.8	73.2
FOTN	0.82	0.8	0.84	0.94	0.85	13	95.53	83.35
TONR	0.83	0.82	0.78	0.91	0.835	168	100.72	84
FOLN	0.76	0.74	0.84	0.92	0.815	7	96.94	84.87
L	0.76	0.74	0.82	0.92	0.81	7	96.52	84.72
SOLN	0.86	0.85	0.85	0.67	0.8075	28	100.92	75.78
SON	0.87	0.86	0.82	0.67	0.805	28	100.16	74.75

Models can be categorized as either simpler or more complex based on the terms they contain and the number of those terms. In this context, opting for models with simpler structures and fewer terms, while still maintaining approximate prediction performance, may seem more reasonable. Regarding the FOTN model, which contains a comparatively fewer number of terms, it might appear more logical to favor it over the SOTN model, which is more complex and comprises a higher number of terms. However, evaluating models solely based on prediction performance is not sufficient. It is more meaningful to assess the stability of the model. This involves systematically varying the values of input parameters at specified increments and observing their impact on the output parameter values. If the resulting output parameter values consistently fall within a realistic and feasible range within the realm of engineering, then it can be concluded that the candidate model has successfully passed the stability test.

Figure 5.2 presents the results of the stability test analysis conducted on the natural frequency output of the SOTN model. During the stability test, changes in the natural frequency output were visually observed by incrementally increasing or decreasing the input parameters by 1%, 5%, 10%, 20%, and 50%. As depicted in Figures 5.2a and 5.2b, altering the input parameters by up to 20% did not lead to any significant deviations in the natural frequency output. However, Figures 5.2c and 5.2d reveal that a 50% change in the inputs resulted in an unreasonable disparity in the natural frequency values. The emergence of negative natural frequencies and values exceeding 150 Hz are unrealistic and implausible in engineering contexts. These findings underscore the importance of assessing model stability to ensure the reliability and practicality of predictions.

Figure 5.3 shows the stability test analysis results for the natural frequency of the FOTN model. Unlike the SOTN model, increasing and decreasing the input values up to 50% did not cause an abnormal change in the natural frequency

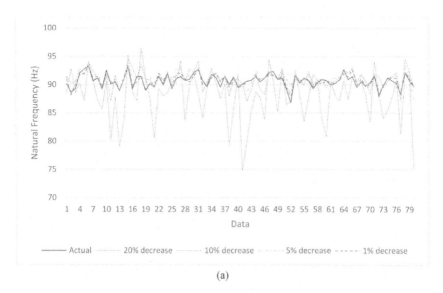

(a)

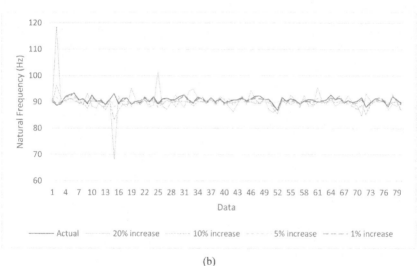

(b)

FIGURE 5.2   Stability analysis of the SOTN model for natural frequency output, when (a) the input parameters are decreased by 1%, 5%, 10%, and 20%, (b) the input parameters are increased by 1%, 5%, 10%, and 20%, (c) the input parameters are decreased by 50%, and (d) the input parameters are increased by 50%

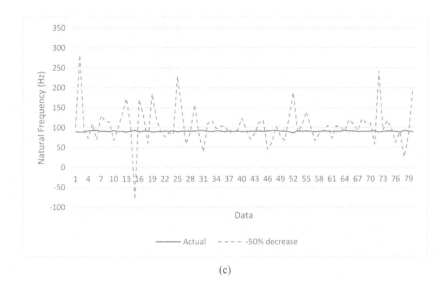

(c)

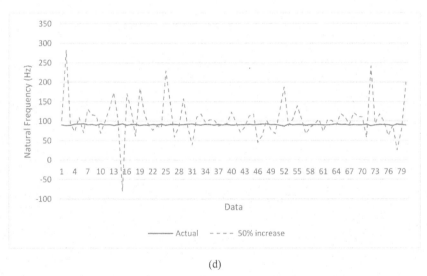

(d)

FIGURE 5.2 *(Continued)*

values, as seen in Figures 5.3a and 5.3b. When the stability test results of the FOTN and SOTN models are examined, the FOTN model is more stable than the SOTN model. In addition, considering the terms and number of terms included in the FOTN model, it has a simpler structure.

Figure 5.4 illustrates various aspects of the FOTN model's performance, including the relationship between predicted and observed values, error rates, their distributions, and the evaluation of predicted values within statistical

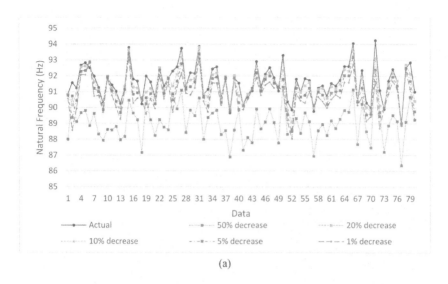

(a)

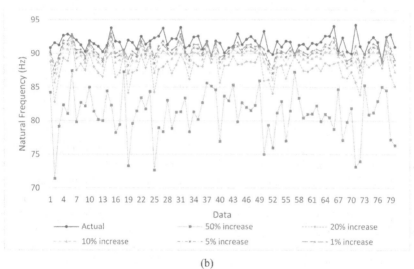

(b)

FIGURE 5.3   Stability analysis of the FOTN model for natural frequency output, when: (a) the input parameters are decreased by 1%, 5%, 10%, 20%, and 50% and (b) the input parameters are increased by 1%, 5%, 10%, 20%, and 50%

prediction bands. In Figure 5.4a, it is evident that the errors resulting from predictions follow a normal distribution, indicating a satisfactory fit of the model to the data. Figure 5.4b demonstrates the ratio of the error, derived from the disparity between actual and predicted values for each output, to the standard error of this error value. Here, the point distributions within the range of +3 to −3

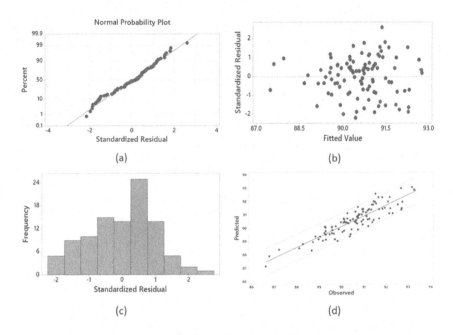

FIGURE 5.4 Statistical analysis of the FOTN model for natural frequency output: (a) normal distribution of errors, (b) standard error, (c) distribution of standard error, and (d) relationship between predicted and observed values

signify statistically significant and acceptable errors. Figure 5.4c highlights that the majority of errors fall within the 0–1 range, indicating a high level of accuracy in prediction. Lastly, in Figure 5.4d, the predicted values are closely clustered around the regression line and within the 95% prediction interval band, affirming the reliability and consistency of the statistical estimates across various graphical representations. These findings collectively suggest that the FOTN model provides reasonable and robust predictions, aligning well with the results depicted in other graphical analyses.

Figure 5.5 shows the relationship between the predicted results obtained using the SOTN model and the actual values and the statistical evaluation of the error values. It is possible to make similar evaluations with the FOTN model. The errors are normally distributed, the estimated values are within the 95% confidence interval, and the errors are concentrated from −1 to 1.

This study revealed that FOTN and SOTN are more suitable models than polynomials for modeling natural frequency, considering model assessment criteria such as boundedness control, stability analysis, and statistical distribution of errors. When FOTN and SOTN models are compared with each other, it would be more meaningful to say that both models are superior to each other in some

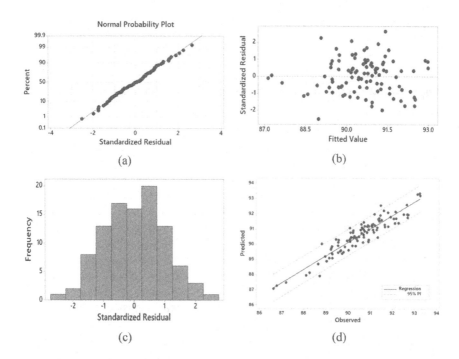

FIGURE 5.5 Statistical analysis of the SOTN model for natural frequency output: (a) normal distribution of errors, (b) standard error, (c) distribution of standard error, and (d) relationship between predicted and observed values

evaluation criteria, and instead of making a clear choice, the FOT model has fewer terms and has higher stability, and the SOT model has a better prediction performance.

REFERENCES

[1] Gerdooei M, Rezaei MJ, Ghaforian Nosrati H. Improving the performance of a multi-layer armored system subjected to shock loading of an underwater explosion. Mechanics of Advanced Materials and Structures 2022; 29(3): 419–428.

[2] Savran M. A new systematic approach for design, modeling and optimization of the engineering problems based on stochastic multiple-nonlinear neuro-regression analysis and non-traditional search algorithms (PhD thesis). İzmir Katip Çelebi University, 2023. https://tez.yok.gov.tr/UlusalTezMerkezi/tezSorguSonucYeni.jsp

[3] Shui L, Chen F, Garg A, Peng X, Bao N, Zhang J. Design optimization of battery pack enclosure for electric vehicle. Structural and Multidisciplinary Optimization 2018; 58: 331–347.

APPENDIX

TABLE A5.1
Proposed Mathematical Models for Stress Output (Related to Table 5.3)

Nomenclature	Models
L	$69.799 + 0.051x_1 - 0.142x_2 - 21.740x_3$
LR	$(14532.900 - 8803.850x_1 - 17492.700x_2 + 750076.000x_3)/(-19670.900 - 273.603x_1 - 398.252x_2 + 33795.500x_3)$
SON	$3167.350 - 5.263x_1 + 0.546x_1^2 - 1.409x_2 + 0.163x_1x_2 + 0.566x_2^2 - 4148.820x_3 - 8.483x_1x_3 - 8.358x_2x_3 + 1448.470x_3^2$
SONR	$(5.178 + 221.225x_1 - 62.185x_1^2 - 196.573x_2 + 8.551x_1x_2 + 64.210x_2^2 + 26.118x_3 + 453.164x_1x_3 - 369.875x_2x_3 + 9.523x_3^2)/(-16.665 + 24.547x_1 - 1.758x_1^2 - 23.916x_2 + 0.091x_1x_2 + 1.817x_2^2 - 30.361x_3 + 1.960x_1x_3 + 3.075x_2x_3 + 17.070x_3^2)$
TON	$-0.176 + 12.300x_1 - 1.687x_1^2 + 0.093x_1^3 + 4.532x_2 + 1.814x_1x_2 - 0.184x_1^2x_2 - 0.848x_2^2 + 0.173x_1x_2^2 - 0.107x_2^3 + 5.628x_3 - 2.826x_1x_3 - 0.446x_1^2x_3 - 2.913x_2x_3 + 0.178x_1x_2x_3 + 0.982x_2^2x_3 + 1.229x_3^2 + 4.626x_1x_3^2 - 6.611x_2x_3^2 - 0.132x_3^3$
TONR	$(5.379 - 10.501x_1 - 3.186x_1^2 + 53.587x_1^3 - 4.435x_2 - 78.205x_1x_2 - 418.985x_1^2x_2 + 1.057x_2^2 - 11.901x_1x_2^2 - 28.653x_2^3 - 10.682x_3 - 73.524x_1x_3 - 50.773x_1^2x_3 - 9.361x_2x_3 - 87.685x_1x_2x_3 - 47.876x_2^2x_3 - 24.186x_3^2 - 24.145x_1x_3^2 - 29.761x_2x_3^2 - 2.008x_3^3)/(-4.761 - 235.002x_1 - 60.790x_1^2 + 5.774x_1^3 - 173.128x_2 + 453.390x_1x_2 - 21.868x_1^2x_2 - 73.899x_2^2 + 9.524x_1x_2^2 - 7.026x_2^3 - 24.441x_3 - 135.613x_1x_3 - 23.394x_1^2x_3 - 489.435x_2x_3 - 229.606x_1x_2x_3 + 66.687x_2^2x_3 + 16.324x_3^2 + 434.088x_1x_3^2 + 95.938x_2x_3^2 + 19.215x_3^3)$
FOT	$-209.930 - 1.362Cos[x_1] - 2.039Cos[x_2] + 43.908Cos[x_3] + 2.743Sin[x_1] - 0.294Sin[x_2] + 243.309Sin[x_3]$
FOTR	$(-3.058 - 22.084Cos[x_1] + 8.622Cos[x_2] + 3.389Cos[x_3] + 21.571Sin[x_1] + 10.509Sin[x_2] - 32.086Sin[x_3])/(-0.077 - 0.603Cos[x_1] + 0.238Cos[x_2] + 0.437Cos[x_3] + 0.603Sin[x_1] + 0.279Sin[x_2] - 0.923Sin[x_3])$
SOT	$8.690 - 8.633Cos[x_1] + 1.357Cos[x_1]^2 - 9.100Cos[x_2] - 9.668Cos[x_1]Cos[x_2] - 11.298Cos[x_2]^2 + 6.384Cos[x_3] - 19.877Cos[x_1]Cos[x_3] + 7.568Cos[x_2]Cos[x_3] + 46.835Cos[x_3]^2 + 6.727Sin[x_1] + 3.746Cos[x_1]Sin[x_1] + 1.294Cos[x_2]Sin[x_1] + 20.612Cos[x_3]Sin[x_1] + 7.633Sin[x_1]^2 - 1.697Sin[x_2] - 3.940Cos[x_1]Sin[x_2] + 2.003Cos[x_2]Sin[x_2] + 7.464Cos[x_3]Sin[x_2] - 0.955Sin[x_1]Sin[x_2] - 5.556Sin[x_2]^2 + 22.573Sin[x_3] + 0.256Cos[x_1]Sin[x_3] + 1.246Cos[x_2]Sin[x_3] - 0.863Cos[x_3]Sin[x_3] - 3.672Sin[x_1]Sin[x_3] - 1.195Sin[x_2]Sin[x_3] + 1.345Sin[x_3]^2$
SOTR	$(-0.962 - 0.213Cos[x_1] - 0.812Cos[x_1]^2 + 4.875Cos[x_2] - 3.209Cos[x_1]Cos[x_2] - 2.192Cos[x_2]^2 - 3.054Cos[x_3] + 0.399Cos[x_1]Cos[x_3] + 1.447Cos[x_2]Cos[x_3] - 2.576Cos[x_3]^2 - 3.477Sin[x_1] + 2.309Cos[x_1]Sin[x_1] - 1.192Cos[x_2]Sin[x_1] - 1.480Cos[x_3]Sin[x_1] - 1.497Sin[x_1]^2 + 0.759Sin[x_2] + 1.521Cos[x_1]Sin[x_2] - 5.049Cos[x_2]Sin[x_2] - 0.058Cos[x_3]Sin[x_2] + 0.056Sin[x_1]Sin[x_2] - 1.440Sin[x_2]^2 -$

(Continued)

TABLE A5.1 *(Continued)*

Proposed Mathematical Models for Stress Output (Related to Table 5.3)

Nomenclature	Models
	$7.417\mathrm{Sin}[x_3] + 1.575\mathrm{Cos}[x_1]\mathrm{Sin}[x_3] + 5.205\mathrm{Cos}[x_2]\mathrm{Sin}[x_3] + 1.608\mathrm{Cos}[x_3]\mathrm{Sin}[x_3] -$
	$4.388\mathrm{Sin}[x_1]\mathrm{Sin}[x_3] - 0.668\mathrm{Sin}[x_2]\mathrm{Sin}[x_3] + 0.143\mathrm{Sin}[x_3]^2)/(0.517 + 1.428\mathrm{Cos}[x_1] -$
	$0.126\mathrm{Cos}[x_1]^2 - 0.988\mathrm{Cos}[x_2] - 0.414\mathrm{Cos}[x_1]\mathrm{Cos}[x_2] + 0.120\mathrm{Cos}[x_2]^2 -$
	$0.690\mathrm{Cos}[x_3] - 2.436\mathrm{Cos}[x_1]\mathrm{Cos}[x_3] + 2.075\mathrm{Cos}[x_2]\mathrm{Cos}[x_3] + 0.260\mathrm{Cos}[x_3]^2 +$
	$0.085\mathrm{Sin}[x_1] + 0.337\mathrm{Cos}[x_1]\mathrm{Sin}[x_1] - 0.109\mathrm{Cos}[x_2]\mathrm{Sin}[x_1] + 2.055\mathrm{Cos}[x_3]\mathrm{Sin}[x_1] +$
	$0.177\mathrm{Sin}[x_1]^2 - 1.483\mathrm{Sin}[x_2] - 1.332\mathrm{Cos}[x_1]\mathrm{Sin}[x_2] - 1.118\mathrm{Cos}[x_2]\mathrm{Sin}[x_2] +$
	$1.424\mathrm{Cos}[x_3]\mathrm{Sin}[x_2] + 0.949\mathrm{Sin}[x_1]\mathrm{Sin}[x_2] - 0.626\mathrm{Sin}[x_2]^2 + 0.300\mathrm{Sin}[x_3] -$
	$2.256\mathrm{Cos}[x_1]\mathrm{Sin}[x_3] + 0.382\mathrm{Cos}[x_2]\mathrm{Sin}[x_3] + 0.372\mathrm{Cos}[x_3]\mathrm{Sin}[x_3] +$
	$0.210\mathrm{Sin}[x_1]\mathrm{Sin}[x_3] + 0.092\mathrm{Sin}[x_2]\mathrm{Sin}[x_3] - 1.920\mathrm{Sin}[x_3]^2)$
FOL	$50.851 + 0.510\mathrm{Log}[x_1] - 1.416\mathrm{Log}[x_2] - 31.288\mathrm{Log}[x_3]$
FOLR	$(-5.166 \times 10^7 + [1.697 \times 10^7]\mathrm{Log}[x_1] + (2.002 \times 10^7)\mathrm{Log}[x_2] -$
	$(6.862 \times 10^6)\mathrm{Log}[x_1]\mathrm{Log}[x_2] - (1.463 \times 10^7)\mathrm{Log}[x_3])/(-1.291 \times 10^6 +$
	$453927.000\mathrm{Log}[x_1] + 524252.000\mathrm{Log}[x_2] - 182455.000\mathrm{Log}[x_1]\mathrm{Log}[x_2] -$
	$589576.000\mathrm{Log}[x_3])$
SOL	$69.941 - 8.312\mathrm{Log}[x_1] + 2.764\mathrm{Log}[x_1]^2 - 3.141\mathrm{Log}[x_2] + 0.324\mathrm{Log}[x_1]\mathrm{Log}[x_2] +$
	$1.206\mathrm{Log}[x_2]^2 - 72.816\mathrm{Log}[x_3] - 15.465\mathrm{Log}[x_1]\mathrm{Log}[x_3] - 9.000\mathrm{Log}[x_2]\mathrm{Log}[x_3] +$
	$128.588\mathrm{Log}[x_3]^2$
SOLR	$(-7.328 - 55.337\mathrm{Log}[x_1] - 18.270\mathrm{Log}[x_1]^2 - 14.315\mathrm{Log}[x_2] +$
	$61.198\mathrm{Log}[x_1]\mathrm{Log}[x_2] + 12.934\mathrm{Log}[x_2]^2 - 6.363\mathrm{Log}[x_3] - 6.376\mathrm{Log}[x_1]\mathrm{Log}[x_3] +$
	$3.227\mathrm{Log}[x_2]\mathrm{Log}[x_3] - 1.104\mathrm{Log}[x_3]^2)/(-1.083 + 0.514\mathrm{Log}[x_1] - 0.752\mathrm{Log}[x_1]^2 -$
	$3.086\mathrm{Log}[x_2] + 1.555\mathrm{Log}[x_1]\mathrm{Log}[x_2] + 0.586\mathrm{Log}[x_2]^2 + 2.950\mathrm{Log}[x_3] -$
	$1.338\mathrm{Log}[x_1]\mathrm{Log}[x_3] + 5.367\mathrm{Log}[x_2]\mathrm{Log}[x_3] - 10.986\mathrm{Log}[x_3]^2)$

TABLE A5.2

Proposed Mathematical Models for Natural Frequency Output (Related to Table 5.5)

Nomenclature	Models
L	$100.293 - 4.045x_1 - 3.229x_2 + 5.343x_3 - 1.861x_4 - 1.035x_5 - 0.012x_6$
LR	$(4731.680 - 911.101x_1 - 193.255x_2 + 1050.120x_3 + 320.721x_4 - 566.614x_5 -$
	$57.088x_6)/(50.701 - 9.541x_1 - 1.614x_2 + 10.902x_3 + 3.819x_4 - 6.092x_5 - 0.629x_6)$
SON	$14.620 + 50.037x_1 - 14.921x_1^2 + 13.243x_2 - 0.539x_1x_2 - 0.268x_2^2 + 2.030x_3 -$
	$0.318x_1x_3 + 6.388x_2x_3 - 4.687x_3^2 - 14.895x_4 + 10.747x_1x_4 + 3.457x_2x_4 -$
	$0.685x_3x_4 - 1.119x_4^2 + 13.479x_5 - 4.204x_1x_5 - 4.595x_2x_5 + 3.757x_3x_5 +$
	$0.029x_4x_5 - 0.256x_5^2 + 1.465x_6 - 0.444x_1x_6 - 0.384x_2x_6 - 0.103x_3x_6 -$
	$0.069x_4x_6 - 0.130x_5x_6 + 0.004x_6^2$

(Continued)

TABLE A5.2 *(Continued)*
Proposed Mathematical Models for Natural Frequency Output
(Related to Table 5.5)

Nomenclature	Models

SONR $(130364.000 - 108020.000x_1 + 21515.100x_1^2 - 3606.660x_2 + 536.269x_1x_2 - 5262.380x_2^2 + 31999.900x_3 - 12459.900x_1x_3 - 7261.010x_2x_3 - 11577.900x_3^2 - 8402.030x_4 - 11301.500x_1x_4 + 4694.350x_2x_4 + 26271.000x_3x_4 + 7492.640x_4^2 - 33289.800x_5 + 18147.000x_1x_5 + 2989.240x_2x_5 - 1299.460x_3x_5 - 5785.840x_4x_5 + 1735.510x_5^2 - 836.303x_6 + 488.612x_1x_6 + 249.930x_2x_6 - 185.085x_3x_6 - 360.068x_4x_6 + 95.515x_5x_6 + 2.478x_6^2)/((1445.150 - 1183.970x_1 + 233.884x_1^2 - 38.048x_2 + 0.797x_1x_2 - 58.581x_2^2 + 343.379x_3 - 130.037x_1x_3 - 76.826x_2x_3 - 129.624x_3^2 - 102.338x_4 - 123.997x_1x_4 + 55.505x_2x_4 + 286.087x_3x_4 + 84.606x_4^2 - 368.742x_5 + 199.460x_1x_5 + 33.113x_2x_5 - 13.364x_3x_5 - 62.766x_4x_5 + 19.143x_5^2 - 9.148x_6 + 5.421x_1x_6 + 2.677x_2x_6 - 1.989x_3x_6 - 3.995x_4x_6 + 1.043x_5x_6 + 0.027x_6^2))$

TON $-198.312 + 1189.480x_1 + 2613.510x_1^2 - 545.717x_1^3 + 255.992x_2 - 666.166x_1x_2 + 329.334x_1^2x_2 + 821.266x_2^2 + 100.982x_1x_2^2 - 178.898x_2^3 + 3554.720x_3 - 3875.730x_1x_3 + 44.438x_1^2x_3 - 1010.170x_2x_3 + 579.824x_1x_2x_3 + 132.244x_2^2x_3 - 654.134x_3^2 + 364.830x_1x_3^2 - 205.505x_2x_3^2 - 132.051x_3^3 - 2218.080x_4 - 2451.690x_1x_4 - 25.039x_1^2x_4 + 314.872x_2x_4 - 749.919x_1x_2x_4 + 4.794x_2^2x_4 + 1496.430x_3x_4 + 1091.580x_1x_3x_4 - 142.688x_2x_3x_4 - 1167.770x_3^2x_4 + 1441.150x_4^2 - 491.930x_1x_4^2 + 314.495x_2x_4^2 - 279.809x_3x_4^2 + 3.003x_4^3 - 2563.280x_5 - 636.376x_1x_5 - 253.850x_1^2x_5 + 307.897x_2x_5 - 59.779x_1x_2x_5 + 10.441x_2^2x_5 + 129.640x_3x_5 + 212.814x_1x_3x_5 - 120.707x_2x_3x_5 + 374.614x_3^2x_5 + 127.792x_4x_5 + 762.718x_1x_4x_5 + 195.250x_2x_4x_5 - 72.297x_3x_4x_5 - 109.312x_4^2x_5 + 651.535x_5^2 + 14.503x_1x_5^2 - 40.619x_2x_5^2 - 64.011x_3x_5^2 - 56.944x_4x_5^2 - 31.122x_5^3 + 131.781x_6 + 56.334x_1x_6 - 5.814x_1^2x_6 - 56.552x_2x_6 - 0.343x_1x_2x_6 - 9.458x_2^2x_6 - 36.500x_3x_6 - 4.137x_1x_3x_6 + 24.941x_2x_3x_6 + 37.782x_3^2x_6 + 23.026x_4x_6 + 37.384x_1x_4x_6 - 16.903x_2x_4x_6 + 28.073x_3x_4x_6 - 16.443x_4^2x_6 + 32.917x_5x_6 - 1.346x_1x_5x_6 - 4.288x_2x_5x_6 - 16.791x_3x_5x_6 - 17.723x_4x_5x_6 - 4.384x_5^2x_6 - 4.372x_6^2 - 1.155x_1x_6^2 + 1.202x_2x_6^2 - 0.981x_3x_6^2 + 0.286x_4x_6^2 + 0.617x_5x_6^2 + 0.026x_6^3$

TONR $(1.001 + 3.011x_1 + 3.006x_1^2 + 0.998x_1^3 + 3.015x_2 + 6.071x_1x_2 + 3.008x_1^2x_2 + 3.024x_2^2 + 3.028x_1x_2^2 + 1.004x_2^3 + 3.009x_3 + 6.052x_1x_3 + 3.007x_1^2x_3 + 6.063x_2x_3 + 6.083x_1x_2x_3 + 3.026x_2^2x_3 + 3.006x_3^2 + 3.011x_1x_3^2 + 3.012x_2x_3^2 + 1.000x_3^3 + 3.012x_4 + 6.065x_1x_4 + 3.011x_1^2x_4 + 6.075x_2x_4 + 6.099x_1x_2x_4 + 3.029x_2^2x_4 + 6.045x_3x_4 + 6.071x_1x_3x_4 + 6.074x_2x_3x_4 + 3.006x_3^2x_4 + 3.017x_4^2 + 3.024x_1x_4^2 + 3.025x_2x_4^2 + 3.015x_3x_4^2 + 1.003x_4^3 + 3.026x_5 + 6.124x_1x_5 + 3.011x_1^2x_5 + 6.153x_2x_5 + 6.174x_1x_2x_5 + 3.056x_2^2x_5 + 6.116x_3x_5 + 6.153x_1x_3x_5 + 6.177x_2x_3x_5 + 3.024x_3^2x_5 + 6.137x_4x_5 + 6.177x_1x_4x_5 + 6.190x_2x_4x_5 + 6.146x_3x_4x_5 + 3.047x_4^2x_5 + 3.079x_5^2 + 3.090x_1x_5^2 + 3.101x_2x_5^2 + 3.098x_3x_5^2 + 3.104x_4x_5^2 + 1.028x_5^3 + 3.257x_6 + 7.300x_1x_6 + 3.086x_1^2x_6 +$

(Continued)

TABLE A5.2 *(Continued)*
**Proposed Mathematical Models for Natural Frequency Output
(Related to Table 5.5)**

Nomenclature	Models

$7.593x_2x_6 + 7.924x_1x_2x_6 + 3.578x_2^2x_6 + 7.144x_3x_6 + 7.713x_1x_3x_6 + 7.910x_2x_3x_6 +$
$3.195x_3^2x_6 + 7.375x_4x_6 + 7.887x_1x_4x_6 + 8.028x_2x_4x_6 + 7.416x_3x_4x_6 + 3.483x_4^2x_6 +$
$9.204x_5x_6 + 9.835x_1x_5x_6 + 10.348x_2x_5x_6 + 9.996x_3x_5x_6 + 10.190x_4x_5x_6 +$
$5.553x_5^2x_6 + 9.998x_6^2 + 11.552x_1x_6^2 + 12.508x_2x_6^2 + 12.043x_3x_6^2 + 12.296x_4x_6^2 +$
$26.937x_5x_6^2 + 21.133x_6^3)/(0.985 + 2.820x_1 + 3.552x_1^2 + 1.314x_1^3 + 2.658x_2 +$
$4.483x_1x_2 + 3.824x_1^2x_2 + 2.335x_2^2 + 2.351x_1x_2^2 + 0.873x_2^3 + 3.034x_3 + 5.565x_1x_3 +$
$3.673x_1^2x_3 + 5.306x_2x_3 + 4.701x_1x_2x_3 + 2.505x_2^2x_3 + 3.570x_3^2 + 3.373x_1x_3^2 +$
$3.520x_2x_3^2 + 1.191x_3^3 + 2.849x_4 + 4.711x_1x_4 + 3.511x_1^2x_4 + 4.591x_2x_4 +$
$3.787x_1x_2x_4 + 2.362x_2^2x_4 + 6.604x_3x_4 + 5.364x_1x_3x_4 + 6.117x_2x_3x_4 + 3.993x_3^2x_4 +$
$2.797x_4^2 + 2.410x_1x_4^2 + 2.632x_2x_4^2 + 3.317x_3x_4^2 + 0.966x_4^3 + 2.763x_5 + 5.375x_1x_5 +$
$5.401x_1^2x_5 + 4.537x_2x_5 + 5.861x_1x_2x_5 + 2.573x_2^2x_5 + 6.095x_3x_5 + 5.440x_1x_3x_5 +$
$5.572x_2x_3x_5 + 4.230x_3^2x_5 + 5.320x_4x_5 + 4.638x_1x_4x_5 + 5.914x_2x_4x_5 + 7.558x_3x_4x_5 +$
$2.932x_4^2x_5 + 2.496x_5^2 + 3.256x_1x_5^2 + 3.596x_2x_5^2 + 2.486x_3x_5^2 + 2.773x_4x_5^2 + 0.780x_5^3 +$
$1.028x_6 - 6.024x_1x_6 + 28.682x_1^2x_6 - 17.002x_2x_6 - 17.271x_1x_2x_6 - 5.034x_2^2x_6 +$
$10.534x_3x_6 - 15.203x_1x_3x_6 - 11.959x_2x_3x_6 + 20.325x_3^2x_6 - 2.581x_4x_6 -$
$22.210x_1x_4x_6 - 13.156x_2x_4x_6 + 25.817x_3x_4x_6 - 0.195x_4^2x_6 - 8.376x_5x_6 +$
$2.960x_1x_5x_6 + 8.351x_2x_5x_6 - 6.990x_3x_5x_6 + 5.151x_4x_5x_6 - 5.901x_5^2x_6 -$
$0.051x_6^2 + 0.434x_1x_6^2 + 2.094x_2x_6^2 - 1.677x_3x_6^2 + 0.385x_4x_6^2 + 0.989x_5x_6^2 + 0.197x_6^3)$

FOTN
$56.766 + 10.060\text{Cos}[x_1] + 3.503\text{Cos}[x_2] - 0.423\text{Cos}[x_3] + 2.582\text{Cos}[x_4] +$
$0.994\text{Cos}[x_5] - 0.002\text{Cos}[x_6] + 20.446\text{Sin}[x_1] - 5.466\text{Sin}[x_2] + 15.647\text{Sin}[x_3] +$
$2.733\text{Sin}[x_4] + 0.775\text{Sin}[x_5] + 0.064\text{Sin}[x_6]$

FOTNR
$(-390.163 + 1668.740\text{Cos}[x_1] - 94.662\text{Cos}[x_2] - 3190.230\text{Cos}[x_3] +$
$512.204\text{Cos}[x_4] + 1861.650\text{Cos}[x_5] + 130.495\text{Cos}[x_6] + 6809.320\text{Sin}[x_1] -$
$5267.950\text{Sin}[x_2] - 4791.960\text{Sin}[x_3] + 6685.110\text{Sin}[x_4] + 241.302\text{Sin}[x_5] -$
$20.362\text{Sin}[x_6])/(-3.792 + 17.699\text{Cos}[x_1] - 1.367\text{Cos}[x_2] - 34.406\text{Cos}[x_3] +$
$5.409\text{Cos}[x_4] + 20.442\text{Cos}[x_5] + 1.443\text{Cos}[x_6] + 73.679\text{Sin}[x_1] - 57.743\text{Sin}[x_2] -$
$52.049\text{Sin}[x_3] + 73.369\text{Sin}[x_4] + 2.614\text{Sin}[x_5] - 0.231\text{Sin}[x_6])$

SOTN
$8.690 - 8.633\text{Cos}[x_1] + 1.357\text{Cos}[x_1]^2 - 9.100\text{Cos}[x_2] - 9.668\text{Cos}[x_1]\text{Cos}[x_2] -$
$11.298\text{Cos}[x_2]^2 + 6.384\text{Cos}[x_3] - 19.877\text{Cos}[x_1]\text{Cos}[x_3] + 7.568\text{Cos}[x_2]\text{Cos}[x_3] +$
$46.835\text{Cos}[x_3]^2 + 6.727\text{Sin}[x_1] + 3.746\text{Cos}[x_1]\text{Sin}[x_1] + 1.294\text{Cos}[x_2]\text{Sin}[x_1] +$
$20.612\text{Cos}[x_3]\text{Sin}[x_1] + 7.633\text{Sin}[x_1]^2 - 1.697\text{Sin}[x_2] - 3.940\text{Cos}[x_1]\text{Sin}[x_2] +$
$2.003\text{Cos}[x_2]\text{Sin}[x_2] + 7.464\text{Cos}[x_3]\text{Sin}[x_2] - 0.955\text{Sin}[x_1]\text{Sin}[x_2] - 5.556\text{Sin}[x_2]^2 +$
$22.573\text{Sin}[x_3] + 0.256\text{Cos}[x_1]\text{Sin}[x_3] + 1.246\text{Cos}[x_2]\text{Sin}[x_3] - 0.863\text{Cos}[x_3]\text{Sin}[x_3] -$
$3.672\text{Sin}[x_1]\text{Sin}[x_3] - 1.195\text{Sin}[x_2]\text{Sin}[x_3] + 1.345\text{Sin}[x_3]^2$

(Continued)

TABLE A5.2 *(Continued)*
Proposed Mathematical Models for Natural Frequency Output
(Related to Table 5.5)

Nomenclature	Models

SOTNR $(25.254 - 24.044\text{Cos}[x_1] - 4.644\text{Cos}[x_1]^2 + 137.325\text{Cos}[x_2] + 44.429\text{Cos}[x_1]\text{Cos}[x_2] +$

$0.757\text{Cos}[x_2]^2 + 15.810\text{Cos}[x_3] - 5.316\text{Cos}[x_1]\text{Cos}[x_3] + 28.802\text{Cos}[x_2]\text{Cos}[x_3] +$

$5.000\text{Cos}[x_3]^2 + 6.754\text{Cos}[x_4] - 8.433\text{Cos}[x_1]\text{Cos}[x_4] + 34.661\text{Cos}[x_2]\text{Cos}[x_4] -$

$0.877\text{Cos}[x_3]\text{Cos}[x_4] + 1.444\text{Cos}[x_4]^2 - 76.413\text{Cos}[x_5] + 32.053\text{Cos}[x_1]\text{Cos}[x_5] -$

$129.668\text{Cos}[x_2]\text{Cos}[x_5] - 3.161\text{Cos}[x_3]\text{Cos}[x_5] + 0.170\text{Cos}[x_4]\text{Cos}[x_5] +$

$15.973\text{Cos}[x_5]^2 + 161.135\text{Cos}[x_6] + 73.351\text{Cos}[x_1]\text{Cos}[x_6] - 57.901\text{Cos}[x_2]\text{Cos}[x_6] +$

$61.481\text{Cos}[x_3]\text{Cos}[x_6] + 41.055\text{Cos}[x_4]\text{Cos}[x_6] - 153.852\text{Cos}[x_5]\text{Cos}[x_6] +$

$22.349\text{Cos}[x_6]^2 + 110.536\text{Sin}[x_1] - 19.696\text{Cos}[x_1]\text{Sin}[x_1] + 129.344\text{Cos}[x_2]\text{Sin}[x_1] +$

$18.492\text{Cos}[x_3]\text{Sin}[x_1] + 10.428\text{Cos}[x_4]\text{Sin}[x_1] - 88.142\text{Cos}[x_5]\text{Sin}[x_1] +$

$146.216\text{Cos}[x_6]\text{Sin}[x_1] + 30.898\text{Sin}[x_1]^2 + 99.493\text{Sin}[x_2] - 24.202\text{Cos}[x_1]\text{Sin}[x_2] +$

$135.422\text{Cos}[x_2]\text{Sin}[x_2] + 15.979\text{Cos}[x_3]\text{Sin}[x_2] + 6.108\text{Cos}[x_4]\text{Sin}[x_2] -$

$76.199\text{Cos}[x_5]\text{Sin}[x_2] + 155.671\text{Cos}[x_6]\text{Sin}[x_2] + 111.180\text{Sin}[x_1]\text{Sin}[x_2] +$

$25.497\text{Sin}[x_2]^2 + 89.407\text{Sin}[x_3] - 26.346\text{Cos}[x_1]\text{Sin}[x_3] + 134.900\text{Cos}[x_2]\text{Sin}[x_3] +$

$7.571\text{Cos}[x_3]\text{Sin}[x_3] + 5.679\text{Cos}[x_4]\text{Sin}[x_3] - 68.752\text{Cos}[x_5]\text{Sin}[x_3] +$

$153.081\text{Cos}[x_6]\text{Sin}[x_3] + 101.231\text{Sin}[x_1]\text{Sin}[x_3] + 90.066\text{Sin}[x_2]\text{Sin}[x_3] +$

$21.254\text{Sin}[x_3]^2 + 97.771\text{Sin}[x_4] - 21.720\text{Cos}[x_1]\text{Sin}[x_4] + 132.522\text{Cos}[x_2]\text{Sin}[x_4] +$

$15.919\text{Cos}[x_3]\text{Sin}[x_4] + 4.185\text{Cos}[x_4]\text{Sin}[x_4] - 75.513\text{Cos}[x_5]\text{Sin}[x_4] +$

$150.845\text{Cos}[x_6]\text{Sin}[x_4] + 108.531\text{Sin}[x_1]\text{Sin}[x_4] + 98.411\text{Sin}[x_2]\text{Sin}[x_4] +$

$88.382\text{Sin}[x_3]\text{Sin}[x_4] + 24.810\text{Sin}[x_4]^2 + 170.695\text{Sin}[x_5] + 42.226\text{Cos}[x_1]\text{Sin}[x_5] +$

$4.759\text{Cos}[x_2]\text{Sin}[x_5] + 61.939\text{Cos}[x_3]\text{Sin}[x_5] + 36.764\text{Cos}[x_4]\text{Sin}[x_5] -$

$162.180\text{Cos}[x_5]\text{Sin}[x_5] - 31.963\text{Cos}[x_6]\text{Sin}[x_5] + 166.754\text{Sin}[x_1]\text{Sin}[x_5] +$

$171.752\text{Sin}[x_2]\text{Sin}[x_5] + 157.908\text{Sin}[x_3]\text{Sin}[x_5] + 166.561\text{Sin}[x_4]\text{Sin}[x_5] +$

$10.281\text{Sin}[x_5]^2 + 155.857\text{Sin}[x_6] + 100.282\text{Cos}[x_1]\text{Sin}[x_6] - 73.022\text{Cos}[x_2]\text{Sin}[x_6] +$

$71.780\text{Cos}[x_3]\text{Sin}[x_6] + 27.912\text{Cos}[x_4]\text{Sin}[x_6] - 145.852\text{Cos}[x_5]\text{Sin}[x_6] -$

$37.306\text{Cos}[x_6]\text{Sin}[x_6] + 129.197\text{Sin}[x_1]\text{Sin}[x_6] + 147.527\text{Sin}[x_2]\text{Sin}[x_6] +$

$143.402\text{Sin}[x_3]\text{Sin}[x_6] + 150.870\text{Sin}[x_4]\text{Sin}[x_6] - 39.175\text{Sin}[x_5]\text{Sin}[x_6] +$

$3.905\text{Sin}[x_6]^2)/(3.427 - 5.337\text{Cos}[x_1] + 21.456\text{Cos}[x_1]^2 - 55.897\text{Cos}[x_2] -$

$29.605\text{Cos}[x_1]\text{Cos}[x_2] - 9.211\text{Cos}[x_2]^2 + 8.531\text{Cos}[x_3] + 21.415\text{Cos}[x_1]\text{Cos}[x_3] -$

$19.854\text{Cos}[x_2]\text{Cos}[x_3] - 9.767\text{Cos}[x_3]^2 - 7.317\text{Cos}[x_4] + 5.162\text{Cos}[x_1]\text{Cos}[x_4] +$

$91.355\text{Cos}[x_2]\text{Cos}[x_4] - 125.160\text{Cos}[x_3]\text{Cos}[x_4] + 20.962\text{Cos}[x_4]^2 + 31.794\text{Cos}[x_5] +$

$66.482\text{Cos}[x_1]\text{Cos}[x_5] - 138.291\text{Cos}[x_2]\text{Cos}[x_5] - 4.608\text{Cos}[x_3]\text{Cos}[x_5] +$

$126.307\text{Cos}[x_4]\text{Cos}[x_5] + 27.236\text{Cos}[x_5]^2 - 77.512\text{Cos}[x_6] - 36.024\text{Cos}[x_1]\text{Cos}[x_6] -$

$11.409\text{Cos}[x_2]\text{Cos}[x_6] + 20.932\text{Cos}[x_3]\text{Cos}[x_6] + 14.168\text{Cos}[x_4]\text{Cos}[x_6] +$

(Continued)

TABLE A5.2 *(Continued)*
Proposed Mathematical Models for Natural Frequency Output
(Related to Table 5.5)

Nomenclature	Models
	$8.522\text{Cos}[x_5]\text{Cos}[x_6] + 1.858\text{Cos}[x_6]^2 - 20.564\text{Sin}[x_1] + 9.083\text{Cos}[x_1]\text{Sin}[x_1] +$
	$5.901\text{Cos}[x_2]\text{Sin}[x_1] - 27.977\text{Cos}[x_3]\text{Sin}[x_1] - 46.282\text{Cos}[x_4]\text{Sin}[x_1] +$
	$24.273\text{Cos}[x_5]\text{Sin}[x_1] - 91.087\text{Cos}[x_6]\text{Sin}[x_1] - 17.030\text{Sin}[x_1]^2 + 27.973\text{Sin}[x_2] +$
	$65.525\text{Cos}[x_1]\text{Sin}[x_2] - 86.672\text{Cos}[x_2]\text{Sin}[x_2] + 33.708\text{Cos}[x_3]\text{Sin}[x_2] +$
	$18.817\text{Cos}[x_4]\text{Sin}[x_2] - 72.680\text{Cos}[x_5]\text{Sin}[x_2] + 68.277\text{Cos}[x_6]\text{Sin}[x_2] -$
	$28.211\text{Sin}[x_1]\text{Sin}[x_2] + 13.638\text{Sin}[x_2]^2 + 21.317\text{Sin}[x_3] - 15.775\text{Cos}[x_1]\text{Sin}[x_3] +$
	$16.597\text{Cos}[x_2]\text{Sin}[x_3] - 21.699\text{Cos}[x_3]\text{Sin}[x_3] + 27.398\text{Cos}[x_4]\text{Sin}[x_3] +$
	$22.332\text{Cos}[x_5]\text{Sin}[x_3] + 60.073\text{Cos}[x_6]\text{Sin}[x_3] + 1.782\text{Sin}[x_1]\text{Sin}[x_3] +$
	$26.502\text{Sin}[x_2]\text{Sin}[x_3] + 14.194\text{Sin}[x_3]^2 + 1.560\text{Sin}[x_4] - 59.620\text{Cos}[x_1]\text{Sin}[x_4] -$
	$10.644\text{Cos}[x_2]\text{Sin}[x_4] + 39.252\text{Cos}[x_3]\text{Sin}[x_4] + 160.140\text{Cos}[x_4]\text{Sin}[x_4] +$
	$30.039\text{Cos}[x_5]\text{Sin}[x_4] + 61.046\text{Cos}[x_6]\text{Sin}[x_4] - 11.002\text{Sin}[x_1]\text{Sin}[x_4] +$
	$12.504\text{Sin}[x_2]\text{Sin}[x_4] - 1.454\text{Sin}[x_3]\text{Sin}[x_4] - 16.535\text{Sin}[x_4]^2 + 64.429\text{Sin}[x_5] +$
	$9.259\text{Cos}[x_1]\text{Sin}[x_5] + 37.979\text{Cos}[x_2]\text{Sin}[x_5] + 54.263\text{Cos}[x_3]\text{Sin}[x_5] -$
	$35.861\text{Cos}[x_4]\text{Sin}[x_5] - 26.875\text{Cos}[x_5]\text{Sin}[x_5] + 1.001\text{Cos}[x_6]\text{Sin}[x_5] -$
	$0.097\text{Sin}[x_1]\text{Sin}[x_5] - 179.619\text{Sin}[x_2]\text{Sin}[x_5] + 65.405\text{Sin}[x_3]\text{Sin}[x_5] +$
	$30.134\text{Sin}[x_4]\text{Sin}[x_5] - 22.809\text{Sin}[x_5]^2 - 176.556\text{Sin}[x_6] - 15.526\text{Cos}[x_1]\text{Sin}[x_6] +$
	$2.287\text{Cos}[x_2]\text{Sin}[x_6] + 28.837\text{Cos}[x_3]\text{Sin}[x_6] + 42.109\text{Cos}[x_4]\text{Sin}[x_6] -$
	$39.258\text{Cos}[x_5]\text{Sin}[x_6] - 0.342\text{Cos}[x_6]\text{Sin}[x_6] - 19.962\text{Sin}[x_1]\text{Sin}[x_6] -$
	$138.278\text{Sin}[x_2]\text{Sin}[x_6] + 84.013\text{Sin}[x_3]\text{Sin}[x_6] + 219.191\text{Sin}[x_4]\text{Sin}[x_6] +$
	$4.261\text{Sin}[x_5]\text{Sin}[x_6] + 2.569\text{Sin}[x_6]^2)$
FOLN	$98.8765 - 4.91267\text{Log}[x_1] - 4.93843\text{Log}[x_2] + 6.72741\text{Log}[x_3] - 2.59446\text{Log}[x_4] -$
	$3.11368\text{Log}[x_5] - 0.587893\text{Log}[x_6]$
FOLNR	$(544.639 - 56.112\text{Log}[x_1] - 35.704\text{Log}[x_2] + 29.052\text{Log}[x_3] + 32.844\text{Log}[x_4] -$
	$72.163\text{Log}[x_5] - 112.603\text{Log}[x_6])/(5.966 - 0.594\text{Log}[x_1] - 0.363\text{Log}[x_2] +$
	$0.289\text{Log}[x_3] + 0.376\text{Log}[x_4] - 0.778\text{Log}[x_5] - 1.241\text{Log}[x_6])$
SOLN	$36.079 + 101.500\text{Log}[x_1] - 24.879\text{Log}[x_1]^2 + 93.853\text{Log}[x_2] -$
	$1.484\text{Log}[x_1]\text{Log}[x_2] - 4.637\text{Log}[x_2]^2 - 1.820\text{Log}[x_3] - 0.287\text{Log}[x_1]\text{Log}[x_3] +$
	$10.707\text{Log}[x_2]\text{Log}[x_3] - 5.821\text{Log}[x_3]^2 + 3.430\text{Log}[x_4] + 18.633\text{Log}[x_1]\text{Log}[x_4] +$
	$8.903\text{Log}[x_2]\text{Log}[x_4] - 1.189\text{Log}[x_3]\text{Log}[x_4] - 2.886\text{Log}[x_4]^2 + 64.651\text{Log}[x_5] -$
	$17.429\text{Log}[x_1]\text{Log}[x_5] - 21.481\text{Log}[x_2]\text{Log}[x_5] + 12.358\text{Log}[x_3]\text{Log}[x_5] -$
	$1.394\text{Log}[x_4]\text{Log}[x_5] - 4.423\text{Log}[x_5]^2 - 4.858\text{Log}[x_6] - 22.327\text{Log}[x_1]\text{Log}[x_6] -$
	$21.301\text{Log}[x_2]\text{Log}[x_6] - 1.629\text{Log}[x_3]\text{Log}[x_6] - 3.125\text{Log}[x_4]\text{Log}[x_6] -$
	$12.655\text{Log}[x_5]\text{Log}[x_6] + 4.650\text{Log}[x_6]^2$

(Continued)

TABLE A5.2 *(Continued)*

Proposed Mathematical Models for Natural Frequency Output (Related to Table 5.5)

Nomenclature	Models
SOLNR	$(-325.724 - 931.482\text{Log}[x_1] + 3443.360\text{Log}[x_1]^2 - 664.887\text{Log}[x_2] - 859.929\text{Log}[x_1]\text{Log}[x_2] - 1055.430\text{Log}[x_2]^2 + 180.848\text{Log}[x_3] + 1309.810\text{Log}[x_1]\text{Log}[x_3] + 1236.540\text{Log}[x_2]\text{Log}[x_3] - 1289.880\text{Log}[x_3]^2 + 1387.780\text{Log}[x_4] - 5133.510\text{Log}[x_1]\text{Log}[x_4] + 3273.440\text{Log}[x_2]\text{Log}[x_4] - 3600.940\text{Log}[x_3]\text{Log}[x_4] - 1721.090\text{Log}[x_4]^2 + 108.601\text{Log}[x_5] - 168.314\text{Log}[x_1]\text{Log}[x_5] + 520.219\text{Log}[x_2]\text{Log}[x_5] + 360.600\text{Log}[x_3]\text{Log}[x_5] + 634.691\text{Log}[x_4]\text{Log}[x_5] - 102.053\text{Log}[x_5]^2 - 4.408\text{Log}[x_6] - 330.622\text{Log}[x_1]\text{Log}[x_6] - 271.750\text{Log}[x_2]\text{Log}[x_6] + 7.417\text{Log}[x_3]\text{Log}[x_6] + 408.998\text{Log}[x_4]\text{Log}[x_6] + 28.629\text{Log}[x_5]\text{Log}[x_6] + 49.710\text{Log}[x_6]^2)/(-2.225 - 13.904\text{Log}[x_1] + 37.882\text{Log}[x_1]^2 - 10.087\text{Log}[x_2] - 12.132\text{Log}[x_1]\text{Log}[x_2] - 11.912\text{Log}[x_2]^2 + 2.134\text{Log}[x_3] + 16.254\text{Log}[x_1]\text{Log}[x_3] + 13.551\text{Log}[x_2]\text{Log}[x_3] - 13.746\text{Log}[x_3]^2 + 14.658\text{Log}[x_4] - 57.577\text{Log}[x_1]\text{Log}[x_4] + 36.328\text{Log}[x_2]\text{Log}[x_4] - 41.244\text{Log}[x_3]\text{Log}[x_4] - 18.363\text{Log}[x_4]^2 - 2.101\text{Log}[x_5] - 1.750\text{Log}[x_1]\text{Log}[x_5] + 6.752\text{Log}[x_2]\text{Log}[x_5] + 3.156\text{Log}[x_3]\text{Log}[x_5] + 7.641\text{Log}[x_4]\text{Log}[x_5] - 1.141\text{Log}[x_5]^2 + 0.752\text{Log}[x_6] - 2.349\text{Log}[x_1]\text{Log}[x_6] - 2.235\text{Log}[x_2]\text{Log}[x_6] + 0.193\text{Log}[x_3]\text{Log}[x_6] + 4.556\text{Log}[x_4]\text{Log}[x_6] + 1.120\text{Log}[x_5]\text{Log}[x_6] + 0.222\text{Log}[x_6]^2)$

6 The Effect of Using the Taguchi Method in Experimental Design on Mathematical Modeling

Taguchi is among the most commonly utilized methods in the literature for experimental design, modeling, and optimization across various fields. Its widespread preference stems from its ability to minimize the number of experiments required during the design process, thereby reducing costs—a critical consideration in problem-solving endeavors. This cost-effectiveness contributes significantly to Taguchi's esteemed status, leading to its adoption in diverse domains such as engineering, physics, chemistry, biology, medicine, and social sciences.

The Taguchi orthogonal array provides researchers with a range of design possibilities when constructing experimental sets. By referencing the Taguchi design table, which is often presented in discussions of experimental design methods, researchers can observe the potential for creating different experimental design sets by considering design parameters and their corresponding levels. While certain Taguchi sequences exhibit similarities to other experimental design methods in terms of the number of experiments and parameters involved, Taguchi's distinctive features and advantages render it a preferred approach in many research contexts.

In contrast, while some Taguchi sequences indeed minimize the number of experiments compared to alternative experimental design methods, this apparent advantage must be considered within the broader context of experimental design, modeling, and optimization as interconnected components of a cohesive process. Any deficiency in one aspect inevitably impacts the others. In essence, Taguchi's capability to model and optimize with fewer experiments may introduce inaccuracies in problem definition and analysis. Hence, experimental design emerges as a critical step in this tripartite process.

It is imperative to thoroughly analyze design parameters, constraints, and their bounds before selecting an experimental design method. Table 6.1 presents examples of the most commonly employed Taguchi orthogonal arrays in the literature along with the studies where they have been applied. This compilation underscores the importance of informed decision-making in experimental design, ensuring robustness and accuracy throughout the entire modeling and optimization process.

DOI: 10.1201/9781003494843-6

TABLE 6.1

Literature Studies Related to Taguchi and the Most Frequently Used Taguchi Arrays

Year	Study	Scope	Number of Design Parameters	Taguchi Orthogonal Array	FFD Experiment	Objective Function	Model
2018	Raju et al. [1]	3D printing	4	$L18(2 \times 3^3)$	54	Surface Roughness; Hardness; Tensile Strength; Flexural Strength	L
2021	Hong et al. [2]	Train brake disc design	18	$L32(2^{18})$	262,144	Temperature	SON
2016	Pahange and Abolbashari [3]	Airplane wing design	6	$L18(2 \times 3^5)$	486	Weight; Deflection	–
2016	Zeng et al. [4]	Automotive composite bumper design	4	$L9(3^4)$	81	Weight; Cost	TON
2021	Fountas and Vaxevanidis [5]	3D printing	3	$L9(3^3)$	27	Printing Time	SON
2019	Kun et al. [6]	Fiber reinforced composite injection molding process	5	$L27(3^5)$	243	Shrinkage; Warpage; Residual Stress	SON
2021	Chaudhari et al. [7]	Electrical discharge machining process	4	$L16(4^4)$	256	Material Removal Rate; Surface Roughness	L
2022	Jafari and Akyüz. [8]	Automobile brake disc design	9	$L16(2^6 \times 3^3)$	1728	Cooling Time	L
2021	Kumar et al. [9]	Drilling process in polymer composites	3	$L16(4^3)$	64	Material Removal Rate; Surface Roughness; Thrust; Torque	L

(Continued)

TABLE 6.1 (Continued)
Literature Studies Related to Taguchi and the Most Frequently Used Taguchi Arrays

Year	Study	Scope	Number of Design Parameters	Taguchi Orthogonal Array	FFD Experiment	Objective Function	Model
2021	Huang et al. [10]	Laser welding process	6	L25(5^6)	15,625	Weld Depth Weld Strength	–
2022	Cheng et al. [11]	Wind turbine blade design	3	L9 (3^3)	27	Efficiency	–
2022	Elsayed et al. [12]	Energy storage systems	6	L25 (5^6)	15,625	Hydrolic Power	–
2020	Biçer et al. [13]	Design of shell-and-tube-type heat exchanger	5	L16(4^5)	1024	Temperature Pressure	–
2021	Maguluri et al. [14]	3D printing	3	L8($2^2 \times 4$)	16	Elastic Modulus Tensile Strength Fracture Strain	–
2019	Lenin et al. [15]	Optimization of building ventilation systems	5	L8($2^4 \times 4$)	64	Temperature Mass Flow Rate	SON
2022	Yuce et al. [16]	Performance analysis of building ventilation systems	5	L25 (5^5)	3125	Air Quality	–

The Taguchi method serves as a versatile tool for experimental design, modeling, and optimization across numerous engineering problems featuring varying numbers of design parameters. Among the most notable Taguchi orthogonal sequences are L8, L9, L16, L18, L25, L27, and L32. These sequences accommodate the utilization of linear, second-, and third-order polynomial mathematical models for modeling purpose

A notable observation arises when comparing the number of experiments recommended by the Taguchi method, as outlined in Table 6.1, with the number of experiments typically required under conventional approaches. Compared to Full Factorial design (FFD), the Taguchi method suggests conducting experiments at a rate of approximately 1%, and in some instances, as low as one in one thousand. Consequently, discussions regarding the reliability of results generated by the Taguchi method have become imperative.

To address this, the efficiency of the Taguchi method was scrutinized by comparing results obtained from experimental design and modeling processes with those derived from FFD. This comparison was facilitated using the most common Taguchi orthogonal arrays in a detailed problem scenario. Such examination is essential for discerning the reliability and efficacy of the Taguchi method in practical engineering applications.

CASE STUDY 6.1

This study aimed to investigate the effect of using Taguchi as an experimental design method on modeling [17]. For this purpose, frequency gap analysis in laminated composite materials was chosen as the reference problem. Choosing such a problem gives us some advantages in terms of experimental design and modeling:

1. Natural frequency gap analysis in laminated composites indeed benefits from having an analytical solution, facilitating the accurate generation of the required results to construct a dataset. This analytical solution enables researchers to obtain precise and reliable data without the need for extensive experimental testing, thereby streamlining the data collection process for subsequent analysis.
2. The objective functions involved in the solution phase of laminated composite mechanics problems often comprise trigonometric expressions characterized by high nonlinearity. Consequently, achieving success in the mathematical modeling process necessitates meticulous attention to detail and precision. Addressing the complexities introduced by these nonlinear expressions is essential for developing accurate and robust models that effectively capture the behavior of laminated composite materials.
3. The widespread use of laminated composite materials across various industries underscores the importance of developing a systematic approach to experimental design and mathematical modeling. By establishing such a systematic framework, researchers can navigate the complexities inherent in composite material mechanics problems without relying solely on complex analytical formulas or differential equation

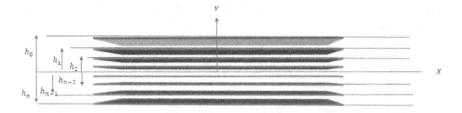

FIGURE 6.1 Eight-layered symmetric graphite/epoxy

solutions. This approach promotes efficiency and accessibility, enabling researchers to provide practical solutions to real-world engineering challenges associated with laminated composite materials.

In line with these evaluations, the eight-layered symmetric graphite/epoxy composite shown in Figure 6.1 was used to model the frequency gap. Here, the fiber angles of each layer were selected as input parameters.

The laminated composite natural frequency gap problem is solved by considering Classical Lamination Theory. The reference study can be examined for detailed information and the assumptions of Classical Lamination Theory [18]. To calculate the natural frequency gap using the analytical formula based on composite material mechanics, the fiber angle-dependent properties of the laminated composite must be determined. The transformed reduced stiffness matrix $[\bar{Q}\_{ij}]$ elements are expressed as follows for angle laminated composite [18]:

$$\bar{Q}_{11} = Q_{11}c^4 + Q_{22}s^4 + 2(Q_{12} + 2Q_{66})s^2c^2$$
$$\bar{Q}_{12} = (Q_{11} + Q_{22} - 4Q_{66})s^2c^2 + Q_{12}(c^4 + s^4)$$
$$\bar{Q}_{22} = Q_{11}s^4 + Q_{22}c^4 + 2(Q_{12} + 2Q_{66})s^2c^2$$
$$\bar{Q}_{16} = (Q_{11} - Q_{12} - 2Q_{66})sc^3 - (Q_{22} - Q_{12} - 2Q_{66})s^3c \tag{6.1}$$
$$\bar{Q}_{26} = (Q_{11} - Q_{12} - 2Q_{66})cs^3 - (Q_{22} - Q_{12} - 2Q_{66})sc^3$$
$$\bar{Q}_{66} = (Q_{11} + Q_{22} - 2Q_{12} - 2Q_{66})s^2c^2 + Q_{66}(c^4 + s^4)$$

The elements in the transformed reduced stiffness matrix can be expressed based on engineering constants.

$$Q_{11} = \frac{E_1}{1 - v_{21}v_{12}} \quad Q_{22} = \frac{E_2}{1 - v_{21}v_{12}} \quad Q_{66} = G_{12}$$
$$Q_{12} = \frac{v_{12}E_2}{1 - v_{21}v_{12}} \quad v_{21} = \frac{E_2}{E_1}v_{12} \tag{6.2}$$

Here, E_1, E_2, v_{12} and G_{12} show the elasticity modules, Poisson ratio, and shear modulus of the laminated composite in the fiber direction and perpendicular to the fiber, respectively. The bending stiffness matrix [D] can be calculated as follows, depending on the transformed reduced stiffness matrix and the thickness of each layer [18]:

$$D_{ij} = \frac{1}{3}\sum_{k=1}^{n}[(\bar{Q}_{ij})]_k (h_k - h_{k-1})^3, \, i,j = 1,2,6 \qquad (6.3)$$

In the problem under consideration, the natural frequency difference between modes (F) selected as the objective function and the natural frequency associated with it (ω) is calculated as follows [17, 19].

$$\omega_{mn}^2 = \frac{\pi^4}{\rho h}\left[D_{11}\left(\frac{m}{a}\right)^4 + 2(D_{12} + 2D_{66})\left(\frac{m}{a}\right)^2\left(\frac{n}{b}\right)^2 + D_{22}\left(\frac{n}{b}\right)^4 \right] \qquad (6.4)$$

$$F = \left(\frac{(\omega_2^{Gl} - \omega_1^{Gl})_{max} - (\omega_2 - \omega_1)}{(\omega_2^{Gl} - \omega_1^{Gl})_{max}} \right)^2 \qquad (6.5)$$

Experimental sets were generated employing Taguchi, one of the most prevalent experimental design and modeling methodologies in the literature. To assess Taguchi's effectiveness in modeling, natural frequency issues within the same composite material were also investigated using the FFD experimental design and modeling approach, and the outcomes were juxtaposed. The focal point lies in determining the significance of employing Taguchi for modeling and optimization with a limited number of experiments. Otherwise, it may give rise to misconceptions, necessitating greater caution in its utilization.

Table 6.2 shows the design parameters and their level values for different Taguchi orthogonal arrays. Experimental sets were created using design parameters for the specified orthogonal arrays. Here, L8 shows that the experiment set consists of eight experiments. Similarly, the experimental numbers of other orthogonal arrays are also indicated.

Table 6.3 shows the number of experiments needed when Taguchi and FFD experimental design methods are used in the same scenario study. For example, if we examine Scenario 1, while it is sufficient to conduct eight experiments when Taguchi is used, FFD recommends 16 experiments for the same problem. Here, it can be seen that there is a significant decrease in the number of experiments compared to FFD, mainly when Taguchi is used in L9, L16, and L25 orthogonal arrays. Therefore, a necessity has arisen to compare Taguchi and FFD experimental design and modeling methods.

Table 6.4 shows the assessment of second and third-order polynomial models created for different orthogonal arrays (L8, L9, L16, L25, L27) when Taguchi is used as the experimental design and modeling method. When R^2 is used as an assessment criterion, the proposed polynomial models for all orthogonal arrays in scenarios (1–5) have high accomplish rates ($R^2 > 0.95$). When the maximum and minimum output values produced by the models are examined, taking into account boundedness control as a second assessment criterion, it is seen that the

TABLE 6.2
Taguchi Orthogonal Arrays, Design Parameters, and Their Levels

Scenario 1	**L8**	**Level 1**	**Level 2**			
	x_1	0	90			
	x_2	0	90			
	x_3	0	90			
	x_4	0	90			
Scenario 2	**L9**	**Level 1**	**Level 2**	**Level 3**		
	x_1	0	45	90		
	x_2	0	45	90		
	x_3	0	45	90		
	x_4	0	45	90		
Scenario 3	**L16**	**Level 1**	**Level 2**	**Level 3**	**Level 4**	
	x_1	0	30	60	90	
	x_2	0	30	60	90	
	x_3	0	30	60	90	
	x_4	0	30	60	90	
Scenario 4	**L25**	**Level 1**	**Level 2**	**Level 3**	**Level 4**	**Level 5**
	x_1	0	22.5	45	67.5	90
	x_2	0	22.5	45	67.5	90
	x_3	0	22.5	45	67.5	90
	x_4	0	22.5	45	67.5	90
Scenario 5	**L27**	**Level 1**	**Level 2**	**Level 3**		
	x_1	0	45	90		
	x_2	0	45	90		
	x_3	0	45	90		
	x_4	0	45	90		

TABLE 6.3
Experiments to be Performed in Taguchi and FFD Designs

Trial	Scenarios	Taguchi	FFD
	Scenario 1*	8 (L8)	16 (L16)
	Scenario 2*	9 (L9)	81 (L81)
	Scenario 3*	16 (L16)	256 (L256)
	Scenario 4*	25 (L25)	625 (L625)
	Scenario 5*	27 (L27)	81 (L81)

* Datasets are given in appendix (Table A6.1–A6.10).

TABLE 6.4
Assessment of Models Created with Neuro Regression Using Taguchi Experimental Set

Scenarios	Design Method	Model*	R^2	R^2_{adj}	Max.	Min.
Scenario 1	Taguchi L8	SON	1	1	1502.68	502.67
		TON	1	1	1502.68	502.67
Scenario 2	Taguchi L9	SON	1	1	1856.76	−338.18
		TON	1	1	2269.76	502.67
Scenario 3	Taguchi L16	SON	0.97	0.95	2696.28	534.75
		TON	1	1	2597.37	461.27
Scenario 4	Taguchi L25	SON	0.95	0.94	2473.11	162.18
		TON	0.99	0.99	2934.85	500.02
Scenario 5	Taguchi L27	SON	1	1	1856.76	−338.18
		TON	1	1	2269.76	502.67

* full form of models are given in appendix (Table A6.11).

SON models in Scenario 2 and Scenario 5 produce meaningless values in terms of engineering. The natural frequency gap can't take a negative value. Apart from these, the maximum and minimum values produced by the models in other scenarios are consistent with those produced using the analytical formula.

In Table 6.5, an assessment similar to the Taguchi method was made when FFD was used. It is seen that the proposed second- and third-order polynomial models successfully meet both R^2 and boundedness check criteria in all scenarios.

TABLE 6.5
Assessment of Models Created with Neuro Regression Using FFD Experimental Set

Scenarios	Design Method	Model*	R^2	R^2_{adj}	Max.	Min.
Scenario 1	FFD L16	SON	0.99	0.99	1770.02	503.78
		TON	0.99	0.99	1770.02	503.78
Scenario 2	FFD L81	SON	0.98	0.98	1732.93	451.13
		TON	0.98	0.98	1789.96	451.13
Scenario 3	FFD L256	SON	0.91	0.91	1817.7	379.8
		TON	0.91	0.91	1817.7	379.8
Scenario 4	FFD L625	SON	0.9	0.9	1846.9	313.01
		TON	0.9	0.9	1846.9	313.01
Scenario 5	FFD L81_2	SON	0.98	0.98	1732.93	451.13
		TON	0.98	0.98	1789.96	451.13

* Full form of models are given in appendix (Table A6.12).

When the results in Table 6.4 and Table 6.5 are examined, it can be thought that Taguchi was entirely accomplished in experimental design and modeling. Still, modeling using only the experimental set data and looking at the model's success only allows us to infer how well the model predicts the actual values. However, the expectation from the proposed models is that they can make accurate predictions when asked for an input set that is not included in the dataset. For this reason, the mathematical models obtained using the experimental sets produced by the Taguchi method were tested using the data lines in the FFD, which are not included in Taguchi. Taguchi and FFD datasets corresponding to each scenario were used while performing this process. For example, if Scenario 1 is considered, it is understood that the model was built using the dataset consisting of eight lines for Taguchi. For the same scenario, FFD recommends a dataset of 16 rows. Therefore, eight lines in FFD are not in Taguchi. Using these eight lines, the mathematical models suggested by Taguchi will be tested. When similar evaluations are carried out for all scenarios, it will be possible to observe the effect of using Taguchi instead of FFD.

Figure 6.2 contrasts the R^2 assessment criteria derived from testing the SON and TON models developed using Taguchi experimental design sets with the data lines encompassed in FFD but not Taguchi. The findings indicate that both SON and TON models exhibited subpar performance during the testing phase, contrary to their high prediction accuracies in Table 6.4. Notably, the highest prediction performances, with R^2 values of 0.25 and 0.32 for the SON and TON models, respectively, were attained when employing Taguchi L16 in Scenario 3. However, these results indicate modest success in achieving the R^2 goodness of fit criterion. For other scenarios, the R^2 values were negative, indicating an entirely unacceptable level of prediction performance.

Overall, the Taguchi method demonstrates limited effectiveness in modeling the natural frequency gap in laminated composite materials. Nonetheless, to comprehensively assess the efficacy of the FFD method and facilitate a comparison with the Taguchi method, it is imperative to subject the mathematical models proposed by both methodologies to testing using the same randomly generated dataset.

Figure 6.3 illustrates the R^2 goodness-of-fit index results obtained from testing the SON and TON models developed using Taguchi experimental design sets with a randomly generated dataset comprising 150 lines. Similar to Figure 6.2, the highest prediction performances were achieved with R^2 values of 0.38 and 0.42 for the SON and TON models respectively when Taguchi L16 was employed in Scenario 3. However, it is worth noting that these success rates are insufficient, and utilizing the proposed models under these circumstances may lead to erroneous designs.

To address whether this failure observed in Taguchi could be rectified by utilizing the FFD design method, the dataset consisting of 150 lines used to test the Taguchi method was also employed to test the FFD method, with the results depicted in Figure 6.4.

According to the findings in Figure 6.4, all models in scenarios other than those proposed using the dataset comprising 16 lines (L16) in Scenario 1 demonstrated

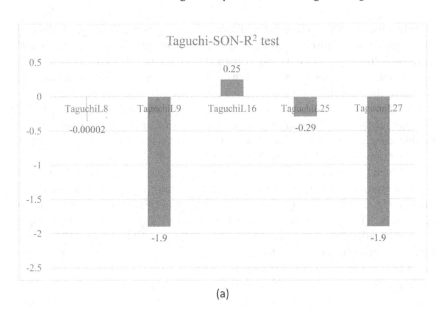

(a)

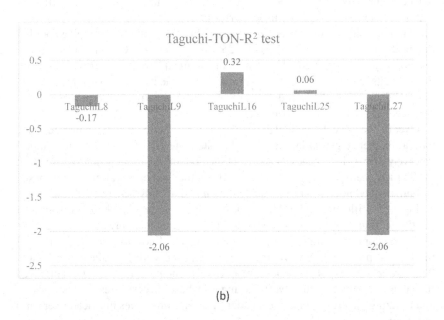

(b)

FIGURE 6.2 R^2 criteria obtained from testing the SON and TON models created using Taguchi experimental design sets: (a) SON and (b) TON

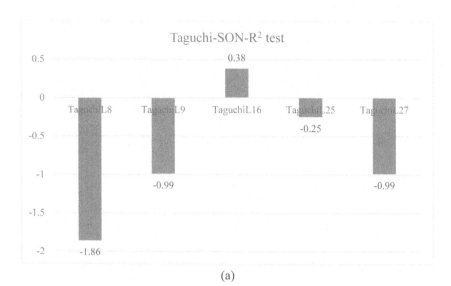

(a)

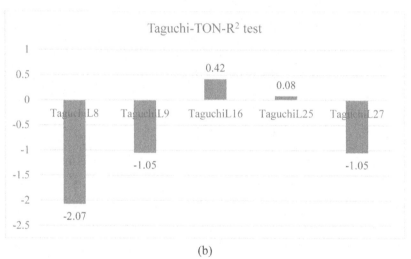

(b)

FIGURE 6.3 R^2 results based on randomly generated 150-row test data: (a) SON and (b) TON (Taguchi)

adequate success in accurately predicting the natural frequency range in terms of the R^2 goodness-of-fit index. Several notable observations emerge: (i) As the number of rows in the datasets increased in FFD, the success rate correspondingly improved. (ii) The most successful design was attained when the L625 dataset was utilized in Scenario 4 for both the SON and TON models. (iii) It is erroneous to draw the same conclusion for Taguchi. In Figure 6.3, the most successful designs were obtained when L16 was used.

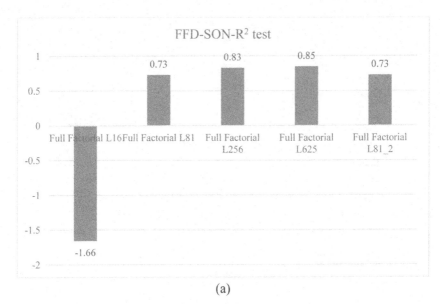

(a)

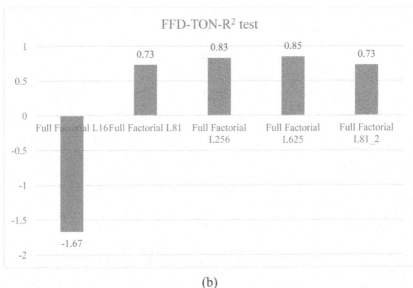

(b)

FIGURE 6.4 R^2 results based on randomly generated 150 row test data: (a) SON and (b) TON (FFD)

This result can be interpreted as the orthogonal array taken into consideration when creating the experimental set in Taguchi, which may have missed important data array lines that could directly impact the design. This situation also raises a question about which orthogonal array should be used to create the dataset. (iv) a typical result observed in FFD and Taguchi is that creating a dataset using two levels is pointless. In this case, both the models proposed by FFD and Taguchi failed. Since there is great flexibility regarding the intermediate values design parameters can take when using two levels, it is tough for the proposed models to accurately determine the natural frequency range behavior intended to be revealed in the problem. (v) When creating an experimental design, it is understood that it is essential and necessary to take into account how many levels of design parameters the design parameters will consist of and the values they will take, as well as the number of experimental lines. In line with all these results, it can be emphasized that one should be careful using Taguchi, which is one of the most preferred methods for experimental design in the literature, and that analyzing the problem well and deciding on the correct experimental design method will have a direct impact on the results.

REFERENCES

[1] Raju M, Gupta MK, Bhanot N, Sharma VS. A hybrid PSO–BFO evolutionary algorithm for optimization of fused deposition modelling process parameters. Journal of Intelligent Manufacturing 2019; 3: 2743–2758.

[2] Hong H, Kim G, Lee H, Kim J, Lee D, Kim M, Lee J ve diğ. Optimal location of brake pad for reduction of temperature deviation on brake disc during high-energy braking. Journal of Mechanical Science and Technology 2021; 35: 1109–1120.

[3] Pahange H, Abolbashari MH. Mass and performance optimization of an airplane wing leading edge structure against bird strike using Taguchi-based grey relational analysis. Chinese Journal of Aeronautics 2016; 29(4): 934–944.

[4] Zeng F, Xie H, Liu Q, Li F, Tan W. Design and optimization of a new composite bumper beam in high-speed frontal crashes. Structural and Multidisciplinary Optimization 2016; 53: 115–122.

[5] Fountas NA, Vaxevanidis NM. Optimization of fused deposition modeling process using a virus-evolutionary genetic algorithm. Computers in Industry 2021; 125: 103371.

[6] Kun L, Shilin Y, Yucheng Z, Wenfeng P, Gang Z. Multi-objective optimization of the fiber-reinforced composite injection molding process using Taguchi method, RSM, and NSGA-II. Simulation Modelling Practice and Theory 2019; 91: 69–82.

[7] Chaudhari R, Khanna S, Vora J, Patel VK, Paneliya S, Pimenov DY, Wojciechowski S, ve diğ. Experimental investigations and optimization of MWCNTs-mixed WEDM process parameters of nitinol shape memory alloy. Journal of Materials Research and Technology 2021; 15: 2152–2169.

[8] Jafari R, Akyüz R. Optimization and thermal analysis of radial ventilated brake disc to enhance the cooling performance. Case Studies in Thermal Engineering 2022; 3: 101731.

[9] Kumar J, Verma RK, Mondal AK. Taguchi-grey theory based harmony search algorithm (GR-HSA) for predictive modeling and multi-objective optimization in drilling of polymer composites. Experimental Techniques 2021; 45: 531–548.

[10] Huang Y, Gao X, Ma B, Liu G, Zhang N, Zhang Y, You D. Optimization of weld strength for laser welding of steel to PMMA using Taguchi design method. Optics & Laser Technology 2021; 136: 106726.

[11] Cheng B, Du J, Yao Y. Power prediction formula for blade design and optimization of Dual Darrieus Wind Turbines based on Taguchi Method and Genetic Expression Programming model. Renewable Energy 2022; 192: 583–605.

[12] Elsayed ME, Abdo S, Attia AA, Attia E, Abd Elrahman MA. Parametric optimisation for the design of gravity energy storage system using Taguchi method. Scientific Reports 2022; 12(1): 1–16.

[13] Biçer N, Engin T, Yaşar H, Büyükkaya E, Aydın A, Topuz A. Design optimization of a shell-and-tube heat exchanger with novel three-zonal baffle by using CFD and Taguchi method. International Journal of Thermal Sciences 2020; 155: 106417.

[14] Maguluri N, Suresh G, Rao KV. Assessing the effect of FDM processing parameters on mechanical properties of PLA parts using Taguchi method. Journal of Thermoplastic Composite Materials 2023; 36(4): 1472–1488.

[15] Lenin VR, Sivalakshmi S, Raja M. Optimization of window type and vent parameters on single-sided natural ventilation buildings. Journal of Thermal Analysis and Calorimetry 2019; 136(1): 367–379.

[16] Yuce BE, Nielsen PV, Wargocki P. The use of Taguchi, ANOVA, and GRA methods to optimize CFD analyses of ventilation performance in buildings. Building and Environment 2022; 225: 109587.

[17] Hosseinzadeh Y, Jalili S, Khani R. Investigating the effects of flax fibers application on multi-objective optimization of laminated composite plates for simultaneous cost minimization and frequency gap maximization. Journal of Building Engineering 2020; 32: 101477.

[18] Kaw AK. Mechanics of Composite Materials. London: CRC, 2006.

[19] Savran M, Aydin L. Stochastic optimization of graphite-flax/epoxy hybrid laminated composite for maximum fundamental frequency and minimum cost. Engineering Structures 2018; 174: 675–687.

APPENDIX

TABLE A6.1

Taguchi Design of Experiment Set (Dataset Defined in Scenario 1 in Table 6.3)

Trial	x_1	x_2	x_3	x_4	Natural Frequency Gap
1	0	0	0	0	502.669
2	0	0	90	90	962.156
3	0	90	0	90	1502.68
4	0	90	90	0	1734.35
5	90	0	0	90	1734.35
6	90	0	90	0	1502.68
7	90	90	0	0	962.156
8	90	90	90	90	502.669

TABLE A6.2
Taguchi Design of Experiment Set (Dataset Defined in Scenario 2 in Table 6.3)

Trial	x_1	x_2	x_3	x_4	Natural Frequency Gap
1	0	0	0	0	502.669
2	0	45	45	45	1203.99
3	0	90	90	90	1771.13
4	45	0	45	90	1563.12
5	45	45	90	0	1758.97
6	45	90	0	45	1676.66
7	90	0	90	45	1480.91
8	90	45	0	90	1336.23
9	90	90	45	0	776.296

TABLE A6.3
Taguchi Design of Experiment Set (Dataset Defined in Scenario 3 in Table 6.3)

Trial	x_1	x_2	x_3	x_4	Natural Frequency Gap
1	0	0	0	0	502.669
2	0	30	30	30	895.171
3	0	60	60	60	1485.13
4	0	90	90	90	1771.13
5	30	0	30	60	1125.14
6	30	30	0	90	1258.19
7	30	60	90	0	1825.62
8	30	90	60	30	1828.55
9	60	0	60	90	1821.18
10	60	30	90	60	1591.42
11	60	60	0	30	1518.63
12	60	90	30	0	1283.93
13	90	0	90	30	1491.42
14	90	30	60	0	1356.31
15	90	60	30	90	1051.66
16	90	90	0	60	923.312

TABLE A6.4

Taguchi Design of Experiment Set (Dataset Defined in Scenario 4 in Table 6.3)

Trial	x_1	x_2	x_3	x_4	Natural Frequency Gap
1	0	0	0	0	502.669
2	0	22.5	22.5	22.5	748.221
3	0	45	45	45	1203.99
4	0	67.5	67.5	67.5	1601.14
5	0	90	90	90	1771.13
6	22.5	0	22.5	45	904.801
7	22.5	22.5	45	67.5	1159.23
8	22.5	45	67.5	90	1525.55
9	22.5	67.5	90	0	1781.77
10	22.5	90	0	22.5	1649.8
11	45	0	45	90	1563.12
12	45	22.5	67.5	0	1706.67
13	45	45	90	22.5	1753.56
14	45	67.5	0	45	1761.25
15	45	90	22.5	67.5	1624.44
16	67.5	0	67.5	22.5	1707.55
17	67.5	22.5	90	45	1547
18	67.5	45	0	67.5	1534.8
19	67.5	67.5	22.5	90	1229.98
20	67.5	90	45	0	1045.91
21	90	0	90	67.5	1467.27
22	90	22.5	0	90	1613.64
23	90	45	22.5	0	1332.01
24	90	67.5	45	22.5	914.513
25	90	90	67.5	45	603.385

TABLE A6.5

Taguchi Design of Experiment Set (Dataset Defined in Scenario 5 in Table 6.3)

No	x_1	x_2	x_3	x_4	Natural Frequency Gap
1	0	0	0	0	502.669
2	0	0	0	0	502.669
3	0	0	0	0	502.669
4	0	45	45	45	1203.99
5	0	45	45	45	1203.99
6	0	45	45	45	1203.99
7	0	90	90	90	1771.13

(Continued)

TABLE A6.5 *(Continued)*
Taguchi Design of Experiment Set (Dataset Defined in Scenario 5 in Table 6.3)

No	x_1	x_2	x_3	x_4	Natural Frequency Gap
8	0	90	90	90	1771.13
9	0	90	90	90	1771.13
10	45	0	45	90	1563.12
11	45	0	45	90	1563.12
12	45	0	45	90	1563.12
13	45	45	90	0	1758.97
14	45	45	90	0	1758.97
15	45	45	90	0	1758.97
16	45	90	0	45	1676.66
17	45	90	0	45	1676.66
18	45	90	0	45	1676.66
19	90	0	90	45	1480.91
20	90	0	90	45	1480.91
21	90	0	90	45	1480.91
22	90	45	0	90	1336.23
23	90	45	0	90	1336.23
24	90	45	0	90	1336.23
25	90	90	45	0	776.296
26	90	90	45	0	776.296
27	90	90	45	0	776.296

TABLE A6.6
FFD Design of Experiment Set (Dataset Defined in Scenario 1 in Table 6.3)

No	x_1	x_2	x_3	x_4	Natural Frequency Gap
1	0	90	90	0	1734.35
2	90	90	0	90	910.514
3	0	0	0	90	567.108
4	90	0	90	0	1502.68
5	90	90	0	0	962.156
6	0	0	90	0	910.514
7	90	90	90	0	567.108
8	0	90	90	90	1771.13
9	90	90	90	90	502.669
10	90	0	0	0	1771.13
11	90	0	90	90	1462.02

(Continued)

TABLE A6.6 *(Continued)*
FFD Design of Experiment Set (Dataset Defined in Scenario 1 in Table 6.3)

No	x_1	x_2	x_3	x_4	Natural Frequency Gap
12	0	0	0	0	502.669
13	90	0	0	90	1734.35
14	0	90	0	0	1462.02
15	0	90	0	90	1502.68
16	0	0	90	90	962.156

TABLE A6.7
FFD Design of Experiment Set (Dataset Defined in Scenario 2 in Table 6.3)

Trial	x_1	x_2	x_3	x_4	Natural Frequency Gap
1	0	90	0	90	1502.68
2	0	45	90	0	1336.23
3	90	90	45	0	776.296
4	45	90	45	0	1563.12
5	90	0	45	0	1628.3
6	45	0	45	45	1544.02
7	45	45	0	45	1741.23
8	90	90	0	0	962.156
9	45	0	90	90	1695.42
10	45	90	90	90	1400.63
11	90	90	90	45	536.224
12	90	45	90	45	1051.47
13	0	90	0	45	1480.91
14	0	45	45	45	1203.99
15	90	45	90	0	1074.99
16	45	45	90	90	1725.5
17	45	90	0	90	1660.18
18	90	45	0	90	1336.23
19	90	45	45	45	1203.99
20	45	45	45	45	1848.19
21	45	45	0	90	1758.97
22	45	90	90	0	1439.23
23	90	0	0	45	1750.86
24	45	0	0	45	1419.16
25	90	45	45	0	1225.94
26	0	45	45	0	1183.04

(Continued)

TABLE A6.7 *(Continued)*
FFD Design of Experiment Set (Dataset Defined in Scenario 2 in Table 6.3)

Trial	x_1	x_2	x_3	x_4	Natural Frequency Gap
27	45	0	0	0	1400.63
28	0	90	90	45	1750.86
29	90	0	90	90	1462.02
30	45	45	0	0	1725.5
31	45	90	0	0	1695.42
32	0	90	90	90	1771.13
33	45	45	45	0	1833.23
34	90	90	90	90	502.669
35	90	0	0	0	1771.13
36	45	0	45	0	1526.67
37	0	0	0	45	536.224
38	0	0	45	90	776.296
39	0	0	90	45	936.15
40	45	90	45	90	1526.67
41	90	90	45	45	748.947
42	90	0	45	45	1607.7
43	90	0	45	90	1590.13
44	90	45	0	45	1355.83
45	45	45	45	90	1833.23
46	90	45	0	0	1377.19
47	0	90	45	90	1628.3
48	0	45	90	90	1377.19
49	0	0	45	45	748.947
50	45	45	90	45	1741.23
51	90	90	0	90	910.514
52	0	45	0	90	1074.99
53	0	90	45	45	1607.7
54	45	45	90	0	1758.97
55	90	90	0	45	936.15
56	0	0	0	90	567.108
57	45	90	0	45	1676.66
58	0	45	0	45	1051.47
59	0	90	45	0	1590.13
60	90	45	45	90	1183.04
61	0	45	90	45	1355.83
62	0	90	0	0	1462.02
63	0	0	90	90	962.156
64	0	0	0	0	502.669
65	0	0	90	0	910.514
66	0	45	45	90	1225.94

(Continued)

TABLE A6.7 *(Continued)*
FFD Design of Experiment Set (Dataset Defined in Scenario 2 in Table 6.3)

Trial	x_1	x_2	x_3	x_4	Natural Frequency Gap
67	45	90	45	45	1544.02
68	45	90	90	45	1419.16
69	90	0	0	90	1734.35
70	90	45	90	90	1028.46
71	90	90	45	90	720.594
72	90	0	90	0	1502.68
73	45	0	90	0	1660.18
74	90	90	90	0	567.108
75	0	45	0	0	1028.46
76	0	0	45	0	720.594
77	45	0	45	90	1563.12
78	90	0	90	45	1480.91
79	45	0	0	90	1439.23
80	0	90	90	0	1734.35
81	45	0	90	45	1676.66

TABLE A6.8
FFD Design of Experiment Set (Dataset Defined in Scenario 3 in Table 6.3)

Trial	x_1	x_2	x_3	x_4	Natural Frequency Gap
1	0	30	0	60	831.093
2	90	90	60	0	676.781
3	60	30	60	0	1674.96
4	60	60	0	30	1518.63
5	90	0	0	0	1771.13
6	90	30	30	60	1465.61
7	90	60	0	60	1145.45
8	0	60	90	60	1559.27
9	0	30	30	90	932.301
10	90	60	90	60	804.841
11	0	60	30	60	1345.35
12	0	60	60	0	1456.64
13	90	90	0	90	910.514
14	90	90	60	60	633.06

(Continued)

TABLE A6.8 *(Continued)*
FFD Design of Experiment Set (Dataset Defined in Scenario 3 in Table 6.3)

Trial	x_1	x_2	x_3	x_4	Natural Frequency Gap
15	30	60	30	90	1674.96
16	60	30	90	30	1609.33
17	60	60	90	60	1227.16
18	60	0	0	90	1786.08
19	30	0	0	30	1025.91
20	60	90	90	0	1060.39
21	0	30	60	60	1085.91
22	0	0	60	0	816.904
23	90	0	30	60	1667.64
24	0	90	90	60	1760.52
25	90	90	30	0	870.187
26	0	60	30	90	1356.31
27	30	60	90	60	1849.49
28	30	90	90	90	1751.36
29	60	60	90	90	1217.21
30	30	0	0	0	1014.52
31	60	0	60	60	1828.55
32	60	0	30	0	1804.22
33	30	30	90	30	1500.14
34	60	60	60	0	1325.16
35	30	30	30	30	1295.16
36	90	30	0	0	1569.85
37	60	0	90	30	1793.59
38	90	60	0	0	1179.55
39	30	90	90	60	1759
40	0	60	60	90	1495.73
41	60	60	0	90	1491.51
42	90	0	60	0	1563.84
43	30	90	90	0	1786.08
44	30	0	30	0	1092.44
45	0	60	30	0	1315.05
46	0	30	60	0	1051.66
47	30	60	30	60	1665.48
48	0	30	30	60	919.953
49	0	90	30	90	1563.84
50	30	90	0	0	1768.77
51	0	90	30	0	1524.59
52	60	90	60	90	1092.44
53	60	90	30	60	1252.57
54	30	60	90	30	1832.94

(Continued)

TABLE A6.8 *(Continued)*
FFD Design of Experiment Set (Dataset Defined in Scenario 3 in Table 6.3)

Trial	x_1	x_2	x_3	x_4	Natural Frequency Gap
55	0	60	60	60	1485.13
56	90	30	30	30	1485.13
57	90	0	30	90	1659.53
58	90	90	90	60	519.835
59	60	0	30	30	1811.66
60	30	60	60	90	1793.99
61	0	90	0	60	1491.42
62	30	30	90	60	1518.63
63	30	0	0	90	1060.39
64	60	60	30	60	1430.41
65	90	90	60	30	662.586
66	30	60	30	0	1640.07
67	60	30	90	90	1583.18
68	60	90	30	30	1273.19
69	0	60	30	30	1324.75
70	90	60	90	0	844.026
71	90	0	60	60	1533.33
72	90	30	60	0	1356.31
73	90	30	90	0	1288.28
74	90	30	60	60	1324.75
75	30	90	60	30	1828.55
76	60	0	0	0	1751.36
77	90	90	60	90	617.563
78	30	0	90	60	1348.24
79	90	0	90	60	1471.12
80	30	30	0	60	1247.62
81	60	0	90	90	1768.77
82	30	60	60	0	1760.45
83	30	60	90	0	1825.62
84	0	0	0	90	567.108
85	90	60	60	30	919.953
86	0	90	90	0	1734.35
87	30	90	0	60	1793.59
88	60	90	30	90	1242.64
89	90	30	60	90	1315.05
90	0	90	60	90	1696.89
91	0	0	90	0	910.514
92	30	90	90	30	1776.35
93	90	90	30	30	856.96
94	60	90	90	60	1025.91
95	90	90	90	30	551.947

(Continued)

TABLE A6.8 *(Continued)*
FFD Design of Experiment Set (Dataset Defined in Scenario 3 in Table 6.3)

Trial	x_1	x_2	x_3	x_4	Natural Frequency Gap
96	0	30	90	30	1145.45
97	0	0	30	60	662.586
98	30	0	60	60	1273.19
99	30	0	60	0	1242.64
100	30	0	30	90	1136.31
101	60	90	0	30	1348.24
102	30	60	60	30	1767.98
103	90	0	0	90	1734.35
104	30	90	30	30	1828.55
105	30	90	30	90	1854.83
106	90	30	30	90	1456.64
107	90	90	90	0	567.108
108	30	90	0	90	1803.29
109	30	90	60	60	1811.66
110	60	0	60	30	1845.36
111	90	0	60	30	1552.93
112	90	30	0	90	1531.43
113	0	0	0	60	551.947
114	60	0	30	60	1828.55
115	30	30	60	90	1459.13
116	60	60	30	0	1459.13
117	90	60	90	90	791.406
118	0	0	30	0	617.563
119	30	0	90	90	1358.93
120	60	90	90	30	1048.81
121	0	90	60	0	1659.53
122	90	30	90	90	1245.38
123	60	90	0	60	1328.01
124	60	0	0	60	1776.35
125	0	30	60	90	1097.66
126	90	60	60	0	932.301
127	90	60	30	0	1097.66
128	0	0	60	30	830.379
129	90	60	30	60	1062.95
130	0	30	90	90	1179.55
131	0	30	0	90	844.026
132	0	60	0	90	1288.28
133	60	30	60	30	1665.48
134	30	90	30	60	1845.36
135	0	90	30	30	1533.33
136	0	60	90	90	1569.85

(Continued)

TABLE A6.8 *(Continued)*
FFD Design of Experiment Set (Dataset Defined in Scenario 3 in Table 6.3)

Trial	x_1	x_2	x_3	x_4	Natural Frequency Gap
137	0	0	30	30	633.06
138	0	0	30	90	676.781
139	60	30	30	0	1793.99
140	60	90	90	90	1014.52
141	0	0	90	90	962.156
142	90	90	0	60	923.312
143	0	90	0	30	1471.12
144	60	60	30	90	1421.52
145	90	0	30	30	1686.3
146	0	90	30	60	1552.93
147	0	90	0	90	1502.68
148	0	90	90	90	1771.13
149	90	30	60	30	1345.35
150	0	90	90	30	1742.14
151	0	30	0	30	804.841
152	30	60	0	90	1619.04
153	30	0	90	30	1328.01
154	0	0	60	60	856.96
155	90	30	0	30	1559.27
156	30	30	90	0	1491.51
157	90	90	90	90	502.669
158	90	0	90	0	1502.68
159	30	30	0	90	1258.19
160	90	0	30	0	1696.89
161	0	90	60	60	1686.3
162	30	90	60	0	1838.05
163	90	0	60	90	1524.59
164	30	60	90	90	1858.74
165	30	90	30	0	1821.18
166	60	60	90	30	1247.62
167	90	60	90	30	831.093
168	0	90	0	0	1462.02
169	30	60	0	60	1609.33
170	60	0	30	90	1838.05
171	60	60	60	30	1314.91
172	0	60	90	0	1531.43
173	30	90	0	30	1776.35
174	60	30	0	60	1832.94
175	30	0	60	90	1283.93
176	30	60	0	0	1583.18

(Continued)

TABLE A6.8 *(Continued)*
FFD Design of Experiment Set (Dataset Defined in Scenario 3 in Table 6.3)

Trial	x_1	x_2	x_3	x_4	Natural Frequency Gap
177	60	90	0	90	1318.42
178	90	30	0	60	1540.08
179	90	60	60	90	882.641
180	60	90	60	30	1125.14
181	30	30	60	60	1449.2
182	90	60	0	30	1167.88
183	30	0	60	30	1252.57
184	60	60	0	60	1500.14
185	30	30	0	0	1217.21
186	60	0	60	0	1854.83
187	90	90	0	0	962.156
188	90	90	30	60	830.379
189	60	60	90	0	1258.19
190	30	60	60	60	1784.74
191	90	60	30	90	1051.66
192	60	0	0	30	1759
193	30	30	60	0	1421.52
194	60	90	60	60	1103.24
195	0	30	0	0	791.406
196	90	30	90	30	1276.95
197	90	0	0	60	1742.14
198	60	90	0	0	1358.93
199	30	90	60	90	1804.22
200	90	0	90	90	1462.02
201	90	0	0	30	1760.52
202	0	0	90	30	923.312
203	60	60	60	90	1285.62
204	0	60	60	30	1465.61
205	0	30	90	60	1167.88
206	0	60	0	60	1276.95
207	60	30	30	90	1760.45
208	30	30	30	60	1314.91
209	0	60	0	30	1255.53
210	90	90	0	30	949.078
211	0	60	90	30	1540.08
212	0	0	0	0	502.669
213	60	60	0	0	1528.53
214	60	0	90	0	1803.29
215	90	0	90	30	1491.42
216	30	30	30	90	1325.16

(Continued)

TABLE A6.8 *(Continued)*
FFD Design of Experiment Set (Dataset Defined in Scenario 3 in Table 6.3)

Trial	x_1	x_2	x_3	x_4	Natural Frequency Gap
217	0	30	90	0	1134.61
218	60	30	90	60	1591.42
219	30	60	30	30	1648.05
220	60	60	30	30	1449.2
221	90	60	30	30	1085.91
222	90	90	30	90	816.904
223	60	30	60	60	1648.05
224	30	60	0	30	1591.42
225	90	60	0	90	1134.61
226	0	60	0	0	1245.38
227	90	30	90	60	1255.53
228	60	30	0	0	1858.74
229	60	0	90	60	1776.35
230	60	30	90	0	1619.04
231	30	30	30	0	1285.62
232	30	30	60	30	1430.41
233	90	30	30	0	1495.73
234	60	30	30	60	1767.98
235	0	0	60	90	870.187
236	30	0	30	30	1103.24
237	0	30	30	30	895.171
238	30	0	30	60	1125.14
239	30	0	0	60	1048.81
240	60	90	60	0	1136.31
241	0	30	30	0	882.641
242	0	0	0	30	519.835
243	30	0	90	0	1318.42
244	30	30	90	90	1528.53
245	60	30	0	30	1849.49
246	60	60	60	60	1295.16
247	0	0	90	60	949.078
248	60	30	0	90	1825.62
249	60	30	30	30	1784.74
250	90	60	60	60	895.171
251	30	30	0	30	1227.16
252	60	90	30	0	1283.93
253	0	30	60	30	1062.95
254	60	30	60	90	1640.07
255	0	90	60	30	1667.64
256	60	0	60	90	1821.18

TABLE A6.9
FFD Design of Experiment Set (Dataset Defined in Scenario 4 in Table 6.3)

Trial	x_1	x_2	x_3	x_4	Natural Frequency Gap
1	22.5	22.5	0	22.5	972.219
2	45	90	22.5	67.5	1624.44
3	0	67.5	67.5	22.5	1574.3
4	67.5	22.5	67.5	67.5	1568.57
5	67.5	0	45	67.5	1764.88
6	22.5	45	0	90	1298.02
7	22.5	22.5	45	90	1165.71
8	22.5	0	22.5	67.5	922.172
9	67.5	90	0	90	1161.98
10	67.5	45	45	22.5	1424.03
11	0	45	0	90	1074.99
12	45	22.5	22.5	22.5	1537.44
13	22.5	67.5	22.5	67.5	1594.55
14	45	0	45	22.5	1531.58
15	0	0	22.5	22.5	581.296
16	0	0	45	67.5	768.357
17	67.5	0	45	22.5	1789.43
18	22.5	90	22.5	22.5	1682.14
19	0	22.5	0	45	709.779
20	90	67.5	0	90	1045.7
21	90	22.5	0	90	1613.64
22	67.5	90	22.5	67.5	1120.6
23	90	90	67.5	90	571.78
24	0	22.5	0	0	680.887
25	22.5	90	45	67.5	1789.43
26	0	90	22.5	90	1538.13
27	90	0	67.5	22.5	1531.57
28	67.5	22.5	90	45	1547
29	22.5	22.5	45	45	1143.85
30	0	45	0	45	1051.47
31	22.5	0	90	0	1161.98
32	22.5	22.5	45	0	1122.56
33	22.5	45	67.5	45	1506.13
34	22.5	0	90	22.5	1168.16
35	22.5	90	45	90	1795.25
36	0	45	22.5	67.5	1113.14
37	0	67.5	90	22.5	1618.41
38	67.5	90	0	67.5	1168.16
39	90	90	90	22.5	558.289
40	0	22.5	22.5	22.5	748.221

(Continued)

TABLE A6.9 *(Continued)*
FFD Design of Experiment Set (Dataset Defined in Scenario 4 in Table 6.3)

Trial	x_1	x_2	x_3	x_4	Natural Frequency Gap
41	90	0	90	45	1480.91
42	67.5	67.5	67.5	67.5	1019.39
43	45	67.5	67.5	0	1568.58
44	90	22.5	45	45	1487.13
45	67.5	22.5	45	90	1649.61
46	67.5	45	67.5	45	1316.22
47	45	0	45	45	1544.02
48	90	90	90	90	502.669
49	45	22.5	90	0	1745.51
50	22.5	67.5	67.5	22.5	1745.71
51	22.5	22.5	90	67.5	1309.85
52	45	67.5	0	22.5	1773.74
53	22.5	67.5	90	67.5	1810.38
54	90	90	0	67.5	918.008
55	67.5	67.5	0	90	1275.04
56	90	67.5	22.5	22.5	1035.72
57	45	90	67.5	67.5	1442.38
58	45	45	67.5	0	1789.08
59	45	45	45	0	1833.23
60	45	22.5	22.5	67.5	1562.88
61	90	45	67.5	45	1097.01
62	67.5	90	45	0	1045.91
63	67.5	67.5	67.5	90	1012.75
64	67.5	22.5	67.5	45	1581.03
65	0	45	22.5	22.5	1081.19
66	90	22.5	22.5	90	1569.41
67	22.5	0	22.5	45	904.801
68	90	90	0	0	962.156
69	90	0	0	67.5	1738.8
70	90	67.5	22.5	67.5	1001.87
71	67.5	67.5	0	22.5	1309.85
72	22.5	90	0	90	1681.72
73	0	45	67.5	90	1332.01
74	90	22.5	90	45	1353.94
75	22.5	45	22.5	90	1336.68
76	90	45	90	22.5	1068.03
77	90	22.5	67.5	22.5	1407.35
78	90	90	45	22.5	768.357
79	90	45	45	0	1225.94
80	0	0	90	45	936.15

(Continued)

TABLE A6.9 *(Continued)*
FFD Design of Experiment Set (Dataset Defined in Scenario 4 in Table 6.3)

Trial	x_1	x_2	x_3	x_4	Natural Frequency Gap
81	22.5	67.5	22.5	22.5	1568.57
82	0	90	0	22.5	1467.27
83	22.5	22.5	22.5	45	1035.43
84	45	90	90	67.5	1405.91
85	67.5	0	90	0	1681.72
86	0	90	0	0	1462.02
87	45	90	90	90	1400.63
88	90	90	0	90	910.514
89	0	90	90	45	1750.86
90	22.5	0	0	45	853.21
91	22.5	45	45	90	1429.97
92	90	45	0	90	1336.23
93	0	0	0	22.5	512.825
94	45	0	0	90	1439.23
95	90	0	22.5	90	1689.74
96	22.5	45	90	45	1547.33
97	67.5	67.5	67.5	45	1035.43
98	0	45	45	45	1203.99
99	22.5	90	0	67.5	1675.66
100	90	0	0	22.5	1764.8
101	45	22.5	90	67.5	1773.74
102	22.5	67.5	22.5	90	1600.49
103	90	67.5	90	67.5	689.52
104	0	22.5	22.5	90	794.202
105	67.5	90	45	90	999.495
106	0	45	45	90	1225.94
107	22.5	22.5	45	22.5	1128.75
108	45	90	0	0	1695.42
109	67.5	90	67.5	0	929.356
110	90	0	90	22.5	1495.99
111	45	90	45	0	1563.12
112	45	22.5	90	45	1761.25
113	67.5	45	0	67.5	1534.8
114	90	0	67.5	45	1516.8
115	90	22.5	22.5	45	1587.04
116	45	22.5	45	45	1634.06
117	22.5	0	0	22.5	835.028
118	0	22.5	22.5	67.5	786.48
119	45	0	90	0	1660.18
120	22.5	22.5	90	22.5	1280.76

(Continued)

TABLE A6.9 *(Continued)*
FFD Design of Experiment Set (Dataset Defined in Scenario 4 in Table 6.3)

Trial	x_1	x_2	x_3	x_4	Natural Frequency Gap
121	0	67.5	45	0	1468.71
122	90	22.5	0	45	1630.96
123	90	0	22.5	45	1706.56
124	22.5	67.5	0	67.5	1560.77
125	45	67.5	67.5	45	1549.75
126	90	67.5	22.5	90	994.957
127	67.5	0	67.5	90	1677.56
128	90	90	45	0	776.296
129	90	45	90	90	1028.46
130	45	90	45	90	1526.67
131	0	90	22.5	67.5	1531.57
132	90	67.5	45	0	921.843
133	45	0	22.5	45	1455.36
134	90	0	67.5	67.5	1503.46
135	67.5	45	90	0	1298.02
136	0	0	67.5	0	855.788
137	45	0	45	0	1526.67
138	0	90	0	45	1480.91
139	0	45	0	22.5	1035.16
140	0	22.5	22.5	0	740.012
141	90	90	45	67.5	729.034
142	22.5	22.5	22.5	90	1058.28
143	45	22.5	67.5	0	1706.67
144	67.5	45	0	90	1529.88
145	45	45	22.5	90	1789.08
146	22.5	22.5	90	90	1316.22
147	90	0	0	45	1750.86
148	22.5	45	67.5	90	1525.55
149	90	67.5	45	22.5	914.513
150	0	67.5	67.5	67.5	1601.14
151	90	22.5	45	67.5	1473.88
152	67.5	22.5	22.5	45	1757.29
153	45	45	0	0	1725.5
154	45	67.5	0	67.5	1749.89
155	0	90	0	90	1502.68
156	22.5	90	90	90	1897.69
157	45	0	67.5	45	1636.46
158	0	0	67.5	22.5	863.518
159	67.5	90	67.5	90	879.942
160	45	90	45	22.5	1557.33

(Continued)

TABLE A6.9 *(Continued)*
FFD Design of Experiment Set (Dataset Defined in Scenario 4 in Table 6.3)

Trial	x_1	x_2	x_3	x_4	Natural Frequency Gap
161	22.5	67.5	45	90	1685.09
162	22.5	45	45	67.5	1424.03
163	67.5	45	90	22.5	1291.79
164	45	67.5	0	90	1745.51
165	90	67.5	0	45	1068.91
166	0	45	0	0	1028.46
167	0	67.5	22.5	67.5	1407.35
168	67.5	22.5	45	22.5	1679.3
169	90	45	67.5	67.5	1081.19
170	90	0	45	45	1607.7
171	90	67.5	67.5	0	794.202
172	0	90	67.5	90	1726.84
173	90	45	0	45	1355.83
174	0	0	45	22.5	729.034
175	22.5	45	22.5	67.5	1330.56
176	45	0	90	45	1676.66
177	90	67.5	67.5	45	767.595
178	0	67.5	90	90	1651.29
179	67.5	45	0	0	1566.69
180	22.5	0	67.5	90	1158.8
181	0	90	90	22.5	1738.8
182	67.5	0	45	45	1776.44
183	90	22.5	22.5	67.5	1574.3
184	90	22.5	90	67.5	1339.67
185	67.5	0	67.5	67.5	1682.14
186	0	45	45	67.5	1219.39
187	22.5	45	0	67.5	1291.79
188	0	0	22.5	45	603.385
189	45	0	90	22.5	1664.78
190	45	0	67.5	67.5	1649.53
191	67.5	90	90	22.5	871.124
192	22.5	22.5	67.5	22.5	1235.82
193	67.5	67.5	90	45	988.761
194	22.5	22.5	90	0	1275.04
195	45	45	22.5	22.5	1760.43
196	0	90	45	90	1628.3
197	22.5	67.5	67.5	45	1757.29
198	45	90	45	67.5	1531.58
199	22.5	90	45	45	1776.44
200	67.5	22.5	22.5	0	1775.91

(Continued)

TABLE A6.9 *(Continued)*
FFD Design of Experiment Set (Dataset Defined in Scenario 4 in Table 6.3)

Trial	x_1	x_2	x_3	x_4	Natural Frequency Gap
201	45	45	90	90	1725.5
202	0	22.5	45	90	921.843
203	90	22.5	22.5	22.5	1601.14
204	67.5	67.5	0	67.5	1280.76
205	45	67.5	0	45	1761.25
206	67.5	67.5	22.5	45	1250.26
207	90	22.5	45	90	1468.71
208	22.5	67.5	90	90	1816.13
209	90	45	90	45	1051.47
210	0	67.5	0	67.5	1369.23
211	45	67.5	22.5	90	1706.67
212	45	22.5	0	45	1515.59
213	67.5	67.5	22.5	67.5	1235.82
214	45	0	22.5	0	1437.22
215	90	22.5	67.5	90	1372.98
216	90	0	67.5	0	1538.13
217	45	90	22.5	22.5	1649.53
218	45	90	90	22.5	1433.18
219	45	67.5	45	67.5	1622.21
220	90	67.5	90	90	680.887
221	0	45	22.5	45	1097.01
222	90	0	90	67.5	1467.27
223	0	90	45	0	1590.13
224	67.5	67.5	22.5	90	1229.98
225	0	90	45	45	1607.7
226	22.5	67.5	0	90	1566.8
227	0	67.5	22.5	22.5	1378.45
228	67.5	22.5	0	45	1797.53
229	45	90	90	45	1419.16
230	67.5	45	45	45	1410.2
231	67.5	45	22.5	22.5	1519.67
232	0	45	22.5	0	1074.71
233	67.5	45	67.5	0	1336.68
234	22.5	67.5	45	67.5	1679.3
235	67.5	22.5	0	67.5	1786.1
236	90	90	67.5	0	632.902
237	22.5	0	45	67.5	1039.04
238	22.5	67.5	90	0	1781.77
239	90	90	22.5	67.5	863.518
240	67.5	67.5	90	0	1012.2

(Continued)

TABLE A6.9 *(Continued)*
FFD Design of Experiment Set (Dataset Defined in Scenario 4 in Table 6.3)

Trial	x_1	x_2	x_3	x_4	Natural Frequency Gap
241	22.5	90	45	22.5	1764.88
242	67.5	67.5	45	45	1143.85
243	90	45	22.5	0	1332.01
244	45	45	90	0	1758.97
245	90	0	0	0	1771.13
246	0	22.5	67.5	90	1042.88
247	90	90	22.5	45	882.074
248	90	67.5	90	0	737.349
249	90	90	0	45	936.15
250	90	45	67.5	22.5	1113.14
251	67.5	0	90	90	1645.13
252	67.5	0	22.5	0	1885.33
253	67.5	0	67.5	45	1694.16
254	0	67.5	0	45	1353.94
255	45	45	90	45	1741.23
256	22.5	0	22.5	90	929.356
257	67.5	45	22.5	90	1488.42
258	22.5	45	45	0	1391.77
259	67.5	45	0	45	1547.33
260	22.5	0	90	67.5	1199.23
261	67.5	22.5	90	67.5	1534.31
262	0	22.5	45	22.5	878.989
263	0	45	67.5	45	1310.56
264	90	45	22.5	90	1290.62
265	0	90	22.5	22.5	1503.46
266	90	22.5	45	22.5	1501.54
267	0	0	67.5	45	882.074
268	0	22.5	67.5	0	994.957
269	67.5	0	22.5	90	1851.4
270	45	45	45	45	1848.19
271	45	90	90	0	1439.23
272	0	45	22.5	90	1119.94
273	0	67.5	90	45	1630.96
274	67.5	0	0	45	1907.31
275	90	90	45	90	720.594
276	67.5	45	67.5	90	1296.86
277	67.5	90	67.5	45	904.801
278	22.5	45	90	22.5	1534.8
279	45	67.5	0	0	1779.25
280	0	90	90	0	1734.35

(Continued)

TABLE A6.9 *(Continued)*
FFD Design of Experiment Set (Dataset Defined in Scenario 4 in Table 6.3)

Trial	x_1	x_2	x_3	x_4	Natural Frequency Gap
281	45	67.5	90	90	1498.07
282	45	0	22.5	22.5	1442.38
283	22.5	0	67.5	0	1114.27
284	0	67.5	45	22.5	1473.88
285	22.5	90	0	45	1662.03
286	0	67.5	90	67.5	1645.01
287	22.5	0	45	22.5	1006.25
288	90	45	45	67.5	1189.08
289	67.5	45	22.5	67.5	1493.44
290	22.5	0	0	0	827.385
291	45	90	0	45	1676.66
292	67.5	0	45	0	1795.25
293	90	22.5	90	0	1375.89
294	67.5	90	0	45	1183.42
295	67.5	90	22.5	45	1136.13
296	45	45	0	90	1758.97
297	45	90	67.5	90	1437.22
298	0	0	90	90	962.156
299	0	90	0	67.5	1495.99
300	67.5	0	45	90	1760.5
301	45	67.5	67.5	90	1532.57
302	45	45	0	22.5	1729.9
303	22.5	22.5	0	0	965.344
304	45	22.5	45	0	1617.55
305	45	45	67.5	45	1771.58
306	90	45	45	90	1183.04
307	22.5	22.5	67.5	45	1250.26
308	22.5	45	90	67.5	1560.81
309	22.5	45	90	0	1529.88
310	67.5	0	90	45	1662.03
311	67.5	90	45	22.5	1039.04
312	22.5	0	22.5	0	879.942
313	0	67.5	22.5	90	1413.87
314	67.5	67.5	90	67.5	972.219
315	22.5	67.5	90	45	1797.53
316	22.5	90	90	0	1892.17
317	45	90	45	45	1544.02
318	0	45	45	22.5	1189.08
319	90	22.5	0	22.5	1645.01
320	90	90	90	0	567.108

(Continued)

TABLE A6.9 *(Continued)*
FFD Design of Experiment Set (Dataset Defined in Scenario 4 in Table 6.3)

Trial	x_1	x_2	x_3	x_4	Natural Frequency Gap
321	22.5	90	45	0	1760.5
322	67.5	67.5	22.5	0	1271.6
323	0	22.5	90	45	1068.91
324	22.5	67.5	45	22.5	1654.23
325	0	90	90	90	1771.13
326	0	0	0	45	536.224
327	67.5	45	90	45	1277.15
328	45	67.5	90	45	1515.59
329	0	67.5	22.5	45	1392.38
330	22.5	22.5	67.5	90	1271.6
331	22.5	67.5	0	22.5	1534.31
332	0	0	22.5	67.5	624.43
333	45	0	90	90	1695.42
334	67.5	22.5	90	0	1566.8
335	22.5	45	0	45	1277.15
336	45	0	45	67.5	1557.33
337	45	67.5	45	90	1617.55
338	67.5	0	0	90	1892.17
339	67.5	67.5	45	22.5	1159.23
340	67.5	22.5	22.5	22.5	1770.17
341	22.5	22.5	0	90	1012.2
342	67.5	22.5	0	0	1816.13
343	45	90	22.5	90	1619.75
344	90	0	90	0	1502.68
345	45	22.5	0	22.5	1503.05
346	22.5	90	67.5	67.5	1879.55
347	90	22.5	67.5	45	1392.38
348	22.5	90	90	67.5	1901.76
349	45	22.5	0	0	1498.07
350	0	0	0	90	567.108
351	67.5	0	0	67.5	1896.25
352	45	22.5	0	67.5	1528.95
353	90	45	90	0	1074.99
354	90	67.5	0	0	1092.95
355	45	22.5	22.5	45	1549.75
356	90	45	45	22.5	1219.39
357	0	0	0	0	502.669
358	45	90	67.5	45	1455.36
359	22.5	90	67.5	22.5	1855.57
360	22.5	90	0	22.5	1649.8

(Continued)

TABLE A6.9 *(Continued)*
FFD Design of Experiment Set (Dataset Defined in Scenario 4 in Table 6.3)

Trial	x_1	x_2	x_3	x_4	Natural Frequency Gap
361	67.5	67.5	45	90	1122.56
362	90	45	0	22.5	1370.74
363	45	0	22.5	67.5	1469.12
364	45	45	45	67.5	1837.4
365	22.5	45	22.5	45	1316.22
366	0	45	0	67.5	1068.03
367	22.5	22.5	45	67.5	1159.23
368	0	45	90	45	1355.83
369	22.5	67.5	45	45	1666.19
370	67.5	90	90	0	878.502
371	0	22.5	67.5	22.5	1001.87
372	0	90	67.5	0	1689.74
373	67.5	22.5	67.5	0	1600.49
374	22.5	67.5	67.5	67.5	1770.17
375	67.5	90	45	45	1022.59
376	0	67.5	45	45	1487.13
377	0	90	90	67.5	1764.8
378	0	67.5	0	90	1375.89
379	90	67.5	45	45	896.814
380	67.5	90	22.5	0	1158.8
381	90	45	22.5	45	1310.56
382	0	22.5	90	90	1092.95
383	90	67.5	22.5	45	1018.68
384	22.5	45	67.5	22.5	1493.44
385	22.5	45	67.5	67.5	1519.67
386	90	0	90	90	1462.02
387	90	22.5	0	67.5	1618.41
388	22.5	0	90	45	1183.42
389	22.5	67.5	90	22.5	1786.1
390	45	0	0	0	1400.63
391	67.5	0	90	67.5	1649.8
392	45	45	45	90	1833.23
393	22.5	45	22.5	22.5	1302.43
394	67.5	67.5	67.5	0	1058.28
395	90	0	22.5	22.5	1720.53
396	0	67.5	67.5	45	1587.04
397	0	22.5	45	67.5	914.513
398	22.5	22.5	22.5	22.5	1019.39
399	90	0	22.5	67.5	1694.31
400	22.5	0	0	90	878.502

(Continued)

TABLE A6.9 *(Continued)*
FFD Design of Experiment Set (Dataset Defined in Scenario 4 in Table 6.3)

Trial	x_1	x_2	x_3	x_4	Natural Frequency Gap
401	0	45	90	90	1377.19
402	22.5	90	90	45	1907.31
403	0	90	22.5	45	1516.8
404	45	0	67.5	90	1655.26
405	90	90	90	45	536.224
406	45	67.5	22.5	67.5	1711.13
407	0	90	22.5	0	1498.34
408	0	0	0	67.5	558.289
409	22.5	45	90	90	1566.69
410	67.5	90	0	22.5	1199.23
411	67.5	90	90	90	827.385
412	45	22.5	45	90	1652.36
413	22.5	0	45	0	999.495
414	0	22.5	90	0	1045.7
415	22.5	0	90	90	1205.97
416	45	45	67.5	67.5	1760.43
417	22.5	0	67.5	45	1136.13
418	90	45	67.5	0	1119.94
419	0	0	67.5	67.5	900.602
420	22.5	90	22.5	90	1713.51
421	22.5	45	45	22.5	1397.04
422	0	0	22.5	90	632.902
423	67.5	67.5	22.5	22.5	1265.23
424	45	67.5	22.5	45	1722.62
425	67.5	67.5	67.5	22.5	1051.55
426	67.5	45	22.5	45	1506.13
427	22.5	0	0	67.5	871.124
428	0	90	67.5	67.5	1720.53
429	22.5	45	22.5	0	1296.86
430	90	90	22.5	22.5	900.602
431	67.5	45	90	67.5	1263.03
432	0	22.5	22.5	45	767.595
433	90	90	22.5	0	908.309
434	45	45	22.5	67.5	1783.74
435	67.5	90	67.5	22.5	922.172
436	67.5	67.5	45	67.5	1128.75
437	45	22.5	22.5	90	1568.58
438	90	67.5	67.5	90	740.012
439	22.5	0	67.5	67.5	1152.06
440	45	22.5	90	22.5	1749.89

(Continued)

TABLE A6.9 *(Continued)*
FFD Design of Experiment Set (Dataset Defined in Scenario 4 in Table 6.3)

Trial	x_1	x_2	x_3	x_4	Natural Frequency Gap
441	0	0	45	45	748.947
442	67.5	22.5	67.5	90	1563.72
443	67.5	0	22.5	45	1866.76
444	67.5	45	22.5	0	1525.55
445	0	67.5	67.5	0	1569.41
446	0	22.5	0	22.5	689.52
447	67.5	67.5	0	45	1294.98
448	0	22.5	67.5	45	1018.68
449	22.5	90	67.5	90	1885.33
450	45	0	67.5	0	1619.75
451	45	67.5	45	45	1634.06
452	0	0	45	0	720.594
453	22.5	45	67.5	0	1488.42
454	45	22.5	67.5	45	1722.62
455	22.5	90	90	22.5	1896.25
456	45	22.5	45	67.5	1646.8
457	90	22.5	0	0	1651.29
458	67.5	22.5	90	22.5	1560.77
459	0	90	67.5	45	1706.56
460	90	0	45	22.5	1621.94
461	67.5	0	67.5	22.5	1707.55
462	90	67.5	45	90	871.537
463	90	0	45	0	1628.3
464	67.5	22.5	0	90	1781.77
465	0	45	67.5	67.5	1325.56
466	90	22.5	45	0	1507.86
467	22.5	67.5	0	0	1529.35
468	67.5	67.5	45	0	1165.71
469	22.5	67.5	0	45	1547
470	45	45	67.5	22.5	1783.74
471	45	45	22.5	45	1771.58
472	67.5	67.5	0	0	1316.22
473	45	90	22.5	45	1636.46
474	0	45	90	0	1336.23
475	90	67.5	22.5	0	1042.88
476	67.5	0	22.5	22.5	1879.55
477	90	0	45	90	1590.13
478	22.5	90	67.5	45	1866.76
479	90	22.5	22.5	0	1607.4
480	22.5	90	22.5	67.5	1707.55

(Continued)

TABLE A6.9 *(Continued)*
FFD Design of Experiment Set (Dataset Defined in Scenario 4 in Table 6.3)

Trial	x_1	x_2	x_3	x_4	Natural Frequency Gap
481	90	22.5	90	22.5	1369.23
482	22.5	45	0	0	1257.32
483	90	22.5	90	90	1334.05
484	90	0	22.5	0	1726.84
485	67.5	0	0	22.5	1901.76
486	0	22.5	45	45	896.814
487	45	67.5	67.5	67.5	1537.44
488	22.5	22.5	67.5	0	1229.98
489	0	22.5	0	90	737.349
490	45	67.5	90	22.5	1528.95
491	90	90	67.5	67.5	581.296
492	90	90	22.5	90	855.788
493	90	45	22.5	22.5	1325.56
494	90	90	90	67.5	512.825
495	90	67.5	0	67.5	1052.43
496	67.5	45	90	90	1257.32
497	22.5	0	67.5	22.5	1120.6
498	45	45	90	67.5	1729.9
499	22.5	90	22.5	0	1677.56
500	45	67.5	90	67.5	1503.05
501	67.5	90	67.5	67.5	887.276
502	67.5	0	67.5	0	1713.51
503	22.5	45	45	45	1410.2
504	90	45	67.5	90	1074.71
505	67.5	90	90	45	853.21
506	22.5	67.5	22.5	45	1581.03
507	90	67.5	90	22.5	729.376
508	45	0	90	67.5	1689.68
509	67.5	45	45	0	1429.97
510	90	90	45	45	748.947
511	67.5	0	22.5	67.5	1855.57
512	45	90	67.5	0	1475.06
513	0	22.5	90	22.5	1052.43
514	22.5	22.5	22.5	67.5	1051.55
515	90	45	22.5	67.5	1296.31
516	67.5	22.5	45	45	1666.19
517	0	67.5	45	90	1507.86
518	45	90	0	67.5	1664.78
519	67.5	90	22.5	22.5	1152.06
520	67.5	22.5	45	0	1685.09

(Continued)

TABLE A6.9 *(Continued)*
FFD Design of Experiment Set (Dataset Defined in Scenario 4 in Table 6.3)

Trial	x_1	x_2	x_3	x_4	Natural Frequency Gap
521	0	45	67.5	0	1290.62
522	45	67.5	22.5	0	1740.66
523	90	22.5	67.5	0	1413.87
524	67.5	45	67.5	67.5	1302.43
525	45	90	22.5	0	1655.26
526	45	22.5	22.5	0	1532.57
527	45	22.5	67.5	90	1740.66
528	0	67.5	45	67.5	1501.54
529	22.5	0	22.5	22.5	887.276
530	22.5	22.5	0	45	988.761
531	67.5	67.5	90	90	965.344
532	90	67.5	67.5	67.5	748.221
533	22.5	45	0	22.5	1263.03
534	90	45	45	45	1203.99
535	45	0	22.5	90	1475.06
536	45	22.5	67.5	67.5	1735.15
537	22.5	22.5	0	67.5	1005.31
538	45	90	67.5	22.5	1469.12
539	67.5	90	90	67.5	835.028
540	22.5	90	22.5	45	1694.16
541	0	0	90	67.5	954.473
542	22.5	67.5	22.5	0	1563.72
543	45	0	67.5	22.5	1624.44
544	45	45	45	22.5	1837.4
545	0	0	90	0	910.514
546	45	45	90	22.5	1753.56
547	45	67.5	67.5	22.5	1562.88
548	90	67.5	0	22.5	1085.8
549	45	0	0	22.5	1405.91
550	45	0	0	67.5	1433.18
551	0	90	45	67.5	1621.94
552	22.5	67.5	67.5	0	1741.3
553	0	0	90	22.5	918.008
554	22.5	22.5	67.5	67.5	1265.23
555	0	67.5	67.5	90	1607.4
556	90	67.5	90	45	709.779
557	90	90	67.5	22.5	624.43
558	67.5	22.5	22.5	90	1741.3
559	45	45	22.5	0	1756.1
560	45	67.5	45	0	1652.36

(Continued)

TABLE A6.9 *(Continued)*
FFD Design of Experiment Set (Dataset Defined in Scenario 4 in Table 6.3)

Trial	x_1	x_2	x_3	x_4	Natural Frequency Gap
561	0	90	45	22.5	1594.97
562	45	22.5	90	90	1779.25
563	67.5	45	45	67.5	1397.04
564	0	45	90	67.5	1370.74
565	90	90	67.5	45	603.385
566	22.5	0	45	90	1045.91
567	90	22.5	67.5	67.5	1378.45
568	67.5	90	22.5	90	1114.27
569	45	22.5	67.5	22.5	1711.13
570	22.5	22.5	90	45	1294.98
571	0	90	67.5	22.5	1694.31
572	45	22.5	45	22.5	1622.21
573	90	45	0	67.5	1341.8
574	90	0	0	90	1734.35
575	67.5	90	45	67.5	1006.25
576	67.5	0	90	22.5	1675.66
577	67.5	22.5	45	67.5	1654.23
578	0	22.5	67.5	67.5	1035.72
579	45	67.5	45	22.5	1646.8
580	90	67.5	67.5	22.5	786.48
581	22.5	0	45	45	1022.59
582	67.5	0	0	0	1897.69
583	67.5	45	45	90	1391.77
584	45	0	45	90	1563.12
585	0	0	67.5	90	908.309
586	45	90	0	22.5	1689.68
587	45	67.5	90	0	1534.73
588	67.5	22.5	0	22.5	1810.38
589	0	0	22.5	0	571.78
590	0	45	67.5	22.5	1296.31
591	90	0	45	67.5	1594.97
592	45	0	0	45	1419.16
593	45	22.5	0	90	1534.73
594	0	67.5	22.5	0	1372.98
595	22.5	67.5	67.5	90	1775.91
596	67.5	22.5	22.5	67.5	1745.71
597	67.5	22.5	67.5	22.5	1594.55
598	90	67.5	45	67.5	878.989
599	0	0	45	90	776.296
600	67.5	67.5	90	22.5	1005.31

(Continued)

TABLE A6.9 *(Continued)*
FFD Design of Experiment Set (Dataset Defined in
Scenario 4 in Table 6.3)

Trial	x_1	x_2	x_3	x_4	Natural Frequency Gap
601	0	22.5	90	67.5	1085.8
602	45	45	67.5	90	1756.1
603	67.5	90	0	0	1205.97
604	22.5	67.5	45	0	1649.61
605	90	0	67.5	90	1498.34
606	0	67.5	90	0	1613.64
607	22.5	22.5	22.5	0	1012.75
608	0	67.5	0	0	1334.05
609	90	45	0	0	1377.19
610	22.5	90	67.5	0	1851.4
611	45	45	0	67.5	1753.56
612	0	22.5	0	67.5	729.376
613	0	45	45	0	1183.04
614	45	67.5	22.5	22.5	1735.15
615	90	45	90	67.5	1035.16
616	0	67.5	0	22.5	1339.67
617	22.5	90	0	0	1645.13
618	45	45	0	45	1741.23
619	67.5	45	0	22.5	1560.81
620	67.5	45	67.5	22.5	1330.56
621	45	90	0	90	1660.18
622	90	90	0	22.5	954.473
623	0	45	90	22.5	1341.8
624	0	22.5	45	0	871.537
625	67.5	22.5	90	90	1529.35

TABLE A6.10
FFD Design of Experiment Set (Dataset Defined in Scenario 5
in Table 6.3)

Trial	x_1	x_2	x_3	x_4	Natural Frequency Gap
1	45	90	45	0	1563.12
2	90	45	45	90	1183.04
3	0	0	45	90	776.296
4	0	0	90	0	910.514
5	45	90	0	0	1695.42
6	0	45	45	0	1183.04

(Continued)

TABLE A6.10 *(Continued)*
FFD Design of Experiment Set (Dataset Defined in Scenario 5 in Table 6.3)

Trial	x_1	x_2	x_3	x_4	Natural Frequency Gap
7	0	45	0	0	1028.46
8	0	0	45	0	720.594
9	0	90	0	0	1462.02
10	90	90	45	45	748.947
11	90	0	0	0	1771.13
12	0	45	90	0	1336.23
13	45	0	0	90	1439.23
14	90	0	45	90	1590.13
15	90	0	0	45	1750.86
16	90	0	45	45	1607.7
17	45	45	0	0	1725.5
18	0	90	45	90	1628.3
19	45	0	90	90	1695.42
20	45	45	0	90	1758.97
21	90	45	45	0	1225.94
22	45	45	0	45	1741.23
23	45	45	45	90	1833.23
24	45	90	90	45	1419.16
25	90	45	90	90	1028.46
26	0	0	0	90	567.108
27	45	0	45	90	1563.12
28	90	90	45	0	776.296
29	90	90	45	90	720.594
30	0	45	45	45	1203.99
31	90	0	45	0	1628.3
32	45	45	90	90	1725.5
33	45	90	0	45	1676.66
34	0	45	90	90	1377.19
35	0	45	45	90	1225.94
36	0	0	45	45	748.947
37	0	90	45	0	1590.13
38	45	90	90	0	1439.23
39	45	0	90	0	1660.18
40	0	45	0	45	1051.47
41	45	0	45	0	1526.67
42	45	0	90	45	1676.66
43	90	90	90	90	502.669
44	0	0	90	45	936.15
45	90	90	0	45	936.15
46	90	0	90	0	1502.68

(Continued)

TABLE A6.10 *(Continued)*
FFD Design of Experiment Set (Dataset Defined in Scenario 5 in Table 6.3)

Trial	x_1	x_2	x_3	x_4	Natural Frequency Gap
47	45	90	90	90	1400.63
48	90	45	0	0	1377.19
49	0	90	0	45	1480.91
50	0	90	90	90	1771.13
51	90	90	0	0	962.156
52	90	45	0	45	1355.83
53	45	90	45	90	1526.67
54	0	45	90	45	1355.83
55	45	45	90	0	1758.97
56	0	90	45	45	1607.7
57	45	0	0	45	1419.16
58	0	0	90	90	962.156
59	45	45	45	45	1848.19
60	90	90	90	45	536.224
61	90	0	90	45	1480.91
62	45	90	45	45	1544.02
63	90	90	90	0	567.108
64	45	45	90	45	1741.23
65	90	45	45	45	1203.99
66	90	90	0	90	910.514
67	0	45	0	90	1074.99
68	45	0	45	45	1544.02
69	0	90	90	45	1750.86
70	90	45	0	90	1336.23
71	90	45	90	45	1051.47
72	0	0	0	45	536.224
73	0	90	90	0	1734.35
74	0	0	0	0	502.669
75	45	0	0	0	1400.63
76	90	0	90	90	1462.02
77	90	0	0	90	1734.35
78	45	90	0	90	1660.18
79	45	45	45	0	1833.23
80	90	45	90	0	1074.99
81	0	90	0	90	1502.68

TABLE A6.11

Models Formed with Neuro Regression Using the Taguchi Experimental Set (Related to Table 6.4)

Scenario	Design Method		Model
1	Taguchi L8	SON	$502.669 + 3.738x_1 + 0.042x_1^2 + 3.738x_2 - 0.109x_1x_2 + 0.042x_2^2 + 3.738x_3 - 0.043x_1x_3 - 0.014x_2x_3 + 0.042x_3^2 + 3.738x_4 - 0.014x_1x_4 - 0.043x_2x_4 - 0.109x_3x_4 + 0.042x_4^2$
		TON	$502.66900 + 1.77988x_1 + 0.01978x_1^2 + 0.00022x_1^3 + 1.77988x_2 - 0.02064x_1x_2 - 0.00023x_1^2x_2 + 0.01978x_2^2 - 0.00023x_1x_2^2 + 0.00022x_2^3 + 1.77988x_3 + 0.00160x_1x_3 + 0.00002x_1^2x_3 + 0.01113x_2x_3 - 0.00053x_1x_2x_3 + 0.00012x_2^2x_3 + 0.01978x_3^2 + 0.00002x_1x_3^2 + 0.00012x_2x_3^2 + 0.00022x_3^3 + 1.77988x_4 + 0.01113x_1x_4 + 0.00012x_1^2x_4 + 0.00160x_2x_4 - 0.00053x_1x_2x_4 + 0.00002x_2^2x_4 - 0.02064x_3x_4 - 0.00053x_1x_3x_4 - 0.00053x_2x_3x_4 - 0.00023x_3^2x_4 + 0.01978x_4^2 + 0.00012x_1x_4^2 + 0.00002x_2x_4^2 - 0.00023x_3x_4^2 + 0.00022x_4^3$
2	Taguchi L9	SON	$502.669 + 2.773x_1 - 0.135x_1^2 + 5.918x_2 - 0.005x_1x_2 + 0.049x_2^2 + 5.871x_3 + 0.043x_1x_3 - 0.098x_2x_3 + 0.092x_3^2 + 5.286x_4 + 0.073x_1x_4 - 0.013x_2x_4 - 0.102x_3x_4 + 0.038x_4^2$
		TON	$502.66900 + 1.34463x_1 - 0.02247x_1^2 - 0.00049x_1^3 + 4.21135x_2 + 0.01058x_1x_2 - 0.00046x_1^2x_2 + 0.02775x_2^2 + 0.00005x_1x_2^2 + 0.00016x_2^3 + 4.27225x_3 + 0.02045x_1x_3 - 0.00027x_1^2x_3 + 0.00780x_2x_3 - 0.00005x_1x_2x_3 - 0.00052x_2^2x_3 + 0.03722x_3^2 + 0.00034x_1x_3^2 + 0.00009x_2x_3^2 + 0.00032x_3^3 + 3.94036x_4 + 0.02275x_1x_4 - 0.00017x_1^2x_4 + 0.02088x_2x_4 + 0.00073x_1x_2x_4 + 0.00017x_2^2x_4 + 0.00400x_3x_4 + 0.00031x_1x_3x_4 - 0.00066x_2x_3x_4 - 0.00044x_3^2x_4 + 0.02430x_4^2 + 0.00020x_1x_4^2 - 0.00034x_2x_4^2 - 0.00006x_3x_4^2 + 0.00012x_4^3$
3	Taguchi L16	SON	$454.371 + 23.700x_1 - 0.059x_1^2 + 16.932x_2 - 0.286x_1x_2 - 0.025x_2^2 - 4.134x_3 - 0.134x_1x_3 + 0.088x_2x_3 + 0.072x_3^2 + 5.032x_4 + 0.062x_1x_4 - 0.138x_2x_4 + 0.017x_3x_4 - 0.043x_4^2$

(Continued)

TABLE A6.11 *(Continued)*
Models Formed with Neuro Regression Using the Taguchi Experimental Set (Related to Table 6.4)

Scenario	Design Method		Model
		TON	$502.66900 + 9.89672x_1 + 0.04625x_1^2 - 0.00014x_1^3 + 5.85707x_2 + 0.05117x_1x_2 - 0.00100x_1^2x_2 + 0.06543x_2^2 - 0.00125x_1x_2^2 - 0.00045x_2^3 - 2.77925x_3 - 0.00837x_1x_3 - 0.00052x_1^2x_3 + 0.08703x_2x_3 - 0.00142x_1x_2x_3 + 0.00017x_2^2x_3 + 0.04949x_3^2 - 0.00005x_1x_3^2 + 0.00061x_2x_3^2 - 0.00009x_3^3 + 1.14334x_4 + 0.06443x_1x_4 - 0.00052x_1^2x_4 + 0.09778x_2x_4 - 0.00086x_1x_2x_4 - 0.00011x_2^2x_4 + 0.05488x_3x_4 + 0.00040x_1x_3x_4 - 0.00095x_2x_3x_4 - 0.00049x_3^2x_4 + 0.03365x_4^2 - 0.00004x_1x_4^2 - 0.00171x_2x_4^2 + 0.00001x_3x_4^2 - 0.00007x_4^3$
4	Taguchi L25	SON	$442.771 + 17.490x_1 - 0.056x_1^2 + 6.877x_2 - 0.099x_1x_2 + 0.040x_2^2 + 4.556x_3 - 0.124x_1x_3 - 0.182x_2x_3 + 0.178x_3^2 + 5.246x_4 - 0.122x_1x_4 + 0.018x_2x_4 - 0.131x_3x_4 + 0.063x_4^2$
		TON	$501.23600 + 16.82070x_1 + 0.07750x_1^2 - 0.00064x_1^3 + 2.84324x_2 - 0.17153x_1x_2 - 0.00102x_1^2x_2 + 0.14303x_2^2 + 0.00148x_1x_2^2 - 0.00125x_2^3 + 1.47856x_3 - 0.13450x_1x_3 - 0.00067x_1^2x_3 - 0.00490x_2x_3 - 0.00118x_1x_2x_3 - 0.00094x_2^2x_3 + 0.15463x_3^2 + 0.00192x_1x_3^2 + 0.00026x_2x_3^2 - 0.00062x_3^3 - 0.52371x_4 - 0.11282x_1x_4 - 0.00058x_1^2x_4 + 0.14133x_2x_4 + 0.00105x_1x_2x_4 + 0.00118x_2^2x_4 - 0.08731x_3x_4 - 0.00166x_1x_3x_4 + 0.00022x_2x_3x_4 - 0.00026x_3^2x_4 + 0.05561x_4^2 + 0.00084x_1x_4^2 - 0.00323x_2x_4^2 + 0.00123x_3x_4^2 + 0.00021x_4^3$
5	Taguchi L27	SON	$502.669 + 2.773x_1 - 0.135x_1^2 + 5.918x_2 - 0.005x_1x_2 + 0.049x_2^2 + 5.871x_3 + 0.043x_1x_3 - 0.098x_2x_3 + 0.092x_3^2 + 5.286x_4 + 0.073x_1x_4 - 0.013x_2x_4 - 0.102x_3x_4 + 0.038x_4^2$
		TON	$502.66900 + 1.34463x_1 - 0.02247x_1^2 - 0.00049x_1^3 + 4.21135x_2 + 0.01058x_1x_2 - 0.00046x_1^2x_2 + 0.02775x_2^2 + 0.00005x_1x_2^2 + 0.00016x_2^3 + 4.27225x_3 + 0.02045x_1x_3 - 0.00027x_1^2x_3 + 0.00780x_2x_3 - 0.00005x_1x_2x_3 - 0.00052x_2^2x_3 + 0.03722x_3^2 + 0.00034x_1x_3^2 + 0.00009x_2x_3^2 + 0.00032x_3^3 + 3.94036x_4 + 0.02275x_1x_4 - 0.00017x_1^2x_4 + 0.02088x_2x_4 + 0.00073x_1x_2x_4 + 0.00017x_2^2x_4 + 0.00400x_3x_4 + 0.00031x_1x_3x_4 - 0.00066x_2x_3x_4 - 0.00044x_3^2x_4 + 0.02430x_4^2 + 0.00020x_1x_4^2 - 0.00034x_2x_4^2 - 0.00006x_3x_4^2 + 0.00012x_4^3$

TABLE A6.12

Models Formed with Neuro Regression Using the FFD Experimental Set (Related to Table 6.5)

Scenario	Design Method		Model
1	FFD L16	SON	$503.784 + 7.035x_1 + 0.078x_1^2 + 5.317x_2 - 0.218x_1x_2 + 0.059x_2^2 + 2.253x_3 - 0.083x_1x_3 - 0.016x_2x_3 + 0.025x_3^2 + 0.346x_4 - 0.012x_1x_4 - 0.002x_2x_4 - 0.001x_3x_4 + 0.004x_4^2$
		TON	$503.78360 + 4.68975x_1 + 0.05211x_1^2 + 0.00058x_1^3 + 3.54490x_2 - 0.07259x_1x_2 - 0.00081x_1^2x_2 + 0.03939x_2^2 - 0.00081x_1x_2^2 + 0.00044x_2^3 + 1.50228x_3 - 0.02765x_1x_3 - 0.00031x_1^2x_3 - 0.00539x_2x_3 + (4.41090 \times 10^{-19})x_1x_2x_3 - 0.00006x_2^2x_3 + 0.01669x_3^2 - 0.00031x_1x_3^2 - 0.00006x_2x_3^2 + 0.00019x_3^3 + 0.23041x_4 - 0.00398x_1x_4 - 0.00004x_1^2x_4 - 0.00080x_2x_4 + (1.25435 \times 10^{-18})x_1x_2x_4 - (8.83425 \times 10^{-6})x_2^2x_4 - 0.00034x_3x_4 + (1.10273 \times 10^{-18})x_1x_3x_4 + (2.75682 \times 10^{-20})x_2x_3x_4 - (3.81276 \times 10^{-6})x_3^2x_4 + 0.00256x_4^2 - 0.00004x_1x_4^2 - (8.83425 \times 10^{-6})x_2x_4^2 - (3.81276 \times 10^{-6})x_3x_4^2 + 0.00003x_4^3$
2	FFD L81	SON	$451.128 + 33.251x_1 - 0.216x_1^2 + 15.516x_2 - 0.216x_1x_2 - 0.046x_2^2 + 5.611x_3 - 0.080x_1x_3 - 0.032x_2x_3 - 0.005x_3^2 + 0.791x_4 - 0.011x_1x_4 - 0.005x_2x_4 - 0.002x_3x_4 + 0.000x_4^2$
		TON	$451.12810 + 26.29822x_1 + 0.01613x_1^2 - 0.00172x_1^3 + 13.32087x_2 - 0.21584x_1x_2 - (8.40114 \times 10^{-19})x_1^2x_2 + 0.02696x_2^2 - (2.05306 \times 10^{-18})x_1x_2^2 - 0.00054x_2^3 + 5.03533x_3 - 0.08033x_1x_3 - (3.60854 \times 10^{-18})x_1^2x_3 - 0.03197x_2x_3 + (2.83472 \times 10^{-18})x_1x_2x_3 - (9.38221 \times 10^{-18})x_2^2x_3 + 0.01379x_3^2 - (5.14217 \times 10^{-18})x_1x_3^2 + (5.05196 \times 10^{-18})x_2x_3^2 - 0.00014x_3^3 + 0.72488x_4 - 0.01146x_1x_4 - (7.57794 \times 10^{-18})x_1^2x_4 - 0.00457x_2x_4 - (1.99719 \times 10^{-18})x_1x_2x_4 + (1.53363 \times 10^{-18})x_2^2x_4 - 0.00162x_3x_4 - (7.34449 \times 10^{-18})x_1x_3x_4 + (6.24926 \times 10^{-18})x_2x_3x_4 - (4.51068 \times 10^{-18})x_3^2x_4 + 0.00223x_4^2 - (3.60854 \times 10^{-18})x_1x_4^2 + (2.43577 \times 10^{-18})x_2x_4^2 - (3.33790 \times 10^{-18})x_3x_4^2 - 0.00002x_4^3$
3	FFD L256	SON	$379.796 + 32.630x_1 - 0.185x_1^2 + 14.513x_2 - 0.251x_1x_2 - 0.020x_2^2 + 5.756x_3 - 0.091x_1x_3 - 0.028x_2x_3 - 0.004x_3^2 + 0.822x_4 - 0.013x_1x_4 - 0.004x_2x_4 - 0.002x_3x_4 + 0.000x_4^2$

(Continued)

TABLE A6.12 *(Continued)*
Models Formed with Neuro Regression Using the FFD Experimental Set (Related to Table 6.5)

Scenario	Design Method		Model
		TON	$379.79630 + 32.63015x_1 - 0.18512x_1^2 + (5.37946 \times 10^{-18})x_1^3 + 14.51322x_2 - 0.25113x_1x_2 - (5.00175 \times 10^{-18})x_1^2x_2 - 0.01981x_2^2 - (6.65005 \times 10^{-18})x_1x_2^2 - (1.50177 \times 10^{-17})x_2^3 + 5.75641x_3 - 0.09074x_1x_3 - (4.77440 \times 10^{-18})x_1^2x_3 - 0.02792x_2x_3 + (4.01906 \times 10^{-19})x_1x_2x_3 - (5.22910 \times 10^{-18})x_2^2x_3 - 0.00362x_3^2 + (1.09129 \times 10^{-17})x_1x_3^2 + (7.27527 \times 10^{-18})x_2x_3^2 + (1.68108 \times 10^{-18})x_3^3 + 0.82240x_4 - 0.01301x_1x_4 - (2.00070 \times 10^{-17})x_1^2x_4 - 0.00383x_2x_4 - (3.93868 \times 10^{-18})x_1x_2x_4 + (6.59322 \times 10^{-18})x_2^2x_4 - 0.00203x_3x_4 - (7.55583 \times 10^{-18})x_1x_3x_4 + (5.46592 \times 10^{-18})x_2x_3x_4 - (7.27527 \times 10^{-18})x_3^2x_4 + 0.00030x_4^2 - (2.02344 \times 10^{-17})x_1x_4^2 - (1.36411 \times 10^{-18})x_2x_4^2 - (7.04792 \times 10^{-18})x_3x_4^2 + (4.70703 \times 10^{-18})x_4^3$
4	FFD L625	SON	$313.012 + 33.720x_1 - 0.185x_1^2 + 15.667x_2 - 0.266x_1x_2 - 0.025x_2^2 + 5.903x_3 - 0.099x_1x_3 - 0.029x_2x_3 - 0.001x_3^2 + 0.870x_4 - 0.014x_1x_4 - 0.004x_2x_4 - 0.001x_3x_4 + 0.000x_4^2$
		TON	$313.01250 + 33.71966x_1 - 0.18529x_1^2 + (3.26808 \times 10^{-18})x_1^3 + 15.66718x_2 - 0.26610x_1x_2 + (4.95873 \times 10^{-18})x_1^2x_2 - 0.02455x_2^2 + (1.47212 \times 10^{-17})x_1x_2^2 + (1.39149 \times 10^{-17})x_2^3 + 5.90312x_3 - 0.09865x_1x_3 - (9.76250 \times 10^{-18})x_1^2x_3 - 0.02866x_2x_3 - (1.11916 \times 10^{-17})x_1x_2x_3 + (1.82853 \times 10^{-17})x_2^2x_3 - 0.00123x_3^2 - (2.85127 \times 10^{-17})x_1x_3^2 + (1.85952 \times 10^{-18})x_2x_3^2 - (1.08255 \times 10^{-17})x_3^3 + 0.87012x_4 - 0.01398x_1x_4 - (6.50833 \times 10^{-18})x_1^2x_4 - 0.00429x_2x_4 + (9.67043 \times 10^{-18})x_1x_2x_4 + (8.83274 \times 10^{-18})x_2^2x_4 - 0.00140x_3x_4 - (1.17349 \times 10^{-17})x_1x_3x_4 + (1.41253 \times 10^{-18})x_2x_3x_4 - (1.16220 \times 10^{-17})x_3^2x_4 + 0.00017x_4^2 + (1.54960 \times 10^{-18})x_1x_4^2 + (4.33889 \times 10^{-18})x_2x_4^2 + (1.11571 \times 10^{-17})x_3x_4^2 + (1.76681 \times 10^{-17})x_4^3$
5	FFD L81	SON	$451.128 + 33.251x_1 - 0.216x_1^2 + 15.516x_2 - 0.216x_1x_2 - 0.046x_2^2 + 5.611x_3 - 0.080x_1x_3 - 0.032x_2x_3 - 0.005x_3^2 + 0.791x_4 - 0.011x_1x_4 - 0.005x_2x_4 - 0.002x_3x_4 + 0.000x_4^2$

(Continued)

TABLE A6.12 *(Continued)*
Models Formed with Neuro Regression Using the FFD Experimental Set (Related to Table 6.5)

Scenario	Design Method	Model
	TON	$451.12810 + 26.29822x_1 + 0.01613x_1^2 - 0.00172x_1^3 + 13.32087x_2 - 0.21584x_1x_2 + (2.00725 \times 10^{-18})x_1^2x_2 + 0.02696x_2^2 + (2.25534 \times 10^{-18})x_1x_2^2 - 0.00054x_2^3 + 5.03533x_3 - 0.08033x_1x_3 + (3.96940 \times 10^{-18})x_1^2x_3 - 0.03197x_2x_3 - (1.21764 \times 10^{-17})x_1x_2x_3 - (6.04431 \times 10^{-18})x_2^2x_3 + 0.01379x_3^2 + (2.52598 \times 10^{-18})x_1x_3^2 - (1.25397 \times 10^{-17})x_2x_3^2 - 0.00014x_3^3 + 0.72488x_4 - 0.01146x_1x_4 - (6.31495 \times 10^{-19})x_1^2x_4 - 0.00457x_2x_4 - (1.15321 \times 10^{-17})x_1x_2x_4 - (6.67581 \times 10^{-18})x_2^2x_4 - 0.00162x_3x_4 - (6.63581 \times 10^{-18})x_1x_3x_4 - (1.61063 \times 10^{-17})x_2x_3x_4 + (2.16513 \times 10^{-18})x_3^2x_4 + 0.00223x_4^2 - (1.44342 \times 10^{-18})x_1x_4^2 - (6.31495 \times 10^{-19})x_2x_4^2 + (1.17278 \times 10^{-18})x_3x_4^2 - 0.00002x_4^3$

7 Comparison of Different Test and Validation Methods Used in Mathematical Modeling

Achievement in mathematical modeling, an essential stage of experimental design, modeling, and optimization processes, is directly linked to the modeling methods used. Many different methods are used in the mathematical modeling process. The most common modeling methods in the literature are Response Surface Method (RSM), Artificial Neural Networks (ANN), Fuzzy Logic method (FL), Adaptive Neural Fuzzy Inference System (ANFIS), regression, Support Vector Machines (SVR), and other methods that work similarly to these mentioned methods. Many modeling and optimization studies using these methods can be found in the literature. It is often preferred to compare the methods with each other, especially by using different modeling methods for the same problem. In this section, the original modeling methods proposed in this book, Neuro Regression (NRM) and Stochastic Neuro Regression (SNRM), are compared according to the models they offer in the reference problem given below. Among these hybrid methods, NRM utilizes ANN and regression, while SNRM employs optimization in addition to these in model determination.

Apart from the modeling method, another parameter that affects the mathematical modeling process is how the dataset is used in the model creation phase. There are different data separation techniques followed here. To make a general classification, data separation can be achieved by using all data for training–testing, training–testing–validation, or without testing and validation. It has been shown in the previous chapters of the book that creating a model using all the data in these three data separation methods and then deciding about the model's success without subjecting it to testing and validation processes is not an acceptable approach. For this reason, in this part of the study, mathematical modeling was carried out using training–test and training–test–validation data separation methods. Here, the processes called testing and verification, which are actively used to evaluate model success, do not differ from each other in terms of their purpose. The aim of the testing and verification steps is to measure the prediction performance of the model created using training data by asking it about data it has never encountered before. In many sources, validation is defined as a part of the training process during the model creation phase. When the trained model shows low prediction performance, it is subjected to a validation process, and it is aimed to improve the success of

 DOI: 10.1201/9781003494843-7

the model by increasing the dataset and diversity it encounters during the training phase. In this chapter, it has been used in an approach that aims to measure the model's performance in predicting the actual values of data that it has not encountered before, completely independent of the training phase, as in the testing phase. In this approach followed throughout the book, the dataset is generally divided into 80% training, 15% testing, and 5% validation.

CASE STUDY 7.1

In the study conducted by Shui et al. [1], the objective was to create an enclosure box for a battery pack with the highest possible natural frequency. The researchers chose the wall thickness and temperature of the battery box as design parameters. They used Central Composite design (CCD) and Latin Hypercube sampling (LHS) experimental methods to form an experimental set and generated a 100-line set of input and output parameters.

This study section compares various test and validation methods used during the modeling phase. Additionally, the evaluation aims to highlight two new modeling techniques proposed in this book. The comparison of the successfulness of the models obtained using NRM and SNRM is another goal of this study.

NRM and SNRM are two methods used to create mathematical models. NRM utilizes artificial intelligence and regression, while SNRM additionally uses stochastic optimization techniques to determine the most appropriate model coefficients and minimize the difference between actual and predicted values. Comparing these two methods can help understand the effectiveness of optimization techniques in model creation.

Table 7.1 presents the evaluation outcomes of models employed to mathematically represent the natural frequency of the battery pack enclosure utilizing the NRM approach. The methodology involved partitioning the data into training, testing, and validation sets, with the performance of each model delineated accordingly. Following experimentation with linear, polynomial, logarithmic, trigonometric, and rational model types, it was ascertained that the SOTN model emerged as the most proficient. This determination was grounded on the average R-squared (R^2) value attained across the training, testing, and validation phases, alongside the extremities of values generated by the model.

Similarly, the L, SON, FOTN, FOLN, SOLN, and TONR models demonstrated success in elucidating the natural frequency. However, other model types were deemed unsuccessful as they failed to satisfy both the R^2 assessment criterion and the boundedness check criteria, which evaluate the maximum and minimum values produced by the model.

Table 7.2 presents the performance of the models when employing SNRM to model natural frequency. In comparison to the results obtained for NRM in Table 7.1, all models obtained using SNRM successfully meet the average R^2 criterion. Moreover, except for SOLNR, the models produce results consistent with the actual values provided in the dataset in the reference study, thereby meeting the boundedness control criterion.

TABLE 7.1

Assessment of Models Obtained for Natural Frequency Using the Neuro Regression Method

	R^2 Train	R^2 Adj	R^2 Test	R^2 Valid	R^2 Avg	Max.	Min.
L	0.76	0.74	0.82	0.92	0.81	96.52	84.72
SON	0.87	0.86	0.82	0.67	0.805	100.16	74.75
TON	1	1	−235	−86	−79.75	513.31	−371.76
FOTN	0.82	0.8	0.84	0.94	0.85	95.53	83.35
SOTN	0.89	0.88	0.89	0.84	0.875	98.8	73.2
FOLN	0.76	0.74	0.84	0.92	0.815	96.94	84.87
SOLN	0.86	0.85	0.85	0.67	0.8075	100.92	75.78
LR	0.81	0.8	0.6	0.7	0.7275	641169	−2.61E+07
SONR	0.98	0.98	−6.2	0.1	−1.035	1.70E+11	−3.50E+07
TONR	0.83	0.82	0.78	0.91	0.835	100.72	84
FOTNR	0.87	0.86	0.7	−0.03	0.6	5.00E+07	−1.17E+06
FOLNR	0.81	0.8	0.6	0.53	0.685	5.12E+04	−1.20E+15
SOLNR	0.92	0.91	−2.11	0.81	0.1325	3381	−21,396

* Full form of models are given in appendix (Table A5.2).

TABLE 7.2

Assessment of Models Obtained for Natural Frequency Using the Stochastic Neuro Regression Method

Model*	R^2 Train	R^2 Adj	R^2 Test	R^2 Valid	R^2 Avg	Max.	Min.
L	0.76	0.74	0.85	0.95	0.825	96.96	84.19
SON	0.86	0.85	0.93	0.85	0.8725	97.52	77.68
TON	0.97	0.96	0.99	0.98	0.975	150.8	26.02
FOTN	0.82	0.81	0.87	0.96	0.865	95.94	83.27
SOTN	0.88	0.87	0.97	0.94	0.915	97.06	78.8
FOLN	0.75	0.74	0.86	0.95	0.825	97.37	84.39
SOLN	0.85	0.84	0.94	0.84	0.8675	97.56	78.64
LR	0.8	0.79	0.87	0.86	0.83	95	74.08
SONR	0.85	0.84	0.92	0.88	0.8725	96.96	81.83
TONR	0.86	0.85	0.93	0.88	0.88	96.58	80.55
FOTNR	0.83	0.82	0.86	0.95	0.865	95.7	71.78
FOLNR	0.78	0.77	0.9	0.87	0.83	95.19	72.9
SOLNR	0.88	0.86	0.96	0.87	0.8925	1222	−249,671

* Full form of models are given in appendix (Table A7.1).

TABLE 7.3

Comparison of the Performance of the Models Obtained Using Neuro Regression and Stochastic Neuro Regression for Two Different Data Separation Methods

	Method*	R^2 Train	R^2 Adj	R^2 Test	R^2 Valid	R^2 Avg.	Max.	Min.
Train–Test	NRM	0.94	0.93	0.97	–	0.947	97.09	78.35
	SNRM	0.94	0.93	0.98	–	0.95	97.19	76.37
Train–Test–	NRM	0.89	0.88	0.89	0.84	0.875	98.8	73.2
Validation	SNRM	0.88	0.87	0.97	0.94	0.915	97.06	78.8

* Full form of models are given in appendix (Table A7.2).

Upon evaluating the results in Tables 7.1 and 7.2 collectively, it can be inferred that SNRM outperforms NRM in modeling the natural frequency. NRM failed to establish successful models for TON, LR, SONR, FOTNR, FOLNR, and SOLNR model types. In contrast, only SOLNR, when employed with SNRM, was considered unsuccessful due to its failure to meet the boundedness control criterion.

Table 7.3 shows the assessment of SOTN models obtained using NRM and SNRM when the dataset is divided into two parts as training–test and three parts as training–test–validation.

In the first case, the dataset was divided into two parts, with 80% of the data used for training and 20% for testing. In the second case, the dataset was divided into three parts, with 80% of the data used for training, 15% for testing, and 5% for validation. When comparing the success rates of SOTN models obtained using NRM and SNRM for these two cases, it was found that the average R^2 values were higher for the first case where the dataset was divided into training–test. In the second case, where the dataset was divided into three parts as training–test–validation in both NRM and SNRM, the average R^2 values were slightly lower. However, in both cases, the success of the models obtained using SNRM was observed to be higher than that of those obtained using the NRM method.

According to the findings, subjecting the model to validation, which can be considered a secondary testing phase, may marginally decrease the average success rate. However, incorporating additional criteria to assess model success enhances the reliability of the model. Thus, the results presented in Table 7.3 suggest that dividing the dataset into three segments—training, testing, and validation—provides a more meaningful approach than dividing it into two segments for more dependable modeling.

The data partitioning method known as bootstrap, as outlined in Table 7.4, segregates the dataset into training, testing, and validation subsets. This technique randomly generates the dataset, allocating 80% for training, 15% for testing, and 5% for validation in each iteration. The performance evaluation of models obtained through 20 repeated runs using both NRM and SNRM with the bootstrap method is provided separately. Following the repeated iterations, the

TABLE 7.4

Comparison of the Performance of the Models Obtained by Neuro Regression and Stochastic Neuro Regression Using the Bootstrap Data Separation Method (80 Training–15 Testing–5 Validation)

	Method*	R^2 Train	R2 Adj	R^2 Test	R^2 Valid	R^2 Avg.	Max.
1	NRM	0.96	0.96	0.74	−0.57	98.96	81.42
	SNRM	0.95	0.95	0.98	0.76	97.1	76.25
2	NRM	0.97	0.97	0.9	0.71	98.16	77.17
	SNRM	0.96	0.96	0.96	0.92	97.1	76.25
3	NRM	0.96	0.96	0.39	0.78	98.37	71.05
	SNRM	0.94	0.94	0.98	0.77	98.12	77.17
4	NRM	0.96	0.96	0.72	0.62	103.94	72.33
	SNRM	0.95	0.95	0.97	0.86	97.23	76.72
5	NRM	0.96	0.95	−0.11	0.86	102.33	78.85
	SNRM	0.95	0.94	0.96	0.89	96.57	76.71
6	NRM	0.95	0.95	0.8	0.96	97.27	77.8
	SNRM	0.94	0.94	0.95	0.96	98.45	74.78
7	NRM	0.96	0.96	0.71	0.84	101.32	77.44
	SNRM	0.95	0.95	0.98	0.84	97.67	75.05
8	NRM	0.96	0.96	0.69	0.76	98.82	75.47
	SNRM	0.95	0.94	0.98	0.89	96.70	76.30
9	NRM	0.96	0.95	0.64	0.27	98.5	78.62
	SNRM	0.95	0.95	0.95	0.71	97.1	76.25
10	NRM	0.95	0.95	0.46	0.92	105.83	78.35
	SNRM	0.95	0.94	0.97	0.94	96.89	76.36
11	NRM	0.98	0.98	−1.44	−0.55	111.38	67.32
	SNRM	0.96	0.96	0.94	0.93	97.29	75.83
12	NRM	0.96	0.95	0.81	0.92	100.99	77.43
	SNRM	0.94	0.94	0.97	0.93	96.65	76.1
13	NRM	0.97	0.97	0.36	0.37	101.66	75.1
	SNRM	0.96	0.96	0.91	0.82	97.1	76.25
14	NRM	0.94	0.94	0.55	0.98	103.17	75.56
	SNRM	0.93	0.93	0.99	0.97	97.69	76.41
15	NRM	0.96	0.96	0.68	0.43	101.08	77.18
	SNRM	0.95	0.95	0.97	0.72	97.11	76.25
16	NRM	0.96	0.95	0.86	0.85	98.59	79.02
	SNRM	0.95	0.95	0.95	0.92	97.57	76.17
17	NRM	0.95	0.95	0.7	0.16	98.4	74.27
	SNRM	0.94	0.94	0.99	0.96	97.95	76.33
18	NRM	0.98	0.98	−0.2	0.21	111.79	74
	SNRM	0.96	0.96	0.94	0.9	96.85	76.16
19	NRM	0.96	0.95	0.77	0.94	97.64	78.21
	SNRM	0.95	0.94	0.95	0.98	97.89	76.31
20	NRM	0.96	0.96	0.34	0.51	102.77	67.77
	SNRM	0.94	0.94	0.92	0.92	98.37	76.55

* Full form of models are given in appendix (Table A7.3).

average R^2 index obtained for the training, testing, and validation phases indicates that SNRM surpasses NRM in modeling the natural frequency of the battery pack enclosure.

Throughout the repeated running process, certain models obtained using NRM faltered during the testing and validation stages. Conversely, models recommended by SNRM consistently exhibited high prediction performance across all 20 iterations. Specifically, models derived from the 1st, 3rd, 5th, 10th, 11th, 13th, 15th, 17th, 18th, and 20th iterations using NRM failed to meet at least one of the R^2 success criteria or demonstrated a performance below 0.5. Conversely, the results indicate that SNRM consistently produced models with exceptionally high prediction performance after each of the 20 repetitions. Even the least successful SOTN model obtained using SNRM achieved an R^2 index of 0.94, 0.93, 0.91, and 0.71 in the training, adjusted, testing, and validation phases, respectively.

The bootstrap data separation method detailed in Table 7.5 was employed to develop a model. Initially, the dataset was partitioned into 80% for training and 20% for testing. Subsequently, the training set was further split into two subsets: 80% for training and 20% for validation. Given that the dataset comprised 100 lines, the initial separation yielded 80 lines for training and 20 for testing. With the subsequent split, 64 lines were allocated for training the model, while 16 lines were reserved for validating the model within the training phase.

TABLE 7.5

Comparison of the Performance of the Models Obtained by Neuro Regression and Stochastic Neuro Regression Using the Bootstrap Data Separation Method (80 Training**–20 Testing)

	Method*	R^2 Train	R^2 Adj	R^2 Test	R^2 Valid	R^2 Avg.	Max.
1	NRM	0.95	0.95	0.71	0.91	100.04	75.2
	SNRM	0.93	0.92	0.97	0.97	96.88	76.49
2	NRM	0.97	0.96	0.34	0.76	99.31	68.9
	SNRM	0.94	0.94	0.96	0.98	97.11	76.25
3	NRM	0.93	0.93	−0.27	0.23	102.73	66.12
	SNRM	0.92	0.91	0.97	0.97	96.98	76.59
4	NRM	0.98	0.97	−0.17	0.92	99.37	74.06
	SNRM	0.95	0.95	0.88	0.97	96.97	76.59
5	NRM	0.96	0.96	0.21	0.93	102.88	71.68
	SNRM	0.93	0.92	0.97	0.98	97.23	76.42
6	NRM	0.97	0.97	0.52	0.91	101.01	76.93
	SNRM	0.94	0.93	0.92	0.97	98.15	78.17
7	NRM	0.96	0.95	−0.19	0.66	100.15	74.9
	SNRM	0.93	0.93	0.94	0.98	97.55	77.15
8	NRM	0.96	0.95	−0.12	0.94	100.64	77.82
	SNRM	0.92	0.92	0.97	0.97	99.83	77.04
9	NRM	0.95	0.94	0.57	0.94	103.17	79.35
	SNRM	0.92	0.91	0.97	0.96	97.25	76.53

(Continued)

TABLE 7.5 *(Continued)*

Comparison of the Performance of the Models Obtained by Neuro Regression and Stochastic Neuro Regression Using the Bootstrap Data Separation Method (80 Training**–20 Testing)

	Method*	R^2 Train	R^2 Adj	R^2 Test	R^2 Valid	R^2 Avg.	Max.
10	NRM	0.95	0.95	0.39	0.93	100.34	75.7
	SNRM	0.94	0.93	0.95	0.97	97.31	76.9
11	NRM	0.97	0.97	−0.84	0.94	101.39	74.97
	SNRM	0.93	0.93	0.95	0.97	97.06	77.08
12	NRM	0.97	0.96	0.53	0.96	103.7	75.46
	SNRM	0.93	0.92	0.94	0.97	96.88	77.35
13	NRM	0.97	0.97	0.003	0.93	102.39	72.93
	SNRM	0.93	0.92	0.95	0.97	97.24	75.83
14	NRM	0.97	0.97	−0.09	0.95	104.14	66.94
	SNRM	0.93	0.93	0.94	0.97	96.92	77.42
15	NRM	0.97	0.96	−0.09	0.94	105.1	61.3
	SNRM	0.93	0.92	0.95	0.97	97.45	76.1
16	NRM	0.95	0.94	0.66	0.93	103.55	77.3
	SNRM	0.93	0.92	0.97	0.98	97	76.57
17	NRM	0.94	0.93	0.77	0.89	96.73	74.85
	SNRM	0.91	0.90	0.98	0.97	97.47	76.44
18	NRM	0.95	0.94	0.71	0.93	101.07	78.33
	SNRM	0.93	0.93	0.96	0.97	96.97	76.59
19	NRM	0.96	0.96	0.41	0.96	101.4	74.79
	SNRM	0.94	0.93	0.93	0.98	97.17	76.96
20	NRM	0.95	0.95	−0.28	0.25	101.56	73.7
	SNRM	0.93	0.93	0.98	0.98	97.34	76.12

* Full form of models are given in appendix (Table A7.4).
** Training was divided into two parts (64 Training–16 Validation).

In this approach, 20% of the test data was predetermined and remained unchanged across subsequent iterations. However, the 80% training set, divided into two parts, underwent random changes in each run. This method diverges from scenarios where the dataset is partitioned into three segments—80% training, 15% testing, and 5% validation—since the training set is also internally tested, with the data used for training undergoing random alterations each time. This setup allows for an investigation into whether the proposed model remains robust when the training data is randomly selected and varied iteratively.

Among the models derived through NRM, those obtained from the 2nd, 3rd, 4th, 5th, 7th, 8th, 10th, 11th, 13th, 14th, 15th, 19th, and 20th iterations failed to meet at least one of the R^2 criteria. In contrast, SNRM consistently produced robust models with high prediction performance across all iteration processes.

Models obtained using SNRM exhibited a minimum R^2 index of 0.92, 0.91, 0.88, and 0.96 in the training, adjusted, testing, and validation phases, respectively. This underscores SNRM's ability to effectively establish a model even amidst randomized changes to the training dataset.

In Table 7.6, the mathematical modeling process was carried out using the bootstrap method. The dataset was initially divided into two parts—training and testing. Then, the training set was further divided into two parts, with 50% for training and 50% for validation. This approach evaluates how the success of modeling methods changes with a reduction in the data used to train the model.

TABLE 7.6

Comparison of the Performance of the Models Obtained by Neuro Regression and Stochastic Neuro Regression Using the Bootstrap Data Separation Method (80 Training**–20 Testing)

	Method*	R² Train	R² Adj	R² Test	R² Valid	R² Avg.	Max.
1	NRM	1	1	−1.56	0.27	123.92	43.85
	SNRM	0.96	0.95	0.9	0.97	96.97	76.59
2	NRM	1	1	−24.98	−2.41	170.25	29.02
	SNRM	0.93	0.92	0.94	0.97	96.97	76.59
3	NRM	1	1	−4.9	−2.44	145.73	25.52
	SNRM	0.93	0.91	0.94	0.97	96.97	76.59
4	NRM	1	1	−4.32	−0.82	149.172	42.27
	SNRM	0.92	0.91	0.94	0.97	97.88	76.36
5	NRM	1	1	−0.31	0.55	112.75	71.97
	SNRM	0.93	0.91	0.94	0.97	97.23	76.38
6	NRM	1	1	−11.82	−1.09	141.61	31.21
	SNRM	0.96	0.95	0.91	0.97	96.97	76.59
7	NRM	1	1	−0.45	−0.26	109.12	57.68
	SNRM	0.91	0.89	0.95	0.98	97.11	76.25
8	NRM	1	1	−4.82	0.2	122.05	58.26
	SNRM	0.94	0.93	0.93	0.97	96.97	76.59
9	NRM	1	1	−33.07	−0.39	186.65	42.82
	SNRM	0.93	0.92	0.93	0.97	96.73	76.16
10	NRM	1	1	−14.87	−6.97	150.08	43.41
	SNRM	0.91	0.90	0.95	0.97	97.57	77.32
11	NRM	1	1	−15.7	−2.97	143.34	48.45
	SNRM	0.9	0.88	0.95	0.97	97.11	76.69
12	NRM	1	1	−43.77	−6.18	159.51	45.66
	SNRM	0.91	0.89	0.96	0.97	98.06	76.05
13	NRM	1	1	−2.41	0.096	146.62	28.8
	SNRM	0.93	0.92	0.93	0.98	97.03	77.20
14	NRM	1	1	−4.21	−2.93	126.15	22.35
	SNRM	0.92	0.91	0.95	0.97	97.46	76.41

(Continued)

TABLE 7.6 *(Continued)*

Comparison of the Performance of the Models Obtained by Neuro Regression and Stochastic Neuro Regression Using the Bootstrap Data Separation Method (80 Training**–20 Testing)

	Method*	R^2 Train	R^2 Adj	R^2 Test	R^2 Valid	R^2 Avg.	Max.
15	NRM	1	1	−87.43	−84.35	213.44	−21.22
	SNRM	0.92	0.90	0.94	0.98	96.96	77
16	NRM	1	1	−13.77	−14.29	139.52	25.59
	SNRM	0.89	0.87	0.96	0.98	97.11	76.25
17	NRM	1	1	−45.82	−24.85	203.16	−1.36
	SNRM	0.92	0.90	0.94	0.98	98.10	77.18
18	NRM	1	1	−249.95	−62.46	256.6	−129.62
	SNRM	0.93	0.92	0.94	0.97	97.86	76.41
19	NRM	1	1	−31.36	−11.61	178.83	−5.53
	SNRM	0.91	0.89	0.96	0.97	97.05	76.06
20	NRM	1	1	−11.59	−5.85	158.76	68.89
	SNRM	0.87	0.85	0.97	0.98	97.37	75.35

* Full form of models are given in appendix (Table A7.5).

** Training was divided into two parts (40 Training–40 Validation).

Based on the results presented in Table 7.6, it was observed that after 20 repeated runs, NRM was unable to propose an appropriate mathematical model. The models put forth by NRM failed to fulfill at least one criterion. On the other hand, the models suggested by SNRM were successful with high prediction performance in all repeated running processes, as previously observed. These results indicate that SNRM can be classified as a highly reliable modeling method.

In Table 7.7, mathematical modeling was conducted with NRM and SNRM using cross-validation, a data separation method. First, the dataset was split into 80% training and 20% testing. The training set was then divided into 10 parts, and in each step, nine parts were used for training, while one part was used for validation. The modeling was performed in 10 repeated runs by changing the dataset allocated for validation in each run. This approach enabled the evaluation of how the training and validation data change affected the model's success.

Like in the previous data separation methods, SNRM suggested SOTN models demonstrating high prediction performance in all 10 repeated runs. On the other hand, NRM could not suggest a model with R^2 indexes of 0.5 or above in any of the repeated runs. Thus, these results are insufficient to conclude that the models proposed by NRM are successful. Table 7.7 demonstrates that the lowest prediction performance of the models recommended by SNRM is R^2 of 0.8 and above.

TABLE 7.7

Comparison of the Performance of the Models Obtained by Neuro Regression and Stochastic Neuro Regression Using the Cross-Validation Data Separation Method (10-Fold Cross-Validation)

	Method*	R² Train	R² Adj	R² Valid	R² Test	Max.	Min.
		%80 Training			%20 Testing		
1	NRM	0.89	0.88	0.37	0.61	97.96	68.63
	SNRM	0.87	0.86	0.93	0.93	98.31	76.66
2	NRM	0.89	0.88	0.4	0.77	102.31	73.84
	SNRM	0.86	0.85	0.95	0.93	96.62	76.37
3	NRM	0.9	0.89	−7.08	−0.83	122.17	81.81
	SNRM	0.85	0.84	0.97	0.93	97.15	75.18
4	NRM	0.89	0.88	0.13	0.56	101.32	76.62
	SNRM	0.87	0.85	0.87	0.93	96.67	74.96
5	NRM	0.9	0.9	0.38	0.84	99.2	68.48
	SNRM	0.88	0.87	0.87	0.93	97.31	77.62
6	NRM	0.89	0.88	0.23	0.83	99.24	67.51
	SNRM	0.87	0.85	0.85	0.94	96.78	78.54
7	NRM	0.91	0.9	−3.5	0.009	97.55	69.73
	SNRM	0.89	0.88	0.82	0.93	98.1	75.63
8	NRM	0.89	0.88	0.36	0.86	97.01	76.32
	SNRM	0.87	0.85	0.98	0.93	97.21	75.46
9	NRM	0.89	0.88	0.33	0.88	99.11	75.39
	SNRM	0.87	0.86	0.8	0.92	97.04	75.53
10	NRM	0.92	0.92	−1.64	0.47	98.63	71.48
	SNRM	0.87	0.86	0.9	0.93	97.12	76.1

* Full form of models are given in appendix (Table A7.6).

For this reason, the modeling process was carried out by dividing the data into five parts using the cross-validation method, and the model performance evaluations are given in Table 7.8. Thus, it was possible to compare the prediction performances of the models obtained when the dataset was divided into 10 and 5 parts. In Table 7.8, for the modeling process, the data was first divided into 80% training and 20% testing, and then the 80% training set was split into five parts. Here, four parts are used for training and one part for validation. By following this procedure, a mathematical model is obtained, and the performance in predicting the actual values is recorded. Afterward, one part of the four-piece training set is transferred to validation, the dataset previously used in validation is transferred to training, and a mathematical model is formed again by following the same modeling procedure. This method obtains an SOTN-type mathematical model in five repetitions, and prediction performance is saved.

TABLE 7.8

Comparison of the Performance of the Models Obtained by Neuro Regression and Stochastic Neuro Regression Using the Cross-Validation Data Separation Method (5-Fold Cross-Validation)

	Method*	%80 Training			%20 Testing		
		R^2 Train	R^2 Adj	r^2 valid	R^2 Test	Max.	Min.
1	NRM	0.94	0.93	−4.55	0.3	105.87	66.14
	SNRM	0.87	0.85	0.91	0.92	97.77	76.18
2	NRM	0.93	0.92	−3.98	−0.63	132.13	80.32
	SNRM	0.83	0.82	0.92	0.96	96.92	78.75
3	NRM	0.92	0.91	−0.44	0.54	103.99	69.33
	SNRM	0.87	0.85	0.88	0.93	96.74	75.55
4	NRM	0.93	0.92	−5.79	−2.65	100.02	51.7
	SNRM	0.87	0.86	0.84	0.93	97.28	76.42
5	NRM	0.89	0.88	0.44	0.84	99.69	80.2
	SNRM	0.86	0.84	0.92	0.92	97.24	75.19

* Full form of models are given in appendix (Table A7.7).

After examining the results in Table 7.8, it becomes clear that SNRM outperforms NRM in modeling. While NRM's models failed to meet the assessment criteria, SNRM's models achieved the best results after the five repeated runs.

A general evaluation is made of the results obtained throughout this study:

1. SNRM has proposed models with a high success rate in creating mathematical models for all data separation methods (training–test, training–test–validation, bootstrap, cross-validation).
2. NRM: While it suggested successful models when training–test and training–test–validation data separation methods were used, it could propose a limited number of successful models in the bootstrap method and could not recommend any successful models in cross-validation data separation methods.
3. Data separation methods, bootstrap, and cross-validation are more effective methods in mathematical modeling. However, this does not mean that dividing the dataset into training–test or training–test–validation will always be insufficient in modeling.
4. Using the entire dataset to create a model during the mathematical modeling process may cause the model to memorize the data. In this case, a model with an excellent performance in predicting the actual values of the output parameters in the dataset may make a meaningless prediction inconsistent with reality for data outside the dataset. This situation

can be seen when the model successes in the tables are examined. In almost all result tables, it is seen that when modeling is performed with both NRM and SNRM, the performance of the models in predicting the actual output values in the training phase gets values very close to 1 in terms of R^2. At the same time, sometimes, they cannot show similar success in the testing and validation stages. This situation indicates that the proposed models must be tested and verified with data they have not encountered before.

REFERENCE

1. Shui L, Chen F, Garg A, Peng X, Bao N, Zhang J. Design optimization of battery pack enclosure for electric vehicle. Structural and Multidisciplinary Optimization 2018; 58: 331–347.

APPENDIX

TABLE A7.1

Models Obtained Using the Stochastic Neuro Regression Method in Modeling the Natural Frequency Output (Related to Table 7.2)

Nomenclature	Model
L	$100.943 - 4.197x_1 - 3.283x_2 + 5.828x_3 - 2.064x_4 - 1.128x_5 - 0.024x_6$
LR	$(-2.719 \times 10^7 + (3.493 \times 10^7)x_1 - (1.204 \times 10^7)x_2 - (9.241 \times 10^7)x_3 -$
	$(1.483 \times 10^7)x_4 + (1.349 \times 10^7)x_5 + 570111.000x_6)/(-203969.000 +$
	$351068.000x_1 - 164554.000x_2 - 971127.000x_3 - 184003.000x_4 + 137879.000x_5 +$
	$6242.810x_6$
SON	$39.514 + 41.864x_1 - 13.593x_1^2 + 6.324x_2 + 0.368x_1x_2 + 1.150x_2^2 + 2.392x_3 -$
	$0.091x_1x_3 + 5.758x_2x_3 - 4.880x_3^2 - 0.486x_4 + 3.043x_1x_4 + 0.494x_2x_4 + 0.843x_3x_4 -$
	$1.629x_4^2 + 8.415x_5 - 2.781x_1x_5 - 3.151x_2x_5 + 1.030x_3x_5 - 0.644x_4x_5 + 0.177x_5^2 +$
	$0.563x_6 - 0.190x_1x_6 - 0.324x_2x_6 + 0.074x_3x_6 - 0.024x_4x_6 - 0.060x_5x_6 + 0.004x_6^2$
SONR	$(-0.068 + 7.250x_1 - 0.266x_1^2 - 2.847x_2 + 0.232x_1x_2 - 0.374x_2^2 + 5.439x_3 +$
	$3.435x_1x_3 - 4.208x_2x_3 + 2.757x_3^2 - 0.186x_4 + 10.401x_1x_4 + 9.268x_2x_4 +$
	$3.375x_3x_4 + 1.354x_4^2 - 4.715x_5 + 9.512x_1x_5 - 1.571x_2x_5 - 2.192x_3x_5 +$
	$0.519x_4x_5 - 4.487x_5^2 - 8.093x_6 + 85.283x_1x_6 + 31.445x_2x_6 - 26.264x_3x_6 +$
	$12.042x_4x_6 - 1.066x_5x_6 - 41.384x_6^2)/(-53.534 + 240.401x_1 - 86.672x_1^2 -$
	$23.373x_2 + 20.773x_1x_2 + 12.666x_2^2 - 205.469x_3 + 19.030x_1x_3 + 63.098x_2x_3 -$
	$26.637x_3^2 - 70.916x_4 + 36.543x_1x_4 + 11.892x_2x_4 + 7.662x_3x_4 + 2.763x_4^2 +$
	$5.244x_5 - 12.630x_1x_5 - 15.307x_2x_5 + 16.804x_3x_5 + 1.020x_4x_5 + 5.009x_5^2 +$
	$6.196x_6 - 2.027x_1x_6 - 2.702x_2x_6 + 3.137x_3x_6 - 0.647x_4x_6 - 0.644x_5x_6 - 0.446x_6^2)$

(Continued)

TABLE A7.1 *(Continued)*

Models Obtained Using the Stochastic Neuro Regression Method in Modeling the Natural Frequency Output (Related to Table 7.2)

Nomenclature	Model

TON

$-49.425 - 229.945x_1 + 686.104x_1^2 - 158.657x_1^3 + 690.610x_2 - 171.617x_1x_2 +$
$92.966x_1^2x_2 - 286.486x_2^2 - 58.811x_1x_2^2 + 35.771x_2^3 - 73.724x_3 - 693.136x_1x_3 -$
$170.654x_1^2x_3 - 186.721x_2x_3 + 325.060x_1x_2x_3 + 123.849x_2^2x_3 + 403.798x_3^2 +$
$44.264x_1x_3^2 - 133.523x_2x_3^2 - 32.773x_3^3 - 528.822x_4 + 236.473x_1x_4 + 137.405x_1^2x_4 +$
$122.252x_2x_4 - 84.980x_1x_2x_4 + 60.131x_2^2x_4 - 43.294x_3x_4 - 29.102x_1x_3x_4 -$
$119.111x_2x_3x_4 + 68.052x_3^2x_4 - 143.574x_4^2 + 24.306x_1x_4^2 - 10.887x_2x_4^2 +$
$42.425x_3x_4^2 + 5.973x_4^3 + 177.009x_5 - 317.795x_1x_5 - 16.156x_1^2x_5 - 143.675x_2x_5 +$
$14.105x_1x_2x_5 + 4.261x_2^2x_5 + 444.710x_3x_5 + 63.843x_1x_3x_5 - 39.601x_2x_3x_5 -$
$130.865x_3^2x_5 + 345.047x_4x_5 - 89.403x_1x_4x_5 - 21.054x_2x_4x_5 - 74.345x_3x_4x_5 +$
$21.338x_4^2x_5 - 120.771x_5^2 + 62.657x_1x_5^2 + 24.665x_2x_5^2 + 6.154x_3x_5^2 - 27.640x_4x_5^2 +$
$2.601x_5^3 - 2.700x_6 + 13.166x_1x_6 - 4.549x_1^2x_6 + 5.233x_2x_6 - 5.200x_1x_2x_6 -$
$1.486x_2^2x_6 - 23.540x_3x_6 + 9.737x_1x_3x_6 + 0.397x_2x_3x_6 + 3.407x_3^2x_6 - 1.297x_4x_6 -$
$5.420x_1x_4x_6 + 1.013x_2x_4x_6 + 5.085x_3x_4x_6 - 0.322x_4^2x_6 - 0.449x_5x_6 + 0.057x_1x_5x_6 +$
$1.094x_2x_5x_6 - 1.731x_3x_5x_6 - 0.046x_4x_5x_6 + 0.223x_5^2x_6 + 0.174x_6^2 + 0.015x_1x_6^2 +$
$0.006x_2x_6^2 + 0.003x_3x_6^2 + 0.028x_4x_6^2 - 0.008x_5x_6^2 - 0.002x_6^3$

TONR

$(3.006 + 9.598x_1 + 2.172x_1^2 - 3.465x_1^3 + 0.144x_2 - 3.615x_1x_2 + 5.755x_1^2x_2 +$
$11.349x_2^2 + 8.538x_1x_2^2 + 1.307x_2^3 + 0.193x_3 + 3.439x_1x_3 - 3.969x_1^2x_3 -$
$7.434x_2x_3 + 0.521x_1x_2x_3 - 21.028x_2^2x_3 + 1.977x_3^2 - 4.277x_1x_3^2 + 0.122x_2x_3^2 -$
$0.362x_3^3 - 3.905x_4 + 2.144x_1x_4 - 1.034x_1^2x_4 - 10.171x_2x_4 - 16.398x_1x_2x_4 -$
$1.207x_2^2x_4 + 20.618x_3x_4 + 22.014x_1x_3x_4 - 12.131x_2x_3x_4 + 3.769x_3^2x_4 - 0.752x_4^2 -$
$18.792x_1x_4^2 - 0.033x_2x_4^2 - 2.072x_3x_4^2 - 3.042x_4^3 - 1.368x_5 - 0.505x_1x_5 -$
$13.483x_1^2x_5 - 3.397x_2x_5 + 15.115x_1x_2x_5 - 10.141x_2^2x_5 - 0.591x_3x_5 + 18.309x_1x_3x_5 -$
$7.051x_2x_3x_5 + 4.642x_3^2x_5 - 4.127x_4x_5 - 0.545x_1x_4x_5 - 4.527x_2x_4x_5 - 16.738x_3x_4x_5 -$
$23.662x_4^2x_5 + 0.474x_5^2 - 7.484x_1x_5^2 - 10.138x_2x_5^2 + 1.749x_3x_5^2 + 1.650x_4x_5^2 -$
$1.121x_5^3 - 2.621x_6 + 17.011x_1x_6 - 2.582x_1^2x_6 - 3.377x_2x_6 - 3.035x_1x_2x_6 +$
$7.416x_2^2x_6 + 26.072x_3x_6 + 13.734x_1x_3x_6 - 10.135x_2x_3x_6 - 1.344x_3^2x_6 - 11.164x_4x_6 +$
$0.938x_1x_4x_6 - 0.712x_2x_4x_6 + 2.161x_3x_4x_6 + 0.547x_4^2x_6 - 2.398x_5x_6 + 17.246x_1x_5x_6 +$
$9.438x_2x_5x_6 - 5.042x_3x_5x_6 - 5.036x_4x_5x_6 - 0.904x_5^2x_6 - 8.334x_6^2 + 25.045x_1x_6^2 -$
$0.392x_2x_6^2 - 9.223x_3x_6^2 + 3.633x_4x_6^2 - 31.382x_5x_6^2 - 8.335x_6^3)/(2.288 - 4.250x_1 +$
$6.287x_1^2 - 0.748x_1^3 - 1.457x_2 - 26.529x_1x_2 - 8.038x_1^2x_2 + 5.679x_2^2 - 6.083x_1x_2^2 +$
$1.436x_2^3 - 5.930x_3 - 3.308x_1x_3 + 2.243x_1^2x_3 - 10.301x_2x_3 + 10.553x_1x_2x_3 -$
$1.808x_2^2x_3 - 3.314x_3^2 + 3.097x_1x_3^2 - 9.197x_2x_3^2 + 1.471x_3^3 + 3.796x_4 - 28.279x_1x_4 -$
$3.323x_1^2x_4 - 17.353x_2x_4 + 24.712x_1x_2x_4 - 0.363x_2^2x_4 - 1.847x_3x_4 + 0.386x_1x_3x_4 -$
$21.694x_2x_3x_4 - 5.855x_3^2x_4 - 2.005x_4^2 - 4.880x_1x_4^2 - 5.481x_2x_4^2 - 9.045x_3x_4^2 +$

(Continued)

TABLE A7.1 *(Continued)*

Models Obtained Using the Stochastic Neuro Regression Method in Modeling the Natural Frequency Output (Related to Table 7.2)

Nomenclature	Model
	$1.717x_4^3 + 7.333x_5 - 5.778x_1x_5 - 10.669x_1^2x_5 - 24.254x_2x_5 + 7.395x_1x_2x_5 -$
	$6.004x_2^2x_5 - 9.112x_3x_5 + 6.988x_1x_3x_5 - 9.229x_2x_3x_5 + 20.359x_3^2x_5 - 14.069x_4x_5 -$
	$5.638x_1x_4x_5 - 83.887x_2x_4x_5 - 25.506x_3x_4x_5 - 15.798x_4^2x_5 + 0.746x_5^2 + 15.116x_1x_5^2 +$
	$3.124x_2x_5^2 + 42.864x_3x_5^2 + 10.338x_4x_5^2 + 3.943x_5^3 - 68.695x_6 + 80.839x_1x_6 -$
	$24.145x_1^2x_6 + 12.299x_2x_6 + 5.025x_1x_2x_6 + 2.264x_2^2x_6 - 55.317x_3x_6 + 2.962x_1x_3x_6 +$
	$18.794x_2x_3x_6 - 6.574x_3^2x_6 + 23.490x_4x_6 + 4.566x_1x_4x_6 + 5.135x_2x_4x_6 +$
	$4.059x_3x_4x_6 - 2.868x_4^2x_6 + 11.390x_5x_6 - 6.681x_1x_5x_6 - 1.825x_2x_5x_6 - 0.563x_3x_5x_6 +$
	$2.416x_4x_5x_6 - 2.518x_5^2x_6 + 1.052x_6^2 - 0.341x_1x_6^2 - 1.123x_2x_6^2 + 0.932x_3x_6^2 -$
	$0.664x_4x_6^2 - 0.295x_5x_6^2 - 0.083x_6^3)$
FOTN	$59.382 + 10.340\mathrm{Cos}[x_1] + 3.449\mathrm{Cos}[x_2] - 1.432\mathrm{Cos}[x_3] + 2.481\mathrm{Cos}[x_4] +$
	$1.358\mathrm{Cos}[x_5] + 0.039\mathrm{Cos}[x_6] + 21.104\mathrm{Sin}[x_1] - 5.589\mathrm{Sin}[x_2] + 13.916\mathrm{Sin}[x_3] +$
	$1.781\mathrm{Sin}[x_4] + 0.809\mathrm{Sin}[x_5] + 0.082\mathrm{Sin}[x_6]$
FOTNR	$(15.128 + 18.151\mathrm{Cos}[x_1] + 40.789\mathrm{Cos}[x_2] - 70.928\mathrm{Cos}[x_3] - 1.542\mathrm{Cos}[x_4] -$
	$13.829\mathrm{Cos}[x_5] + 9.247\mathrm{Cos}[x_6] + 58.501\mathrm{Sin}[x_1] - 38.502\mathrm{Sin}[x_2] + 40.948\mathrm{Sin}[x_3] -$
	$7.431\mathrm{Sin}[x_4] + 13.212\mathrm{Sin}[x_5] - 2.851\mathrm{Sin}[x_6])/(0.280 + 0.122\mathrm{Cos}[x_1] +$
	$0.427\mathrm{Cos}[x_2] - 0.730\mathrm{Cos}[x_3] - 0.041\mathrm{Cos}[x_4] - 0.161\mathrm{Cos}[x_5] + 0.102\mathrm{Cos}[x_6] +$
	$0.487\mathrm{Sin}[x_1] - 0.359\mathrm{Sin}[x_2] + 0.469\mathrm{Sin}[x_3] - 0.125\mathrm{Sin}[x_4] + 0.140\mathrm{Sin}[x_5] -$
	$0.033\mathrm{Sin}[x_6])$
SOTN	$74.691 - 288.534\mathrm{Cos}[x_1] - 193.450\mathrm{Cos}[x_1]^2 + 10.589\mathrm{Cos}[x_2] + 41.619\mathrm{Cos}[x_2]^2 +$
	$99.522\mathrm{Cos}[x_3] + 36.990\mathrm{Cos}[x_1]\mathrm{Cos}[x_3] + 5.655\mathrm{Cos}[x_2]\mathrm{Cos}[x_3] + 140.268\mathrm{Cos}[x_3]^2 +$
	$20.571\mathrm{Cos}[x_4] + 0.880\mathrm{Cos}[x_2]\mathrm{Cos}[x_4] + 64.206\mathrm{Cos}[x_4]^2 - 63.044\mathrm{Cos}[x_5] +$
	$45.745\mathrm{Cos}[x_5]^2 + 1.621\mathrm{Cos}[x_6] + 36.057\mathrm{Cos}[x_6]^2 - 979.897\mathrm{Sin}[x_1] +$
	$263.152\mathrm{Cos}[x_1]\mathrm{Sin}[x_1] - 7.613\mathrm{Cos}[x_2]\mathrm{Sin}[x_1] + 92.294\mathrm{Cos}[x_3]\mathrm{Sin}[x_1] -$
	$10.724\mathrm{Cos}[x_4]\mathrm{Sin}[x_1] + 45.966\mathrm{Cos}[x_5]\mathrm{Sin}[x_1] + 3.400\mathrm{Cos}[x_6]\mathrm{Sin}[x_1] +$
	$280.355\mathrm{Sin}[x_1]^2 + 41.110\mathrm{Sin}[x_2]^2 + 565.733\mathrm{Sin}[x_3] + 25.349\mathrm{Cos}[x_1]\mathrm{Sin}[x_3] -$
	$2.281\mathrm{Cos}[x_2]\mathrm{Sin}[x_3] - 189.384\mathrm{Cos}[x_3]\mathrm{Sin}[x_3] - 8.311\mathrm{Cos}[x_4]\mathrm{Sin}[x_3] +$
	$27.491\mathrm{Cos}[x_5]\mathrm{Sin}[x_3] - 5.176\mathrm{Cos}[x_6]\mathrm{Sin}[x_3] - 77.457\mathrm{Sin}[x_3]^2 + 102.473\mathrm{Sin}[x_4] +$
	$11.227\mathrm{Sin}[x_4]^2 + 1.129\mathrm{Sin}[x_5] + 42.551\mathrm{Sin}[x_5]^2 + 33.507\mathrm{Sin}[x_6] +$
	$0.935\mathrm{Cos}[x_1]\mathrm{Sin}[x_6] - 0.748\mathrm{Cos}[x_2]\mathrm{Sin}[x_6] - 6.071\mathrm{Cos}[x_3]\mathrm{Sin}[x_6] -$
	$1.293\mathrm{Cos}[x_4]\mathrm{Sin}[x_6] - 1.395\mathrm{Cos}[x_5]\mathrm{Sin}[x_6] - 0.242\mathrm{Cos}[x_6]\mathrm{Sin}[x_6] -$
	$0.577\mathrm{Sin}[x_1]\mathrm{Sin}[x_6] - 12.755\mathrm{Sin}[x_2]\mathrm{Sin}[x_6] - 17.112\mathrm{Sin}[x_3]\mathrm{Sin}[x_6] -$
	$3.741\mathrm{Sin}[x_4]\mathrm{Sin}[x_6] + 0.230\mathrm{Sin}[x_5]\mathrm{Sin}[x_6] + 35.863\mathrm{Sin}[x_6]^2$
FOLN	$100.408 - 5.10259\mathrm{Log}[x_1] - 5.01257\mathrm{Log}[x_2] + 7.27561\mathrm{Log}[x_3] - 2.88494\mathrm{Log}[x_4] -$
	$3.42764\mathrm{Log}[x_5] - 0.907352\mathrm{Log}[x_6]$
FOLNR	$-457.926 + 1129.160\mathrm{Log}[x_1] + 250.311\mathrm{Log}[x_2] - 567.336\mathrm{Log}[x_3] + 55.789\mathrm{Log}[x_4] +$
	$135.374\mathrm{Log}[x_5] - 210.170\mathrm{Log}[x_6]$

(Continued)

TABLE A7.1 *(Continued)*

Models Obtained Using the Stochastic Neuro Regression Method in Modeling the Natural Frequency Output (Related to Table 7.2)

Nomenclature	Model
SOLN	$113.213 + 49.015\text{Log}[x_1] - 22.758\text{Log}[x_1]^2 + 69.338\text{Log}[x_2] + 0.803\text{Log}[x_1]\text{Log}[x_2] - 1.425\text{Log}[x_2]^2 - 13.125\text{Log}[x_3] - 0.813\text{Log}[x_1]\text{Log}[x_3] + 9.933\text{Log}[x_2]\text{Log}[x_3] - 5.496\text{Log}[x_3]^2 + 3.492\text{Log}[x_4] + 5.613\text{Log}[x_1]\text{Log}[x_4] + 1.908\text{Log}[x_2]\text{Log}[x_4] + 2.128\text{Log}[x_3]\text{Log}[x_4] - 4.734\text{Log}[x_4]^2 + 25.590\text{Log}[x_5] - 11.964\text{Log}[x_1]\text{Log}[x_5] - 14.567\text{Log}[x_2]\text{Log}[x_5] + 3.455\text{Log}[x_3]\text{Log}[x_5] - 4.120\text{Log}[x_4]\text{Log}[x_5] - 0.427\text{Log}[x_5]^2 - 28.243\text{Log}[x_6] - 8.844\text{Log}[x_1]\text{Log}[x_6] - 16.714\text{Log}[x_2]\text{Log}[x_6] + 3.934\text{Log}[x_3]\text{Log}[x_6] - 0.437\text{Log}[x_4]\text{Log}[x_6] - 4.805\text{Log}[x_5]\text{Log}[x_6] + 5.640\text{Log}[x_6]^2$

TABLE A7.2

Models Obtained Using Neuro Regression and Stochastic Neuro Regression for Two Different Data Separation Methods (Related to Table 7.3)

Data Separation Method	Modeling Method	Model
Training–Testing	NRM	$187.200 - 355.232\text{Cos}[x_1] - 418.079\text{Cos}[x_1]^2 - 43.855\text{Cos}[x_2] + 198.075\text{Cos}[x_2]^2 - 103.864\text{Cos}[x_3] + 46.847\text{Cos}[x_1]\text{Cos}[x_3] + 17.074\text{Cos}[x_2]\text{Cos}[x_3] + 100.964\text{Cos}[x_3]^2 + 12.849\text{Cos}[x_4] + 0.143\text{Cos}[x_2]\text{Cos}[x_4] + 234.900\text{Cos}[x_4]^2 - 78.045\text{Cos}[x_5] + 223.558\text{Cos}[x_5]^2 + 4.916\text{Cos}[x_6] + 250.406\text{Cos}[x_6]^2 - 1385.570\text{Sin}[x_1] + 364.990\text{Cos}[x_1]\text{Sin}[x_1] + 2.413\text{Cos}[x_2]\text{Sin}[x_1] + 165.522\text{Cos}[x_3]\text{Sin}[x_1] - 9.238\text{Cos}[x_4]\text{Sin}[x_1] + 41.584\text{Cos}[x_5]\text{Sin}[x_1] + 2.780\text{Cos}[x_6]\text{Sin}[x_1] + 235.064\text{Sin}[x_1]^2 + 195.630\text{Sin}[x_2]^2 - 113.911\text{Sin}[x_3] - 12.651\text{Cos}[x_1]\text{Sin}[x_3] + 42.406\text{Cos}[x_2]\text{Sin}[x_3] - 50.038\text{Cos}[x_3]\text{Sin}[x_3] - 0.946\text{Cos}[x_4]\text{Sin}[x_3] + 47.408\text{Cos}[x_5]\text{Sin}[x_3] - 8.057\text{Cos}[x_6]\text{Sin}[x_3] + 237.883\text{Sin}[x_3]^2 + 65.832\text{Sin}[x_4] + 203.085\text{Sin}[x_4]^2 + 1.215\text{Sin}[x_5] + 220.636\text{Sin}[x_5]^2 + 27.716\text{Sin}[x_6] - 1.746\text{Cos}[x_1]\text{Sin}[x_6] + 0.740\text{Cos}[x_2]\text{Sin}[x_6] - 8.374\text{Cos}[x_3]\text{Sin}[x_6] - 0.293\text{Cos}[x_4]\text{Sin}[x_6] - 1.713\text{Cos}[x_5]\text{Sin}[x_6] - 0.311\text{Cos}[x_6]\text{Sin}[x_6] - 8.100\text{Sin}[x_1]\text{Sin}[x_6] - 5.249\text{Sin}[x_2]\text{Sin}[x_6] - 23.326\text{Sin}[x_3]\text{Sin}[x_6] + 8.941\text{Sin}[x_4]\text{Sin}[x_6] + 0.475\text{Sin}[x_5]\text{Sin}[x_6] + 249.910\text{Sin}[x_6]^2$
	SNRM	$187.200 - 355.232\text{Cos}[x_1] - 418.079\text{Cos}[x_1]^2 - 43.855\text{Cos}[x_2] + 198.075\text{Cos}[x_2]^2 - 103.864\text{Cos}[x_3] + 46.847\text{Cos}[x_1]\text{Cos}[x_3] + 17.074\text{Cos}[x_2]\text{Cos}[x_3] + 100.964\text{Cos}[x_3]^2 + 12.849\text{Cos}[x_4] + 0.143\text{Cos}[x_2]\text{Cos}[x_4] + 234.900\text{Cos}[x_4]^2 - 78.045\text{Cos}[x_5] + 223.558\text{Cos}[x_5]^2 + 4.916\text{Cos}[x_6] + 250.406\text{Cos}[x_6]^2 - 1385.570\text{Sin}[x_1] + 364.990\text{Cos}[x_1]\text{Sin}[x_1] + 2.413\text{Cos}[x_2]\text{Sin}[x_1] +$

(Continued)

TABLE A7.2 *(Continued)*

Models Obtained Using Neuro Regression and Stochastic Neuro Regression for Two Different Data Separation Methods (Related to Table 7.3)

Data Separation Method	Modeling Method	Model
		$165.522\mathrm{Cos}[x_3]\mathrm{Sin}[x_1] - 9.238\mathrm{Cos}[x_4]\mathrm{Sin}[x_1] + 41.584\mathrm{Cos}[x_5]\mathrm{Sin}[x_1] +$ $2.780\mathrm{Cos}[x_6]\mathrm{Sin}[x_1] + 235.064\mathrm{Sin}[x_1]^2 + 195.630\mathrm{Sin}[x_2]^2 -$ $113.911\mathrm{Sin}[x_3] - 12.651\mathrm{Cos}[x_1]\mathrm{Sin}[x_3] + 42.406\mathrm{Cos}[x_2]\mathrm{Sin}[x_3] -$ $50.038\mathrm{Cos}[x_3]\mathrm{Sin}[x_3] - 0.946\mathrm{Cos}[x_4]\mathrm{Sin}[x_3] + 47.408\mathrm{Cos}[x_5]\mathrm{Sin}[x_3] -$ $8.057\mathrm{Cos}[x_6]\mathrm{Sin}[x_3] + 237.883\mathrm{Sin}[x_3]^2 + 65.832\mathrm{Sin}[x_4] +$ $203.085\mathrm{Sin}[x_4]^2 + 1.215\mathrm{Sin}[x_5] + 220.636\mathrm{Sin}[x_5]^2 + 27.716\mathrm{Sin}[x_6] -$ $1.746\mathrm{Cos}[x_1]\mathrm{Sin}[x_6] + 0.740\mathrm{Cos}[x_2]\mathrm{Sin}[x_6] - 8.374\mathrm{Cos}[x_3]\mathrm{Sin}[x_6] -$ $0.293\mathrm{Cos}[x_4]\mathrm{Sin}[x_6] - 1.713\mathrm{Cos}[x_5]\mathrm{Sin}[x_6] - 0.311\mathrm{Cos}[x_6]\mathrm{Sin}[x_6] -$ $8.100\mathrm{Sin}[x_1]\mathrm{Sin}[x_6] - 5.249\mathrm{Sin}[x_2]\mathrm{Sin}[x_6] - 23.326\mathrm{Sin}[x_3]\mathrm{Sin}[x_6] +$ $8.941\mathrm{Sin}[x_4]\mathrm{Sin}[x_6] + 0.475\mathrm{Sin}[x_5]\mathrm{Sin}[x_6] + 249.910\mathrm{Sin}[x_6]^2$
Training–Testing–Validation	NRM	$8.690 - 8.633\mathrm{Cos}[x_1] + 1.357\mathrm{Cos}[x_1]^2 - 9.100\mathrm{Cos}[x_2] -$ $9.668\mathrm{Cos}[x_1]\mathrm{Cos}[x_2] - 11.298\mathrm{Cos}[x_2]^2 + 6.384\mathrm{Cos}[x_3] -$ $19.877\mathrm{Cos}[x_1]\mathrm{Cos}[x_3] + 7.568\mathrm{Cos}[x_2]\mathrm{Cos}[x_3] + 46.835\mathrm{Cos}[x_3]^2 +$ $6.727\mathrm{Sin}[x_1] + 3.746\mathrm{Cos}[x_1]\mathrm{Sin}[x_1] + 1.294\mathrm{Cos}[x_2]\mathrm{Sin}[x_1] +$ $20.612\mathrm{Cos}[x_3]\mathrm{Sin}[x_1] + 7.633\mathrm{Sin}[x_1]^2 - 1.697\mathrm{Sin}[x_2] -$ $3.940\mathrm{Cos}[x_1]\mathrm{Sin}[x_2] + 2.003\mathrm{Cos}[x_2]\mathrm{Sin}[x_2] + 7.464\mathrm{Cos}[x_3]\mathrm{Sin}[x_2] -$ $0.955\mathrm{Sin}[x_1]\mathrm{Sin}[x_2] - 5.556\mathrm{Sin}[x_2]^2 + 22.573\mathrm{Sin}[x_3] +$ $0.256\mathrm{Cos}[x_1]\mathrm{Sin}[x_3] + 1.246\mathrm{Cos}[x_2]\mathrm{Sin}[x_3] - 0.863\mathrm{Cos}[x_3]\mathrm{Sin}[x_3] -$ $3.672\mathrm{Sin}[x_1]\mathrm{Sin}[x_3] - 1.195\mathrm{Sin}[x_2]\mathrm{Sin}[x_3] + 1.345\mathrm{Sin}[x_3]^2$
	SNRM	$74.691 - 288.534\mathrm{Cos}[x_1] - 193.450\mathrm{Cos}[x_1]^2 + 10.589\mathrm{Cos}[x_2] +$ $41.619\mathrm{Cos}[x_2]^2 + 99.522\mathrm{Cos}[x_3] + 36.990\mathrm{Cos}[x_1]\mathrm{Cos}[x_3] +$ $5.655\mathrm{Cos}[x_2]\mathrm{Cos}[x_3] + 140.268\mathrm{Cos}[x_3]^2 + 20.571\mathrm{Cos}[x_4] +$ $0.880\mathrm{Cos}[x_2]\mathrm{Cos}[x_4] + 64.206\mathrm{Cos}[x_4]^2 - 63.044\mathrm{Cos}[x_5] +$ $45.745\mathrm{Cos}[x_5]^2 + 1.621\mathrm{Cos}[x_6] + 36.057\mathrm{Cos}[x_6]^2 - 979.897\mathrm{Sin}[x_1] +$ $263.152\mathrm{Cos}[x_1]\mathrm{Sin}[x_1] - 7.613\mathrm{Cos}[x_2]\mathrm{Sin}[x_1] + 92.294\mathrm{Cos}[x_3]\mathrm{Sin}[x_1] -$ $10.724\mathrm{Cos}[x_4]\mathrm{Sin}[x_1] + 45.966\mathrm{Cos}[x_5]\mathrm{Sin}[x_1] + 3.400\mathrm{Cos}[x_6]\mathrm{Sin}[x_1] +$ $280.355\mathrm{Sin}[x_1]^2 + 41.110\mathrm{Sin}[x_2]^2 + 565.733\mathrm{Sin}[x_3] +$ $25.349\mathrm{Cos}[x_1]\mathrm{Sin}[x_3] - 2.281\mathrm{Cos}[x_2]\mathrm{Sin}[x_3] - 189.384\mathrm{Cos}[x_3]\mathrm{Sin}[x_3] -$ $8.311\mathrm{Cos}[x_4]\mathrm{Sin}[x_3] + 27.491\mathrm{Cos}[x_5]\mathrm{Sin}[x_3] - 5.176\mathrm{Cos}[x_6]\mathrm{Sin}[x_3] -$ $77.457\mathrm{Sin}[x_3]^2 + 102.473\mathrm{Sin}[x_4] + 11.227\mathrm{Sin}[x_4]^2 + 1.129\mathrm{Sin}[x_5] +$ $42.551\mathrm{Sin}[x_5]^2 + 33.507\mathrm{Sin}[x_6] + 0.935\mathrm{Cos}[x_1]\mathrm{Sin}[x_6] -$ $0.748\mathrm{Cos}[x_2]\mathrm{Sin}[x_6] - 6.071\mathrm{Cos}[x_3]\mathrm{Sin}[x_6] - 1.293\mathrm{Cos}[x_4]\mathrm{Sin}[x_6] -$ $1.395\mathrm{Cos}[x_5]\mathrm{Sin}[x_6] - 0.242\mathrm{Cos}[x_6]\mathrm{Sin}[x_6] - 0.577\mathrm{Sin}[x_1]\mathrm{Sin}[x_6] -$ $12.755\mathrm{Sin}[x_2]\mathrm{Sin}[x_6] - 17.112\mathrm{Sin}[x_3]\mathrm{Sin}[x_6] - 3.741\mathrm{Sin}[x_4]\mathrm{Sin}[x_6] +$ $0.230\mathrm{Sin}[x_5]\mathrm{Sin}[x_6] + 35.863\mathrm{Sin}[x_6]^2$

TABLE A7.3

Models Obtained by NRM and SNRM Using the Bootstrap Data Separation Method (80 Training–15 Testing–5 Validation) (*Models Obtained Using the NRM and SNRM Methods*)(*(Related to Table 7.4)*)

(*Models proposed by NRM*)

(*1*)

Out[•]//NumberForm=

$259.596 - 520.513$ Cos $[x_1] - 513.732$ Cos $[x_1]^2 - 45.462$ Cos $[x_2] + 275.590$ Cos $[x_2]^2 -$
28.827 Cos $[x_3] + 9.342$ Cos $[x_1]$ Cos $[x_3] + 10.782$ Cos $[x_2]$ Cos $[x_3] + 101.404$ Cos $[x_3]^2 +$
28.446 Cos $[x_4] + 5.112$ Cos $[x_2]$ Cos $[x_4] + 323.565$ Cos $[x_4]^2 - 64.141$ Cos $[x_5] + 314.768$ Cos $[x_5]^2 +$
5.603 Cos $[x_6] + 356.394$ Cos $[x_6]^2 - 1831.991$ Sin $[x_1] + 551.649$ Cos $[x_1]$ Sin $[x_1] +$
9.048 Cos $[x_2]$ Sin $[x_1] + 71.476$ Cos $[x_3]$ Sin $[x_1] - 10.046$ Cos $[x_4]$ Sin $[x_1] + 31.878$ Cos $[x_5]$ Sin $[x_1] +$
2.881 Cos $[x_6]$ Sin $[x_1] + 316.297$ Sin $[x_1]^2 + 271.274$ Sin $[x_2]^2 - 280.423$ Sin $[x_3] -$
36.804 Cos $[x_1]$ Sin $[x_3] + 38.850$ Cos $[x_2]$ Sin $[x_3] - 20.924$ Cos $[x_3]$ Sin $[x_3] - 17.304$ Cos $[x_4]$ Sin $[x_3] +$
50.983 Cos $[x_5]$ Sin $[x_3] - 9.030$ Cos $[x_6]$ Sin $[x_3] + 332.424$ Sin $[x_3]^2 + 87.398$ Sin $[x_4] +$
280.356 Sin $[x_4]^2 + 1.055$ Sin $[x_5] + 307.804$ Sin $[x_5]^2 + 19.943$ Sin $[x_6] + 2.153$ Cos $[x_1]$ Sin $[x_6] +$
1.280 Cos $[x_2]$ Sin $[x_6] - 10.850$ Cos $[x_3]$ Sin $[x_6] - 0.288$ Cos $[x_4]$ Sin $[x_6] + 0.725$ Cos $[x_5]$ Sin $[x_6] -$
0.446 Cos $[x_6]$ Sin $[x_6] + 2.112$ Sin $[x_1]$ Sin $[x_6] - 4.589$ Sin $[x_2]$ Sin $[x_6] - 30.497$ Sin $[x_3]$ Sin $[x_6] +$
15.235 Sin $[x_4]$ Sin $[x_6] + 0.315$ Sin $[x_5]$ Sin $[x_6] + 355.622$ Sin $[x_6]^2$

(*2*)

Out[•]//NumberForm=

$453.081 - 467.818$ Cos $[x_1] - 300.362$ Cos $[x_1]^2 - 30.149$ Cos $[x_2] + 476.387$ Cos $[x_2]^2 -$
716.987 Cos $[x_3] + 47.881$ Cos $[x_1]$ Cos $[x_3] + 10.861$ Cos $[x_2]$ Cos $[x_3] - 349.444$ Cos $[x_3]^2 +$
21.296 Cos $[x_4] + 4.053$ Cos $[x_2]$ Cos $[x_4] + 585.359$ Cos $[x_4]^2 - 89.783$ Cos $[x_5] + 550.306$ Cos $[x_5]^2 +$
6.226 Cos $[x_6] + 603.603$ Cos $[x_6]^2 - 1822.227$ Sin $[x_1] + 484.832$ Cos $[x_1]$ Sin $[x_1] -$
3.656 Cos $[x_2]$ Sin $[x_1] + 163.930$ Cos $[x_3]$ Sin $[x_1] - 1.614$ Cos $[x_4]$ Sin $[x_1] + 53.592$ Cos $[x_5]$ Sin $[x_1] +$
3.181 Cos $[x_6]$ Sin $[x_1] + 553.888$ Sin $[x_1]^2 + 474.540$ Sin $[x_2]^2 - 1967.542$ Sin $[x_3] -$
24.258 Cos $[x_1]$ Sin $[x_3] + 35.632$ Cos $[x_2]$ Sin $[x_3] + 542.958$ Cos $[x_3]$ Sin $[x_3] - 18.060$ Cos $[x_4]$ Sin $[x_3] +$
54.455 Cos $[x_5]$ Sin $[x_3] - 9.858$ Cos $[x_6]$ Sin $[x_3] + 589.291$ Sin $[x_3]^2 + 181.726$ Sin $[x_4] +$
491.183 Sin $[x_4]^2 + 1.247$ Sin $[x_5] + 544.076$ Sin $[x_5]^2 + 35.534$ Sin $[x_6] - 2.588$ Cos $[x_1]$ Sin $[x_6] +$
1.148 Cos $[x_2]$ Sin $[x_6] - 8.280$ Cos $[x_3]$ Sin $[x_6] - 1.259$ Cos $[x_4]$ Sin $[x_6] - 0.588$ Cos $[x_5]$ Sin $[x_6] -$
0.393 Cos $[x_6]$ Sin $[x_6] - 8.674$ Sin $[x_1]$ Sin $[x_6] - 1.988$ Sin $[x_2]$ Sin $[x_6] - 31.079$ Sin $[x_3]$ Sin $[x_6] +$
7.155 Sin $[x_4]$ Sin $[x_6] + 0.243$ Sin $[x_5]$ Sin $[x_6] + 602.867$ Sin $[x_6]^2$

(*3*)

Out[•]//NumberForm=

$32.863 - 156.683$ Cos $[x_1] - 349.704$ Cos $[x_1]^2 + 20.271$ Cos $[x_2] + 34.256$ Cos $[x_2]^2 + 15.459$ Cos $[x_3] +$
77.721 Cos $[x_1]$ Cos $[x_3] + 4.833$ Cos $[x_2]$ Cos $[x_3] + 131.979$ Cos $[x_3]^2 - 45.961$ Cos $[x_4] +$
0.041 Cos $[x_2]$ Cos $[x_4] + 161.079$ Cos $[x_4]^2 + 12.352$ Cos $[x_5] + 39.598$ Cos $[x_5]^2 + 3.420$ Cos $[x_6] +$
45.561 Cos $[x_6]^2 - 734.165$ Sin $[x_1] + 76.260$ Cos $[x_1]$ Sin $[x_1] + 17.245$ Cos $[x_2]$ Sin $[x_1] +$
168.403 Cos $[x_3]$ Sin $[x_1] + 22.195$ Cos $[x_4]$ Sin $[x_1] + 12.628$ Cos $[x_5]$ Sin $[x_1] + 1.959$ Cos $[x_6]$ Sin $[x_1] +$
44.584 Sin $[x_1]^2 + 34.505$ Sin $[x_2]^2 + 304.535$ Sin $[x_3] + 91.527$ Cos $[x_1]$ Sin $[x_3] - 36.119$ Cos $[x_2]$ Sin $[x_3] -$
175.684 Cos $[x_3]$ Sin $[x_3] + 29.981$ Cos $[x_4]$ Sin $[x_3] - 19.180$ Cos $[x_5]$ Sin $[x_3] - 5.637$ Cos $[x_6]$ Sin $[x_3] +$
38.751 Sin $[x_3]^2 + 252.259$ Sin $[x_4] + 35.005$ Sin $[x_4]^2 + 1.251$ Sin $[x_5] + 38.035$ Sin $[x_5]^2 + 71.331$ Sin $[x_6] -$
10.147 Cos $[x_1]$ Sin $[x_6] - 0.242$ Cos $[x_2]$ Sin $[x_6] - 15.141$ Cos $[x_3]$ Sin $[x_6] + 0.187$ Cos $[x_4]$ Sin $[x_6] -$
2.878 Cos $[x_5]$ Sin $[x_6] - 0.344$ Cos $[x_6]$ Sin $[x_6] - 35.984$ Sin $[x_1]$ Sin $[x_6] - 9.774$ Sin $[x_2]$ Sin $[x_6] -$
38.737 Sin $[x_3]$ Sin $[x_6] + 14.541$ Sin $[x_4]$ Sin $[x_6] + 0.276$ Sin $[x_5]$ Sin $[x_6] + 44.995$ Sin $[x_6]^2$

(Continued)

TABLE A7.3 *(Continued)*

Models Obtained by NRM and SNRM Using the Bootstrap Data Separation Method (80 Training–15 Testing–5 Validation) (*Models Obtained Using the NRM and SNRM Methods*)(*(Related to Table 7.4)*)

(*4*)

Out[•]//NumberForm=

$-179.109 + 126.940$ Cos $[x_1] - 155.675$ Cos $[x_1]^2 - 92.636$ Cos $[x_2] - 179.898$ Cos $[x_2]^2 +$
377.115 Cos $[x_3] + 75.539$ Cos $[x_1]$ Cos $[x_3] + 25.220$ Cos $[x_2]$ Cos $[x_3] + 283.157$ Cos $[x_3]^2 +$
9.266 Cos $[x_4] + 5.440$ Cos $[x_2]$ Cos $[x_4] - 185.787$ Cos $[x_4]^2 + 69.563$ Cos $[x_5] - 215.466$ Cos $[x_5]^2 +$
2.950 Cos $[x_6] - 239.442$ Cos $[x_6]^2 + 140.511$ Sin $[x_1] - 102.861$ Cos $[x_1]$ Sin $[x_1] +$
24.846 Cos $[x_2]$ Sin $[x_1] + 284.666$ Cos $[x_3]$ Sin $[x_1] - 3.725$ Cos $[x_4]$ Sin $[x_1] - 10.241$ Cos $[x_5]$ Sin $[x_1] +$
3.348 Cos $[x_6]$ Sin $[x_1] - 213.738$ Sin $[x_1]^2 - 186.708$ Sin $[x_2]^2 + 1430.232$ Sin $[x_3] -$
19.627 Cos $[x_1]$ Sin $[x_3] + 66.682$ Cos $[x_2]$ Sin $[x_3] - 624.415$ Cos $[x_3]$ Sin $[x_3] - 3.007$ Cos $[x_4]$ Sin $[x_3] -$
18.497 Cos $[x_5]$ Sin $[x_3] - 6.563$ Cos $[x_6]$ Sin $[x_3] - 236.401$ Sin $[x_3]^2 + 22.092$ Sin $[x_4] -$
194.308 Sin $[x_4]^2 + 1.764$ Sin $[x_5] - 239.541$ Sin $[x_5]^2 + 66.235$ Sin $[x_6] - 6.445$ Cos $[x_1]$ Sin $[x_6] -$
1.554 Cos $[x_2]$ Sin $[x_6] - 10.701$ Cos $[x_3]$ Sin $[x_6] - 0.004$ Cos $[x_4]$ Sin $[x_6] + 2.303$ Cos $[x_5]$ Sin $[x_6] -$
0.460 Cos $[x_6]$ Sin $[x_6] - 25.942$ Sin $[x_1]$ Sin $[x_6] - 17.245$ Sin $[x_2]$ Sin $[x_6] - 28.676$ Sin $[x_3]$ Sin $[x_6] +$
10.766 Sin $[x_4]$ Sin $[x_6] - 0.562$ Sin $[x_5]$ Sin $[x_6] - 240.158$ Sin $[x_6]^2$

(*5*)

Out[•]//NumberForm=

$132.796 - 478.908$ Cos $[x_1] - 913.668$ Cos $[x_1]^2 - 115.554$ Cos $[x_2] + 139.894$ Cos $[x_2]^2 +$
339.238 Cos $[x_3] + 20.699$ Cos $[x_1]$ Cos $[x_3] + 37.535$ Cos $[x_2]$ Cos $[x_3] + 616.114$ Cos $[x_3]^2 +$
29.870 Cos $[x_4] - 1.100$ Cos $[x_2]$ Cos $[x_4] + 185.162$ Cos $[x_4]^2 - 59.641$ Cos $[x_5] + 160.166$ Cos $[x_5]^2 +$
0.554 Cos $[x_6] + 176.099$ Cos $[x_6]^2 - 2254.889$ Sin $[x_1] + 526.588$ Cos $[x_1]$ Sin $[x_1] +$
19.500 Cos $[x_2]$ Sin $[x_1] + 129.128$ Cos $[x_3]$ Sin $[x_1] - 20.815$ Cos $[x_4]$ Sin $[x_1] + 59.096$ Cos $[x_5]$ Sin $[x_1] +$
6.689 Cos $[x_6]$ Sin $[x_1] + 171.791$ Sin $[x_1]^2 + 138.754$ Sin $[x_2]^2 + 1205.944$ Sin $[x_3] -$
44.713 Cos $[x_1]$ Sin $[x_3] + 94.122$ Cos $[x_2]$ Sin $[x_3] - 447.013$ Cos $[x_3]$ Sin $[x_3] - 8.106$ Cos $[x_4]$ Sin $[x_3] +$
21.672 Cos $[x_5]$ Sin $[x_3] - 7.386$ Cos $[x_6]$ Sin $[x_3] + 154.445$ Sin $[x_3]^2 + 80.153$ Sin $[x_4] +$
143.537 Sin $[x_4]^2 + 1.164$ Sin $[x_5] + 150.543$ Sin $[x_5]^2 + 4.246$ Sin $[x_6] + 3.301$ Cos $[x_1]$ Sin $[x_6] +$
0.137 Cos $[x_2]$ Sin $[x_6] - 4.056$ Cos $[x_3]$ Sin $[x_6] - 0.249$ Cos $[x_4]$ Sin $[x_6] + 1.172$ Cos $[x_5]$ Sin $[x_6] -$
0.328 Cos $[x_6]$ Sin $[x_6] + 8.868$ Sin $[x_1]$ Sin $[x_6] - 4.371$ Sin $[x_2]$ Sin $[x_6] - 13.104$ Sin $[x_3]$ Sin $[x_6] +$
5.795 Sin $[x_4]$ Sin $[x_6] + 0.080$ Sin $[x_5]$ Sin $[x_6] + 175.835$ Sin $[x_6]^2$

(*6*)

Out[•]//NumberForm=

$187.014 - 370.002$ Cos $[x_1] - 368.828$ Cos $[x_1]^2 - 31.334$ Cos $[x_2] + 199.727$ Cos $[x_2]^2 - 142.242$ Cos $[x_3] +$
74.025 Cos $[x_1]$ Cos $[x_3] + 9.668$ Cos $[x_2]$ Cos $[x_3] + 80.932$ Cos $[x_3]^2 + 6.923$ Cos $[x_4] -$
0.671 Cos $[x_2]$ Cos $[x_4] + 205.806$ Cos $[x_4]^2 - 69.750$ Cos $[x_5] + 224.385$ Cos $[x_5]^2 + 5.194$ Cos $[x_6] +$
251.784 Cos $[x_6]^2 - 1277.350$ Sin $[x_1] + 328.575$ Cos $[x_1]$ Sin $[x_1] + 9.755$ Cos $[x_2]$ Sin $[x_1] +$
196.329 Cos $[x_3]$ Sin $[x_1] - 2.801$ Cos $[x_4]$ Sin $[x_1] + 36.167$ Cos $[x_5]$ Sin $[x_1] + 3.101$ Cos $[x_6]$ Sin $[x_1] +$
232.408 Sin $[x_1]^2 + 195.792$ Sin $[x_2]^2 - 168.112$ Sin $[x_3] + 35.822$ Cos $[x_1]$ Sin $[x_3] + 24.084$ Cos $[x_2]$ Sin $[x_3] -$
47.973 Cos $[x_3]$ Sin $[x_3] - 0.965$ Cos $[x_4]$ Sin $[x_3] + 43.636$ Cos $[x_5]$ Sin $[x_3] - 8.662$ Cos $[x_6]$ Sin $[x_3] +$
238.569 Sin $[x_3]^2 + 11.123$ Sin $[x_4] + 203.414$ Sin $[x_4]^2 + 0.915$ Sin $[x_5] + 222.026$ Sin $[x_5]^2 +$
25.667 Sin $[x_6] - 1.377$ Cos $[x_1]$ Sin $[x_6] + 1.071$ Cos $[x_2]$ Sin $[x_6] - 9.903$ Cos $[x_3]$ Sin $[x_6] +$
0.030 Cos $[x_4]$ Sin $[x_6] - 0.197$ Cos $[x_5]$ Sin $[x_6] - 0.158$ Cos $[x_6]$ Sin $[x_6] - 5.901$ Sin $[x_1]$ Sin $[x_6] -$
4.680 Sin $[x_2]$ Sin $[x_6] - 26.585$ Sin $[x_3]$ Sin $[x_6] + 13.407$ Sin $[x_4]$ Sin $[x_6] + 0.047$ Sin $[x_5]$ Sin $[x_6] +$
251.430 Sin $[x_6]^2$

(Continued)

TABLE A7.3 *(Continued)*

Models Obtained by NRM and SNRM Using the Bootstrap Data Separation Method (80 Training–15 Testing–5 Validation) (*Models Obtained Using the NRM and SNRM Methods*)(*(Related to Table 7.4)*)

(*7*)

Out[•]//NumberForm=

342.535 − 580.676 Cos $[x_1]$ − 586.246 Cos $[x_1]^2$ − 58.582 Cos $[x_2]$ + 359.698 Cos $[x_2]^2$ −
268.466 Cos $[x_3]$ + 58.048 Cos $[x_1]$ Cos $[x_3]$ + 13.128 Cos $[x_2]$ Cos $[x_3]$ + 124.553 Cos $[x_3]^2$ +
27.163 Cos $[x_4]$ + 6.660 Cos $[x_2]$ Cos $[x_4]$ + 395.463 Cos $[x_4]^2$ − 27.114 Cos $[x_5]$ + 414.510 Cos $[x_5]^2$ +
4.828 Cos $[x_6]$ + 459.533 Cos $[x_6]^2$ − 2185.305 Sin $[x_1]$ + 561.540 Cos $[x_1]$ Sin $[x_1]$ +
14.925 Cos $[x_2]$ Sin $[x_1]$ + 186.394 Cos $[x_3]$ Sin $[x_1]$ − 12.439 Cos $[x_4]$ Sin $[x_1]$ +
25.980 Cos $[x_5]$ Sin $[x_1]$ + 2.963 Cos $[x_6]$ Sin $[x_1]$ + 417.437 Sin $[x_1]^2$ + 357.597 Sin $[x_2]^2$ −
538.950 Sin $[x_3]$ + 7.269 Cos $[x_1]$ Sin $[x_3]$ + 45.710 Cos $[x_2]$ Sin $[x_3]$ + 87.061 Cos $[x_3]$ Sin $[x_3]$ −
13.520 Cos $[x_4]$ Sin $[x_3]$ + 38.995 Cos $[x_5]$ Sin $[x_3]$ − 8.166 Cos $[x_6]$ Sin $[x_3]$ + 437.360 Sin $[x_3]^2$ +
53.985 Sin $[x_4]$ + 368.385 Sin $[x_4]^2$ + 1.856 Sin $[x_5]$ + 394.674 Sin $[x_5]^2$ + 19.640 Sin $[x_6]$ −
0.319 Cos $[x_1]$ Sin $[x_6]$ + 1.242 Cos $[x_2]$ Sin $[x_6]$ − 7.101 Cos $[x_3]$ Sin $[x_6]$ − 0.522 Cos $[x_4]$ Sin $[x_6]$ +
3.107 Cos $[x_5]$ Sin $[x_6]$ − 0.396 Cos $[x_6]$ Sin $[x_6]$ − 4.314 Sin $[x_1]$ Sin $[x_6]$ + 0.673 Sin $[x_2]$ Sin $[x_6]$ −
22.397 Sin $[x_3]$ Sin $[x_6]$ + 10.740 Sin $[x_4]$ Sin $[x_6]$ − 0.485 Sin $[x_5]$ Sin $[x_6]$ + 458.897 Sin $[x_6]^2$)

(*8*)

Out[•]//NumberForm=

133.381 − 655.681 Cos $[x_1]$ − 993.521 Cos $[x_1]^2$ + 17.692 Cos $[x_2]$ + 141.542 Cos $[x_2]^2$ + 459.771 Cos $[x_3]$ +
52.534 Cos $[x_1]$ Cos $[x_3]$ + 1.815 Cos $[x_2]$ Cos $[x_3]$ + 773.791 Cos $[x_3]^2$ + 0.845 Cos $[x_4]$ −
4.342 Cos $[x_2]$ Cos $[x_4]$ + 73.877 Cos $[x_4]^2$ − 97.579 Cos $[x_5]$ + 162.591 Cos $[x_5]^2$ + 3.826 Cos $[x_6]$ +
179.780 Cos $[x_6]^2$ − 2469.675 Sin $[x_1]$ + 606.264 Cos $[x_1]$ Sin $[x_1]$ + 5.939 Cos $[x_2]$ Sin $[x_1]$ +
117.107 Cos $[x_3]$ Sin $[x_1]$ − 13.478 Cos $[x_4]$ Sin $[x_1]$ + 34.761 Cos $[x_5]$ Sin $[x_1]$ + 4.145 Cos $[x_6]$ Sin $[x_1]$ +
174.527 Sin $[x_1]^2$ + 139.368 Sin $[x_2]^2$ + 1589.722 Sin $[x_3]$ + 45.373 Cos $[x_1]$ Sin $[x_3]$ −
20.987 Cos $[x_2]$ Sin $[x_3]$ − 555.995 Cos $[x_3]$ Sin $[x_3]$ + 16.538 Cos $[x_4]$ Sin $[x_3]$ + 51.891 Cos $[x_5]$ Sin $[x_3]$ −
8.166 Cos $[x_6]$ Sin $[x_3]$ + 150.842 Sin $[x_3]^2$ − 128.602 Sin $[x_4]$ + 145.149 Sin $[x_4]^2$ + 1.171 Sin $[x_5]$ +
172.263 Sin $[x_5]^2$ − 18.751 Sin $[x_6]$ + 0.873 Cos $[x_1]$ Sin $[x_6]$ + 0.474 Cos $[x_2]$ Sin $[x_6]$ −
2.237 Cos $[x_3]$ Sin $[x_6]$ + 0.754 Cos $[x_4]$ Sin $[x_6]$ − 3.822 Cos $[x_5]$ Sin $[x_6]$ − 0.435 Cos $[x_6]$ Sin $[x_6]$ +
0.219 Sin $[x_1]$ Sin $[x_6]$ − 2.672 Sin $[x_2]$ Sin $[x_6]$ − 1.851 Sin $[x_3]$ Sin $[x_6]$ + 20.056 Sin $[x_4]$ Sin $[x_6]$ +
0.689 Sin $[x_5]$ Sin $[x_6]$ + 179.698 Sin $[x_6]^2$

(*9*)

Out[•]//NumberForm=

214.219 − 496.828 Cos $[x_1]$ − 607.884 Cos $[x_1]^2$ + 14.636 Cos $[x_2]$ + 228.395 Cos $[x_2]^2$ −
9.040 Cos $[x_3]$ + 52.724 Cos $[x_1]$ Cos $[x_3]$ − 4.239 Cos $[x_2]$ Cos $[x_3]$ + 186.940 Cos $[x_3]^2$ +
6.958 Cos $[x_4]$ + 1.132 Cos $[x_2]$ Cos $[x_4]$ + 328.551 Cos $[x_4]^2$ − 41.316 Cos $[x_5]$ + 260.935 Cos $[x_5]^2$ +
7.587 Cos $[x_6]$ + 293.053 Cos $[x_6]^2$ − 1878.914 Sin $[x_1]$ + 492.155 Cos $[x_1]$ Sin $[x_1]$ −
5.580 Cos $[x_2]$ Sin $[x_1]$ + 172.293 Cos $[x_3]$ Sin $[x_1]$ + 0.943 Cos $[x_4]$ Sin $[x_1]$ + 53.308 Cos $[x_5]$ Sin $[x_1]$ +
2.120 Cos $[x_6]$ Sin $[x_1]$ + 271.538 Sin $[x_1]^2$ + 224.300 Sin $[x_2]^2$ + 42.723 Sin $[x_3]$ − 3.858 Cos $[x_1]$ Sin $[x_3]$ −
4.606 Cos $[x_2]$ Sin $[x_3]$ − 141.260 Cos $[x_3]$ Sin $[x_3]$ − 5.073 Cos $[x_4]$ Sin $[x_3]$ + 22.156 Cos $[x_5]$ Sin $[x_3]$ −
10.204 Cos $[x_6]$ Sin $[x_3]$ + 269.940 Sin $[x_3]^2$ + 194.188 Sin $[x_4]$ + 229.819 Sin $[x_4]^2$ + 1.316 Sin $[x_5]$ +
244.538 Sin $[x_5]^2$ + 15.128 Sin $[x_6]$ + 2.703 Cos $[x_1]$ Sin $[x_6]$ + 1.713 Cos $[x_2]$ Sin $[x_6]$ −
12.447 Cos $[x_3]$ Sin $[x_6]$ − 0.528 Cos $[x_4]$ Sin $[x_6]$ − 1.748 Cos $[x_5]$ Sin $[x_6]$ − 0.231 Cos $[x_6]$ Sin $[x_6]$ +
9.504 Sin $[x_1]$ Sin $[x_6]$ + 2.307 Sin $[x_2]$ Sin $[x_6]$ − 37.313 Sin $[x_3]$ Sin $[x_6]$ + 10.545 Sin $[x_4]$ Sin $[x_6]$ +
0.546 Sin $[x_5]$ Sin $[x_6]$ + 292.478 Sin $[x_6]^2$

(Continued)

TABLE A7.3 *(Continued)*

Models Obtained by NRM and SNRM Using the Bootstrap Data Separation Method (80 Training–15 Testing–5 Validation) (*Models Obtained Using the NRM and SNRM Methods*)(*(Related to Table 7.4)*)

(*10*)

Out[•]//NumberForm=

$180.615 - 201.960 \cos [x_1] - 347.966 \cos [x_1]^2 - 157.158 \cos [x_2] + 193.946 \cos [x_2]^2 - 161.279 \cos [x_3] + 14.721 \cos [x_1] \cos [x_3] + 61.860 \cos [x_2] \cos [x_3] + 90.116 \cos [x_3]^2 + 12.840 \cos [x_4] + 1.604 \cos [x_2] \cos [x_4] + 216.916 \cos [x_4]^2 - 7.940 \cos [x_5] + 215.514 \cos [x_5]^2 + 3.730 \cos [x_6] + 241.232 \cos [x_6]^2 - 1260.534 \sin [x_1] + 290.684 \cos [x_1] \sin [x_1] + 2.429 \cos [x_2] \sin [x_1] + 172.717 \cos [x_3] \sin [x_1] - 7.105 \cos [x_4] \sin [x_1] - 26.202 \cos [x_5] \sin [x_1] + 1.439 \cos [x_6] \sin [x_1] + 223.960 \sin [x_1]^2 + 188.959 \sin [x_2]^2 - 154.546 \sin [x_3] - 83.916 \cos [x_1] \sin [x_3] + 146.457 \cos [x_2] \sin [x_3] - 3.381 \cos [x_3] \sin [x_3] - 3.520 \cos [x_4] \sin [x_3] + 51.455 \cos [x_5] \sin [x_3] - 5.430 \cos [x_6] \sin [x_3] + 230.355 \sin [x_3]^2 + 44.238 \sin [x_4] + 195.977 \sin [x_4]^2 + 1.098 \sin [x_5] + 206.815 \sin [x_5]^2 - 9.360 \sin [x_6] + 1.223 \cos [x_1] \sin [x_6] + 0.561 \cos [x_2] \sin [x_6] - 0.053 \cos [x_3] \sin [x_6] + 0.204 \cos [x_4] \sin [x_6] - 4.074 \cos [x_5] \sin [x_6] - 0.528 \cos [x_6] \sin [x_6] - 0.012 \sin [x_1] \sin [x_6] - 2.651 \sin [x_2] \sin [x_6] - 3.175 \sin [x_3] \sin [x_6] + 10.801 \sin [x_4] \sin [x_6] + 1.163 \sin [x_5] \sin [x_6] + 240.735 \sin [x_6]^2)$

(*11*)

Out[•]//NumberForm=

$-132.654 + 774.329 \cos [x_1] + 894.854 \cos [x_1]^2 - 153.212 \cos [x_2] - 136.330 \cos [x_2]^2 - 1133.252 \cos [x_3] + 167.660 \cos [x_1] \cos [x_3] + 75.600 \cos [x_2] \cos [x_3] - 830.413 \cos [x_3]^2 + 22.966 \cos [x_4] - 7.039 \cos [x_2] \cos [x_4] + 40.579 \cos [x_4]^2 - 63.638 \cos [x_5] - 160.366 \cos [x_5]^2 + 3.849 \cos [x_6] - 181.849 \cos [x_6]^2 + 2355.862 \sin [x_1] - 804.463 \cos [x_1] \sin [x_1] - 20.808 \cos [x_2] \sin [x_1] + 625.246 \cos [x_3] \sin [x_1] - 3.336 \cos [x_4] \sin [x_1] + 40.431 \cos [x_5] \sin [x_1] + 5.979 \cos [x_6] \sin [x_1] - 170.497 \sin [x_1]^2 - 138.391 \sin [x_2]^2 - 1529.374 \sin [x_3] + 19.654 \cos [x_1] \sin [x_3] + 161.510 \cos [x_2] \sin [x_3] + 469.979 \cos [x_3] \sin [x_3] - 20.672 \cos [x_4] \sin [x_3] + 23.177 \cos [x_5] \sin [x_3] - 10.204 \cos [x_6] \sin [x_3] - 149.309 \sin [x_3]^2 + 344.271 \sin [x_4] - 143.128 \sin [x_4]^2 + 0.338 \sin [x_5] - 156.522 \sin [x_5]^2 + 48.133 \sin [x_6] + 0.627 \cos [x_1] \sin [x_6] - 1.575 \cos [x_2] \sin [x_6] - 15.591 \cos [x_3] \sin [x_6] + 0.956 \cos [x_4] \sin [x_6] - 2.898 \cos [x_5] \sin [x_6] - 0.200 \cos [x_6] \sin [x_6] - 6.459 \sin [x_1] \sin [x_6] - 0.650 \sin [x_2] \sin [x_6] - 47.936 \sin [x_3] \sin [x_6] + 5.908 \sin [x_4] \sin [x_6] + 1.764 \sin [x_5] \sin [x_6] - 182.229 \sin [x_6]^2$

(*12*)

Out[•]//NumberForm=

$157.820 - 224.464 \cos [x_1] - 335.525 \cos [x_1]^2 - 111.673 \cos [x_2] + 167.917 \cos [x_2]^2 - 111.111 \cos [x_3] + 43.705 \cos [x_1] \cos [x_3] + 38.431 \cos [x_2] \cos [x_3] + 40.910 \cos [x_3]^2 + 20.981 \cos [x_4] - 3.538 \cos [x_2] \cos [x_4] + 230.651 \cos [x_4]^2 - 23.859 \cos [x_5] + 189.937 \cos [x_5]^2 + 3.166 \cos [x_6] + 217.096 \cos [x_6]^2 - 1128.858 \sin [x_1] + 255.478 \cos [x_1] \sin [x_1] - 4.329 \cos [x_2] \sin [x_1] + 161.406 \cos [x_3] \sin [x_1] - 12.261 \cos [x_4] \sin [x_1] - 3.109 \cos [x_5] \sin [x_1] + 7.040 \cos [x_6] \sin [x_1] + 194.512 \sin [x_1]^2 + 164.933 \sin [x_2]^2 - 173.612 \sin [x_3] - 29.955 \cos [x_1] \sin [x_3] + 114.343 \cos [x_2] \sin [x_3] - 41.541 \cos [x_3] \sin [x_3] - 6.908 \cos [x_4] \sin [x_3] + 50.382 \cos [x_5] \sin [x_3] - 10.506 \cos [x_6] \sin [x_3] + 205.301 \sin [x_3]^2 + 113.765 \sin [x_4] + 171.072 \sin [x_4]^2 + 1.105 \sin [x_5] + 179.198 \sin [x_5]^2 + 37.233 \sin [x_6] - 1.604 \cos [x_1] \sin [x_6] - 0.829 \cos [x_2] \sin [x_6] - 10.922 \cos [x_3] \sin [x_6] + 0.702 \cos [x_4] \sin [x_6] - 1.865 \cos [x_5] \sin [x_6] - 0.298 \cos [x_6] \sin [x_6] - 6.598 \sin [x_1] \sin [x_6] - 10.977 \sin [x_2] \sin [x_6] - 34.162 \sin [x_3] \sin [x_6] + 14.615 \sin [x_4] \sin [x_6] + 0.558 \sin [x_5] \sin [x_6] + 216.777 \sin [x_6]^2$

(Continued)

TABLE A7.3 *(Continued)*

Models Obtained by NRM and SNRM Using the Bootstrap Data Separation Method (80 Training–15 Testing–5 Validation) (*Models Obtained Using the NRM and SNRM Methods*)(*(Related to Table 7.4)*)

(*13*)

Out[•]//NumberForm=

$525.339 - 751.233$ Cos $[x_1] - 610.608$ Cos $[x_1]^2 + 93.513$ Cos $[x_2] + 548.508$ Cos $[x_2]^2 -$
715.810 Cos $[x_3] + 78.997$ Cos $[x_1]$ Cos $[x_3] - 23.884$ Cos $[x_2]$ Cos $[x_3] - 30.129$ Cos $[x_3]^2 -$
4.554 Cos $[x_4] - 3.826$ Cos $[x_2]$ Cos $[x_4] + 562.686$ Cos $[x_4]^2 - 114.243$ Cos $[x_5] + 634.611$ Cos $[x_5]^2 +$
5.696 Cos $[x_6] + 714.610$ Cos $[x_6]^2 - 2658.311$ Sin $[x_1] + 680.935$ Cos $[x_1]$ Sin $[x_1] +$
6.329 Cos $[x_2]$ Sin $[x_1] + 212.029$ Cos $[x_3]$ Sin $[x_1] + 0.966$ Cos $[x_4]$ Sin $[x_1] + 63.746$ Cos $[x_5]$ Sin $[x_1] +$
1.666 Cos $[x_6]$ Sin $[x_1] + 631.528$ Sin $[x_1]^2 + 547.648$ Sin $[x_2]^2 - 1529.232$ Sin $[x_3] +$
55.603 Cos $[x_1]$ Sin $[x_3] - 92.615$ Cos $[x_2]$ Sin $[x_3] + 482.282$ Cos $[x_3]$ Sin $[x_3] + 7.920$ Cos $[x_4]$ Sin $[x_3] +$
69.999 Cos $[x_5]$ Sin $[x_3] - 7.801$ Cos $[x_6]$ Sin $[x_3] + 678.034$ Sin $[x_3]^2 - 4.528$ Sin $[x_4] +$
570.524 Sin $[x_4]^2 + 1.339$ Sin $[x_5] + 627.938$ Sin $[x_5]^2 - 31.808$ Sin $[x_6] + 0.031$ Cos $[x_1]$ Sin $[x_6] +$
0.788 Cos $[x_2]$ Sin $[x_6] + 3.592$ Cos $[x_3]$ Sin $[x_6] + 0.841$ Cos $[x_4]$ Sin $[x_6] - 2.452$ Cos $[x_5]$ Sin $[x_6] -$
0.387 Cos $[x_6]$ Sin $[x_6] - 0.909$ Sin $[x_1]$ Sin $[x_6] + 3.933$ Sin $[x_2]$ Sin $[x_6] + 7.215$ Sin $[x_3]$ Sin $[x_6] +$
18.630 Sin $[x_4]$ Sin $[x_6] + 1.016$ Sin $[x_5]$ Sin $[x_6] + 714.326$ Sin $[x_6]^2$

(*14*)

Out[•]//NumberForm=

$371.546 - 72.755$ Cos $[x_1] - 53.258$ Cos $[x_1]^2 - 157.037$ Cos $[x_2] + 389.693$ Cos $[x_2]^2 -$
855.719 Cos $[x_3] + 21.167$ Cos $[x_1]$ Cos $[x_3] + 52.913$ Cos $[x_2]$ Cos $[x_3] - 439.345$ Cos $[x_3]^2 +$
5.302 Cos $[x_4] - 0.408$ Cos $[x_2]$ Cos $[x_4] + 436.706$ Cos $[x_4]^2 - 93.263$ Cos $[x_5] + 451.921$ Cos $[x_5]^2 +$
3.508 Cos $[x_6] + 504.156$ Cos $[x_6]^2 - 1023.704$ Sin $[x_1] + 224.783$ Cos $[x_1]$ Sin $[x_1] +$
9.000 Cos $[x_2]$ Sin $[x_1] + 228.535$ Cos $[x_3]$ Sin $[x_1] - 7.181$ Cos $[x_4]$ Sin $[x_1] + 56.988$ Cos $[x_5]$ Sin $[x_1] +$
2.926 Cos $[x_6]$ Sin $[x_1] + 452.099$ Sin $[x_1]^2 + 385.808$ Sin $[x_2]^2 - 2022.640$ Sin $[x_3] -$
147.738 Cos $[x_1]$ Sin $[x_3] + 143.494$ Cos $[x_2]$ Sin $[x_3] + 611.941$ Cos $[x_3]$ Sin $[x_3] + 5.251$ Cos $[x_4]$ Sin $[x_3] +$
31.336 Cos $[x_5]$ Sin $[x_3] - 6.717$ Cos $[x_6]$ Sin $[x_3] + 491.572$ Sin $[x_3]^2 + 72.330$ Sin $[x_4] +$
401.365 Sin $[x_4]^2 + 1.269$ Sin $[x_5] + 457.575$ Sin $[x_5]^2 + 41.514$ Sin $[x_6] - 3.826$ Cos $[x_1]$ Sin $[x_6] +$
0.678 Cos $[x_2]$ Sin $[x_6] - 13.002$ Cos $[x_3]$ Sin $[x_6] + 0.737$ Cos $[x_4]$ Sin $[x_6] - 3.631$ Cos $[x_5]$ Sin $[x_6] -$
0.276 Cos $[x_6]$ Sin $[x_6] - 12.487$ Sin $[x_1]$ Sin $[x_6] - 6.833$ Sin $[x_2]$ Sin $[x_6] - 38.090$ Sin $[x_3]$ Sin $[x_6] +$
15.019 Sin $[x_4]$ Sin $[x_6] + 0.807$ Sin $[x_5]$ Sin $[x_6] + 503.761$ Sin $[x_6]^2$

(*15*)

Out[•]//NumberForm=

$137.756 - 545.943$ Cos $[x_1] - 857.371$ Cos $[x_1]^2 - 60.810$ Cos $[x_2] + 145.822$ Cos $[x_2]^2 +$
368.955 Cos $[x_3] + 37.634$ Cos $[x_1]$ Cos $[x_3] + 22.621$ Cos $[x_2]$ Cos $[x_3] + 608.951$ Cos $[x_3]^2 +$
20.041 Cos $[x_4] - 0.651$ Cos $[x_2]$ Cos $[x_4] + 135.986$ Cos $[x_4]^2 - 36.172$ Cos $[x_5] + 168.433$ Cos $[x_5]^2 +$
1.861 Cos $[x_6] + 186.042$ Cos $[x_6]^2 - 2205.490$ Sin $[x_1] + 549.720$ Cos $[x_1]$ Sin $[x_1] +$
9.048 Cos $[x_2]$ Sin $[x_1] + 128.753$ Cos $[x_3]$ Sin $[x_1] - 32.425$ Cos $[x_4]$ Sin $[x_1] + 24.536$ Cos $[x_5]$ Sin $[x_1] +$
7.782 Cos $[x_6]$ Sin $[x_1] + 178.001$ Sin $[x_1]^2 + 143.887$ Sin $[x_2]^2 + 1217.361$ Sin $[x_3] -$
5.776 Cos $[x_1]$ Sin $[x_3] + 51.226$ Cos $[x_2]$ Sin $[x_3] - 472.497$ Cos $[x_3]$ Sin $[x_3] + 14.696$ Cos $[x_4]$ Sin $[x_3] +$
37.812 Cos $[x_5]$ Sin $[x_3] - 9.877$ Cos $[x_6]$ Sin $[x_3] + 161.290$ Sin $[x_3]^2 - 22.866$ Sin $[x_4] +$
149.216 Sin $[x_4]^2 + 2.199$ Sin $[x_5] + 154.195$ Sin $[x_5]^2 + 24.122$ Sin $[x_6] - 1.105$ Cos $[x_1]$ Sin $[x_6] +$
1.086 Cos $[x_2]$ Sin $[x_6] - 9.574$ Cos $[x_3]$ Sin $[x_6] + 0.308$ Cos $[x_4]$ Sin $[x_6] + 2.814$ Cos $[x_5]$ Sin $[x_6] -$
0.362 Cos $[x_6]$ Sin $[x_6] - 6.051$ Sin $[x_1]$ Sin $[x_6] - 1.645$ Sin $[x_2]$ Sin $[x_6] - 25.776$ Sin $[x_3]$ Sin $[x_6] +$
14.162 Sin $[x_4]$ Sin $[x_6] - 1.014$ Sin $[x_5]$ Sin $[x_6] + 185.717$ Sin $[x_6]^2$

(Continued)

TABLE A7.3 *(Continued)*

Models Obtained by NRM and SNRM Using the Bootstrap Data Separation Method (80 Training–15 Testing–5 Validation) (*Models Obtained Using the NRM and SNRM Methods*)(*(Related to Table 7.4)*)

(*16*)

Out[•]//NumberForm=

$347.315 - 417.912 \cos [x_1] - 400.499 \cos [x_1]^2 - 28.851 \cos [x_2] + 364.569 \cos [x_2]^2 -$
$448.957 \cos [x_3] + 28.340 \cos [x_1] \cos [x_3] + 19.294 \cos [x_2] \cos [x_3] - 49.589 \cos [x_3]^2 +$
$22.055 \cos [x_4] + 0.387 \cos [x_2] \cos [x_4] + 385.910 \cos [x_4]^2 - 90.307 \cos [x_5] + 421.971 \cos [x_5]^2 +$
$5.057 \cos [x_6] + 466.989 \cos [x_6]^2 - 1741.228 \sin [x_1] + 461.853 \cos [x_1] \sin [x_1] -$
$8.544 \cos [x_2] \sin [x_1] + 142.454 \cos [x_3] \sin [x_1] - 15.769 \cos [x_4] \sin [x_1] + 58.913 \cos [x_5] \sin [x_1] +$
$1.629 \cos [x_6] \sin [x_1] + 422.746 \sin [x_1]^2 + 362.304 \sin [x_2]^2 - 1026.872 \sin [x_3] -$
$47.039 \cos [x_1] \sin [x_3] + 36.958 \cos [x_2] \sin [x_3] + 299.714 \cos [x_3] \sin [x_3] - 3.964 \cos [x_4] \sin [x_3] +$
$43.834 \cos [x_5] \sin [x_3] - 6.984 \cos [x_6] \sin [x_3] + 448.708 \sin [x_3]^2 + 23.227 \sin [x_4] +$
$376.962 \sin [x_4]^2 + 1.274 \sin [x_5] + 418.019 \sin [x_5]^2 + 11.917 \sin [x_6] - 0.047 \cos [x_1] \sin [x_6] +$
$0.534 \cos [x_2] \sin [x_6] - 6.283 \cos [x_3] \sin [x_6] + 0.080 \cos [x_4] \sin [x_6] - 3.599 \cos [x_5] \sin [x_6] -$
$0.318 \cos [x_6] \sin [x_6] + 0.114 \sin [x_1] \sin [x_6] - 6.142 \sin [x_2] \sin [x_6] - 19.679 \sin [x_3] \sin [x_6] +$
$11.258 \sin [x_4] \sin [x_6] + 0.935 \sin [x_5] \sin [x_6] + 466.416 \sin [x_6]^2$

(*17*)

Out[•]//NumberForm=

$229.030 - 568.620 \cos [x_1] - 403.434 \cos [x_1]^2 + 22.348 \cos [x_2] + 242.100 \cos [x_2]^2 -$
$306.747 \cos [x_3] + 111.663 \cos [x_1] \cos [x_3] + 2.164 \cos [x_2] \cos [x_3] + 148.146 \cos [x_3]^2 +$
$6.400 \cos [x_4] + 2.036 \cos [x_2] \cos [x_4] + 238.025 \cos [x_4]^2 - 41.101 \cos [x_5] + 282.302 \cos [x_5]^2 +$
$5.418 \cos [x_6] + 312.077 \cos [x_6]^2 - 1483.095 \sin [x_1] + 385.073 \cos [x_1] \sin [x_1] -$
$5.622 \cos [x_2] \sin [x_1] + 209.843 \cos [x_3] \sin [x_1] - 15.706 \cos [x_4] \sin [x_1] + 23.537 \cos [x_5] \sin [x_1] +$
$2.482 \cos [x_6] \sin [x_1] + 280.717 \sin [x_1]^2 + 239.837 \sin [x_2]^2 - 282.906 \sin [x_3] +$
$169.802 \cos [x_1] \sin [x_3] - 15.215 \cos [x_2] \sin [x_3] + 82.558 \cos [x_3] \sin [x_3] + 13.130 \cos [x_4] \sin [x_3] +$
$32.753 \cos [x_5] \sin [x_3] - 8.327 \cos [x_6] \sin [x_3] + 289.053 \sin [x_3]^2 - 8.656 \sin [x_4] +$
$248.025 \sin [x_4]^2 + 1.073 \sin [x_5] + 276.558 \sin [x_5]^2 + 16.251 \sin [x_6] + 1.071 \cos [x_1] \sin [x_6] +$
$0.606 \cos [x_2] \sin [x_6] - 10.512 \cos [x_3] \sin [x_6] + 2.556 \cos [x_4] \sin [x_6] - 1.367 \cos [x_5] \sin [x_6] -$
$0.227 \cos [x_6] \sin [x_6] - 1.980 \sin [x_1] \sin [x_6] - 4.307 \sin [x_2] \sin [x_6] - 30.454 \sin [x_3] \sin [x_6] +$
$20.310 \sin [x_4] \sin [x_6] + 0.484 \sin [x_5] \sin [x_6] + 311.479 \sin [x_6]^2$

(*18*)

Out[•]//NumberForm=

$666.180 - 344.548 \cos [x_1] + 292.899 \cos [x_1]^2 + 125.870 \cos [x_2] + 699.000 \cos [x_2]^2 -$
$1908.801 \cos [x_3] + 137.745 \cos [x_1] \cos [x_3] - 28.982 \cos [x_2] \cos [x_3] - 1012.112 \cos [x_3]^2 +$
$6.990 \cos [x_4] - 1.377 \cos [x_2] \cos [x_4] + 794.442 \cos [x_4]^2 - 59.303 \cos [x_5] + 802.032 \cos [x_5]^2 +$
$3.420 \cos [x_6] + 921.581 \cos [x_6]^2 - 1170.660 \sin [x_1] + 255.120 \cos [x_1] \sin [x_1] -$
$15.822 \cos [x_2] \sin [x_1] + 426.860 \cos [x_3] \sin [x_1] - 6.245 \cos [x_4] \sin [x_1] - 11.780 \cos [x_5] \sin [x_1] +$
$0.995 \cos [x_6] \sin [x_1] + 791.649 \sin [x_1]^2 + 696.756 \sin [x_2]^2 - 4341.122 \sin [x_3] +$
$63.017 \cos [x_1] \sin [x_3] - 103.169 \cos [x_2] \sin [x_3] + 1414.867 \cos [x_3] \sin [x_3] +$
$0.648 \cos [x_4] \sin [x_3] + 106.692 \cos [x_5] \sin [x_3] - 4.759 \cos [x_6] \sin [x_3] + 901.243 \sin [x_3]^2 +$
$143.304 \sin [x_4] + 719.765 \sin [x_4]^2 + 1.226 \sin [x_5] + 786.124 \sin [x_5]^2 - 25.239 \sin [x_6] -$
$1.990 \cos [x_1] \sin [x_6] + 1.563 \cos [x_2] \sin [x_6] + 7.131 \cos [x_3] \sin [x_6] + 1.228 \cos [x_4] \sin [x_6] -$
$4.766 \cos [x_5] \sin [x_6] - 0.216 \cos [x_6] \sin [x_6] - 10.191 \sin [x_1] \sin [x_6] + 6.443 \sin [x_2] \sin [x_6] +$
$8.333 \sin [x_3] \sin [x_6] + 14.219 \sin [x_4] \sin [x_6] + 1.432 \sin [x_5] \sin [x_6] + 920.897 \sin [x_6]^2$

(Continued)

TABLE A7.3 *(Continued)*

Models Obtained by NRM and SNRM Using the Bootstrap Data Separation Method (80 Training–15 Testing–5 Validation) (*Models Obtained Using the NRM and SNRM Methods*)(*(Related to Table 7.4)*)

(***19***)

Out[•]//NumberForm=

$296.239 - 278.280 \cos [x_1] - 181.942 \cos [x_1]^2 - 30.623 \cos [x_2] + 312.717 \cos [x_2]^2 - 497.619 \cos [x_3] + 41.939 \cos [x_1] \cos [x_3] + 10.725 \cos [x_2] \cos [x_3] - 190.745 \cos [x_3]^2 + 20.942 \cos [x_4] + 1.406 \cos [x_2] \cos [x_4] + 311.209 \cos [x_4]^2 - 77.595 \cos [x_5] + 362.563 \cos [x_5]^2 + 6.377 \cos [x_6] + 396.728 \cos [x_6]^2 - 1132.048 \sin [x_1] + 298.711 \cos [x_1] \sin [x_1] + 8.765 \cos [x_2] \sin [x_1] + 162.288 \cos [x_3] \sin [x_1] - 12.061 \cos [x_4] \sin [x_1] + 44.164 \cos [x_5] \sin [x_1] + 1.610 \cos [x_6] \sin [x_1] + 357.768 \sin [x_1]^2 + 309.152 \sin [x_2]^2 - 1180.422 \sin [x_3] - 21.261 \cos [x_1] \sin [x_3] + 23.759 \cos [x_2] \sin [x_3] + 331.103 \cos [x_3] \sin [x_3] - 6.437 \cos [x_4] \sin [x_3] + 53.671 \cos [x_5] \sin [x_3] - 8.452 \cos [x_6] \sin [x_3] + 390.116 \sin [x_3]^2 - 10.783 \sin [x_4] + 319.948 \sin [x_4]^2 + 1.205 \sin [x_5] + 354.621 \sin [x_5]^2 + 18.042 \sin [x_6] - 4.078 \cos [x_1] \sin [x_6] + 0.654 \cos [x_2] \sin [x_6] - 3.798 \cos [x_3] \sin [x_6] + 0.718 \cos [x_4] \sin [x_6] - 1.573 \cos [x_5] \sin [x_6] - 0.392 \cos [x_6] \sin [x_6] - 17.615 \sin [x_1] \sin [x_6] - 3.717 \sin [x_2] \sin [x_6] - 13.140 \sin [x_3] \sin [x_6] + 15.820 \sin [x_4] \sin [x_6] + 0.666 \sin [x_5] \sin [x_6] + 395.965 \sin [x_6]^2$

(***20***)

Out[•]//NumberForm=

$505.095 - 120.186 \cos [x_1] + 159.981 \cos [x_1]^2 - 245.166 \cos [x_2] + 530.943 \cos [x_2]^2 - 1318.046 \cos [x_3] + 78.507 \cos [x_1] \cos [x_3] + 72.890 \cos [x_2] \cos [x_3] - 826.232 \cos [x_3]^2 - 2.703 \cos [x_4] + 5.027 \cos [x_2] \cos [x_4] + 658.418 \cos [x_4]^2 - 41.849 \cos [x_5] + 610.847 \cos [x_5]^2 + 3.190 \cos [x_6] + 683.174 \cos [x_6]^2 - 935.393 \sin [x_1] + 186.028 \cos [x_1] \sin [x_1] + 1.936 \cos [x_2] \sin [x_1] + 293.344 \cos [x_3] \sin [x_1] + 3.196 \cos [x_4] \sin [x_1] + 25.306 \cos [x_5] \sin [x_1] + 2.607 \cos [x_6] \sin [x_1] + 609.435 \sin [x_1]^2 + 527.835 \sin [x_2]^2 - 3319.803 \sin [x_3] - 69.049 \cos [x_1] \sin [x_3] + 236.323 \cos [x_2] \sin [x_3] + 984.270 \cos [x_3] \sin [x_3] + 2.575 \cos [x_4] \sin [x_3] + 33.898 \cos [x_5] \sin [x_3] - 6.078 \cos [x_6] \sin [x_3] + 667.569 \sin [x_3]^2 + 214.086 \sin [x_4] + 546.563 \sin [x_4]^2 + 1.028 \sin [x_5] + 603.767 \sin [x_5]^2 + 72.700 \sin [x_6] - 5.269 \cos [x_1] \sin [x_6] + 0.930 \cos [x_2] \sin [x_6] - 16.328 \cos [x_3] \sin [x_6] - 1.515 \cos [x_4] \sin [x_6] - 0.144 \cos [x_5] \sin [x_6] - 0.349 \cos [x_6] \sin [x_6] - 17.010 \sin [x_1] \sin [x_6] - 2.329 \sin [x_2] \sin [x_6] - 54.264 \sin [x_3] \sin [x_6] + 3.994 \sin [x_4] \sin [x_6] + 0.083 \sin [x_5] \sin [x_6] + 682.612 \sin [x_6]^2$

(***Models proposed by SNRM***)

(***1***)

Out[•]//NumberForm=

$47.442 - 71.611 \cos [x_1] - 84.857 \cos [x_1]^2 - 46.061 \cos [x_2] + 25.895 \cos [x_2]^2 - 56.170 \cos [x_3] + 63.236 \cos [x_1] \cos [x_3] + 16.522 \cos [x_2] \cos [x_3] + 3.186 \cos [x_3]^2 + 9.800 \cos [x_4] + 0.923 \cos [x_2] \cos [x_4] + 44.084 \cos [x_4]^2 - 69.122 \cos [x_5] + 25.374 \cos [x_5]^2 + 5.053 \cos [x_6] + 21.297 \cos [x_6]^2 - 395.248 \sin [x_1] + 80.015 \cos [x_1] \sin [x_1] + 8.629 \cos [x_2] \sin [x_1] + 221.574 \cos [x_3] \sin [x_1] - 7.362 \cos [x_4] \sin [x_1] + 42.674 \cos [x_5] \sin [x_1] + 3.363 \cos [x_6] \sin [x_1] + 118.803 \sin [x_1]^2 + 22.651 \sin [x_2]^2 + 138.274 \sin [x_3] - 8.598 \cos [x_1] \sin [x_3] + 38.175 \cos [x_2] \sin [x_3] - 149.554 \cos [x_3] \sin [x_3] + 0.211 \cos [x_4] \sin [x_3] + 42.158 \cos [x_5] \sin [x_3] - 8.804 \cos [x_6] \sin [x_3] + 38.937 \sin [x_3]^2 + 84.241 \sin [x_4] + 1.930 \sin [x_4]^2 + 1.196 \sin [x_5] + 19.789 \sin [x_5]^2 + 37.038 \sin [x_6] - 4.061 \cos [x_1] \sin [x_6] + 0.206 \cos [x_2] \sin [x_6] - 7.655 \cos [x_3] \sin [x_6] - 0.082 \cos [x_4] \sin [x_6] - 0.998 \cos [x_5] \sin [x_6] - 0.358 \cos [x_6] \sin [x_6] - 17.338 \sin [x_1] \sin [x_6] - 7.121 \sin [x_2] \sin [x_6] - 22.138 \sin [x_3] \sin [x_6] + 10.386 \sin [x_4] \sin [x_6] + 0.297 \sin [x_5] \sin [x_6] + 20.790 \sin [x_6]^2$

(Continued)

TABLE A7.3 *(Continued)*
Models Obtained by NRM and SNRM Using the Bootstrap Data Separation Method (80 Training–15 Testing–5 Validation) (*Models Obtained Using the NRM and SNRM Methods*)(*(Related to Table 7.4)*)

(*2*)

Out[•]//NumberForm=

$47.442 - 71.611 \cos [x_1] - 84.857 \cos [x_1]^2 - 46.061 \cos [x_2] + 25.895 \cos [x_2]^2 - 56.170 \cos [x_3] + 63.236 \cos [x_1] \cos [x_3] + 16.522 \cos [x_2] \cos [x_3] + 3.186 \cos [x_3]^2 + 9.800 \cos [x_4] + 0.923 \cos [x_2] \cos [x_4] + 44.084 \cos [x_4]^2 - 69.122 \cos [x_5] + 25.374 \cos [x_5]^2 + 5.053 \cos [x_6] + 21.297 \cos [x_6]^2 - 395.248 \sin [x_1] + 80.015 \cos [x_1] \sin [x_1] + 8.629 \cos [x_2] \sin [x_1] + 221.574 \cos [x_3] \sin [x_1] - 7.362 \cos [x_4] \sin [x_1] + 42.674 \cos [x_5] \sin [x_1] + 3.363 \cos [x_6] \sin [x_1] + 118.803 \sin [x_1]^2 + 22.651 \sin [x_2]^2 + 138.274 \sin [x_3] - 8.598 \cos [x_1] \sin [x_3] + 38.175 \cos [x_2] \sin [x_3] - 149.554 \cos [x_3] \sin [x_3] + 0.211 \cos [x_4] \sin [x_3] + 42.158 \cos [x_5] \sin [x_3] - 8.804 \cos [x_6] \sin [x_3] + 38.937 \sin [x_3]^2 + 84.241 \sin [x_4] + 1.930 \sin [x_4]^2 + 1.196 \sin [x_5] + 19.789 \sin [x_5]^2 + 37.038 \sin [x_6] - 4.061 \cos [x_1] \sin [x_6] + 0.206 \cos [x_2] \sin [x_6] - 7.655 \cos [x_3] \sin [x_6] - 0.082 \cos [x_4] \sin [x_6] - 0.998 \cos [x_5] \sin [x_6] - 0.358 \cos [x_6] \sin [x_6] - 17.338 \sin [x_1] \sin [x_6] - 7.121 \sin [x_2] \sin [x_6] - 22.138 \sin [x_3] \sin [x_6] + 10.386 \sin [x_4] \sin [x_6] + 0.297 \sin [x_5] \sin [x_6] + 20.790 \sin [x_6]^2$

(*3*)

Out[•]//NumberForm=

$60.381 - 119.991 \cos [x_1] - 98.415 \cos [x_1]^2 - 79.045 \cos [x_2] + 34.666 \cos [x_2]^2 - 69.617 \cos [x_3] + 70.939 \cos [x_1] \cos [x_3] + 22.959 \cos [x_2] \cos [x_3] + 27.317 \cos [x_3]^2 + 19.527 \cos [x_4] + 2.178 \cos [x_2] \cos [x_4] + 51.235 \cos [x_4]^2 - 63.903 \cos [x_5] + 30.039 \cos [x_5]^2 + 4.326 \cos [x_6] + 31.389 \cos [x_6]^2 - 506.963 \sin [x_1] + 119.005 \cos [x_1] \sin [x_1] + 12.123 \cos [x_2] \sin [x_1] + 241.052 \cos [x_3] \sin [x_1] - 11.933 \cos [x_4] \sin [x_1] + 35.761 \cos [x_5] \sin [x_1] + 3.947 \cos [x_6] \sin [x_1] + 143.650 \sin [x_1]^2 + 29.960 \sin [x_2]^2 + 192.240 \sin [x_3] - 3.675 \cos [x_1] \sin [x_3] + 66.947 \cos [x_2] \sin [x_3] - 159.016 \cos [x_3] \sin [x_3] - 5.909 \cos [x_4] \sin [x_3] + 47.074 \cos [x_5] \sin [x_3] - 8.592 \cos [x_6] \sin [x_3] + 34.936 \sin [x_3]^2 + 83.629 \sin [x_4] + 8.412 \sin [x_4]^2 + 1.125 \sin [x_5] + 22.488 \sin [x_5]^2 + 33.900 \sin [x_6] - 2.162 \cos [x_1] \sin [x_6] - 0.107 \cos [x_2] \sin [x_6] - 6.786 \cos [x_3] \sin [x_6] - 0.479 \cos [x_4] \sin [x_6] - 0.452 \cos [x_5] \sin [x_6] - 0.402 \cos [x_6] \sin [x_6] - 12.316 \sin [x_1] \sin [x_6] - 8.064 \sin [x_2] \sin [x_6] - 20.780 \sin [x_3] \sin [x_6] + 8.156 \sin [x_4] \sin [x_6] + 0.269 \sin [x_5] \sin [x_6] + 30.950 \sin [x_6]^2$

(*4*)

Out[•]//NumberForm=

$53.037 - 90.490 \cos [x_1] - 92.725 \cos [x_1]^2 - 46.674 \cos [x_2] + 29.053 \cos [x_2]^2 - 82.649 \cos [x_3] + 64.087 \cos [x_1] \cos [x_3] + 17.588 \cos [x_2] \cos [x_3] + 6.272 \cos [x_3]^2 + 9.986 \cos [x_4] + 0.257 \cos [x_2] \cos [x_4] + 62.427 \cos [x_4]^2 - 80.720 \cos [x_5] + 29.424 \cos [x_5]^2 + 5.508 \cos [x_6] + 31.693 \cos [x_6]^2 - 436.467 \sin [x_1] + 93.089 \cos [x_1] \sin [x_1] + 5.921 \cos [x_2] \sin [x_1] + 225.690 \cos [x_3] \sin [x_1] - 7.370 \cos [x_4] \sin [x_1] + 51.081 \cos [x_5] \sin [x_1] + 3.271 \cos [x_6] \sin [x_1] + 129.019 \sin [x_1]^2 + 26.039 \sin [x_2]^2 + 124.726 \sin [x_3] - 5.077 \cos [x_1] \sin [x_3] + 41.340 \cos [x_2] \sin [x_3] - 129.922 \cos [x_3] \sin [x_3] + 0.058 \cos [x_4] \sin [x_3] + 48.919 \cos [x_5] \sin [x_3] - 9.200 \cos [x_6] \sin [x_3] + 42.451 \sin [x_3]^2 + 91.220 \sin [x_4] + 16.498 \sin [x_4]^2 + 1.293 \sin [x_5] + 21.929 \sin [x_5]^2 + 26.231 \sin [x_6] - 1.972 \cos [x_1] \sin [x_6] + 0.573 \cos [x_2] \sin [x_6] - 6.132 \cos [x_3] \sin [x_6] - 0.407 \cos [x_4] \sin [x_6] + 0.302 \cos [x_5] \sin [x_6] - 0.355 \cos [x_6] \sin [x_6] - 9.427 \sin [x_1] \sin [x_6] - 4.449 \sin [x_2] \sin [x_6] - 17.936 \sin [x_3] \sin [x_6] + 7.194 \sin [x_4] \sin [x_6] + 0.132 \sin [x_5] \sin [x_6] + 31.232 \sin [x_6]^2$

(Continued)

TABLE A7.3 *(Continued)*

Models Obtained by NRM and SNRM Using the Bootstrap Data Separation Method (80 Training–15 Testing–5 Validation) (*Models Obtained Using the NRM and SNRM Methods*)(*(Related to Table 7.4)*)

(*5*)

Out[•]//NumberForm=

$39.943 - 80.506 \text{ Cos } [x_1] - 82.880 \text{ Cos } [x_1]^2 - 37.489 \text{ Cos } [x_2] + 31.628 \text{ Cos } [x_2]^2 - 15.122 \text{ Cos } [x_3] +$
$61.733 \text{ Cos } [x_1] \text{ Cos } [x_3] + 14.156 \text{ Cos } [x_2] \text{ Cos } [x_3] + 3.723 \text{ Cos } [x_3]^2 + 8.415 \text{ Cos } [x_4] -$
$0.769 \text{ Cos } [x_2] \text{ Cos } [x_4] + 38.024 \text{ Cos } [x_4]^2 - 66.529 \text{ Cos } [x_5] + 22.557 \text{ Cos } [x_5]^2 + 4.727 \text{ Cos } [x_6] +$
$22.622 \text{ Cos } [x_6]^2 - 355.296 \text{ Sin } [x_1] + 65.915 \text{ Cos } [x_1] \text{ Sin } [x_1] + 10.453 \text{ Cos } [x_2] \text{ Sin } [x_1] +$
$177.012 \text{ Cos } [x_3] \text{ Sin } [x_1] - 1.947 \text{ Cos } [x_4] \text{ Sin } [x_1] + 32.112 \text{ Cos } [x_5] \text{ Sin } [x_1] + 3.606 \text{ Cos } [x_6] \text{ Sin } [x_1] +$
$105.044 \text{ Sin } [x_1]^2 + 27.256 \text{ Sin } [x_2]^2 + 139.981 \text{ Sin } [x_3] + 16.519 \text{ Cos } [x_1] \text{ Sin } [x_3] +$
$28.058 \text{ Cos } [x_2] \text{ Sin } [x_3] - 149.515 \text{ Cos } [x_3] \text{ Sin } [x_3] - 3.834 \text{ Cos } [x_4] \text{ Sin } [x_3] + 35.795 \text{ Cos } [x_5] \text{ Sin } [x_3] -$
$8.626 \text{ Cos } [x_6] \text{ Sin } [x_3] + 29.654 \text{ Sin } [x_3]^2 + 38.983 \text{ Sin } [x_4] + 18.969 \text{ Sin } [x_4]^2 + 0.871 \text{ Sin } [x_5] +$
$25.339 \text{ Sin } [x_5]^2 + 40.710 \text{ Sin } [x_6] - 3.699 \text{ Cos } [x_1] \text{ Sin } [x_6] + 0.150 \text{ Cos } [x_2] \text{ Sin } [x_6] -$
$10.638 \text{ Cos } [x_3] \text{ Sin } [x_6] - 0.059 \text{ Cos } [x_4] \text{ Sin } [x_6] - 1.681 \text{ Cos } [x_5] \text{ Sin } [x_6] - 0.378 \text{ Cos } [x_6] \text{ Sin } [x_6] -$
$15.755 \text{ Sin } [x_1] \text{ Sin } [x_6] - 8.109 \text{ Sin } [x_2] \text{ Sin } [x_6] - 29.262 \text{ Sin } [x_3] \text{ Sin } [x_6] + 13.107 \text{ Sin } [x_4] \text{ Sin } [x_6] +$
$0.374 \text{ Sin } [x_5] \text{ Sin } [x_6] + 22.233 \text{ Sin } [x_6]^2$

(*6*)

Out[•]//NumberForm=

$2.737 + 30.668 \text{ Cos } [x_1] - 35.502 \text{ Cos } [x_1]^2 - 68.283 \text{ Cos } [x_2] + 2.093 \text{ Cos } [x_2]^2 - 118.647 \text{ Cos } [x_3] +$
$78.827 \text{ Cos } [x_1] \text{ Cos } [x_3] + 21.486 \text{ Cos } [x_2] \text{ Cos } [x_3] - 36.099 \text{ Cos } [x_3]^2 + 5.816 \text{ Cos } [x_4] +$
$1.203 \text{ Cos } [x_2] \text{ Cos } [x_4] + 15.296 \text{ Cos } [x_4]^2 - 47.339 \text{ Cos } [x_5] + 6.770 \text{ Cos } [x_5]^2 + 5.046 \text{ Cos } [x_6] -$
$0.329 \text{ Cos } [x_6]^2 - 92.627 \text{ Sin } [x_1] - 34.328 \text{ Cos } [x_1] \text{ Sin } [x_1] + 13.984 \text{ Cos } [x_2] \text{ Sin } [x_1] +$
$250.989 \text{ Cos } [x_3] \text{ Sin } [x_1] - 5.703 \text{ Cos } [x_4] \text{ Sin } [x_1] + 29.931 \text{ Cos } [x_5] \text{ Sin } [x_1] + 3.845 \text{ Cos } [x_6] \text{ Sin } [x_1] +$
$36.204 \text{ Sin } [x_1]^2 - 1.886 \text{ Sin } [x_2]^2 + 50.468 \text{ Sin } [x_3] + 5.938 \text{ Cos } [x_1] \text{ Sin } [x_3] + 54.543 \text{ Cos } [x_2] \text{ Sin } [x_3] -$
$121.623 \text{ Cos } [x_3] \text{ Sin } [x_3] + 2.846 \text{ Cos } [x_4] \text{ Sin } [x_3] + 37.080 \text{ Cos } [x_5] \text{ Sin } [x_3] - 9.292 \text{ Cos } [x_6] \text{ Sin } [x_3] +$
$30.431 \text{ Sin } [x_3]^2 + 124.777 \text{ Sin } [x_4] - 48.520 \text{ Sin } [x_4]^2 + 1.254 \text{ Sin } [x_5] - 1.580 \text{ Sin } [x_5]^2 + 49.574 \text{ Sin } [x_6] -$
$6.681 \text{ Cos } [x_1] \text{ Sin } [x_6] - 0.108 \text{ Cos } [x_2] \text{ Sin } [x_6] - 8.107 \text{ Cos } [x_3] \text{ Sin } [x_6] - 0.294 \text{ Cos } [x_4] \text{ Sin } [x_6] -$
$1.608 \text{ Cos } [x_5] \text{ Sin } [x_6] - 0.435 \text{ Cos } [x_6] \text{ Sin } [x_6] - 25.082 \text{ Sin } [x_1] \text{ Sin } [x_6] - 9.728 \text{ Sin } [x_2] \text{ Sin } [x_6] -$
$23.889 \text{ Sin } [x_3] \text{ Sin } [x_6] + 9.793 \text{ Sin } [x_4] \text{ Sin } [x_6] + 0.474 \text{ Sin } [x_5] \text{ Sin } [x_6] - 0.857 \text{ Sin } [x_6]^2$

(*7*)

Out[•]//NumberForm=

$28.263 - 15.073 \text{ Cos } [x_1] - 72.350 \text{ Cos } [x_1]^2 - 57.439 \text{ Cos } [x_2] + 18.197 \text{ Cos } [x_2]^2 - 48.551 \text{ Cos } [x_3] +$
$64.271 \text{ Cos } [x_1] \text{ Cos } [x_3] + 16.456 \text{ Cos } [x_2] \text{ Cos } [x_3] - 40.828 \text{ Cos } [x_3]^2 + 18.956 \text{ Cos } [x_4] +$
$5.299 \text{ Cos } [x_2] \text{ Cos } [x_4] + 53.677 \text{ Cos } [x_4]^2 - 72.633 \text{ Cos } [x_5] + 18.099 \text{ Cos } [x_5]^2 + 6.287 \text{ Cos } [x_6] +$
$5.462 \text{ Cos } [x_6]^2 - 300.130 \text{ Sin } [x_1] + 54.376 \text{ Cos } [x_1] \text{ Sin } [x_1] + 9.009 \text{ Cos } [x_2] \text{ Sin } [x_1] +$
$254.151 \text{ Cos } [x_3] \text{ Sin } [x_1] - 7.577 \text{ Cos } [x_4] \text{ Sin } [x_1] + 52.232 \text{ Cos } [x_5] \text{ Sin } [x_1] + 3.373 \text{ Cos } [x_6] \text{ Sin } [x_1] +$
$89.632 \text{ Sin } [x_1]^2 + 15.023 \text{ Sin } [x_2]^2 + 68.586 \text{ Sin } [x_3] - 39.907 \text{ Cos } [x_1] \text{ Sin } [x_3] + 48.944 \text{ Cos } [x_2] \text{ Sin } [x_3] -$
$180.318 \text{ Cos } [x_3] \text{ Sin } [x_3] - 9.419 \text{ Cos } [x_4] \text{ Sin } [x_3] + 37.767 \text{ Cos } [x_5] \text{ Sin } [x_3] - 10.137 \text{ Cos } [x_6] \text{ Sin } [x_3] +$
$56.807 \text{ Sin } [x_3]^2 + 142.458 \text{ Sin } [x_4] - 19.528 \text{ Sin } [x_4]^2 + 1.105 \text{ Sin } [x_5] + 11.797 \text{ Sin } [x_5]^2 +$
$44.556 \text{ Sin } [x_6] - 4.853 \text{ Cos } [x_1] \text{ Sin } [x_6] + 0.439 \text{ Cos } [x_2] \text{ Sin } [x_6] - 9.041 \text{ Cos } [x_3] \text{ Sin } [x_6] +$
$0.171 \text{ Cos } [x_4] \text{ Sin } [x_6] - 0.596 \text{ Cos } [x_5] \text{ Sin } [x_6] - 0.352 \text{ Cos } [x_6] \text{ Sin } [x_6] - 20.823 \text{ Sin } [x_1] \text{ Sin } [x_6] -$
$5.081 \text{ Sin } [x_2] \text{ Sin } [x_6] - 28.244 \text{ Sin } [x_3] \text{ Sin } [x_6] + 11.054 \text{ Sin } [x_4] \text{ Sin } [x_6] + 0.201 \text{ Sin } [x_5] \text{ Sin } [x_6] +$
$4.750 \text{ Sin } [x_6]^2$

(Continued)

TABLE A7.3 *(Continued)*

Models Obtained by NRM and SNRM Using the Bootstrap Data Separation Method (80 Training–15 Testing–5 Validation) (*Models Obtained Using the NRM and SNRM Methods*)(*(Related to Table 7.4)*)

(*8*)

Out[•]//NumberForm=

$45.520 - 68.013$ Cos $[x_1] - 73.867$ Cos $[x_1]^2 - 51.033$ Cos $[x_2] + 25.151$ Cos $[x_2]^2 - 60.442$ Cos $[x_3] +$
66.871 Cos $[x_1]$ Cos $[x_3] + 19.104$ Cos $[x_2]$ Cos $[x_3] - 9.480$ Cos $[x_3]^2 + 16.071$ Cos $[x_4] +$
1.357 Cos $[x_2]$ Cos $[x_4] + 42.304$ Cos $[x_4]^2 - 65.455$ Cos $[x_5] + 13.115$ Cos $[x_5]^2 + 4.701$ Cos $[x_6] +$
20.329 Cos $[x_6]^2 - 320.732$ Sin $[x_1] + 58.943$ Cos $[x_1]$ Sin $[x_1] + 5.212$ Cos $[x_2]$ Sin $[x_1] +$
209.265 Cos $[x_3]$ Sin $[x_1] - 4.900$ Cos $[x_4]$ Sin $[x_1] + 41.916$ Cos $[x_5]$ Sin $[x_1] + 3.789$ Cos $[x_6]$ Sin $[x_1] +$
97.150 Sin $[x_1]^2 + 22.442$ Sin $[x_2]^2 + 81.904$ Sin $[x_3] + 8.975$ Cos $[x_1]$ Sin $[x_3] + 45.897$ Cos $[x_2]$ Sin $[x_3] -$
135.946 Cos $[x_3]$ Sin $[x_3] - 9.421$ Cos $[x_4]$ Sin $[x_3] + 35.757$ Cos $[x_5]$ Sin $[x_3] - 8.824$ Cos $[x_6]$ Sin $[x_3]$
$+ 45.775$ Sin $[x_3]^2 + 110.536$ Sin $[x_4] - 15.411$ Sin $[x_4]^2 + 1.057$ Sin $[x_5] + 9.616$ Sin $[x_5]^2 + 42.140$ Sin $[x_6] -$
3.332 Cos $[x_1]$ Sin $[x_6] + 0.011$ Cos $[x_2]$ Sin $[x_6] - 9.690$ Cos $[x_3]$ Sin $[x_6] - 0.299$ Cos $[x_4]$ Sin $[x_6] -$
0.840 Cos $[x_5]$ Sin $[x_6] - 0.373$ Cos $[x_6]$ Sin $[x_6] - 14.702$ Sin $[x_1]$ Sin $[x_6] - 7.109$ Sin $[x_2]$ Sin $[x_6] -$
28.756 Sin $[x_3]$ Sin $[x_6] + 9.635$ Sin $[x_4]$ Sin $[x_6] + 0.230$ Sin $[x_5]$ Sin $[x_6] + 19.915$ Sin $[x_6]^2$

(*9*)

Out[•]//NumberForm=

$47.442 - 71.611$ Cos $[x_1] - 84.857$ Cos $[x_1]^2 - 46.061$ Cos $[x_2] + 25.895$ Cos $[x_2]^2 - 56.170$ Cos $[x_3] +$
63.236 Cos $[x_1]$ Cos $[x_3] + 16.522$ Cos $[x_2]$ Cos $[x_3] + 3.186$ Cos $[x_3]^2 + 9.800$ Cos $[x_4] +$
0.923 Cos $[x_2]$ Cos $[x_4] + 44.084$ Cos $[x_4]^2 - 69.122$ Cos $[x_5] + 25.374$ Cos $[x_5]^2 + 5.053$ Cos $[x_6] +$
21.297 Cos $[x_6]^2 - 395.248$ Sin $[x_1] + 80.015$ Cos $[x_1]$ Sin $[x_1] + 8.629$ Cos $[x_2]$ Sin $[x_1] +$
221.574 Cos $[x_3]$ Sin $[x_1] - 7.362$ Cos $[x_4]$ Sin $[x_1] + 42.674$ Cos $[x_5]$ Sin $[x_1] + 3.363$ Cos $[x_6]$ Sin $[x_1] +$
118.803 Sin $[x_1]^2 + 22.651$ Sin $[x_2]^2 + 138.274$ Sin $[x_3] - 8.598$ Cos $[x_1]$ Sin $[x_3] + 38.175$ Cos $[x_2]$ Sin $[x_3] -$
149.554 Cos $[x_3]$ Sin $[x_3] + 0.211$ Cos $[x_4]$ Sin $[x_3] + 42.158$ Cos $[x_5]$ Sin $[x_3] - 8.804$ Cos $[x_6]$ Sin $[x_3] +$
38.937 Sin $[x_3]^2 + 84.241$ Sin $[x_4] + 1.930$ Sin $[x_4]^2 + 1.196$ Sin $[x_5] + 19.789$ Sin $[x_5]^2 + 37.038$ Sin $[x_6] -$
4.061 Cos $[x_1]$ Sin $[x_6] + 0.206$ Cos $[x_2]$ Sin $[x_6] - 7.655$ Cos $[x_3]$ Sin $[x_6] - 0.082$ Cos $[x_4]$ Sin $[x_6] -$
0.998 Cos $[x_5]$ Sin $[x_6] - 0.358$ Cos $[x_6]$ Sin $[x_6] - 17.338$ Sin $[x_1]$ Sin $[x_6] - 7.121$ Sin $[x_2]$ Sin $[x_6] -$
22.138 Sin $[x_3]$ Sin $[x_6] + 10.386$ Sin $[x_4]$ Sin $[x_6] + 0.297$ Sin $[x_5]$ Sin $[x_6] + 20.790$ Sin $[x_6]^2$

(*10*)

Out[•]//NumberForm=

$48.588 - 98.200$ Cos $[x_1] - 90.581$ Cos $[x_1]^2 - 51.384$ Cos $[x_2] + 29.790$ Cos $[x_2]^2 - 68.617$ Cos $[x_3] +$
70.398 Cos $[x_1]$ Cos $[x_3] + 18.437$ Cos $[x_2]$ Cos $[x_3] + 16.891$ Cos $[x_3]^2 + 10.872$ Cos $[x_4] +$
1.273 Cos $[x_2]$ Cos $[x_4] + 47.472$ Cos $[x_4]^2 - 72.968$ Cos $[x_5] + 34.176$ Cos $[x_5]^2 + 4.357$ Cos $[x_6] +$
26.242 Cos $[x_6]^2 - 445.132$ Sin $[x_1] + 93.243$ Cos $[x_1]$ Sin $[x_1] + 8.786$ Cos $[x_2]$ Sin $[x_1] +$
229.094 Cos $[x_3]$ Sin $[x_1] - 9.837$ Cos $[x_4]$ Sin $[x_1] + 41.928$ Cos $[x_5]$ Sin $[x_1] + 3.619$ Cos $[x_6]$ Sin $[x_1] +$
135.158 Sin $[x_1]^2 + 26.895$ Sin $[x_2]^2 + 154.731$ Sin $[x_3] + 3.311$ Cos $[x_1]$ Sin $[x_3] + 43.028$ Cos $[x_2]$ Sin $[x_3] -$
147.880 Cos $[x_3]$ Sin $[x_3] + 1.672$ Cos $[x_4]$ Sin $[x_3] + 44.038$ Cos $[x_5]$ Sin $[x_3] - 8.307$ Cos $[x_6]$ Sin $[x_3] +$
39.824 Sin $[x_3]^2 + 65.587$ Sin $[x_4] + 15.429$ Sin $[x_4]^2 + 1.193$ Sin $[x_5] + 30.050$ Sin $[x_5]^2 + 37.655$ Sin $[x_6] -$
4.188 Cos $[x_1]$ Sin $[x_6] + 0.271$ Cos $[x_2]$ Sin $[x_6] - 8.174$ Cos $[x_3]$ Sin $[x_6] - 0.079$ Cos $[x_4]$ Sin $[x_6] -$
1.965 Cos $[x_5]$ Sin $[x_6] - 0.385$ Cos $[x_6]$ Sin $[x_6] - 18.173$ Sin $[x_1]$ Sin $[x_6] - 7.683$ Sin $[x_2]$ Sin $[x_6] -$
23.846 Sin $[x_3]$ Sin $[x_6] + 11.991$ Sin $[x_4]$ Sin $[x_6] + 0.385$ Sin $[x_5]$ Sin $[x_6] + 25.712$ Sin $[x_6]^2$

(Continued)

TABLE A7.3 *(Continued)*

Models Obtained by NRM and SNRM Using the Bootstrap Data Separation Method (80 Training–15 Testing–5 Validation) (*Models Obtained Using the NRM and SNRM Methods*)(*(Related to Table 7.4)*)

(*11*)

Out[•]//NumberForm=

$50.990 - 58.698$ Cos $[x_1] - 76.090$ Cos $[x_1]^2 - 32.642$ Cos $[x_2] + 32.120$ Cos $[x_2]^2 - 61.975$ Cos $[x_3] +$
60.042 Cos $[x_1]$ Cos $[x_3] + 12.586$ Cos $[x_2]$ Cos $[x_3] - 11.203$ Cos $[x_3]^2 + 5.837$ Cos $[x_4] +$
2.579 Cos $[x_2]$ Cos $[x_4] + 46.685$ Cos $[x_4]^2 - 69.869$ Cos $[x_5] + 24.288$ Cos $[x_5]^2 + 4.844$ Cos $[x_6] +$
29.018 Cos $[x_6]^2 - 364.004$ Sin $[x_1] + 76.240$ Cos $[x_1]$ Sin $[x_1] + 10.998$ Cos $[x_2]$ Sin $[x_1] +$
220.385 Cos $[x_3]$ Sin $[x_1] - 7.253$ Cos $[x_4]$ Sin $[x_1] + 41.884$ Cos $[x_5]$ Sin $[x_1] + 3.447$ Cos $[x_6]$ Sin $[x_1] +$
112.790 Sin $[x_1]^2 + 28.877$ Sin $[x_2]^2 + 86.762$ Sin $[x_3] - 16.580$ Cos $[x_1]$ Sin $[x_3] + 22.584$ Cos $[x_2]$ Sin $[x_3] -$
139.535 Cos $[x_3]$ Sin $[x_3] + 4.734$ Cos $[x_4]$ Sin $[x_3] + 45.622$ Cos $[x_5]$ Sin $[x_3] - 8.717$ Cos $[x_6]$ Sin $[x_3] +$
55.466 Sin $[x_3]^2 + 64.904$ Sin $[x_4] + 16.024$ Sin $[x_4]^2 + 1.251$ Sin $[x_5] + 17.665$ Sin $[x_5]^2 + 38.997$ Sin $[x_6] -$
5.069 Cos $[x_1]$ Sin $[x_6] + 0.325$ Cos $[x_2]$ Sin $[x_6] - 7.485$ Cos $[x_3]$ Sin $[x_6] + 0.028$ Cos $[x_4]$ Sin $[x_6] -$
1.709 Cos $[x_5]$ Sin $[x_6] - 0.378$ Cos $[x_6]$ Sin $[x_6] - 21.029$ Sin $[x_1]$ Sin $[x_6] - 7.834$ Sin $[x_2]$ Sin $[x_6] -$
21.131 Sin $[x_3]$ Sin $[x_6] + 11.240$ Sin $[x_4]$ Sin $[x_6] + 0.283$ Sin $[x_5]$ Sin $[x_6] + 28.449$ Sin $[x_6]^2$

(*12*)

Out[•]//NumberForm=

$50.124 - 84.199$ Cos $[x_1] - 99.984$ Cos $[x_1]^2 - 5.708$ Cos $[x_2] + 28.687$ Cos $[x_2]^2 - 24.667$ Cos $[x_3] +$
61.290 Cos $[x_1]$ Cos $[x_3] + 5.368$ Cos $[x_2]$ Cos $[x_3] + 17.783$ Cos $[x_3]^2 + 6.800$ Cos $[x_4] -$
0.223 Cos $[x_2]$ Cos $[x_4] + 37.260$ Cos $[x_4]^2 - 92.178$ Cos $[x_5] + 25.028$ Cos $[x_5]^2 + 5.373$ Cos $[x_6] +$
18.074 Cos $[x_6]^2 - 405.348$ Sin $[x_1] + 80.446$ Cos $[x_1]$ Sin $[x_1] + 8.120$ Cos $[x_2]$ Sin $[x_1] +$
196.538 Cos $[x_3]$ Sin $[x_1] - 1.978$ Cos $[x_4]$ Sin $[x_1] + 58.547$ Cos $[x_5]$ Sin $[x_1] + 3.068$ Cos $[x_6]$ Sin $[x_1] +$
120.482 Sin $[x_1]^2 + 25.122$ Sin $[x_2]^2 + 160.693$ Sin $[x_3] + 4.672$ Cos $[x_1]$ Sin $[x_3] - 0.357$ Cos $[x_2]$ Sin $[x_3] -$
155.921 Cos $[x_3]$ Sin $[x_3] - 1.887$ Cos $[x_4]$ Sin $[x_3] + 39.359$ Cos $[x_5]$ Sin $[x_3] - 8.833$ Cos $[x_6]$ Sin $[x_3] +$
41.615 Sin $[x_3]^2 + 37.630$ Sin $[x_4] + 20.237$ Sin $[x_4]^2 + 1.094$ Sin $[x_5] + 25.709$ Sin $[x_5]^2 + 33.988$ Sin $[x_6] -$
3.405 Cos $[x_1]$ Sin $[x_6] + 0.478$ Cos $[x_2]$ Sin $[x_6] - 8.906$ Cos $[x_3]$ Sin $[x_6] + 0.065$ Cos $[x_4]$ Sin $[x_6] -$
1.133 Cos $[x_5]$ Sin $[x_6] - 0.280$ Cos $[x_6]$ Sin $[x_6] - 13.716$ Sin $[x_1]$ Sin $[x_6] - 6.744$ Sin $[x_2]$ Sin $[x_6] -$
24.006 Sin $[x_3]$ Sin $[x_6] + 11.412$ Sin $[x_4]$ Sin $[x_6] + 0.365$ Sin $[x_5]$ Sin $[x_6] + 17.626$ Sin $[x_6]^2$

(*13*)

Out[•]//NumberForm=

$47.442 - 71.611$ Cos $[x_1] - 84.857$ Cos $[x_1]^2 - 46.061$ Cos $[x_2] + 25.895$ Cos $[x_2]^2 - 56.170$ Cos $[x_3] +$
63.236 Cos $[x_1]$ Cos $[x_3] + 16.522$ Cos $[x_2]$ Cos $[x_3] + 3.186$ Cos $[x_3]^2 + 9.800$ Cos $[x_4] +$
0.923 Cos $[x_2]$ Cos $[x_4] + 44.084$ Cos $[x_4]^2 - 69.122$ Cos $[x_5] + 25.374$ Cos $[x_5]^2 + 5.053$ Cos $[x_6] +$
21.297 Cos $[x_6]^2 - 395.248$ Sin $[x_1] + 80.015$ Cos $[x_1]$ Sin $[x_1] + 8.629$ Cos $[x_2]$ Sin $[x_1] +$
221.574 Cos $[x_3]$ Sin $[x_1] - 7.362$ Cos $[x_4]$ Sin $[x_1] + 42.674$ Cos $[x_5]$ Sin $[x_1] + 3.363$ Cos $[x_6]$ Sin $[x_1] +$
118.803 Sin $[x_1]^2 + 22.651$ Sin $[x_2]^2 + 138.274$ Sin $[x_3] - 8.598$ Cos $[x_1]$ Sin $[x_3] + 38.175$ Cos $[x_2]$ Sin $[x_3] -$
149.554 Cos $[x_3]$ Sin $[x_3] + 0.211$ Cos $[x_4]$ Sin $[x_3] + 42.158$ Cos $[x_5]$ Sin $[x_3] - 8.804$ Cos $[x_6]$ Sin $[x_3] +$
38.937 Sin $[x_3]^2 + 84.241$ Sin $[x_4] + 1.930$ Sin $[x_4]^2 + 1.196$ Sin $[x_5] + 19.789$ Sin $[x_5]^2 + 37.038$ Sin $[x_6] -$
4.061 Cos $[x_1]$ Sin $[x_6] + 0.206$ Cos $[x_2]$ Sin $[x_6] - 7.655$ Cos $[x_3]$ Sin $[x_6] - 0.082$ Cos $[x_4]$ Sin $[x_6] -$
0.998 Cos $[x_5]$ Sin $[x_6] - 0.358$ Cos $[x_6]$ Sin $[x_6] - 17.338$ Sin $[x_1]$ Sin $[x_6] - 7.121$ Sin $[x_2]$ Sin $[x_6] -$
22.138 Sin $[x_3]$ Sin $[x_6] + 10.386$ Sin $[x_4]$ Sin $[x_6] + 0.297$ Sin $[x_5]$ Sin $[x_6] + 20.790$ Sin $[x_6]^2$

(Continued)

TABLE A7.3 *(Continued)*
Models Obtained by NRM and SNRM Using the Bootstrap Data Separation Method (80 Training–15 Testing–5 Validation) (*Models Obtained Using the NRM and SNRM Methods*)(*(Related to Table 7.4)*)

(*14*)
Out[•]//NumberForm=

$45.115 - 90.193 \text{ Cos } [x_1] - 80.819 \text{ Cos } [x_1]^2 - 33.190 \text{ Cos } [x_2] + 27.220 \text{ Cos } [x_2]^2 - 36.893 \text{ Cos } [x_3] +$
$68.497 \text{ Cos } [x_1] \text{ Cos } [x_3] + 11.414 \text{ Cos } [x_2] \text{ Cos } [x_3] - 1.136 \text{ Cos } [x_3]^2 + 12.807 \text{ Cos } [x_4] +$
$0.647 \text{ Cos } [x_2] \text{ Cos } [x_4] + 32.525 \text{ Cos } [x_4]^2 - 66.757 \text{ Cos } [x_5] + 27.103 \text{ Cos } [x_5]^2 + 4.740 \text{ Cos } [x_6] +$
$27.210 \text{ Cos } [x_6]^2 - 382.495 \text{ Sin } [x_1] + 81.601 \text{ Cos } [x_1] \text{ Sin } [x_1] + 6.970 \text{ Cos } [x_2] \text{ Sin } [x_1] +$
$223.615 \text{ Cos } [x_3] \text{ Sin } [x_1] - 8.788 \text{ Cos } [x_4] \text{ Sin } [x_1] + 39.890 \text{ Cos } [x_5] \text{ Sin } [x_1] + 3.894 \text{ Cos } [x_6] \text{ Sin } [x_1] +$
$113.161 \text{ Sin } [x_1]^2 + 24.002 \text{ Sin } [x_2]^2 + 154.236 \text{ Sin } [x_3] + 7.205 \text{ Cos } [x_1] \text{ Sin } [x_3] + 27.977 \text{ Cos } [x_2] \text{ Sin } [x_3] -$
$169.561 \text{ Cos } [x_3] \text{ Sin } [x_3] - 1.523 \text{ Cos } [x_4] \text{ Sin } [x_3] + 45.607 \text{ Cos } [x_5] \text{ Sin } [x_3] - 8.995 \text{ Cos } [x_6] \text{ Sin } [x_3] +$
$35.121 \text{ Sin } [x_3]^2 + 59.812 \text{ Sin } [x_4] + 3.462 \text{ Sin } [x_4]^2 + 1.201 \text{ Sin } [x_5] + 19.799 \text{ Sin } [x_5]^2 + 39.369 \text{ Sin } [x_6] -$
$4.823 \text{ Cos } [x_1] \text{ Sin } [x_6] + 0.298 \text{ Cos } [x_2] \text{ Sin } [x_6] - 8.356 \text{ Cos } [x_3] \text{ Sin } [x_6] - 0.056 \text{ Cos } [x_4] \text{ Sin } [x_6] -$
$0.714 \text{ Cos } [x_5] \text{ Sin } [x_6] - 0.356 \text{ Cos } [x_6] \text{ Sin } [x_6] - 19.339 \text{ Sin } [x_1] \text{ Sin } [x_6] - 5.490 \text{ Sin } [x_2] \text{ Sin } [x_6] -$
$24.810 \text{ Sin } [x_3] \text{ Sin } [x_6] + 11.563 \text{ Sin } [x_4] \text{ Sin } [x_6] + 0.401 \text{ Sin } [x_5] \text{ Sin } [x_6] + 26.687 \text{ Sin } [x_6]^2$

(*15*)
Out[•]//NumberForm=

$47.442 - 71.611 \text{ Cos } [x_1] - 84.857 \text{ Cos } [x_1]^2 - 46.061 \text{ Cos } [x_2] + 25.895 \text{ Cos } [x_2]^2 - 56.170 \text{ Cos } [x_3] +$
$63.236 \text{ Cos } [x_1] \text{ Cos } [x_3] + 16.522 \text{ Cos } [x_2] \text{ Cos } [x_3] + 3.186 \text{ Cos } [x_3]^2 + 9.800 \text{ Cos } [x_4] +$
$0.923 \text{ Cos } [x_2] \text{ Cos } [x_4] + 44.084 \text{ Cos } [x_4]^2 - 69.122 \text{ Cos } [x_5] + 25.374 \text{ Cos } [x_5]^2 + 5.053 \text{ Cos } [x_6] +$
$21.297 \text{ Cos } [x_6]^2 - 395.248 \text{ Sin } [x_1] + 80.015 \text{ Cos } [x_1] \text{ Sin } [x_1] + 8.629 \text{ Cos } [x_2] \text{ Sin } [x_1] +$
$221.574 \text{ Cos } [x_3] \text{ Sin } [x_1] - 7.362 \text{ Cos } [x_4] \text{ Sin } [x_1] + 42.674 \text{ Cos } [x_5] \text{ Sin } [x_1] + 3.363 \text{ Cos } [x_6] \text{ Sin } [x_1] +$
$118.803 \text{ Sin } [x_1]^2 + 22.651 \text{ Sin } [x_2]^2 + 138.274 \text{ Sin } [x_3] - 8.598 \text{ Cos } [x_1] \text{ Sin } [x_3] + 38.175 \text{ Cos } [x_2] \text{ Sin } [x_3] -$
$149.554 \text{ Cos } [x_3] \text{ Sin } [x_3] + 0.211 \text{ Cos } [x_4] \text{ Sin } [x_3] + 42.158 \text{ Cos } [x_5] \text{ Sin } [x_3] - 8.804 \text{ Cos } [x_6] \text{ Sin } [x_3] +$
$38.937 \text{ Sin } [x_3]^2 + 84.241 \text{ Sin } [x_4] + 1.930 \text{ Sin } [x_4]^2 + 1.196 \text{ Sin } [x_5] + 19.789 \text{ Sin } [x_5]^2 + 37.038 \text{ Sin } [x_6] -$
$4.061 \text{ Cos } [x_1] \text{ Sin } [x_6] + 0.206 \text{ Cos } [x_2] \text{ Sin } [x_6] - 7.655 \text{ Cos } [x_3] \text{ Sin } [x_6] - 0.082 \text{ Cos } [x_4] \text{ Sin } [x_6] -$
$0.998 \text{ Cos } [x_5] \text{ Sin } [x_6] - 0.358 \text{ Cos } [x_6] \text{ Sin } [x_6] - 17.338 \text{ Sin } [x_1] \text{ Sin } [x_6] - 7.121 \text{ Sin } [x_2] \text{ Sin } [x_6] -$
$22.138 \text{ Sin } [x_3] \text{ Sin } [x_6] + 10.386 \text{ Sin } [x_4] \text{ Sin } [x_6] + 0.297 \text{ Sin } [x_5] \text{ Sin } [x_6] + 20.790 \text{ Sin } [x_6]^2$

(*16*)
Out[•]//NumberForm=

$66.508 - 48.184 \text{ Cos } [x_1] - 79.202 \text{ Cos } [x_1]^2 - 64.357 \text{ Cos } [x_2] + 40.749 \text{ Cos } [x_2]^2 - 110.814 \text{ Cos } [x_3] +$
$57.706 \text{ Cos } [x_1] \text{ Cos } [x_3] + 21.977 \text{ Cos } [x_2] \text{ Cos } [x_3] - 10.336 \text{ Cos } [x_3]^2 + 6.390 \text{ Cos } [x_4] +$
$0.764 \text{ Cos } [x_2] \text{ Cos } [x_4] + 56.606 \text{ Cos } [x_4]^2 - 73.251 \text{ Cos } [x_5] + 38.513 \text{ Cos } [x_5]^2 + 4.234 \text{ Cos } [x_6] +$
$46.043 \text{ Cos } [x_6]^2 - 411.674 \text{ Sin } [x_1] + 81.510 \text{ Cos } [x_1] \text{ Sin } [x_1] + 10.898 \text{ Cos } [x_2] \text{ Sin } [x_1] +$
$232.187 \text{ Cos } [x_3] \text{ Sin } [x_1] - 6.581 \text{ Cos } [x_4] \text{ Sin } [x_1] + 48.284 \text{ Cos } [x_5] \text{ Sin } [x_1] + 3.674 \text{ Cos } [x_6] \text{ Sin } [x_1] +$
$134.336 \text{ Sin } [x_1]^2 + 38.195 \text{ Sin } [x_2]^2 + 29.965 \text{ Sin } [x_3] - 32.772 \text{ Cos } [x_1] \text{ Sin } [x_3] + 53.479 \text{ Cos } [x_2] \text{ Sin } [x_3] -$
$107.433 \text{ Cos } [x_3] \text{ Sin } [x_3] + 3.157 \text{ Cos } [x_4] \text{ Sin } [x_3] + 39.465 \text{ Cos } [x_5] \text{ Sin } [x_3] - 8.242 \text{ Cos } [x_6] \text{ Sin } [x_3] +$
$67.892 \text{ Sin } [x_3]^2 + 71.126 \text{ Sin } [x_4] + 21.693 \text{ Sin } [x_4]^2 + 1.195 \text{ Sin } [x_5] + 33.496 \text{ Sin } [x_5]^2 + 36.414 \text{ Sin } [x_6] -$
$3.972 \text{ Cos } [x_1] \text{ Sin } [x_6] + 0.253 \text{ Cos } [x_2] \text{ Sin } [x_6] - 8.370 \text{ Cos } [x_3] \text{ Sin } [x_6] - 0.113 \text{ Cos } [x_4] \text{ Sin } [x_6] -$
$1.298 \text{ Cos } [x_5] \text{ Sin } [x_6] - 0.336 \text{ Cos } [x_6] \text{ Sin } [x_6] - 16.454 \text{ Sin } [x_1] \text{ Sin } [x_6] - 5.801 \text{ Sin } [x_2] \text{ Sin } [x_6] -$
$23.762 \text{ Sin } [x_3] \text{ Sin } [x_6] + 10.303 \text{ Sin } [x_4] \text{ Sin } [x_6] + 0.328 \text{ Sin } [x_5] \text{ Sin } [x_6] + 45.553 \text{ Sin } [x_6]^2$

(Continued)

TABLE A7.3 *(Continued)*

Models Obtained by NRM and SNRM Using the Bootstrap Data Separation Method (80 Training–15 Testing–5 Validation) (*Models Obtained Using the NRM and SNRM Methods*)(*(Related to Table 7.4)*)

(*17*)

Out[•]//NumberForm=

$43.208 - 46.746 \, \text{Cos} \, [x_1] - 90.206 \, \text{Cos} \, [x_1]^2 - 76.706 \, \text{Cos} \, [x_2] + 28.439 \, \text{Cos} \, [x_2]^2 - 70.340 \, \text{Cos} \, [x_3] +$
$61.568 \, \text{Cos} \, [x_1] \, \text{Cos} \, [x_3] + 25.726 \, \text{Cos} \, [x_2] \, \text{Cos} \, [x_3] + 3.690 \, \text{Cos} \, [x_3]^2 + 6.829 \, \text{Cos} \, [x_4] +$
$1.246 \, \text{Cos} \, [x_2] \, \text{Cos} \, [x_4] + 54.141 \, \text{Cos} \, [x_4]^2 - 56.334 \, \text{Cos} \, [x_5] + 25.681 \, \text{Cos} \, [x_5]^2 + 5.007 \, \text{Cos} \, [x_6] +$
$20.584 \, \text{Cos} \, [x_6]^2 - 397.229 \, \text{Sin} \, [x_1] + 69.692 \, \text{Cos} \, [x_1] \, \text{Sin} \, [x_1] + 9.046 \, \text{Cos} \, [x_2] \, \text{Sin} \, [x_1] +$
$230.393 \, \text{Cos} \, [x_3] \, \text{Sin} \, [x_1] - 7.033 \, \text{Cos} \, [x_4] \, \text{Sin} \, [x_1] + 34.528 \, \text{Cos} \, [x_5] \, \text{Sin} \, [x_1] + 3.503 \, \text{Cos} \, [x_6] \, \text{Sin} \, [x_1] +$
$113.205 \, \text{Sin} \, [x_1]^2 + 24.809 \, \text{Sin} \, [x_2]^2 + 120.292 \, \text{Sin} \, [x_3] - 22.363 \, \text{Cos} \, [x_1] \, \text{Sin} \, [x_3] + 67.051 \, \text{Cos} \, [x_2] \, \text{Sin} \, [x_3] -$
$144.300 \, \text{Cos} \, [x_3] \, \text{Sin} \, [x_3] + 2.945 \, \text{Cos} \, [x_4] \, \text{Sin} \, [x_3] + 39.051 \, \text{Cos} \, [x_5] \, \text{Sin} \, [x_3] - 8.871 \, \text{Cos} \, [x_6] \, \text{Sin} \, [x_3] +$
$46.250 \, \text{Sin} \, [x_3]^2 + 110.728 \, \text{Sin} \, [x_4] - 2.289 \, \text{Sin} \, [x_4]^2 + 1.189 \, \text{Sin} \, [x_5] + 18.875 \, \text{Sin} \, [x_5]^2 + 39.834 \, \text{Sin} \, [x_6] -$
$3.684 \, \text{Cos} \, [x_1] \, \text{Sin} \, [x_6] + 0.041 \, \text{Cos} \, [x_2] \, \text{Sin} \, [x_6] - 8.121 \, \text{Cos} \, [x_3] \, \text{Sin} \, [x_6] - 0.379 \, \text{Cos} \, [x_4] \, \text{Sin} \, [x_6] -$
$0.998 \, \text{Cos} \, [x_5] \, \text{Sin} \, [x_6] - 0.399 \, \text{Cos} \, [x_6] \, \text{Sin} \, [x_6] - 15.957 \, \text{Sin} \, [x_1] \, \text{Sin} \, [x_6] - 7.112 \, \text{Sin} \, [x_2] \, \text{Sin} \, [x_6] -$
$24.287 \, \text{Sin} \, [x_3] \, \text{Sin} \, [x_6] + 8.376 \, \text{Sin} \, [x_4] \, \text{Sin} \, [x_6] + 0.288 \, \text{Sin} \, [x_5] \, \text{Sin} \, [x_6] + 20.014 \, \text{Sin} \, [x_6]^2$

(*18*)

Out[•]//NumberForm=

$53.832 - 74.198 \, \text{Cos} \, [x_1] - 106.860 \, \text{Cos} \, [x_1]^2 - 40.713 \, \text{Cos} \, [x_2] + 29.252 \, \text{Cos} \, [x_2]^2 + 10.456 \, \text{Cos} \, [x_3] +$
$46.503 \, \text{Cos} \, [x_1] \, \text{Cos} \, [x_3] + 15.789 \, \text{Cos} \, [x_2] \, \text{Cos} \, [x_3] + 10.924 \, \text{Cos} \, [x_3]^2 + 6.112 \, \text{Cos} \, [x_4] +$
$0.162 \, \text{Cos} \, [x_2] \, \text{Cos} \, [x_4] + 44.172 \, \text{Cos} \, [x_4]^2 - 70.149 \, \text{Cos} \, [x_5] + 34.789 \, \text{Cos} \, [x_5]^2 + 5.397 \, \text{Cos} \, [x_6] +$
$25.718 \, \text{Cos} \, [x_6]^2 - 479.607 \, \text{Sin} \, [x_1] + 104.778 \, \text{Cos} \, [x_1] \, \text{Sin} \, [x_1] + 9.536 \, \text{Cos} \, [x_2] \, \text{Sin} \, [x_1] +$
$180.388 \, \text{Cos} \, [x_3] \, \text{Sin} \, [x_1] - 3.115 \, \text{Cos} \, [x_4] \, \text{Sin} \, [x_1] + 45.172 \, \text{Cos} \, [x_5] \, \text{Sin} \, [x_1] + 2.953 \, \text{Cos} \, [x_6] \, \text{Sin} \, [x_1] +$
$142.759 \, \text{Sin} \, [x_1]^2 + 25.689 \, \text{Sin} \, [x_2]^2 + 182.072 \, \text{Sin} \, [x_3] - 26.762 \, \text{Cos} \, [x_1] \, \text{Sin} \, [x_3] + 31.781 \, \text{Cos} \, [x_2] \, \text{Sin} \, [x_3] -$
$170.723 \, \text{Cos} \, [x_3] \, \text{Sin} \, [x_3] + 0.290 \, \text{Cos} \, [x_4] \, \text{Sin} \, [x_3] + 39.297 \, \text{Cos} \, [x_5] \, \text{Sin} \, [x_3] - 8.755 \, \text{Cos} \, [x_6] \, \text{Sin} \, [x_3] +$
$32.919 \, \text{Sin} \, [x_3]^2 + 75.863 \, \text{Sin} \, [x_4] + 7.568 \, \text{Sin} \, [x_4]^2 + 1.196 \, \text{Sin} \, [x_5] + 29.917 \, \text{Sin} \, [x_5]^2 + 36.044 \, \text{Sin} \, [x_6] -$
$3.425 \, \text{Cos} \, [x_1] \, \text{Sin} \, [x_6] - 0.069 \, \text{Cos} \, [x_2] \, \text{Sin} \, [x_6] - 8.597 \, \text{Cos} \, [x_3] \, \text{Sin} \, [x_6] + 0.143 \, \text{Cos} \, [x_4] \, \text{Sin} \, [x_6] -$
$1.464 \, \text{Cos} \, [x_5] \, \text{Sin} \, [x_6] - 0.393 \, \text{Cos} \, [x_6] \, \text{Sin} \, [x_6] - 15.142 \, \text{Sin} \, [x_1] \, \text{Sin} \, [x_6] - 8.320 \, \text{Sin} \, [x_2] \, \text{Sin} \, [x_6] -$
$23.711 \, \text{Sin} \, [x_3] \, \text{Sin} \, [x_6] + 11.684 \, \text{Sin} \, [x_4] \, \text{Sin} \, [x_6] + 0.371 \, \text{Sin} \, [x_5] \, \text{Sin} \, [x_6] + 25.264 \, \text{Sin} \, [x_6]^2$

(*19*)

Out[•]//NumberForm=

$47.014 - 69.596 \, \text{Cos} \, [x_1] - 96.224 \, \text{Cos} \, [x_1]^2 - 43.942 \, \text{Cos} \, [x_2] + 25.631 \, \text{Cos} \, [x_2]^2 - 29.140 \, \text{Cos} \, [x_3] +$
$63.526 \, \text{Cos} \, [x_1] \, \text{Cos} \, [x_3] + 15.135 \, \text{Cos} \, [x_2] \, \text{Cos} \, [x_3] + 6.565 \, \text{Cos} \, [x_3]^2 + 7.704 \, \text{Cos} \, [x_4] +$
$0.123 \, \text{Cos} \, [x_2] \, \text{Cos} \, [x_4] + 40.630 \, \text{Cos} \, [x_4]^2 - 62.718 \, \text{Cos} \, [x_5] + 27.068 \, \text{Cos} \, [x_5]^2 + 4.554 \, \text{Cos} \, [x_6] +$
$23.213 \, \text{Cos} \, [x_6]^2 - 434.584 \, \text{Sin} \, [x_1] + 81.645 \, \text{Cos} \, [x_1] \, \text{Sin} \, [x_1] + 7.901 \, \text{Cos} \, [x_2] \, \text{Sin} \, [x_1] +$
$227.081 \, \text{Cos} \, [x_3] \, \text{Sin} \, [x_1] - 6.312 \, \text{Cos} \, [x_4] \, \text{Sin} \, [x_1] + 36.199 \, \text{Cos} \, [x_5] \, \text{Sin} \, [x_1] + 3.783 \, \text{Cos} \, [x_6] \, \text{Sin} \, [x_1] +$
$124.790 \, \text{Sin} \, [x_1]^2 + 21.908 \, \text{Sin} \, [x_2]^2 + 178.013 \, \text{Sin} \, [x_3] - 11.753 \, \text{Cos} \, [x_1] \, \text{Sin} \, [x_3] + 37.226 \, \text{Cos} \, [x_2] \, \text{Sin} \, [x_3] -$
$178.812 \, \text{Cos} \, [x_3] \, \text{Sin} \, [x_3] + 1.323 \, \text{Cos} \, [x_4] \, \text{Sin} \, [x_3] + 43.042 \, \text{Cos} \, [x_5] \, \text{Sin} \, [x_3] - 8.698 \, \text{Cos} \, [x_6] \, \text{Sin} \, [x_3] +$
$34.799 \, \text{Sin} \, [x_3]^2 + 83.200 \, \text{Sin} \, [x_4] - 1.282 \, \text{Sin} \, [x_4]^2 + 1.214 \, \text{Sin} \, [x_5] + 20.896 \, \text{Sin} \, [x_5]^2 + 35.745 \, \text{Sin} \, [x_6] -$
$3.444 \, \text{Cos} \, [x_1] \, \text{Sin} \, [x_6] + 0.196 \, \text{Cos} \, [x_2] \, \text{Sin} \, [x_6] - 8.089 \, \text{Cos} \, [x_3] \, \text{Sin} \, [x_6] - 0.066 \, \text{Cos} \, [x_4] \, \text{Sin} \, [x_6] -$
$0.780 \, \text{Cos} \, [x_5] \, \text{Sin} \, [x_6] - 0.321 \, \text{Cos} \, [x_6] \, \text{Sin} \, [x_6] - 14.638 \, \text{Sin} \, [x_1] \, \text{Sin} \, [x_6] - 6.707 \, \text{Sin} \, [x_2] \, \text{Sin} \, [x_6] -$
$23.183 \, \text{Sin} \, [x_3] \, \text{Sin} \, [x_6] + 9.893 \, \text{Sin} \, [x_4] \, \text{Sin} \, [x_6] + 0.289 \, \text{Sin} \, [x_5] \, \text{Sin} \, [x_6] + 22.756 \, \text{Sin} \, [x_6]^2$

(Continued)

TABLE A7.3 *(Continued)*

Models Obtained by NRM and SNRM Using the Bootstrap Data Separation Method (80 Training–15 Testing–5 Validation) (*Models Obtained Using the NRM and SNRM Methods*)(*(Related to Table 7.4)*)

(*20*)

Out[•]//NumberForm=

$60.644 - 52.141 \cos [x_1] - 83.523 \cos [x_1]^2 - 52.110 \cos [x_2] + 36.548 \cos [x_2]^2 - 106.536 \cos [x_3] + 50.756 \cos [x_1] \cos [x_3] + 21.437 \cos [x_2] \cos [x_3] - 0.972 \cos [x_3]^2 + 14.654 \cos [x_4] - 0.498 \cos [x_2] \cos [x_4] + 52.972 \cos [x_4]^2 - 90.091 \cos [x_5] + 28.506 \cos [x_5]^2 + 4.476 \cos [x_6] + 30.722 \cos [x_6]^2 - 370.155 \sin [x_1] + 77.403 \cos [x_1] \sin [x_1] + 10.715 \cos [x_2] \sin [x_1] + 206.541 \cos [x_3] \sin [x_1] - 8.475 \cos [x_4] \sin [x_1] + 61.834 \cos [x_5] \sin [x_1] + 2.926 \cos [x_6] \sin [x_1] + 118.027 \sin [x_1]^2 + 33.486 \sin [x_2]^2 + 49.906 \sin [x_3] - 23.365 \cos [x_1] \sin [x_3] + 40.969 \cos [x_2] \sin [x_3] - 89.920 \cos [x_3] \sin [x_3] - 3.858 \cos [x_4] \sin [x_3] + 34.691 \cos [x_5] \sin [x_3] - 7.741 \cos [x_6] \sin [x_3] + 51.927 \sin [x_3]^2 + 77.926 \sin [x_4] + 13.820 \sin [x_4]^2 + 1.167 \sin [x_5] + 28.501 \sin [x_5]^2 + 32.203 \sin [x_6] - 2.751 \cos [x_1] \sin [x_6] + 0.001 \cos [x_2] \sin [x_6] - 8.735 \cos [x_3] \sin [x_6] + 0.504 \cos [x_4] \sin [x_6] - 1.524 \cos [x_5] \sin [x_6] - 0.362 \cos [x_6] \sin [x_6] - 13.185 \sin [x_1] \sin [x_6] - 7.766 \sin [x_2] \sin [x_6] - 24.094 \sin [x_3] \sin [x_6] + 13.178 \sin [x_4] \sin [x_6] + 0.465 \sin [x_5] \sin [x_6] + 30.387 \sin [x_6]^2$

TABLE A7.4

Models Obtained by NRM and SNRM Using the Bootstrap Data Separation Method*) (*(80 Training–20 Testing) Training Was Divided into Two Parts–64 Training–16 Validation*) (*Models Obtained Using the NRM and SNRM Methods*)(*(Related to Table 7.5)*)

(*Models proposed by NRM*)

*In[•]:=(*1*)*

Out[•]//NumberForm=

$234.988 - 579.729 \cos [x_1] - 717.050 \cos [x_1]^2 - 87.588 \cos [x_2] + 246.736 \cos [x_2]^2 - 49.056 \cos [x_3] + 79.210 \cos [x_1] \cos [x_3] + 39.101 \cos [x_2] \cos [x_3] + 370.302 \cos [x_3]^2 + 18.491 \cos [x_4] + 0.533 \cos [x_2] \cos [x_4] + 252.935 \cos [x_4]^2 - 88.023 \cos [x_5] + 284.657 \cos [x_5]^2 + 5.285 \cos [x_6] + 307.100 \cos [x_6]^2 - 2205.990 \sin [x_1] + 559.708 \cos [x_1] \sin [x_1] - 7.412 \cos [x_2] \sin [x_1] + 254.513 \cos [x_3] \sin [x_1] - 31.440 \cos [x_4] \sin [x_1] + 37.791 \cos [x_5] \sin [x_1] + 4.779 \cos [x_6] \sin [x_1] + 305.056 \sin [x_1]^2 + 245.473 \sin [x_2]^2 + 382.768 \sin [x_3] + 6.751 \cos [x_1] \sin [x_3] + 91.479 \cos [x_2] \sin [x_3] - 194.686 \cos [x_3] \sin [x_3] + 16.337 \cos [x_4] \sin [x_3] + 66.790 \cos [x_5] \sin [x_3] - 10.441 \cos [x_6] \sin [x_3] + 291.848 \sin [x_3]^2 + 5.492 \sin [x_4] + 254.307 \sin [x_4]^2 + 1.467 \sin [x_5] + 278.026 \sin [x_5]^2 + 12.258 \sin [x_6] + 1.457 \cos [x_1] \sin [x_6] + 1.523 \cos [x_2] \sin [x_6] - 10.078 \cos [x_3] \sin [x_6] - 0.735 \cos [x_4] \sin [x_6] - 4.656 \cos [x_5] \sin [x_6] - 0.408 \cos [x_6] \sin [x_6] + 4.768 \sin [x_1] \sin [x_6] - 4.047 \sin [x_2] \sin [x_6] - 25.107 \sin [x_3] \sin [x_6] + 9.166 \sin [x_4] \sin [x_6] + 1.683 \sin [x_5] \sin [x_6] + 306.587 \sin [x_6]^2$

(Continued)

TABLE A7.4 *(Continued)*

Models Obtained by NRM and SNRM Using the Bootstrap Data Separation Method*) (*(80 Training–20 Testing) Training Was Divided into Two Parts–64 Training–16 Validation*) (*Models Obtained Using the NRM and SNRM Methods*)(*(Related to Table 7.5)*)

In[•]:=(*2*)

Out[•]//NumberForm=

– 102.579 + 103.029 Cos [x_1] – 460.093 Cos [x_1]^2 – 172.677 Cos [x_2] – 102.867 Cos [x_2]^2 + 328.331 Cos [x_3] + 45.757 Cos [x_1] Cos [x_3] + 54.995 Cos [x_2] Cos [x_3] + 337.628 Cos [x_3]^2 – 6.207 Cos [x_4] – 5.355 Cos [x_2] Cos [x_4] – 49.300 Cos [x_4]^2 – 96.932 Cos [x_5] – 120.257 Cos [x_5]^2 + 7.298 Cos [x_6] – 137.382 Cos [x_6]^2 – 579.774 Sin [x_1] + 14.350 Cos [x_1] Sin [x_1] – 6.967 Cos [x_2] Sin [x_1] + 245.778 Cos [x_3] Sin [x_1] + 5.399 Cos [x_4] Sin [x_1] + 39.430 Cos [x_5] Sin [x_1] – 2.346 Cos [x_6] Sin [x_1] – 118.507 Sin [x_1]^2 – 107.038 Sin [x_2]^2 + 1372.673 Sin [x_3] – 101.080 Cos [x_1] Sin [x_3] + 175.594 Cos [x_2] Sin [x_3] – 538.359 Cos [x_3] Sin [x_3] + 5.498 Cos [x_4] Sin [x_3] + 55.936 Cos [x_5] Sin [x_3] – 5.589 Cos [x_6] Sin [x_3] – 139.000 Sin [x_3]^2 + 127.669 Sin [x_4] – 112.040 Sin [x_4]^2 + 1.443 Sin [x_5] – 115.220 Sin [x_5]^2 + 51.950 Sin [x_6] – 12.277 Cos [x_1] Sin [x_6] + 0.889 Cos [x_2] Sin [x_6] – 9.747 Cos [x_3] Sin [x_6] + 0.375 Cos [x_4] Sin [x_6] – 7.680 Cos [x_5] Sin [x_6] – 0.217 Cos [x_6] Sin [x_6] – 35.826 Sin [x_1] Sin [x_6] – 8.078 Sin [x_2] Sin [x_6] – 24.946 Sin [x_3] Sin [x_6] + 12.931 Sin [x_4] Sin [x_6] + 1.834 Sin [x_5] Sin [x_6] – 137.614 Sin [x_6]^2

In[•]:=(*3*)

Out[•]//NumberForm=

– 304.139 – 207.644 Cos [x_1] – 1203.740 Cos [x_1]^2 – 131.656 Cos [x_2] – 313.593 Cos [x_2]^2 + 1221.300 Cos [x_3] + 49.818 Cos [x_1] Cos [x_3] + 36.844 Cos [x_2] Cos [x_3] + 1239.234 Cos [x_3]^2 + 28.470 Cos [x_4] – 2.987 Cos [x_2] Cos [x_4] – 228.928 Cos [x_4]^2 – 46.909 Cos [x_5] – 362.064 Cos [x_5]^2 + 2.708 Cos [x_6] – 402.660 Cos [x_6]^2 – 1745.810 Sin [x_1] + 323.669 Cos [x_1] Sin [x_1] + 12.621 Cos [x_2] Sin [x_1] + 286.209 Cos [x_3] Sin [x_1] – 6.716 Cos [x_4] Sin [x_1] + 27.539 Cos [x_5] Sin [x_1] + 2.682 Cos [x_6] Sin [x_1] – 351.276 Sin [x_1]^2 – 316.122 Sin [x_2]^2 + 4139.004 Sin [x_3] – 109.556 Cos [x_1] Sin [x_3] + 119.322 Cos [x_2] Sin [x_3] – 1433.624 Cos [x_3] Sin [x_3] – 20.391 Cos [x_4] Sin [x_3] + 39.623 Cos [x_5] Sin [x_3] – 5.556 Cos [x_6] Sin [x_3] – 418.683 Sin [x_3]^2 + 198.470 Sin [x_4] – 332.914 Sin [x_4]^2 + 1.176 Sin [x_5] – 370.945 Sin [x_5]^2 + 74.358 Sin [x_6] – 10.287 Cos [x_1] Sin [x_6] – 0.178 Cos [x_2] Sin [x_6] – 12.902 Cos [x_3] Sin [x_6] – 0.182 Cos [x_4] Sin [x_6] – 1.570 Cos [x_5] Sin [x_6] – 0.203 Cos [x_6] Sin [x_6] – 32.681 Sin [x_1] Sin [x_6] – 14.993 Sin [x_2] Sin [x_6] – 36.295 Sin [x_3] Sin [x_6] + 11.784 Sin [x_4] Sin [x_6] + 0.466 Sin [x_5] Sin [x_6] – 402.953 Sin [x_6]^2

In[•]:=(*4*)

Out[•]//NumberForm=

489.685 – 227.283 Cos [x_1] + 31.811 Cos [x_1]^2 – 210.943 Cos [x_2] + 519.493 Cos [x_2]^2 – 1150.314 Cos [x_3] + 46.973 Cos [x_1] Cos [x_3] + 49.679 Cos [x_2] Cos [x_3] – 598.981 Cos [x_3]^2 + 34.646 Cos [x_4] – 3.056 Cos [x_2] Cos [x_4] + 559.722 Cos [x_4]^2 – 102.066 Cos [x_5] + 592.258 Cos [x_5]^2 + 2.297 Cos [x_6] + 648.214 Cos [x_6]^2 – 1127.851 Sin [x_1] + 252.812 Cos [x_1] Sin [x_1] + 41.854 Cos [x_2] Sin [x_1] + 164.262 Cos [x_3] Sin [x_1] – 5.742 Cos [x_4] Sin [x_1] + 47.379 Cos [x_5] Sin [x_1] + 4.890 Cos [x_6] Sin [x_1] + 599.714 Sin [x_1]^2 + 512.047 Sin [x_2]^2 – 2861.299 Sin [x_3] – 16.590 Cos [x_1] Sin [x_3] + 170.489 Cos [x_2] Sin [x_3] + 943.193 Cos [x_3] Sin [x_3] – 28.513 Cos [x_4] Sin [x_3] + 62.065 Cos [x_5] Sin [x_3] – 7.255 Cos [x_6] Sin [x_3] + 659.038 Sin [x_3]^2 + 55.214 Sin [x_4] + 530.984 Sin [x_4]^2 + 1.431 Sin [x_5] + 592.201 Sin [x_5]^2 + 47.485 Sin [x_6] – 6.831 Cos [x_1] Sin [x_6] + 0.902 Cos [x_2] Sin [x_6] –

(Continued)

TABLE A7.4 *(Continued)*

Models Obtained by NRM and SNRM Using the Bootstrap Data Separation Method*) (*(80 Training–20 Testing) Training Was Divided into Two Parts–64 Training–16 Validation*) (*Models Obtained Using the NRM and SNRM Methods*)(*(Related to Table 7.5)*)

6.394 Cos $[x_3]$ Sin $[x_6]$ – 0.447 Cos $[x_4]$ Sin $[x_6]$ – 1.584 Cos $[x_5]$ Sin $[x_6]$ – 0.598 Cos $[x_6]$ Sin $[x_6]$ – 23.508 Sin $[x_1]$ Sin $[x_6]$ – 11.926 Sin $[x_2]$ Sin $[x_6]$ – 25.519 Sin $[x_3]$ Sin $[x_6]$ + 13.402 Sin $[x_4]$ Sin $[x_6]$ + 0.675 Sin $[x_5]$ Sin $[x_6]$ + 647.699 Sin $[x_6]^2$

$In[\bullet]:=(*5*)$

$Out[\bullet]//NumberForm=$

226.804 – 548.744 Cos $[x_1]$ – 832.121 Cos $[x_1]^2$ + 16.635 Cos $[x_2]$ + 238.308 Cos $[x_2]^2$ – 56.769 Cos $[x_3]$ + 58.458 Cos $[x_1]$ Cos $[x_3]$ – 2.234 Cos $[x_2]$ Cos $[x_3]$ + 553.751 Cos $[x_3]^2$ – 24.190 Cos $[x_4]$ – 1.023 Cos $[x_2]$ Cos $[x_4]$ + 167.548 Cos $[x_4]^2$ – 156.156 Cos $[x_5]$ + 270.079 Cos $[x_5]^2$ + 5.612 Cos $[x_6]$ + 299.917 Cos $[x_6]^2$ – 2350.177 Sin $[x_1]$ + 569.308 Cos $[x_1]$ Sin $[x_1]$ + 16.072 Cos $[x_2]$ Sin $[x_1]$ + 279.645 Cos $[x_3]$ Sin $[x_1]$ – 8.850 Cos $[x_4]$ Sin $[x_1]$ + 97.691 Cos $[x_5]$ Sin $[x_1]$ + 1.880 Cos $[x_6]$ Sin $[x_1]$ + 284.505 Sin $[x_1]^2$ + 238.364 Sin $[x_2]^2$ + 745.330 Sin $[x_3]$ – 31.221 Sin $[x_1]$ Sin $[x_3]$ – 29.112 Cos $[x_2]$ Sin $[x_3]$ – 216.640 Cos $[x_3]$ Sin $[x_3]$ + 39.313 Cos $[x_4]$ Sin $[x_3]$ + 72.715 Cos $[x_5]$ Sin $[x_3]$ – 7.856 Cos $[x_6]$ Sin $[x_3]$ + 276.688 Sin $[x_3]^2$ – 134.950 Sin $[x_4]$ + 246.460 Sin $[x_4]^2$ + 1.308 Sin $[x_5]$ + 266.077 Sin $[x_5]^2$ – 16.225 Sin $[x_6]$ + 0.408 Cos $[x_1]$ Sin $[x_6]$ + 2.014 Cos $[x_2]$ Sin $[x_6]$ + 0.778 Cos $[x_3]$ Sin $[x_6]$ – 1.003 Cos $[x_4]$ Sin $[x_6]$ + 0.181 Cos $[x_5]$ Sin $[x_6]$ – 0.132 Cos $[x_6]$ Sin $[x_6]$ – 0.399 Sin $[x_1]$ Sin $[x_6]$ + 6.181 Sin $[x_2]$ Sin $[x_6]$ + 8.496 Sin $[x_3]$ Sin $[x_6]$ + 2.537 Sin $[x_4]$ Sin $[x_6]$ – 0.401 Sin $[x_5]$ Sin $[x_6]$ + 299.613 Sin $[x_6]^2$

$In[\bullet]:=(*6*)$

$Out[\bullet]//NumberForm=$

272.717 – 506.669 Cos $[x_1]$ – 630.240 Cos $[x_1]^2$ + 33.179 Cos $[x_2]$ + 282.330 Cos $[x_2]^2$ – 31.464 Cos $[x_3]$ + 37.616 Cos $[x_1]$ Cos $[x_3]$ + 4.182 Cos $[x_2]$ Cos $[x_3]$ + 74.177 Cos $[x_3]^2$ + 12.543 Cos $[x_4]$ + 10.323 Cos $[x_2]$ Cos $[x_4]$ + 416.433 Cos $[x_4]^2$ – 100.741 Cos $[x_5]$ + 327.758 Cos $[x_5]^2$ + 8.200 Cos $[x_6]$ + 368.452 Cos $[x_6]^2$ – 2018.056 Sin $[x_1]$ + 517.629 Cos $[x_1]$ Sin $[x_1]$ – 29.372 Cos $[x_2]$ Sin $[x_1]$ + 90.384 Cos $[x_3]$ Sin $[x_1]$ – 2.938 Cos $[x_4]$ Sin $[x_1]$ + 68.561 Cos $[x_5]$ Sin $[x_1]$ + 2.376 Cos $[x_6]$ Sin $[x_1]$ + 344.163 Sin $[x_1]^2$ + 283.690 Sin $[x_2]^2$ – 355.317 Sin $[x_3]$ – 11.380 Cos $[x_1]$ Sin $[x_3]$ – 4.332 Cos $[x_2]$ Sin $[x_3]$ – 40.750 Cos $[x_3]$ Sin $[x_3]$ – 7.661 Cos $[x_4]$ Sin $[x_3]$ + 41.272 Cos $[x_5]$ Sin $[x_3]$ – 11.137 Cos $[x_6]$ Sin $[x_3]$ + 347.855 Sin $[x_3]^2$ +)230.283 Sin $[x_4]$ + 295.255 Sin $[x_4]^2$ + 1.417 Sin $[x_5]$ + 325.941 Sin $[x_5]^2$ + 48.360 Sin $[x_6]$ + 1.417 Cos $[x_1]$ Sin $[x_6]$ + 1.502 Cos $[x_2]$ Sin $[x_6]$ – 17.621 Cos $[x_3]$ Sin $[x_6]$ – 2.627 Cos $[x_4]$ Sin $[x_6]$ – 4.218 Cos $[x_5]$ Sin $[x_6]$ – 0.349 Cos $[x_6]$ Sin $[x_6]$ + 8.460 Sin $[x_1]$ Sin $[x_6]$ – 5.828 Sin $[x_2]$ Sin $[x_6]$ – 49.123 Sin $[x_3]$ Sin $[x_6]$ – 2.803 Sin $[x_4]$ Sin $[x_6]$ + 0.485 Sin $[x_5]$ Sin $[x_6]$ + 367.664 Sin $[x_6]^2$

$In[\bullet]:=(*7*)$

$Out[\bullet]//NumberForm=$

152.046 – 119.152 Cos $[x_1]$ – 236.848 Cos $[x_1]^2$ – 192.646 Cos $[x_2]$ + 168.930 Cos $[x_2]^2$ – 190.359 Cos $[x_3]$ + 48.932 Cos $[x_1]$ Cos $[x_3]$ + 73.179 Cos $[x_2]$ Cos $[x_3]$ – 113.630 Cos $[x_3]^2$ + 44.313 Cos $[x_4]$ + 2.630 Cos $[x_2]$ Cos $[x_4]$ + 213.823 Cos $[x_4]^2$ – 122.081 Cos $[x_5]$ + 181.091 Cos $[x_5]^2$ + 6.509 Cos $[x_6]$ + 204.778 Cos $[x_6]^2$ – 828.754 Sin $[x_1]$ + 184.704 Cos $[x_1]$ Sin $[x_1]$ – 2.486 Cos $[x_2]$ Sin $[x_1]$ + 218.332 Cos $[x_3]$ Sin $[x_1]$ – 45.395 Cos $[x_4]$ Sin $[x_1]$ + 67.730 Cos $[x_5]$ Sin $[x_1]$ + 3.586 Cos $[x_6]$ Sin $[x_1]$ + 188.492 Sin $[x_1]^2$ + 158.919 Sin $[x_2]^2$ – 436.065 Sin $[x_3]$ – 63.218 Cos $[x_1]$ Sin $[x_3]$ +

(Continued)

TABLE A7.4 *(Continued)*

Models Obtained by NRM and SNRM Using the Bootstrap Data Separation Method*) (*(80 Training–20 Testing) Training Was Divided into Two Parts–64 Training–16 Validation*) (*Models Obtained Using the NRM and SNRM Methods*)(*(Related to Table 7.5)*)

185.921 Cos $[x_1]$ Sin $[x_3]$ − 11.275 Cos $[x_3]$ Sin $[x_3]$ + 2.958 Cos $[x_4]$ Sin $[x_3]$ + 51.030 Cos $[x_5]$ Sin $[x_3]$ − 10.640 Cos $[x_6]$ Sin $[x_3]$ + 200.598 Sin $[x_3]^2$ + 99.717 Sin $[x_4]$ + 163.573 Sin $[x_4]^2$ + 1.534 Sin $[x_5]$ + 186.244 Sin $[x_5]^2$ + 46.662 Sin $[x_6]$ − 2.398 Cos $[x_1]$ Sin $[x_6]$ − 0.077 Cos $[x_2]$ Sin $[x_6]$ − 15.181 Cos $[x_3]$ Sin $[x_6]$ − 2.724 Cos $[x_4]$ Sin $[x_6]$ − 1.092 Cos $[x_5]$ Sin $[x_6]$ − 0.462 Cos $[x_6]$ Sin $[x_6]$ − 11.483 Sin $[x_1]$ Sin $[x_6]$ + 14.691 Sin $[x_2]$ Sin $[x_6]$ − 44.773 Sin $[x_3]$ Sin $[x_6]$ − 3.208 Sin $[x_4]$ Sin $[x_6]$ + 0.329 Sin $[x_5]$ Sin $[x_6]$ + 204.310 Sin $[x_6]^2$

In[•]:=(*8*)

Out[•]//NumberForm=

455.399 − 1034.240 Cos $[x_1]$ − 1237.654 Cos $[x_1]^2$ − 159.896 Cos $[x_2]$ + 477.708 Cos $[x_2]^2$ + 60.957 Cos $[x_3]$ + 7.492 Cos $[x_1]$ Cos $[x_3]$ + 57.061 Cos $[x_2]$ Cos $[x_3]$ + 583.979 Cos $[x_3]^2$ − 35.350 Cos $[x_4]$ + 0.961 Cos $[x_2]$ Cos $[x_4]$ + 519.734 Cos $[x_4]^2$ − 119.236 Cos $[x_5]$ + 544.223 Cos $[x_5]^2$ + 7.639 Cos $[x_6]$ + 612.631 Cos $[x_6]^2$ − 3786.085 Sin $[x_1]$ + 962.297 Cos $[x_1]$ Sin $[x_1]$ + 20.161 Cos $[x_2]$ Sin $[x_1]$ − 36.670 Cos $[x_3]$ Sin $[x_1]$ + 0.635 Cos $[x_4]$ Sin $[x_1]$ + 75.419 Cos $[x_5]$ Sin $[x_1]$ − 1.028 Cos $[x_6]$ Sin $[x_1]$ + 570.961 Sin $[x_1]^2$ + 478.323 Sin $[x_2]^2$ + 88.743 Sin $[x_3]$ + 76.382 Cos $[x_1]$ Sin $[x_3]$ + 133.757 Cos $[x_2]$ Sin $[x_3]$ − 21.180 Cos $[x_3]$ Sin $[x_3]$ + 40.911 Cos $[x_4]$ Sin $[x_3]$ + 45.047 Cos $[x_5]$ Sin $[x_3]$ − 7.149 Cos $[x_6]$ Sin $[x_3]$ + 572.522 Sin $[x_3]^2$ + 56.515 Sin $[x_4]$ + 495.437 Sin $[x_4]^2$ + 1.929 Sin $[x_5]$ + 545.970 Sin $[x_5]^2$ + 48.583 Sin $[x_6]$ − 2.018 Cos $[x_1]$ Sin $[x_6]$ + 0.689 Cos $[x_2]$ Sin $[x_6]$ − 15.627 Cos $[x_3]$ Sin $[x_6]$ − 1.279 Cos $[x_4]$ Sin $[x_6]$ − 6.205 Cos $[x_5]$ Sin $[x_6]$ − 0.500 Cos $[x_6]$ Sin $[x_6]$ − 7.941 Sin $[x_1]$ Sin $[x_6]$ − 0.979 Sin $[x_2]$ Sin $[x_6]$ − 44.351 Sin $[x_3]$ Sin $[x_6]$ + 1.532 Sin $[x_4]$ Sin $[x_6]$ + 0.866 Sin $[x_5]$ Sin $[x_6]$ + 611.807 Sin $[x_6]^2$

In[•]:=(*9*)

Out[•]//NumberForm=

425.617 − 668.460 Cos $[x_1]$ − 859.066 Cos $[x_1]^2$ − 180.065 Cos $[x_2]$ + 446.929 Cos $[x_2]^2$ − 117.921 Cos $[x_3]$ + 3.253 Cos $[x_1]$ Cos $[x_3]$ + 55.086 Cos $[x_2]$ Cos $[x_3]$ + 305.547 Cos $[x_3]^2$ + 12.695 Cos $[x_4]$ − 4.940 Cos $[x_2]$ Cos $[x_4]$ + 464.047 Cos $[x_4]^2$ − 3.540 Cos $[x_5]$ + 504.841 Cos $[x_5]^2$ − 2.278 Cos $[x_6]$ + 578.327 Cos $[x_6]^2$ − 2975.465 Sin $[x_1]$ + 738.992 Cos $[x_1]$ Sin $[x_1]$ + 9.144 Cos $[x_2]$ Sin $[x_1]$ + 52.685 Cos $[x_3]$ Sin $[x_1]$ − 15.523 Cos $[x_4]$ Sin $[x_1]$ − 6.198 Cos $[x_5]$ Sin $[x_1]$ + 5.738 Cos $[x_6]$ Sin $[x_1]$ + 526.815 Sin $[x_1]^2$ + 445.663 Sin $[x_2]^2$ − 406.121 Sin $[x_3]$ − 68.497 Cos $[x_1]$ Sin $[x_3]$ + 168.288 Cos $[x_2]$ Sin $[x_3]$ + 66.952 Cos $[x_3]$ Sin $[x_3]$ + 5.838 Cos $[x_4]$ Sin $[x_3]$ + 24.688 Cos $[x_5]$ Sin $[x_3]$ − 3.375 Cos $[x_6]$ Sin $[x_3]$ + 536.149 Sin $[x_3]^2$ + 3.772 Sin $[x_4]$ + 464.125 Sin $[x_4]^2$ + 1.258 Sin $[x_5]$ + 497.532 Sin $[x_5]^2$ + 20.802 Sin $[x_6]$ − 2.359 Cos $[x_1]$ Sin $[x_6]$ − 0.075 Cos $[x_2]$ Sin $[x_6]$ − 8.629 Cos $[x_3]$ Sin $[x_6]$ + 0.739 Cos $[x_4]$ Sin $[x_6]$ − 2.405 Cos $[x_5]$ Sin $[x_6]$ − 0.313 Cos $[x_6]$ Sin $[x_6]$ − 6.186 Sin $[x_1]$ Sin $[x_6]$ − 5.686 Sin $[x_2]$ Sin $[x_6]$ − 27.295 Sin $[x_3]$ Sin $[x_6]$ + 17.781 Sin $[x_4]$ Sin $[x_6]$ + 1.006 Sin $[x_5]$ Sin $[x_6]$ + 578.092 Sin $[x_6]^2$

In[•]:=(*10*)

Out[•]//NumberForm=

397.342 − 122.170 Cos $[x_1]$ − 30.383 Cos $[x_1]^2$ − 91.763 Cos $[x_2]$ + 414.299 Cos $[x_2]^2$ − 961.765 Cos $[x_3]$ + 34.836 Cos $[x_1]$ Cos $[x_3]$ + 34.593 Cos $[x_2]$ Cos $[x_3]$ − 519.215 Cos $[x_3]^2$ +

(Continued)

TABLE A7.4 *(Continued)*

Models Obtained by NRM and SNRM Using the Bootstrap Data Separation Method*) (*(80 Training–20 Testing) Training Was Divided into Two Parts–64 Training–16 Validation*) (*Models Obtained Using the NRM and SNRM Methods*)(*(Related to Table 7.5)*)

1.497 Cos $[x_4]$ + 1.154 Cos $[x_2]$ Cos $[x_4]$ + 513.006 Cos $[x_4]^2$ − 98.813 Cos $[x_5]$ + 479.358 Cos $[x_5]^2$ + 4.855 Cos $[x_6]$ + 532.841 Cos $[x_6]^2$ − 1038.582 Sin $[x_1]$ + 240.300 Cos $[x_1]$ Sin $[x_1]$ − 6.648 Cos $[x_2]$ Sin $[x_1]$ + 222.389 Cos $[x_3]$ Sin $[x_1]$ − 14.232 Cos $[x_4]$ Sin $[x_1]$ + 48.044 Cos $[x_5]$ Sin $[x_1]$ + 2.448 Cos $[x_6]$ Sin $[x_1]$ + 476.860 Sin $[x_1]^2$ + 414.248 Sin $[x_2]^2$ − 2295.015 Sin $[x_3]$ − 116.267 Cos $[x_1]$ Sin $[x_3]$ + 96.223 Cos $[x_2]$ Sin $[x_3]$ + 718.542 Cos $[x_3]$ Sin $[x_3]$ + 16.786 Cos $[x_4]$ Sin $[x_3]$ + 52.709 Cos $[x_5]$ Sin $[x_3]$ − 7.756 Cos $[x_6]$ Sin $[x_3]$ + 530.134 Sin $[x_3]^2$ + 164.690 Sin $[x_4]$ + 429.810 Sin $[x_4]^2$ + 1.256 Sin $[x_5]$ + 481.775 Sin $[x_5]^2$ + 39.552 Sin $[x_6]$ − 3.051 Cos $[x_1]$ Sin $[x_6]$ + 0.701 Cos $[x_2]$ Sin $[x_6]$ − 11.410 Cos $[x_3]$ Sin $[x_6]$ + 0.257 Cos $[x_4]$ Sin $[x_6]$ − 3.705 Cos $[x_5]$ Sin $[x_6]$ − 0.306 Cos $[x_6]$ Sin $[x_6]$ − 13.243 Sin $[x_1]$ Sin $[x_6]$ − 4.892 Sin $[x_2]$ Sin $[x_6]$ − 32.938 Sin $[x_3]$ Sin $[x_6]$ + 10.144 Sin $[x_4]$ Sin $[x_6]$ + 0.327 Sin $[x_5]$ Sin $[x_6]$ + 532.396 Sin $[x_6]^2$

In[•]:=(*11*)

Out[•]//NumberForm=

−177.454 − 451.284 Cos $[x_1]$ − 1014.302 Cos $[x_1]^2$ + 97.389 Cos $[x_2]$ − 184.326 Cos $[x_2]^2$ + 1254.411 Cos $[x_3]$ + 28.465 Cos $[x_1]$ Cos $[x_3]$ − 20.040 Cos $[x_2]$ Cos $[x_3]$ + 1034.003 Cos $[x_3]^2$ + 15.302 Cos $[x_4]$ + 15.021 Cos $[x_2]$ Cos $[x_4]$ − 202.285 Cos $[x_4]^2$ + 3.040 Cos $[x_5]$ − 209.388 Cos $[x_5]^2$ + 0.715 Cos $[x_6]$ − 239.570 Cos $[x_6]^2$ − 1748.116 Sin $[x_1]$ + 497.551 Cos $[x_1]$ Sin $[x_1]$ − 33.906 Cos $[x_2]$ Sin $[x_1]$ + 139.352 Cos $[x_3]$ Sin $[x_1]$ + 0.304 Cos $[x_4]$ Sin $[x_1]$ + 63.675 Cos $[x_5]$ Sin $[x_1]$ + 7.432 Cos $[x_6]$ Sin $[x_1]$ − 198.198 Sin $[x_1]^2$ − 184.366 Sin $[x_2]^2$ + 3334.002 Sin $[x_3]$ − 51.680 Cos $[x_1]$ Sin $[x_3]$ − 60.103 Cos $[x_2]$ Sin $[x_3]$ − 1308.753 Cos $[x_3]$ Sin $[x_3]$ − 14.056 Cos $[x_4]$ Sin $[x_3]$ − 34.036 Cos $[x_5]$ Sin $[x_3]$ − 8.455 Cos $[x_6]$ Sin $[x_3]$ − 255.285 Sin $[x_3]^2$ − 12. 771 Sin $[x_4]$ − 192.576 Sin $[x_4]^2$ + 1.149 Sin $[x_5]$ − 226.047 Sin $[x_5]^2$ + 45.293 Sin $[x_6]$ + 1.520 Cos $[x_1]$ Sin $[x_6]$ + 2.775 Cos $[x_2]$ Sin $[x_6]$ − 21.134 Cos $[x_3]$ Sin $[x_6]$ − 1.520 Cos $[x_4]$ Sin $[x_6]$ − 2.579 Cos $[x_5]$ Sin $[x_6]$ − 0.352 Cos $[x_6]$ Sin $[x_6]$ + 4.849 Sin $[x_1]$ Sin $[x_6]$ − 3.657 Sin $[x_2]$ Sin $[x_6]$ − 56.494 Sin $[x_3]$ Sin $[x_6]$ + 11.202 Sin $[x_4]$ Sin $[x_6]$ + 0.262 Sin $[x_5]$ Sin $[x_6]$ − 240.241 Sin $[x_6]^2$

In[•]:=(*12*)

Out[•]//NumberForm=

354.947 − 484.300 Cos $[x_1]$ − 703.639 Cos $[x_1]^2$ − 44.138 Cos $[x_2]$ + 373.419 Cos $[x_2]^2$ − 203.975 Cos $[x_3]$ + 16.341 Cos $[x_1]$ Cos $[x_3]$ + 27.322 Cos $[x_2]$ Cos $[x_3]$ + 259.929 Cos $[x_3]^2$ − 42.916 Cos $[x_4]$ + 2.064 Cos $[x_2]$ Cos $[x_4]$ + 368.735 Cos $[x_4]^2$ + 16.647 Cos $[x_5]$ + 420.506 Cos $[x_5]^2$ + 7.506 Cos $[x_6]$ + 472.750 Cos $[x_6]^2$ − 2463.459 Sin $[x_1]$ + 567.819 Cos $[x_1]$ Sin $[x_1]$ − 0.917 Cos $[x_2]$ Sin $[x_1]$ + 161.349 Cos $[x_3]$ Sin $[x_1]$ + 4.640 Cos $[x_4]$ Sin $[x_1]$ − 17.796 Cos $[x_5]$ Sin $[x_1]$ + 0.335 Cos $[x_6]$ Sin $[x_1]$ + 437.953 Sin $[x_1]^2$ + 371.386 Sin $[x_2]^2$ − 294.681 Sin $[x_3]$ − 82.220 Cos $[x_1]$ Sin $[x_3]$ + 41.877 Cos $[x_2]$ Sin $[x_3]$ + 54.399 Cos $[x_3]$ Sin $[x_3]$ + 45.098 Cos $[x_4]$ Sin $[x_3]$ + 24.763 Cos $[x_5]$ Sin $[x_3]$ − 8.473 Cos $[x_6]$ Sin $[x_3]$ + 447.207 Sin $[x_3]^2$ − 22.826 Sin $[x_4]$ + 387.165 Sin $[x_4]^2$ + 1.810 Sin $[x_5]$ + 407.937 Sin $[x_5]^2$ + 18.587 Sin $[x_6]$ + 0.306 Cos $[x_1]$ Sin $[x_6]$ + 0.594 Cos $[x_2]$ Sin $[x_6]$ − 5.172 Cos $[x_3]$ Sin $[x_6]$ − 0.452 Cos $[x_4]$ Sin $[x_6]$ − 2.951 Cos $[x_5]$ Sin $[x_6]$ − 0.350 Cos $[x_6]$ Sin $[x_6]$ + 0.514 Sin $[x_1]$ Sin $[x_6]$ − 6.777 Sin $[x_2]$ Sin $[x_6]$ − 16.050 Sin $[x_3]$ Sin $[x_6]$ + 1.565 Sin $[x_4]$ Sin $[x_6]$ + 0.387 Sin $[x_5]$ Sin $[x_6]$ + 471.808 Sin $[x_6]^2$

(Continued)

TABLE A7.4 *(Continued)*

Models Obtained by NRM and SNRM Using the Bootstrap Data Separation Method*) (*(80 Training–20 Testing) Training Was Divided into Two Parts–64 Training–16 Validation*) (*Models Obtained Using the NRM and SNRM Methods*)(*(Related to Table 7.5)*)

In[•]:=(*13*)

Out[•]//NumberForm=

$-87.418 - 147.666 \text{ Cos } [x_1] - 495.837 \text{ Cos } [x_1]^2 - 60.009 \text{ Cos } [x_2] - 89.001 \text{ Cos } [x_2]^2 + 259.833 \text{ Cos } [x_3] + 58.167 \text{ Cos } [x_1] \text{ Cos } [x_3] + 31.906 \text{ Cos } [x_2] \text{ Cos } [x_3] + 452.716 \text{ Cos } [x_3]^2 + 48.917 \text{ Cos } [x_4] - 3.638 \text{ Cos } [x_2] \text{ Cos } [x_4] + 4.927 \text{ Cos } [x_4]^2 - 34.682 \text{ Cos } [x_5] - 104.521 \text{ Cos } [x_5]^2 + 8.877 \text{ Cos } [x_6] - 115.081 \text{ Cos } [x_6]^2 - 858.729 \text{ Sin } [x_1] + 204.166 \text{ Cos } [x_1] \text{ Sin } [x_1] - 22.030 \text{ Cos } [x_2] \text{ Sin } [x_1] + 288.573 \text{ Cos } [x_3] \text{ Sin } [x_1] - 54.545 \text{ Cos } [x_4] \text{ Sin } [x_1] + 28.832 \text{ Cos } [x_5] \text{ Sin } [x_1] + 1.569 \text{ Cos } [x_6] \text{ Sin } [x_1] - 98.118 \text{ Sin } [x_1]^2 - 91.181 \text{ Sin } [x_2]^2 + 1494.782 \text{ Sin } [x_3] - 62.004 \text{ Cos } [x_1] \text{ Sin } [x_3] + 80.097 \text{ Cos } [x_2] \text{ Sin } [x_3] - 524.003 \text{ Cos } [x_3] \text{ Sin } [x_3] + 7.692 \text{ Cos } [x_4] \text{ Sin } [x_3] + 41.949 \text{ Cos } [x_5] \text{ Sin } [x_3] - 10.931 \text{ Cos } [x_6] \text{ Sin } [x_3] - 122.744 \text{ Sin } [x_3]^2 + 202.123 \text{ Sin } [x_4] - 94.545 \text{ Sin } [x_4]^2 + 0.803 \text{ Sin } [x_5] - 122.579 \text{ Sin } [x_5]^2 - 27.855 \text{ Sin } [x_6] + 4.815 \text{ Cos } [x_1] \text{ Sin } [x_6] - 0.222 \text{ Cos } [x_2] \text{ Sin } [x_6] - 0.126 \text{ Cos } [x_3] \text{ Sin } [x_6] + 1.369 \text{ Cos } [x_4] \text{ Sin } [x_6] - 3.375 \text{ Cos } [x_5] \text{ Sin } [x_6] - 0.317 \text{ Cos } [x_6] \text{ Sin } [x_6] + 16.249 \text{ Sin } [x_1] \text{ Sin } [x_6] - 6.165 \text{ Sin } [x_2] \text{ Sin } [x_6] - 0.971 \text{ Sin } [x_3] \text{ Sin } [x_6] + 14.753 \text{ Sin } [x_4] \text{ Sin } [x_6] + 1.274 \text{ Sin } [x_5] \text{ Sin } [x_6] - 115.662 \text{ Sin } [x_6]^2$

In[•]:=(*14*)

Out[•]//NumberForm=

$-34.354 + 247.045 \text{ Cos } [x_1] - 3.032 \text{ Cos } [x_1]^2 - 183.373 \text{ Cos } [x_2] - 30.569 \text{ Cos } [x_2]^2 - 184.075 \text{ Cos } [x_3] + 65.263 \text{ Cos } [x_1] \text{ Cos } [x_3] + 55.825 \text{ Cos } [x_2] \text{ Cos } [x_3] - 244.106 \text{ Cos } [x_3]^2 - 14.034 \text{ Cos } [x_4] + 5.512 \text{ Cos } [x_2] \text{ Cos } [x_4] + 122.037 \text{ Cos } [x_4]^2 - 120.650 \text{ Cos } [x_5] - 41.200 \text{ Cos } [x_5]^2 + 9.084 \text{ Cos } [x_6] - 45.406 \text{ Cos } [x_6]^2 + 264.375 \text{ Sin } [x_1] - 166.117 \text{ Cos } [x_1] \text{ Sin } [x_1] + 28.907 \text{ Cos } [x_2] \text{ Sin } [x_1] + 288.640 \text{ Cos } [x_3] \text{ Sin } [x_1] + 28.200 \text{ Cos } [x_4] \text{ Sin } [x_1] + 100.076 \text{ Cos } [x_5] \text{ Sin } [x_1] + 2.353 \text{ Cos } [x_6] \text{ Sin } [x_1] - 41.613 \text{ Sin } [x_1]^2 - 35.925 \text{ Sin } [x_2]^2 - 212.868 \text{ Sin } [x_3] - 67.673 \text{ Cos } [x_1] \text{ Sin } [x_3] + 148.770 \text{ Cos } [x_2] \text{ Sin } [x_3] - 85.716 \text{ Cos } [x_3] \text{ Sin } [x_3] - 9.605 \text{ Cos } [x_4] \text{ Sin } [x_3] + 20.308 \text{ Cos } [x_5] \text{ Sin } [x_3] - 12.136 \text{ Cos } [x_6] \text{ Sin } [x_3] - 37.653 \text{ Sin } [x_3]^2 + 308.911 \text{ Sin } [x_4] - 38.078 \text{ Sin } [x_4]^2 + 1.694 \text{ Sin } [x_5] - 37.121 \text{ Sin } [x_5]^2 + 92.194 \text{ Sin } [x_6] - 11.017 \text{ Cos } [x_1] \text{ Sin } [x_6] + 0.138 \text{ Cos } [x_2] \text{ Sin } [x_6] - 12.278 \text{ Cos } [x_3] \text{ Sin } [x_6] - 2.660 \text{ Cos } [x_4] \text{ Sin } [x_6] - 0.607 \text{ Cos } [x_5] \text{ Sin } [x_6] - 0.263 \text{ Cos } [x_6] \text{ Sin } [x_6] - 34.634 \text{ Sin } [x_1] \text{ Sin } [x_6] - 14.964 \text{ Sin } [x_2] \text{ Sin } [x_6] - 36.216 \text{ Sin } [x_3] \text{ Sin } [x_6] - 2.981 \text{ Sin } [x_4] \text{ Sin } [x_6] - 0.025 \text{ Sin } [x_5] \text{ Sin } [x_6] - 46.164 \text{ Sin } [x_6]^2$

In[•]:=(*15*)

Out[•]//NumberForm=

$504.483 - 1150.798 \text{ Cos } [x_1] - 698.489 \text{ Cos } [x_1]^2 + 121.954 \text{ Cos } [x_2] + 518.616 \text{ Cos } [x_2]^2 - 680.086 \text{ Cos } [x_3] + 111.610 \text{ Cos } [x_1] \text{ Cos } [x_3] - 8.652 \text{ Cos } [x_2] \text{ Cos } [x_3] + 243.844 \text{ Cos } [x_3]^2 + 43.860 \text{ Cos } [x_4] - 4.495 \text{ Cos } [x_2] \text{ Cos } [x_4] + 535.026 \text{ Cos } [x_4]^2 - 120.360 \text{ Cos } [x_5] + 619.814 \text{ Cos } [x_5]^2 - 2.423 \text{ Cos } [x_6] + 684.306 \text{ Cos } [x_6]^2 - 2809.205 \text{ Sin } [x_1] + 796.893 \text{ Cos } [x_1] \text{ Sin } [x_1] - 15.148 \text{ Cos } [x_2] \text{ Sin } [x_1] - 36.934 \text{ Cos } [x_3] \text{ Sin } [x_1] - 20.277 \text{ Cos } [x_4] \text{ Sin } [x_1] + 43.004 \text{ Cos } [x_5] \text{ Sin } [x_1] + 5.495 \text{ Cos } [x_6] \text{ Sin } [x_1] + 621.742 \text{ Sin } [x_1]^2 + 525.429 \text{ Sin } [x_2]^2 - 1261.844 \text{ Sin } [x_3] + 340.113 \text{ Cos } [x_1] \text{ Sin } [x_3] - 106.248 \text{ Cos } [x_2] \text{ Sin } [x_3] + 629.649 \text{ Cos } [x_3] \text{ Sin } [x_3] - 23.548 \text{ Cos } [x_4] \text{ Sin } [x_3] + 23.155 \text{ Cos } [x_5] \text{ Sin } [x_3] - 2.814 \text{ Cos } [x_6] \text{ Sin } [x_3] + 638.092 \text{ Sin } [x_3]^2 - 28.140 \text{ Sin } [x_4] + 549.392 \text{ Sin } [x_4]^2 + 0.845 \text{ Sin } [x_5] + 654.265 \text{ Sin } [x_5]^2 - 46.335 \text{ Sin } [x_6] + 5.752 \text{ Cos } [x_1] \text{ Sin } [x_6] + 0.522 \text{ Cos } [x_2] \text{ Sin } [x_6] - 5.608 \text{ Cos } [x_3] \text{ Sin } [x_6] + 1.199 \text{ Cos } [x_4] \text{ Sin } [x_6] - 9.926 \text{ Cos } [x_5] \text{ Sin } [x_6] - 0.469 \text{ Cos } [x_6] \text{ Sin } [x_6] + 16.814 \text{ Sin } [x_1] \text{ Sin } [x_6] - 1.729 \text{ Sin } [x_2] \text{ Sin } [x_6] - 9.214 \text{ Sin } [x_3] \text{ Sin } [x_6] + 31.583 \text{ Sin } [x_4] \text{ Sin } [x_6] + 1.717 \text{ Sin } [x_5] \text{ Sin } [x_6] + 684.866 \text{ Sin } [x_6]^2$

(Continued)

TABLE A7.4 *(Continued)*

Models Obtained by NRM and SNRM Using the Bootstrap Data Separation Method*) (*(80 Training–20 Testing) Training Was Divided into Two Parts–64 Training–16 Validation*) (*Models Obtained Using the NRM and SNRM Methods*)(*(Related to Table 7.5)*)

In[•]:=(*16*)

Out[•]//NumberForm=

$445.878 - 577.794$ Cos $[x_1] - 410.370$ Cos $[x_1]^2 - 102.141$ Cos $[x_2] + 471.882$ Cos $[x_2]^2 -$
691.488 Cos $[x_3] + 75.890$ Cos $[x_1]$ Cos $[x_3] + 32.235$ Cos $[x_2]$ Cos $[x_3] - 52.569$ Cos $[x_3]^2 -$
15.311 Cos $[x_4] - 1.724$ Cos $[x_2]$ Cos $[x_4] + 438.872$ Cos $[x_4]^2 - 82.239$ Cos $[x_5] + 525.387$ Cos $[x_5]^2 +$
4.429 Cos $[x_6] + 599.154$ Cos $[x_6]^2 - 2146.814$ Sin $[x_1] + 544.108$ Cos $[x_1]$ Sin $[x_1] +$
5.636 Cos $[x_2]$ Sin $[x_1] + 247.362$ Cos $[x_3]$ Sin $[x_1] + 10.250$ Cos $[x_4]$ Sin $[x_1] + 4.131$ Cos $[x_5]$ Sin $[x_1] +$
0.572 Cos $[x_6]$ Sin $[x_1] + 555.865$ Sin $[x_1]^2 + 467.511$ Sin $[x_2]^2 - 1322.901$ Sin $[x_3] +$
16.560 Cos $[x_1]$ Sin $[x_3] + 96.090$ Cos $[x_2]$ Sin $[x_3] + 422.720$ Cos $[x_3]$ Sin $[x_3] + 10.395$ Cos $[x_4]$ Sin $[x_3] +$
97.154 Cos $[x_5]$ Sin $[x_3] - 5.392$ Cos $[x_6]$ Sin $[x_3] + 575.489$ Sin $[x_3]^2 - 71.432$ Sin $[x_4] +$
482.148 Sin $[x_4]^2 + 1.779$ Sin $[x_5] + 517.470$ Sin $[x_5]^2 - 19.443$ Sin $[x_6] + 0.962$ Cos $[x_1]$ Sin $[x_6] +$
1.845 Cos $[x_2]$ Sin $[x_6] + 1.630$ Cos $[x_3]$ Sin $[x_6] - 0.343$ Cos $[x_4]$ Sin $[x_6] - 3.657$ Cos $[x_5]$ Sin $[x_6] -$
0.146 Cos $[x_6]$ Sin $[x_6] + 2.666$ Sin $[x_1]$ Sin $[x_6] + 0.091$ Sin $[x_2]$ Sin $[x_6] + 4.768$ Sin $[x_3]$ Sin $[x_6] +$
7.836 Sin $[x_4]$ Sin $[x_6] + 1.682$ Sin $[x_5]$ Sin $[x_6] + 598.703$ Sin $[x_6]^2$

In[•]:=(*17*)

Out[•]//NumberForm=

$32.240 - 389.785$ Cos $[x_1] - 464.548$ Cos $[x_1]^2 + 96.310$ Cos $[x_2] + 33.507$ Cos $[x_2]^2 + 238.556$ Cos $[x_3] +$
44.090 Cos $[x_1]$ Cos $[x_3] - 23.212$ Cos $[x_2]$ Cos $[x_3] + 289.977$ Cos $[x_3]^2 + 27.577$ Cos $[x_4] -$
1.368 Cos $[x_2]$ Cos $[x_4] + 143.157$ Cos $[x_4]^2 - 18.698$ Cos $[x_5] + 38.298$ Cos $[x_5]^2 + 2.048$ Cos $[x_6] +$
43.084 Cos $[x_6]^2 - 1112.475$ Sin $[x_1] + 321.284$ Cos $[x_1]$ Sin $[x_1] - 0.933$ Cos $[x_2]$ Sin $[x_1] +$
65.742 Cos $[x_3]$ Sin $[x_1] - 4.512$ Cos $[x_4]$ Sin $[x_1] - 2.959$ Cos $[x_5]$ Sin $[x_1] + 4.894$ Cos $[x_6]$ Sin $[x_1] +$
47.624 Sin $[x_1]^2 + 33.720$ Sin $[x_2]^2 + 727.355$ Sin $[x_3] + 70.237$ Cos $[x_1]$ Sin $[x_3] - 88.808$ Cos $[x_2]$ Sin $[x_3] -$
286.184 Cos $[x_3]$ Sin $[x_3] - 22.504$ Cos $[x_4]$ Sin $[x_3] + 18.857$ Cos $[x_5]$ Sin $[x_3] - 7.076$ Cos $[x_6]$ Sin $[x_3] +$
34.365 Sin $[x_3]^2 + 205.986$ Sin $[x_4] + 34.626$ Sin $[x_4]^2 + 0.923$ Sin $[x_5] + 41.815$ Sin $[x_5]^2 + 37.462$ Sin $[x_6] -$
2.764 Cos $[x_1]$ Sin $[x_6] - 0.177$ Cos $[x_2]$ Sin $[x_6] - 11.463$ Cos $[x_3]$ Sin $[x_6] + 0.614$ Cos $[x_4]$ Sin $[x_6] -$
3.590 Cos $[x_5]$ Sin $[x_6] - 0.455$ Cos $[x_6]$ Sin $[x_6] - 15.832$ Sin $[x_1]$ Sin $[x_6] - 10.343$ Sin $[x_2]$ Sin $[x_6] -$
29.187 Sin $[x_3]$ Sin $[x_6] + 16.632$ Sin $[x_4]$ Sin $[x_6] + 0.436$ Sin $[x_5]$ Sin $[x_6] + 43.068$ Sin $[x_6]^2$

In[•]:=(*18*)

Out[•]//NumberForm=

$140.423 - 3.059$ Cos $[x_1] - 20.687$ Cos $[x_1]^2 - 24.776$ Cos $[x_2] + 152.350$ Cos $[x_2]^2 - 88.213$ Cos $[x_3] -$
18.276 Cos $[x_1]$ Cos $[x_3] + 4.311$ Cos $[x_2]$ Cos $[x_3] - 193.865$ Cos $[x_3]^2 + 9.544$ Cos $[x_4] -$
4.756 Cos $[x_2]$ Cos $[x_4] + 132.087$ Cos $[x_4]^2 + 2.223$ Cos $[x_5] + 166.967$ Cos $[x_5]^2 + 3.221$ Cos $[x_6] +$
191.464 Cos $[x_6]^2 - 372.858$ Sin $[x_1] + 73.492$ Cos $[x_1]$ Sin $[x_1] - 0.165$ Cos $[x_2]$ Sin $[x_1] -$
2.792 Cos $[x_3]$ Sin $[x_1] + 10.542$ Cos $[x_4]$ Sin $[x_1] - 20.243$ Cos $[x_5]$ Sin $[x_1] + 3.451$ Cos $[x_6]$ Sin $[x_1] +$
172.752 Sin $[x_1]^2 + 147.374$ Sin $[x_2]^2 - 648.075$ Sin $[x_3] - 50.524$ Cos $[x_1]$ Sin $[x_3] +$
29.447 Cos $[x_2]$ Sin $[x_3] + 109.622$ Cos $[x_3]$ Sin $[x_3] - 15.147$ Cos $[x_4]$ Sin $[x_3] + 40.146$ Cos $[x_5]$ Sin $[x_3] -$
6.926 Cos $[x_6]$ Sin $[x_3] + 182.890$ Sin $[x_3]^2 - 26.111$ Sin $[x_4] + 153.059$ Sin $[x_4]^2 + 1.107$ Sin $[x_5] +$
156.479 Sin $[x_5]^2 + 32.547$ Sin $[x_6] - 8.550$ Cos $[x_1]$ Sin $[x_6] - 0.827$ Cos $[x_2]$ Sin $[x_6] -$
7.079 Cos $[x_3]$ Sin $[x_6] + 3.040$ Cos $[x_4]$ Sin $[x_6] + 0.519$ Cos $[x_5]$ Sin $[x_6] - 0.478$ Cos $[x_6]$ Sin $[x_6] -$
26.450 Sin $[x_1]$ Sin $[x_6] - 8.465$ Sin $[x_2]$ Sin $[x_6] - 20.621$ Sin $[x_3]$ Sin $[x_6] + 25.857$ Sin $[x_4]$ Sin $[x_6] +$
0.460 Sin $[x_5]$ Sin $[x_6] + 191.046$ Sin $[x_6]^2$

(Continued)

TABLE A7.4 *(Continued)*

Models Obtained by NRM and SNRM Using the Bootstrap Data Separation Method*) (*(80 Training–20 Testing) Training Was Divided into Two Parts–64 Training–16 Validation*) (*Models Obtained Using the NRM and SNRM Methods*)(*(Related to Table 7.5)*)

In[•]:=(*19*)

Out[•]//NumberForm=

127.388 − 613.951 Cos $[x_1]$ − 530.437 Cos $[x_1]^2$ − 63.404 Cos $[x_2]$ + 134.581 Cos $[x_2]^2$ − 51.879 Cos $[x_3]$ + 130.430 Cos $[x_1]$ Cos $[x_3]$ + 19.835 Cos $[x_2]$ Cos $[x_3]$ + 342.180 Cos $[x_3]^2$ + 0.150 Cos $[x_4]$ − 0.644 Cos $[x_2]$ Cos $[x_4]$ + 176.755 Cos $[x_4]^2$ − 58.388 Cos $[x_5]$ + 155.130 Cos $[x_5]^2$ + 7.036 Cos $[x_6]$ + 169.490 Cos $[x_6]^2$ − 1650.580 Sin $[x_1]$ + 516.518 Cos $[x_1]$ Sin $[x_1]$ + 2.273 Cos $[x_2]$ Sin $[x_1]$ + 368.875 Cos $[x_3]$ Sin $[x_1]$ − 1.110 Cos $[x_4]$ Sin $[x_1]$ + 26.921 Cos $[x_5]$ Sin $[x_1]$ + 4.894 Cos $[x_6]$ Sin $[x_1]$ + 165.427 Sin $[x_1]^2$ + 133.538 Sin $[x_2]^2$ + 644.207 Sin $[x_3]$ + 62.715 Cos $[x_1]$ Sin $[x_3]$ + 62.385 Cos $[x_2]$ Sin $[x_3]$ − 307.442 Cos $[x_3]$ Sin $[x_3]$ + 4.357 Cos $[x_4]$ Sin $[x_3]$ + 56.710 Cos $[x_5]$ Sin $[x_3]$ − 12.375 Cos $[x_6]$ Sin $[x_3]$ + 154.096 Sin $[x_3]^2$ + 79.022 Sin $[x_4]$ + 138.543 Sin $[x_4]^2$ + 0.597 Sin $[x_5]$ + 144.692 Sin $[x_5]^2$ + 48.578 Sin $[x_6]$ − 3.759 Cos $[x_1]$ Sin $[x_6]$ + 1.001 Cos $[x_2]$ Sin $[x_6]$ − 9.669 Cos $[x_3]$ Sin $[x_6]$ − 0.958 Cos $[x_4]$ Sin $[x_6]$ + 1.800 Cos $[x_5]$ Sin $[x_6]$ − 0.077 Cos $[x_6]$ Sin $[x_6]$ − 18.385 Sin $[x_1]$ Sin $[x_6]$ − 4.488 Sin $[x_2]$ Sin $[x_6]$ − 25.134 Sin $[x_3]$ Sin $[x_6]$ + 3.298 Sin $[x_4]$ Sin $[x_6]$ − 0.471 Sin $[x_5]$ Sin $[x_6]$ + 168.927 Sin $[x_6]^2$

In[•]:=(*20*)

Out[•]//NumberForm=

52.693 − 577.866 Cos $[x_1]$ − 1079.151 Cos $[x_1]^2$ − 100.770 Cos $[x_2]$ + 59.440 Cos $[x_2]^2$ + 579.550 Cos $[x_3]$ + 30.925 Cos $[x_1]$ Cos $[x_3]$ + 31.036 Cos $[x_2]$ Cos $[x_3]$ + 942.840 Cos $[x_3]^2$ + 6.834 Cos $[x_4]$ + 1.062 Cos $[x_2]$ Cos $[x_4]$ + 42.621 Cos $[x_4]^2$ − 54.232 Cos $[x_5]$ + 63.493 Cos $[x_5]^2$ + 3.287 Cos $[x_6]$ + 69.945 Cos $[x_6]^2$ − 2482.173 Sin $[x_1]$ + 620.321 Cos $[x_1]$ Sin $[x_1]$ + 1.264 Cos $[x_2]$ Sin $[x_1]$ + 115.386 Cos $[x_3]$ Sin $[x_1]$ − 44.359 Cos $[x_4]$ Sin $[x_1]$ + 18. 204 Cos $[x_5]$ Sin $[x_1]$ + 0.745 Cos $[x_6]$ Sin $[x_1]$ + 83.045 Sin $[x_1]^2$ + 55.225 Sin $[x_2]^2$ + 2183.072 Sin $[x_3]$ − 47.067 Cos $[x_1]$ Sin $[x_3]$ + 98.448 Cos $[x_2]$ Sin $[x_3]$ − 666.175 Cos $[x_3]$ Sin $[x_3]$ + 40.433 Cos $[x_4]$ Sin $[x_3]$ + 52.768 Cos $[x_5]$ Sin $[x_3]$ − 4.381 Cos $[x_6]$ Sin $[x_3]$ + 47.082 Sin $[x_3]^2$ − 24.624 Sin $[x_4]$ + 56.676 Sin $[x_4]^2$ + 0.992 Sin $[x_5]$ + 56.782 Sin $[x_5]^2$ + 31.624 Sin $[x_6]$ + 3.439 Cos $[x_1]$ Sin $[x_6]$ + 1.924 Cos $[x_2]$ Sin $[x_6]$ − 18.619 Cos $[x_3]$ Sin $[x_6]$ + 1.170 Cos $[x_4]$ Sin $[x_6]$ − 2.452 Cos $[x_5]$ Sin $[x_6]$ − 0.167 Cos $[x_6]$ Sin $[x_6]$ + 11.996 Sin $[x_1]$ Sin $[x_6]$ − 0.184 Sin $[x_2]$ Sin $[x_6]$ − 55.220 Sin $[x_3]$ Sin $[x_6]$ + 11.687 Sin $[x_4]$ Sin $[x_6]$ + 0.969 Sin $[x_5]$ Sin $[x_6]$ + 69.668 Sin $[x_6]^2$

(*Models proposed by SNRM*)

In[•]:=(*1*)

Out[•]//NumberForm=

46.291 − 75.115 Cos $[x_1]$ − 73.824 Cos $[x_1]^2$ − 33.971 Cos $[x_2]$ + 25.550 Cos $[x_2]^2$ − 7.837 Cos $[x_3]$ + 55.520 Cos $[x_1]$ Cos $[x_3]$ + 11.662 Cos $[x_2]$ Cos $[x_3]$ − 11.320 Cos $[x_3]^2$ + 10.062 Cos $[x_4]$ + 1.100 Cos $[x_2]$ Cos $[x_4]$ + 45.623 Cos $[x_4]^2$ − 61.390 Cos $[x_5]$ + 30.615 Cos $[x_5]^2$ + 4.437 Cos $[x_6]$ + 24.971 Cos $[x_6]^2$ − 397.609 Sin $[x_1]$ + 86.284 Cos $[x_1]$ Sin $[x_1]$ + 5.882 Cos $[x_2]$ Sin $[x_1]$ + 189.364 Cos $[x_3]$ Sin $[x_1]$ − 1.533 Cos $[x_4]$ Sin $[x_1]$ + 36.461 Cos $[x_5]$ Sin $[x_1]$ + 3.122 Cos $[x_6]$ Sin $[x_1]$ + 129.819 Sin $[x_1]^2$ + 22.721 Sin $[x_2]^2$ + 98.983 Sin $[x_3]$ − 9.843 Cos $[x_1]$ Sin $[x_3]$ + 29.835 Cos $[x_2]$ Sin $[x_3]$ − 161.893 Cos $[x_3]$ Sin $[x_3]$ − 6.155 Cos $[x_4]$ Sin $[x_3]$ + 36.405 Cos $[x_5]$ Sin $[x_3]$ − 7.890 Cos $[x_6]$ Sin $[x_3]$ + 52.099 Sin $[x_3]^2$ + 92.719 Sin $[x_4]$ − 1.464 Sin $[x_4]^2$ + 1.125 Sin $[x_5]$ + 27.292 Sin $[x_5]^2$ + 43.094 Sin $[x_6]$ −

(Continued)

TABLE A7.4 *(Continued)*

Models Obtained by NRM and SNRM Using the Bootstrap Data Separation Method*) (*(80 Training–20 Testing) Training Was Divided into Two Parts–64 Training–16 Validation*) (*Models Obtained Using the NRM and SNRM Methods*)(*(Related to Table 7.5)*)

3.920 Cos $[x_1]$ Sin $[x_6]$ + 0.379 Cos $[x_2]$ Sin $[x_6]$ – 9.652 Cos $[x_3]$ Sin $[x_6]$ – 0.136 Cos $[x_4]$ Sin $[x_6]$ – 0.609 Cos $[x_5]$ Sin $[x_6]$ – 0.331 Cos $[x_6]$ Sin $[x_6]$ – 17.212 Sin $[x_1]$ Sin $[x_6]$ – 5.796 Sin $[x_2]$ Sin $[x_6]$ – 27.744 Sin $[x_3]$ Sin $[x_6]$ + 9.165 Sin $[x_4]$ Sin $[x_6]$ + 0.083 Sin $[x_5]$ Sin $[x_6]$ + 24.474 Sin $[x_6]^2$

In[•]:=(*2*)

Out[•]//NumberForm=

47.442 – 71.611 Cos $[x_1]$ – 84.857 Cos $[x_1]^2$ – 46.061 Cos $[x_2]$ + 25.895 Cos $[x_2]^2$ – 56.170 Cos $[x_3]$ + 63.236 Cos $[x_1]$ Cos $[x_3]$ + 16.522 Cos $[x_2]$ Cos $[x_3]$ + 3.186 Cos $[x_3]^2$ + 9.800 Cos $[x_4]$ + 0.923 Cos $[x_2]$ Cos $[x_4]$ + 44.084 Cos $[x_4]^2$ – 69.122 Cos $[x_5]$ + 25.374 Cos $[x_5]^2$ + 5.053 Cos $[x_6]$ + 21.297 Cos $[x_6]^2$ – 395.248 Sin $[x_1]$ + 80.015 Cos $[x_1]$ Sin $[x_1]$ + 8.629 Cos $[x_2]$ Sin $[x_1]$ + 221.574 Cos $[x_3]$ Sin $[x_1]$ – 7.362 Cos $[x_4]$ Sin $[x_1]$ + 42.674 Cos $[x_5]$ Sin $[x_1]$ + 3.363 Cos $[x_6]$ Sin $[x_1]$ + 118.803 Sin $[x_1]^2$ + 22.651 Sin $[x_2]^2$ + 138.274 Sin $[x_3]$ – 8.598 Cos $[x_1]$ Sin $[x_3]$ + 38.175 Cos $[x_2]$ Sin $[x_3]$ – 149.554 Cos $[x_3]$ Sin $[x_3]$ + 0.211 Cos $[x_4]$ Sin $[x_3]$ + 42.158 Cos $[x_5]$ Sin $[x_3]$ – 8.804 Cos $[x_6]$ Sin $[x_3]$ + 38.937 Sin $[x_3]^2$ + 84.241 Sin $[x_4]$ + 1.930 Sin $[x_4]^2$ + 1.196 Sin $[x_5]$ + 19.789 Sin $[x_5]^2$ + 37.038 Sin $[x_6]$ – 4.061 Cos $[x_1]$ Sin $[x_6]$ + 0.206 Cos $[x_2]$ Sin $[x_6]$ – 7.655 Cos $[x_3]$ Sin $[x_6]$ – 0.082 Cos $[x_4]$ Sin $[x_6]$ – 0.998 Cos $[x_5]$ Sin $[x_6]$ – 0.358 Cos $[x_6]$ Sin $[x_6]$ – 17.338 Sin $[x_1]$ Sin $[x_6]$ – 7.121 Sin $[x_2]$ Sin $[x_6]$ – 22.138 Sin $[x_3]$ Sin $[x_6]$ + 10.386 Sin $[x_4]$ Sin $[x_6]$ + 0.297 Sin $[x_5]$ Sin $[x_6]$ + 20.790 Sin $[x_6]^2$

In[•]:=(*3*)

Out[•]//NumberForm=

43.198 – 69.999 Cos $[x_1]$ – 90.196 Cos $[x_1]^2$ – 55.131 Cos $[x_2]$ + 27.875 Cos $[x_2]^2$ – 52.742 Cos $[x_3]$ + 62.051 Cos $[x_1]$ Cos $[x_3]$ + 19.220 Cos $[x_2]$ Cos $[x_3]$ + 3.577 Cos $[x_3]^2$ + 9.389 Cos $[x_4]$ + 0.824 Cos $[x_2]$ Cos $[x_4]$ + 45.357 Cos $[x_4]^2$ – 78.860 Cos $[x_5]$ + 24.914 Cos $[x_5]^2$ + 4.947 Cos $[x_6]$ + 24.411 Cos $[x_6]^2$ – 400.666 Sin $[x_1]$ + 81.609 Cos $[x_1]$ Sin $[x_1]$ + 6.877 Cos $[x_2]$ Sin $[x_1]$ + 218.528 Cos $[x_3]$ Sin $[x_1]$ – 7.102 Cos $[x_4]$ Sin $[x_1]$ + 45.678 Cos $[x_5]$ Sin $[x_1]$ + 2.981 Cos $[x_6]$ Sin $[x_1]$ + 118.031 Sin $[x_1]^2$ + 25.145 Sin $[x_2]^2$ + 140.449 Sin $[x_3]$ – 11.628 Cos $[x_1]$ Sin $[x_3]$ + 48.842 Cos $[x_2]$ Sin $[x_3]$ – 149.721 Cos $[x_3]$ Sin $[x_3]$ + 0.598 Cos $[x_4]$ Sin $[x_3]$ + 45.554 Cos $[x_5]$ Sin $[x_3]$ – 8.303 Cos $[x_6]$ Sin $[x_3]$ + 40.369 Sin $[x_3]^2$ + 76.815 Sin $[x_4]$ + 7.699 Sin $[x_4]^2$ + 1.153 Sin $[x_5]$ + 21.395 Sin $[x_5]^2$ + 37.892 Sin $[x_6]$ – 4.113 Cos $[x_1]$ Sin $[x_6]$ + 0.448 Cos $[x_2]$ Sin $[x_6]$ – 7.931 Cos $[x_3]$ Sin $[x_6]$ – 0.328 Cos $[x_4]$ Sin $[x_6]$ – 1.153 Cos $[x_5]$ Sin $[x_6]$ – 0.345 Cos $[x_6]$ Sin $[x_6]$ – 17.361 Sin $[x_1]$ Sin $[x_6]$ – 5.464 Sin $[x_2]$ Sin $[x_6]$ – 22.164 Sin $[x_3]$ Sin $[x_6]$ + 7.821 Sin $[x_4]$ Sin $[x_6]$ + 0.339 Sin $[x_5]$ Sin $[x_6]$ + 23.908 Sin $[x_6]^2$

In[•]:=(*4*)

Out[•]//NumberForm=

43.198 – 69.999 Cos $[x_1]$ – 90.196 Cos $[x_1]^2$ – 55.131 Cos $[x_2]$ + 27.875 Cos $[x_2]^2$ – 52.742 Cos $[x_3]$ + 62.051 Cos $[x_1]$ Cos $[x_3]$ + 19.220 Cos $[x_2]$ Cos $[x_3]$ + 3.577 Cos $[x_3]^2$ + 9.389 Cos $[x_4]$ + 0.824 Cos $[x_2]$ Cos $[x_4]$ + 45.357 Cos $[x_4]^2$ – 78.860 Cos $[x_5]$ + 24.914 Cos $[x_5]^2$ + 4.947 Cos $[x_6]$ + 24.411 Cos $[x_6]^2$ – 400.666 Sin $[x_1]$ + 81.609 Cos $[x_1]$ Sin $[x_1]$ + 6.877 Cos $[x_2]$ Sin $[x_1]$ + 218.528 Cos $[x_3]$ Sin $[x_1]$ – 7.102 Cos $[x_4]$ Sin $[x_1]$ + 45.678 Cos $[x_5]$ Sin $[x_1]$ + 2.981 Cos $[x_6]$ Sin $[x_1]$ + 118.031 Sin $[x_1]^2$ + 25.145 Sin $[x_2]^2$ + 140.449 Sin $[x_3]$ – 11.628 Cos $[x_1]$ Sin $[x_3]$ + 48.842 Cos $[x_2]$ Sin $[x_3]$ – 149.721 Cos $[x_3]$ Sin $[x_3]$ + 0.598 Cos $[x_4]$ Sin $[x_3]$ + 45.554 Cos $[x_5]$ Sin $[x_3]$ – 8.303 Cos $[x_6]$ Sin $[x_3]$ + 40.369 Sin $[x_3]^2$ + 76.815 Sin $[x_4]$ + 7.699 Sin $[x_4]^2$ + 1.153 Sin $[x_5]$ + 21.395 Sin $[x_5]^2$ + 37.892 Sin $[x_6]$ –

(Continued)

TABLE A7.4 *(Continued)*
Models Obtained by NRM and SNRM Using the Bootstrap Data Separation Method*) (*(80 Training–20 Testing) Training Was Divided into Two Parts–64 Training–16 Validation*) (*Models Obtained Using the NRM and SNRM Methods*)(*(Related to Table 7.5)*)

4.113 Cos $[x_1]$ Sin $[x_6]$ + 0.448 Cos $[x_2]$ Sin $[x_6]$ − 7.931 Cos $[x_3]$ Sin $[x_6]$ − 0.328 Cos $[x_4]$ Sin $[x_6]$ −
1.153 Cos $[x_5]$ Sin $[x_6]$ − 0.345 Cos $[x_6]$ Sin $[x_6]$ − 17.361 Sin $[x_1]$ Sin $[x_6]$ − 5.464 Sin $[x_2]$ Sin $[x_6]$ −
22.164 Sin $[x_3]$ Sin $[x_6]$ + 7.821 Sin $[x_4]$ Sin $[x_6]$ + 0.339 Sin $[x_5]$ Sin $[x_6]$ + 23.908 Sin $[x_6]^2$

In[•]:=(*5*)

Out[•]//NumberForm=

45.283 − 50.605 Cos $[x_1]$ − 88.091 Cos $[x_1]^2$ − 66.976 Cos $[x_2]$ + 34.057 Cos $[x_2]^2$ − 18.900 Cos $[x_3]$ +
50.261 Cos $[x_1]$ Cos $[x_3]$ + 22.438 Cos $[x_2]$ Cos $[x_3]$ − 18.499 Cos $[x_3]^2$ + 14.770 Cos $[x_4]$ +
2.076 Cos $[x_2]$ Cos $[x_4]$ + 54.969 Cos $[x_4]^2$ − 64.678 Cos $[x_5]$ + 30.186 Cos $[x_5]^2$ + 5.162 Cos $[x_6]$ +
19.331 Cos $[x_6]^2$ − 398.713 Sin $[x_1]$ + 82.667 Cos $[x_1]$ Sin $[x_1]$ + 7.640 Cos $[x_2]$ Sin $[x_1]$ +
196.992 Cos $[x_3]$ Sin $[x_1]$ − 6.277 Cos $[x_4]$ Sin $[x_1]$ + 43.583 Cos $[x_5]$ Sin $[x_1]$ + 3.003 Cos $[x_6]$ Sin $[x_1]$ +
120.802 Sin $[x_1]^2$ + 30.988 Sin $[x_2]^2$ + 81.130 Sin $[x_3]$ − 29.540 Cos $[x_1]$ Sin $[x_3]$ + 59.288 Cos $[x_2]$ Sin $[x_3]$ −
156.628 Cos $[x_3]$ Sin $[x_3]$ − 6.416 Cos $[x_4]$ Sin $[x_3]$ + 36.188 Cos $[x_5]$ Sin $[x_3]$ − 8.585 Cos $[x_6]$ Sin $[x_3]$ +
55.494 Sin $[x_3]^2$ + 129.905 Sin $[x_4]$ − 11.960 Sin $[x_4]^2$ + 1.192 Sin $[x_5]$ + 24.914 Sin $[x_5]^2$ + 50.687 Sin $[x_6]$ −
4.784 Cos $[x_1]$ Sin $[x_6]$ + 0.137 Cos $[x_2]$ Sin $[x_6]$ − 10.073 Cos $[x_3]$ Sin $[x_6]$ − 0.452 Cos $[x_4]$ Sin $[x_6]$ −
0.966 Cos $[x_5]$ Sin $[x_6]$ − 0.361 Cos $[x_6]$ Sin $[x_6]$ − 19.637 Sin $[x_1]$ Sin $[x_6]$ − 7.511 Sin $[x_2]$ Sin $[x_6]$ −
29.688 Sin $[x_3]$ Sin $[x_6]$ + 7.411 Sin $[x_4]$ Sin $[x_6]$ + 0.379 Sin $[x_5]$ Sin $[x_6]$ + 18.769 Sin $[x_6]^2$

In[•]:=(*6*)

Out[•]//NumberForm=

114.617 − 184.901 Cos $[x_1]$ − 143.369 Cos $[x_1]^2$ − 94.522 Cos $[x_2]$ + 66.014 Cos $[x_2]^2$ − 72.247 Cos $[x_3]$ +
62.747 Cos $[x_1]$ Cos $[x_3]$ + 28.478 Cos $[x_2]$ Cos $[x_3]$ + 79.999 Cos $[x_3]^2$ + 6.244 Cos $[x_4]$ −
0.950 Cos $[x_2]$ Cos $[x_4]$ + 64.597 Cos $[x_4]^2$ − 62.669 Cos $[x_5]$ + 61.641 Cos $[x_5]^2$ + 3.953 Cos $[x_6]$ +
62.293 Cos $[x_6]^2$ − 814.430 Sin $[x_1]$ + 193.670 Cos $[x_1]$ Sin $[x_1]$ + 13.688 Cos $[x_2]$ Sin $[x_1]$ +
232.371 Cos $[x_3]$ Sin $[x_1]$ − 10.686 Cos $[x_4]$ Sin $[x_1]$ + 28.463 Cos $[x_5]$ Sin $[x_1]$ + 2.662 Cos $[x_6]$ Sin $[x_1]$ +
239.893 Sin $[x_1]^2$ + 62.325 Sin $[x_2]^2$ + 256.335 Sin $[x_3]$ − 12.377 Cos $[x_1]$ Sin $[x_3]$ + 80.969 Cos $[x_2]$ Sin $[x_3]$ −
149.089 Cos $[x_3]$ Sin $[x_3]$ + 7.917 Cos $[x_4]$ Sin $[x_3]$ + 49.854 Cos $[x_5]$ Sin $[x_3]$ − 6.922 Cos $[x_6]$ Sin $[x_3]$ +
47.023 Sin $[x_3]^2$ + 11.925 Sin $[x_4]$ + 62.652 Sin $[x_4]^2$ + 1.289 Sin $[x_5]$ + 55.569 Sin $[x_5]^2$ + 21.202 Sin $[x_6]$ −
3.757 Cos $[x_1]$ Sin $[x_6]$ + 0.861 Cos $[x_2]$ Sin $[x_6]$ − 4.471 Cos $[x_3]$ Sin $[x_6]$ + 0.176 Cos $[x_4]$ Sin $[x_6]$ −
0.754 Cos $[x_5]$ Sin $[x_6]$ − 0.312 Cos $[x_6]$ Sin $[x_6]$ − 16.242 Sin $[x_1]$ Sin $[x_6]$ − 5.213 Sin $[x_2]$ Sin $[x_6]$ −
11.539 Sin $[x_3]$ Sin $[x_6]$ + 12.285 Sin $[x_4]$ Sin $[x_6]$ + 0.585 Sin $[x_5]$ Sin $[x_6]$ + 61.785 Sin $[x_6]^2$

In[•]:=(*7*)

Out[•]//NumberForm=

69.372 − 137.929 Cos $[x_1]$ − 110.655 Cos $[x_1]^2$ − 78.552 Cos $[x_2]$ + 37.167 Cos $[x_2]^2$ − 82.321 Cos $[x_3]$ +
67.647 Cos $[x_1]$ Cos $[x_3]$ + 22.780 Cos $[x_2]$ Cos $[x_3]$ + 32.754 Cos $[x_3]^2$ + 10.857 Cos $[x_4]$ +
0.468 Cos $[x_2]$ Cos $[x_4]$ + 67.026 Cos $[x_4]^2$ − 65.591 Cos $[x_5]$ + 41.942 Cos $[x_5]^2$ + 4.028 Cos $[x_6]$ +
40.406 Cos $[x_6]^2$ − 563.987 Sin $[x_1]$ + 135.985 Cos $[x_1]$ Sin $[x_1]$ + 12.010 Cos $[x_2]$ Sin $[x_1]$ +
234.132 Cos $[x_3]$ Sin $[x_1]$ − 5.245 Cos $[x_4]$ Sin $[x_1]$ + 37.909 Cos $[x_5]$ Sin $[x_1]$ + 3.142 Cos $[x_6]$ Sin $[x_1]$ +
159.478 Sin $[x_1]^2$ + 34.158 Sin $[x_2]^2$ + 172.302 Sin $[x_3]$ − 1.475 Cos $[x_1]$ Sin $[x_3]$ + 67.383 Cos $[x_2]$ Sin $[x_3]$ −
140.709 Cos $[x_3]$ Sin $[x_3]$ − 3.164 Cos $[x_4]$ Sin $[x_3]$ + 42.896 Cos $[x_5]$ Sin $[x_3]$ − 7.474 Cos $[x_6]$ Sin $[x_3]$ +
40.398 Sin $[x_3]^2$ + 85.168 Sin $[x_4]$ + 24.575 Sin $[x_4]^2$ + 1.080 Sin $[x_5]$ + 36.644 Sin $[x_5]^2$ + 37.455 Sin $[x_6]$ −

(Continued)

TABLE A7.4 *(Continued)*

Models Obtained by NRM and SNRM Using the Bootstrap Data Separation Method*) (*(80 Training–20 Testing) Training Was Divided into Two Parts–64 Training–16 Validation*) (*Models Obtained Using the NRM and SNRM Methods*)(*(Related to Table 7.5)*)

$4.436 \cos [x_1] \sin [x_6] + 0.580 \cos [x_2] \sin [x_6] - 6.393 \cos [x_3] \sin [x_6] - 0.494 \cos [x_4] \sin [x_6] - 0.085 \cos [x_5] \sin [x_6] - 0.283 \cos [x_6] \sin [x_6] - 18.380 \sin [x_1] \sin [x_6] - 6.468 \sin [x_2] \sin [x_6] - 18.575 \sin [x_3] \sin [x_6] + 7.495 \sin [x_4] \sin [x_6] + 0.165 \sin [x_5] \sin [x_6] + 39.860 \sin [x_6]^2$

In[•]:=(*8*)

Out[•]//NumberForm=

$75.162 - 154.409 \cos [x_1] - 127.574 \cos [x_1]^2 - 79.987 \cos [x_2] + 45.691 \cos [x_2]^2 - 66.013 \cos [x_3] + 82.291 \cos [x_1] \cos [x_3] + 23.890 \cos [x_2] \cos [x_3] + 52.032 \cos [x_3]^2 + 12.295 \cos [x_4] + 1.030 \cos [x_2] \cos [x_4] + 52.198 \cos [x_4]^2 - 57.912 \cos [x_5] + 33.086 \cos [x_5]^2 + 4.513 \cos [x_6] + 33.428 \cos [x_6]^2 - 673.859 \sin [x_1] + 149.040 \cos [x_1] \sin [x_1] + 6.728 \cos [x_2] \sin [x_1] + 283.677 \cos [x_3] \sin [x_1] - 11.549 \cos [x_4] \sin [x_1] + 28.185 \cos [x_5] \sin [x_1] + 3.510 \cos [x_6] \sin [x_1] + 184.625 \sin [x_1]^2 + 42.348 \sin [x_2]^2 + 308.428 \sin [x_3] - 2.866 \cos [x_1] \sin [x_3] + 73.883 \cos [x_2] \sin [x_3] - 203.282 \cos [x_3] \sin [x_3] + 1.811 \cos [x_4] \sin [x_3] + 50.707 \cos [x_5] \sin [x_3] - 8.342 \cos [x_6] \sin [x_3] + 15.471 \sin [x_3]^2 + 77.072 \sin [x_4] + 14.001 \sin [x_4]^2 + 1.005 \sin [x_5] + 24.566 \sin [x_5]^2 + 26.639 \sin [x_6] - 1.916 \cos [x_1] \sin [x_6] + 0.787 \cos [x_2] \sin [x_6] - 5.903 \cos [x_3] \sin [x_6] - 0.808 \cos [x_4] \sin [x_6] + 0.691 \cos [x_5] \sin [x_6] - 0.263 \cos [x_6] \sin [x_6] - 9.934 \sin [x_1] \sin [x_6] - 2.754 \sin [x_2] \sin [x_6] - 17.044 \sin [x_3] \sin [x_6] + 4.991 \sin [x_4] \sin [x_6] + 0.247 \sin [x_5] \sin [x_6] + 32.949 \sin [x_6]^2$

In[•]:=(*9*)

Out[•]//NumberForm=

$57.389 - 99.477 \cos [x_1] - 95.354 \cos [x_1]^2 - 30.560 \cos [x_2] + 33.730 \cos [x_2]^2 - 75.852 \cos [x_3] + 63.046 \cos [x_1] \cos [x_3] + 13.228 \cos [x_2] \cos [x_3] - 0.040 \cos [x_3]^2 + 5.696 \cos [x_4] + 3.188 \cos [x_2] \cos [x_4] + 49.517 \cos [x_4]^2 - 103.371 \cos [x_5] + 33.876 \cos [x_5]^2 + 5.804 \cos [x_6] + 29.639 \cos [x_6]^2 - 462.706 \sin [x_1] + 106.516 \cos [x_1] \sin [x_1] + 5.178 \cos [x_2] \sin [x_1] + 222.063 \cos [x_3] \sin [x_1] - 7.621 \cos [x_4] \sin [x_1] + 63.222 \cos [x_5] \sin [x_1] + 2.660 \cos [x_6] \sin [x_1] + 146.900 \sin [x_1]^2 + 31.668 \sin [x_2]^2 + 120.142 \sin [x_3] - 7.896 \cos [x_1] \sin [x_3] + 26.070 \cos [x_2] \sin [x_3] - 130.090 \cos [x_3] \sin [x_3] + 5.196 \cos [x_4] \sin [x_3] + 53.315 \cos [x_5] \sin [x_3] - 8.929 \cos [x_6] \sin [x_3] + 45.344 \sin [x_3]^2 + 96.193 \sin [x_4] + 2.578 \sin [x_4]^2 + 1.251 \sin [x_5] + 30.214 \sin [x_5]^2 + 36.330 \sin [x_6] - 3.951 \cos [x_1] \sin [x_6] + 0.792 \cos [x_2] \sin [x_6] - 7.152 \cos [x_3] \sin [x_6] - 0.759 \cos [x_4] \sin [x_6] - 1.535 \cos [x_5] \sin [x_6] - 0.291 \cos [x_6] \sin [x_6] - 17.110 \sin [x_1] \sin [x_6] - 3.788 \sin [x_2] \sin [x_6] - 19.082 \sin [x_3] \sin [x_6] + 3.891 \sin [x_4] \sin [x_6] + 0.270 \sin [x_5] \sin [x_6] + 29.069 \sin [x_6]^2$

In[•]:=(*10*)

Out[•]//NumberForm=

$66.991 - 150.362 \cos [x_1] - 120.854 \cos [x_1]^2 - 16.850 \cos [x_2] + 36.674 \cos [x_2]^2 - 1.142 \cos [x_3] + 65.790 \cos [x_1] \cos [x_3] + 5.476 \cos [x_2] \cos [x_3] + 39.537 \cos [x_3]^2 + 13.744 \cos [x_4] - 0.055 \cos [x_2] \cos [x_4] + 47.854 \cos [x_4]^2 - 77.815 \cos [x_5] + 40.629 \cos [x_5]^2 + 4.726 \cos [x_6] + 36.732 \cos [x_6]^2 - 589.768 \sin [x_1] + 137.372 \cos [x_1] \sin [x_1] + 8.271 \cos [x_2] \sin [x_1] + 208.247 \cos [x_3] \sin [x_1] - 7.417 \cos [x_4] \sin [x_1] + 44.999 \cos [x_5] \sin [x_1] + 3.019 \cos [x_6] \sin [x_1] + 170.667 \sin [x_1]^2 + 33.068 \sin [x_2]^2 + 245.007 \sin [x_3] + 10.640 \cos [x_1] \sin [x_3] + 11.775 \cos [x_2] \sin [x_3] - 190.211 \cos [x_3] \sin [x_3] - 3.874 \cos [x_4] \sin [x_3] + 46.839 \cos [x_5] \sin [x_3] - 8.074 \cos [x_6] \sin [x_3] + 32.930 \sin [x_3]^2 + 32.237 \sin [x_4] + 33.651 \sin [x_4]^2 + 1.100 \sin [x_5] + 36.377 \sin [x_5]^2 + 34.968 \sin [x_6] -$

(Continued)

TABLE A7.4 *(Continued)*

Models Obtained by NRM and SNRM Using the Bootstrap Data Separation Method*) (*(80 Training–20 Testing) Training Was Divided into Two Parts–64 Training–16 Validation*) (*Models Obtained Using the NRM and SNRM Methods*)(*(Related to Table 7.5)*)

4.011 Cos $[x_1]$ Sin $[x_6]$ + 0.494 Cos $[x_2]$ Sin $[x_6]$ – 7.558 Cos $[x_3]$ Sin $[x_6]$ – 0.300 Cos $[x_4]$ Sin $[x_6]$ – 1.362 Cos $[x_5]$ Sin $[x_6]$ – 0.305 Cos $[x_6]$ Sin $[x_6]$ – 15.802 Sin $[x_1]$ Sin $[x_6]$ – 5.942 Sin $[x_2]$ Sin $[x_6]$ – 21.036 Sin $[x_3]$ Sin $[x_6]$ + 8.249 Sin $[x_4]$ Sin $[x_6]$ + 0.604 Sin $[x_5]$ Sin $[x_6]$ + 36.277 Sin $[x_6]^2$

In[•]:=(*11*)

Out[•]//NumberForm=

71.207 – 73.334 Cos $[x_1]$ – 107.893 Cos $[x_1]^2$ – 68.225 Cos $[x_2]$ + 36.792 Cos $[x_2]^2$ – 99.629 Cos $[x_3]$ + 58.473 Cos $[x_1]$ Cos $[x_3]$ + 24.787 Cos $[x_2]$ Cos $[x_3]$ + 15.469 Cos $[x_3]^2$ + 3.615 Cos $[x_4]$ – 1.331 Cos $[x_2]$ Cos $[x_4]$ + 56.060 Cos $[x_4]^2$ – 96.425 Cos $[x_5]$ + 33.304 Cos $[x_5]^2$ + 5.175 Cos $[x_6]$ + 30.024 Cos $[x_6]^2$ – 484.321 Sin $[x_1]$ + 93.570 Cos $[x_1]$ Sin $[x_1]$ + 4.538 Cos $[x_2]$ Sin $[x_1]$ + 221.719 Cos $[x_3]$ Sin $[x_1]$ – 7.961 Cos $[x_4]$ Sin $[x_1]$ + 51.112 Cos $[x_5]$ Sin $[x_1]$ + 2.874 Cos $[x_6]$ Sin $[x_1]$ + 144.536 Sin $[x_1]^2$ + 34.254 Sin $[x_2]^2$ + 140.800 Sin $[x_3]$ – 19.504 Cos $[x_1]$ Sin $[x_3]$ + 63.522 Cos $[x_2]$ Sin $[x_3]$ – 110.765 Cos $[x_3]$ Sin $[x_3]$ + 8.053 Cos $[x_4]$ Sin $[x_3]$ + 56.983 Cos $[x_5]$ Sin $[x_3]$ – 8.439 Cos $[x_6]$ Sin $[x_3]$ + 38.223 Sin $[x_3]^2$ + 67.718 Sin $[x_4]$ + 24.105 Sin $[x_4]^2$ + 1.193 Sin $[x_5]$ + 30.548 Sin $[x_5]^2$ + 21.644 Sin $[x_6]$ – 3.649 Cos $[x_1]$ Sin $[x_6]$ + 0.448 Cos $[x_2]$ Sin $[x_6]$ – 4.174 Cos $[x_3]$ Sin $[x_6]$ + 0.011 Cos $[x_4]$ Sin $[x_6]$ – 1.520 Cos $[x_5]$ Sin $[x_6]$ – 0.370 Cos $[x_6]$ Sin $[x_6]$ – 14.619 Sin $[x_1]$ Sin $[x_6]$ – 4.798 Sin $[x_2]$ Sin $[x_6]$ – 10.714 Sin $[x_3]$ Sin $[x_6]$ + 8.230 Sin $[x_4]$ Sin $[x_6]$ + 0.626 Sin $[x_5]$ Sin $[x_6]$ + 29.589 Sin $[x_6]^2$

In[•]:=(*12*)

Out[•]//NumberForm=

45.366 – 122.419 Cos $[x_1]$ – 92.332 Cos $[x_1]^2$ – 77.430 Cos $[x_2]$ + 37.176 Cos $[x_2]^2$ – 64.368 Cos $[x_3]$ + 70.363 Cos $[x_1]$ Cos $[x_3]$ + 23.980 Cos $[x_2]$ Cos $[x_3]$ + 15.976 Cos $[x_3]^2$ + 14.255 Cos $[x_4]$ – 0.953 Cos $[x_2]$ Cos $[x_4]$ + 57.515 Cos $[x_4]^2$ – 71.133 Cos $[x_5]$ + 32.336 Cos $[x_5]^2$ + 3.967 Cos $[x_6]$ + 24.849 Cos $[x_6]^2$ – 447.576 Sin $[x_1]$ + 101.234 Cos $[x_1]$ Sin $[x_1]$ + 11.339 Cos $[x_2]$ Sin $[x_1]$ + 216.534 Cos $[x_3]$ Sin $[x_1]$ – 5.218 Cos $[x_4]$ Sin $[x_1]$ + 39.522 Cos $[x_5]$ Sin $[x_1]$ + 3.408 Cos $[x_6]$ Sin $[x_1]$ + 130.778 Sin $[x_1]^2$ + 33.863 Sin $[x_2]^2$ + 138.647 Sin $[x_3]$ + 19.380 Cos $[x_1]$ Sin $[x_3]$ + 66.714 Cos $[x_2]$ Sin $[x_3]$ – 141.465 Cos $[x_3]$ Sin $[x_3]$ – 6.695 Cos $[x_4]$ Sin $[x_3]$ + 43.934 Cos $[x_5]$ Sin $[x_3]$ – 7.636 Cos $[x_6]$ Sin $[x_3]$ + 42.260 Sin $[x_3]^2$ + 86.047 Sin $[x_4]$ + 14.379 Sin $[x_4]^2$ + 0.958 Sin $[x_5]$ + 28.918 Sin $[x_5]^2$ + 35.730 Sin $[x_6]$ – 3.632 Cos $[x_1]$ Sin $[x_6]$ + 0.170 Cos $[x_2]$ Sin $[x_6]$ – 8.280 Cos $[x_3]$ Sin $[x_6]$ – 0.147 Cos $[x_4]$ Sin $[x_6]$ – 0.658 Cos $[x_5]$ Sin $[x_6]$ – 0.310 Cos $[x_6]$ Sin $[x_6]$ – 15.767 Sin $[x_1]$ Sin $[x_6]$ – 5.997 Sin $[x_2]$ Sin $[x_6]$ – 22.687 Sin $[x_3]$ Sin $[x_6]$ + 9.970 Sin $[x_4]$ Sin $[x_6]$ + 0.408 Sin $[x_5]$ Sin $[x_6]$ + 24.480 Sin $[x_6]^2$

In[•]:=(*13*)

Out[•]//NumberForm=

36.667 – 74.885 Cos $[x_1]$ – 90.656 Cos $[x_1]^2$ – 79.597 Cos $[x_2]$ + 28.085 Cos $[x_2]^2$ – 38.575 Cos $[x_3]$ + 71.408 Cos $[x_1]$ Cos $[x_3]$ + 23.272 Cos $[x_2]$ Cos $[x_3]$ – 21.943 Cos $[x_3]^2$ + 11.812 Cos $[x_4]$ + 2.321 Cos $[x_2]$ Cos $[x_4]$ + 51.536 Cos $[x_4]^2$ – 82.575 Cos $[x_5]$ + 19.989 Cos $[x_5]^2$ + 4.314 Cos $[x_6]$ + 31.712 Cos $[x_6]^2$ – 381.196 Sin $[x_1]$ + 73.078 Cos $[x_1]$ Sin $[x_1]$ + 10.556 Cos $[x_2]$ Sin $[x_1]$ + 225.713 Cos $[x_3]$ Sin $[x_1]$ – 3.799 Cos $[x_4]$ Sin $[x_1]$ + 46.347 Cos $[x_5]$ Sin $[x_1]$ + 3.698 Cos $[x_6]$ Sin $[x_1]$ + 110.581 Sin $[x_1]^2$ + 24.590 Sin $[x_2]^2$ + 103.371 Sin $[x_3]$ + 0.909 Cos $[x_1]$ Sin $[x_3]$ + 69.294 Cos $[x_2]$ Sin $[x_3]$ – 169.474 Cos $[x_3]$ Sin $[x_3]$ – 5.905 Cos $[x_4]$ Sin $[x_3]$ + 43.074 Cos $[x_5]$ Sin $[x_3]$ – 8.390 Cos $[x_6]$ Sin $[x_3]$ + 43.702 Sin $[x_3]^2$ + 111.565 Sin $[x_4]$ – 6.124 Sin $[x_4]^2$ + 1.316 Sin $[x_5]$ + 19.604 Sin $[x_5]^2$ + 54.309 Sin $[x_6]$ –

(Continued)

TABLE A7.4 *(Continued)*

Models Obtained by NRM and SNRM Using the Bootstrap Data Separation Method*) (*(80 Training–20 Testing) Training Was Divided into Two Parts–64 Training–16 Validation*) (*Models Obtained Using the NRM and SNRM Methods*)(*(Related to Table 7.5)*)

4.700 Cos $[x_1]$ Sin $[x_6]$ + 0.690 Cos $[x_2]$ Sin $[x_6]$ − 11.516 Cos $[x_3]$ Sin $[x_6]$ − 0.736 Cos $[x_4]$ Sin $[x_6]$ − 0.573 Cos $[x_5]$ Sin $[x_6]$ − 0.382 Cos $[x_6]$ Sin $[x_6]$ − 19.609 Sin $[x_1]$ Sin $[x_6]$ − 5.444 Sin $[x_2]$ Sin $[x_6]$ − 33.059 Sin $[x_3]$ Sin $[x_6]$ + 5.735 Sin $[x_4]$ Sin $[x_6]$ + 0.244 Sin $[x_5]$ Sin $[x_6]$ + 31.252 Sin $[x_6]^2$

In[•]:=(***14***)

Out[•]//NumberForm=

70.194 − 142.297 Cos $[x_1]$ − 90.550 Cos $[x_1]^2$ − 27.991 Cos $[x_2]$ + 45.258 Cos $[x_2]^2$ − 59.905 Cos $[x_3]$ + 68.216 Cos $[x_1]$ Cos $[x_3]$ + 10.094 Cos $[x_2]$ Cos $[x_3]$ + 42.045 Cos $[x_3]^2$ + 6.525 Cos $[x_4]$ − 1.591 Cos $[x_2]$ Cos $[x_4]$ + 26.850 Cos $[x_4]^2$ − 73.881 Cos $[x_5]$ + 45.848 Cos $[x_5]^2$ + 3.996 Cos $[x_6]$ + 38.424 Cos $[x_6]^2$ − 504.651 Sin $[x_1]$ + 120.374 Cos $[x_1]$ Sin $[x_1]$ + 8.088 Cos $[x_2]$ Sin $[x_1]$ + 210.381 Cos $[x_3]$ Sin $[x_1]$ − 6.898 Cos $[x_4]$ Sin $[x_1]$ + 35.831 Cos $[x_5]$ Sin $[x_1]$ + 2.769 Cos $[x_6]$ Sin $[x_1]$ + 154.126 Sin $[x_1]^2$ + 42.298 Sin $[x_2]^2$ + 191.383 Sin $[x_3]$ + 18.800 Cos $[x_1]$ Sin $[x_3]$ + 22.518 Cos $[x_2]$ Sin $[x_3]$ − 140.008 Cos $[x_3]$ Sin $[x_3]$ + 3.652 Cos $[x_4]$ Sin $[x_3]$ + 48.877 Cos $[x_5]$ Sin $[x_3]$ − 7.021 Cos $[x_6]$ Sin $[x_3]$ + 40.272 Sin $[x_3]^2$ − 19.124 Sin $[x_4]$ + 40.755 Sin $[x_4]^2$ + 1.014 Sin $[x_5]$ + 43.166 Sin $[x_5]^2$ + 24.947 Sin $[x_6]$ − 3.662 Cos $[x_1]$ Sin $[x_6]$ + 0.625 Cos $[x_2]$ Sin $[x_6]$ − 6.231 Cos $[x_3]$ Sin $[x_6]$ + 0.000 Cos $[x_4]$ Sin $[x_6]$ − 1.398 Cos $[x_5]$ Sin $[x_6]$ − 0.290 Cos $[x_6]$ Sin $[x_6]$ − 15.357 Sin $[x_1]$ Sin $[x_6]$ − 3.689 Sin $[x_2]$ Sin $[x_6]$ − 16.061 Sin $[x_3]$ Sin $[x_6]$ + 10.345 Sin $[x_4]$ Sin $[x_6]$ + 0.417 Sin $[x_5]$ Sin $[x_6]$ + 37.973 Sin $[x_6]^2$

In[•]:=(***15***)

Out[•]//NumberForm=

64.116 − 67.410 Cos $[x_1]$ − 105.787 Cos $[x_1]^2$ − 73.512 Cos $[x_2]$ + 39.005 Cos $[x_2]^2$ − 43.223 Cos $[x_3]$ + 58.027 Cos $[x_1]$ Cos $[x_3]$ + 23.875 Cos $[x_2]$ Cos $[x_3]$ + 8.054 Cos $[x_3]^2$ − 0.323 Cos $[x_4]$ + 1.662 Cos $[x_2]$ Cos $[x_4]$ + 45.941 Cos $[x_4]^2$ − 64.954 Cos $[x_5]$ + 37.375 Cos $[x_5]^2$ + 5.965 Cos $[x_6]$ + 37.008 Cos $[x_6]^2$ − 513.159 Sin $[x_1]$ + 102.812 Cos $[x_1]$ Sin $[x_1]$ + 8.946 Cos $[x_2]$ Sin $[x_1]$ + 233.252 Cos $[x_3]$ Sin $[x_1]$ − 2.849 Cos $[x_4]$ Sin $[x_1]$ + 41.677 Cos $[x_5]$ Sin $[x_1]$ + 2.437 Cos $[x_6]$ Sin $[x_1]$ + 152.269 Sin $[x_1]^2$ + 35.905 Sin $[x_2]^2$ + 151.198 Sin $[x_3]$ − 35.206 Cos $[x_1]$ Sin $[x_3]$ + 64.727 Cos $[x_2]$ Sin $[x_3]$ − 169.416 Cos $[x_3]$ Sin $[x_3]$ + 6.668 Cos $[x_4]$ Sin $[x_3]$ + 43.687 Cos $[x_5]$ Sin $[x_3]$ − 8.861 Cos $[x_6]$ Sin $[x_3]$ + 50.774 Sin $[x_3]^2$ + 91.310 Sin $[x_4]$ + 1.125 Sin $[x_4]^2$ + 1.331 Sin $[x_5]$ + 28.748 Sin $[x_5]^2$ + 42.667 Sin $[x_6]$ − 4.699 Cos $[x_1]$ Sin $[x_6]$ + 0.732 Cos $[x_2]$ Sin $[x_6]$ − 8.174 Cos $[x_3]$ Sin $[x_6]$ − 0.812 Cos $[x_4]$ Sin $[x_6]$ − 0.506 Cos $[x_5]$ Sin $[x_6]$ − 0.297 Cos $[x_6]$ Sin $[x_6]$ − 18.048 Sin $[x_1]$ Sin $[x_6]$ − 4.968 Sin $[x_2]$ Sin $[x_6]$ − 22.846 Sin $[x_3]$ Sin $[x_6]$ + 4.740 Sin $[x_4]$ Sin $[x_6]$ + 0.235 Sin $[x_5]$ Sin $[x_6]$ + 36.336 Sin $[x_6]^2$

In[•]:=(***16***)

Out[•]//NumberForm=

61.617 − 138.698 Cos $[x_1]$ − 146.604 Cos $[x_1]^2$ − 49.003 Cos $[x_2]$ + 32.428 Cos $[x_2]^2$ + 42.612 Cos $[x_3]$ + 53.720 Cos $[x_1]$ Cos $[x_3]$ + 19.420 Cos $[x_2]$ Cos $[x_3]$ + 44.690 Cos $[x_3]^2$ + 15.136 Cos $[x_4]$ + 1.063 Cos $[x_2]$ Cos $[x_4]$ + 48.182 Cos $[x_4]^2$ − 64.664 Cos $[x_5]$ + 35.768 Cos $[x_5]^2$ + 4.792 Cos $[x_6]$ + 31.514 Cos $[x_6]^2$ − 640.457 Sin $[x_1]$ + 153.699 Cos $[x_1]$ Sin $[x_1]$ + 4.542 Cos $[x_2]$ Sin $[x_1]$ + 188.470 Cos $[x_3]$ Sin $[x_1]$ − 11.454 Cos $[x_4]$ Sin $[x_1]$ + 41.203 Cos $[x_5]$ Sin $[x_1]$ + 3.613 Cos $[x_6]$ Sin $[x_1]$ + 171.313 Sin $[x_1]^2$ + 30.057 Sin $[x_2]^2$ + 307.416 Sin $[x_3]$ − 13.858 Cos $[x_1]$ Sin $[x_3]$ + 44.578 Cos $[x_2]$ Sin $[x_3]$ − 211.774 Cos $[x_3]$ Sin $[x_3]$ − 1.416 Cos $[x_4]$ Sin $[x_3]$ + 34.639 Cos $[x_5]$ Sin $[x_3]$ − 8.750 Cos $[x_6]$ Sin $[x_3]$ + 7.969 Sin $[x_3]^2$ + 92.368 Sin $[x_4]$ + 1.374 Sin $[x_4]^2$ + 1.093 Sin $[x_5]$ + 32.545 Sin $[x_5]^2$ + 44.441 Sin $[x_6]$ −

(Continued)

TABLE A7.4 *(Continued)*

Models Obtained by NRM and SNRM Using the Bootstrap Data Separation Method*) (*(80 Training–20 Testing) Training Was Divided into Two Parts–64 Training–16 Validation*) (*Models Obtained Using the NRM and SNRM Methods*)(*(Related to Table 7.5)*)

3.682 Cos $[x_1]$ Sin $[x_6]$ + 0.315 Cos $[x_2]$ Sin $[x_6]$ − 10.300 Cos $[x_3]$ Sin $[x_6]$ − 0.276 Cos $[x_4]$ Sin $[x_6]$ − 1.291 Cos $[x_5]$ Sin $[x_6]$ − 0.342 Cos $[x_6]$ Sin $[x_6]$ − 16.060 Sin $[x_1]$ Sin $[x_6]$ − 7.086 Sin $[x_2]$ Sin $[x_6]$ − 29.023 Sin $[x_3]$ Sin $[x_6]$ + 8.662 Sin $[x_4]$ Sin $[x_6]$ + 0.200 Sin $[x_5]$ Sin $[x_6]$ + 30.991 Sin $[x_6]^2$

In[•]:=(***17***)

Out[•]//NumberForm=

49.283 − 48.325 Cos $[x_1]$ − 89.565 Cos $[x_1]^2$ − 79.411 Cos $[x_2]$ + 33.762 Cos $[x_2]^2$ − 95.738 Cos $[x_3]$ + 63.468 Cos $[x_1]$ Cos $[x_3]$ + 26.422 Cos $[x_2]$ Cos $[x_3]$ + 1.680 Cos $[x_3]^2$ + 7.419 Cos $[x_4]$ + 1.364 Cos $[x_2]$ Cos $[x_4]$ + 37.271 Cos $[x_4]^2$ − 90.619 Cos $[x_5]$ + 28.247 Cos $[x_5]^2$ + 5.362 Cos $[x_6]$ + 32.772 Cos $[x_6]^2$ − 419.913 Sin $[x_1]$ + 80.365 Cos $[x_1]$ Sin $[x_1]$ + 8.732 Cos $[x_2]$ Sin $[x_1]$ + 252.848 Cos $[x_3]$ Sin $[x_1]$ − 9.071 Cos $[x_4]$ Sin $[x_1]$ + 57.849 Cos $[x_5]$ Sin $[x_1]$ + 2.846 Cos $[x_6]$ Sin $[x_1]$ + 130.622 Sin $[x_1]^2$ + 30.417 Sin $[x_2]^2$ + 134.615 Sin $[x_3]$ − 33.192 Cos $[x_1]$ Sin $[x_3]$ + 70.144 Cos $[x_2]$ Sin $[x_3]$ − 141.195 Cos $[x_3]$ Sin $[x_3]$ + 4.818 Cos $[x_4]$ Sin $[x_3]$ + 47.907 Cos $[x_5]$ Sin $[x_3]$ − 8.645 Cos $[x_6]$ Sin $[x_3]$ + 40.766 Sin $[x_3]^2$ + 75.684 Sin $[x_4]$ + 0.636 Sin $[x_4]^2$ + 1.272 Sin $[x_5]$ + 23.338 Sin $[x_5]^2$ + 33.592 Sin $[x_6]$ − 4.482 Cos $[x_1]$ Sin $[x_6]$ + 0.577 Cos $[x_2]$ Sin $[x_6]$ − 5.915 Cos $[x_3]$ Sin $[x_6]$ − 0.353 Cos $[x_4]$ Sin $[x_6]$ − 0.846 Cos $[x_5]$ Sin $[x_6]$ − 0.339 Cos $[x_6]$ Sin $[x_6]$ − 17.592 Sin $[x_1]$ Sin $[x_6]$ − 4.952 Sin $[x_2]$ Sin $[x_6]$ − 17.692 Sin $[x_3]$ Sin $[x_6]$ + 7.363 Sin $[x_4]$ Sin $[x_6]$ + 0.409 Sin $[x_5]$ Sin $[x_6]$ + 32.155 Sin $[x_6]^2$

In[•]:=(***18***)

Out[•]//NumberForm=

62.183 − 176.417 Cos $[x_1]$ − 143.587 Cos $[x_1]^2$ − 73.148 Cos $[x_2]$ + 43.930 Cos $[x_2]^2$ − 31.515 Cos $[x_3]$ + 68.165 Cos $[x_1]$ Cos $[x_3]$ + 25.641 Cos $[x_2]$ Cos $[x_3]$ + 50.167 Cos $[x_3]^2$ + 15.499 Cos $[x_4]$ + 2.454 Cos $[x_2]$ Cos $[x_4]$ + 50.652 Cos $[x_4]^2$ − 93.620 Cos $[x_5]$ + 40.280 Cos $[x_5]^2$ + 5.011 Cos $[x_6]$ + 36.483 Cos $[x_6]^2$ − 687.499 Sin $[x_1]$ + 175.186 Cos $[x_1]$ Sin $[x_1]$ + 7.432 Cos $[x_2]$ Sin $[x_1]$ + 232.180 Cos $[x_3]$ Sin $[x_1]$ − 16.108 Cos $[x_4]$ Sin $[x_1]$ + 56.453 Cos $[x_5]$ Sin $[x_1]$ + 3.280 Cos $[x_6]$ Sin $[x_1]$ + 192.569 Sin $[x_1]^2$ + 41.786 Sin $[x_2]^2$ + 298.105 Sin $[x_3]$ − 2.636 Cos $[x_1]$ Sin $[x_3]$ + 64.979 Cos $[x_2]$ Sin $[x_3]$ − 186.401 Cos $[x_3]$ Sin $[x_3]$ + 2.788 Cos $[x_4]$ Sin $[x_3]$ + 49.651 Cos $[x_5]$ Sin $[x_3]$ − 8.692 Cos $[x_6]$ Sin $[x_3]$ + 11.804 Sin $[x_3]^2$ + 110.975 Sin $[x_4]$ − 5.787 Sin $[x_4]^2$ + 1.218 Sin $[x_5]$ + 36.767 Sin $[x_5]^2$ + 35.197 Sin $[x_6]$ − 2.994 Cos $[x_1]$ Sin $[x_6]$ + 0.641 Cos $[x_2]$ Sin $[x_6]$ − 7.885 Cos $[x_3]$ Sin $[x_6]$ − 0.592 Cos $[x_4]$ Sin $[x_6]$ − 1.566 Cos $[x_5]$ Sin $[x_6]$ − 0.334 Cos $[x_6]$ Sin $[x_6]$ − 14.602 Sin $[x_1]$ Sin $[x_6]$ − 4.890 Sin $[x_2]$ Sin $[x_6]$ − 22.341 Sin $[x_3]$ Sin $[x_6]$ + 6.802 Sin $[x_4]$ Sin $[x_6]$ + 0.241 Sin $[x_5]$ Sin $[x_6]$ + 35.941 Sin $[x_6]^2$

In[•]:=(***19***)

Out[•]//NumberForm=

75.137 − 95.388 Cos $[x_1]$ − 118.848 Cos $[x_1]^2$ − 75.605 Cos $[x_2]$ + 43.595 Cos $[x_2]^2$ − 47.181 Cos $[x_3]$ + 54.891 Cos $[x_1]$ Cos $[x_3]$ + 26.809 Cos $[x_2]$ Cos $[x_3]$ + 16.463 Cos $[x_3]^2$ + 10.380 Cos $[x_4]$ + 2.435 Cos $[x_2]$ Cos $[x_4]$ + 57.651 Cos $[x_4]^2$ − 80.227 Cos $[x_5]$ + 50.740 Cos $[x_5]^2$ + 4.847 Cos $[x_6]$ + 38.368 Cos $[x_6]^2$ − 575.659 Sin $[x_1]$ + 123.646 Cos $[x_1]$ Sin $[x_1]$ + 6.942 Cos $[x_2]$ Sin $[x_1]$ + 213.758 Cos $[x_3]$ Sin $[x_1]$ − 6.868 Cos $[x_4]$ Sin $[x_1]$ + 51.215 Cos $[x_5]$ Sin $[x_1]$ + 2.856 Cos $[x_6]$ Sin $[x_1]$ + 173.996 Sin $[x_1]^2$ + 40.192 Sin $[x_2]^2$ + 142.810 Sin $[x_3]$ − 27.707 Cos $[x_1]$ Sin $[x_3]$ + 67.544 Cos $[x_2]$ Sin $[x_3]$ − 149.710 Cos $[x_3]$ Sin $[x_3]$ − 0.708 Cos $[x_4]$ Sin $[x_3]$ + 38.364 Cos $[x_5]$ Sin $[x_3]$ − 8.105 Cos $[x_6]$ Sin $[x_3]$ + 49.532 Sin $[x_3]^2$ + 88.253 Sin $[x_4]$ + 13.857 Sin $[x_4]^2$ + 1.339 Sin $[x_5]$ + 48.669 Sin $[x_5]^2$ + 34.875 Sin $[x_6]$ −

(Continued)

TABLE A7.4 *(Continued)*
Models Obtained by NRM and SNRM Using the Bootstrap Data Separation Method*) (*(80 Training–20 Testing) Training Was Divided into Two Parts–64 Training–16 Validation*) (*Models Obtained Using the NRM and SNRM Methods*)(*(Related to Table 7.5)*)

$3.575 \cos [x_1] \sin [x_6] + 0.308 \cos [x_2] \sin [x_6] - 8.020 \cos [x_3] \sin [x_6] - 0.172 \cos [x_4] \sin [x_6] -$
$1.588 \cos [x_5] \sin [x_6] - 0.398 \cos [x_6] \sin [x_6] - 14.765 \sin [x_1] \sin [x_6] - 6.223 \sin [x_2] \sin [x_6] -$
$23.614 \sin [x_3] \sin [x_6] + 9.966 \sin [x_4] \sin [x_6] + 0.520 \sin [x_5] \sin [x_6] + 37.847 \sin [x_6]^2$

In[•]:=(***20***)
Out[•]//NumberForm=

$79.374 - 116.790 \cos [x_1] - 117.638 \cos [x_1]^2 - 51.206 \cos [x_2] + 47.537 \cos [x_2]^2 - 67.022 \cos [x_3] +$
$62.975 \cos [x_1] \cos [x_3] + 17.561 \cos [x_2] \cos [x_3] + 17.036 \cos [x_3]^2 + 9.353 \cos [x_4] +$
$1.640 \cos [x_2] \cos [x_4] + 69.906 \cos [x_4]^2 - 74.089 \cos [x_5] + 45.711 \cos [x_5]^2 + 5.244 \cos [x_6] +$
$43.642 \cos [x_6]^2 - 585.223 \sin [x_1] + 135.369 \cos [x_1] \sin [x_1] + 7.833 \cos [x_2] \sin [x_1] +$
$234.800 \cos [x_3] \sin [x_1] - 7.110 \cos [x_4] \sin [x_1] + 47.071 \cos [x_5] \sin [x_1] + 3.220 \cos [x_6] \sin [x_1] +$
$171.025 \sin [x_1]^2 + 44.695 \sin [x_2]^2 + 142.392 \sin [x_3] - 20.365 \cos [x_1] \sin [x_3] + 44.267 \cos [x_2] \sin [x_3] -$
$150.808 \cos [x_3] \sin [x_3] + 0.682 \cos [x_4] \sin [x_3] + 45.388 \cos [x_5] \sin [x_3] - 8.872 \cos [x_6] \sin [x_3] +$
$54.599 \sin [x_3]^2 + 78.276 \sin [x_4] + 31.793 \sin [x_4]^2 + 1.248 \sin [x_5] + 38.603 \sin [x_5]^2 + 34.831 \sin [x_6] -$
$4.108 \cos [x_1] \sin [x_6] + 0.329 \cos [x_2] \sin [x_6] - 6.745 \cos [x_3] \sin [x_6] - 0.265 \cos [x_4] \sin [x_6] -$
$1.283 \cos [x_5] \sin [x_6] - 0.387 \cos [x_6] \sin [x_6] - 16.935 \sin [x_1] \sin [x_6] - 6.242 \sin [x_2] \sin [x_6] -$
$19.770 \sin [x_3] \sin [x_6] + 8.521 \sin [x_4] \sin [x_6] + 0.342 \sin [x_5] \sin [x_6] + 43.052 \sin [x_6]^2$

TABLE A7.5
Models Obtained by NRM and SNRM Using the Bootstrap Data Separation Method (80 Training–15 Testing–5 Validation*) (*Models Obtained Using the NRM and SNRM Methods*)(*Related to Table 7.6*)

(*Models proposed by NRM*)
In[•]:= (*1*)
Out[•]//NumberForm=

$-7.555 - 745.970 \cos [x1] - 32.241 \cos [x1]^2 + 543.555 \cos [x2] - 20.436 \cos [x2]^2 -$
$364.413 \cos [x3] + 213.925 \cos [x1] \cos [x3] - 167.953 \cos [x2] \cos [x3] + 175.331 \cos [x3]^2 +$
$95.863 \cos [x4] - 8.347 \cos [x2] \cos [x4] + 109.936 \cos [x4]^2 - 5.362 \cos [x5] - 7.813 \cos [x5]^2 +$
$5.988 \cos [x6] - 10.117 \cos [x6]^2 - 3.382 \sin [x1] + 110.211 \cos [x1] \sin [x1] +$
$30.314 \cos [x2] \sin [x1] + 68.670 \cos [x3] \sin [x1] + 4.459 \cos [x4] \sin [x1] +$
$109.850 \cos [x5] \sin [x1] + 2.319 \cos [x6] \sin [x1] - 8.697 \sin [x1]^2 - 7.864 \sin [x2]^2 + 6.117 \sin [x3] +$
$608.735 \cos [x1] \sin [x3] - 541.937 \cos [x2] \sin [x3] + 209.002 \cos [x3] \sin [x3] -$
$103.427 \cos [x4] \sin [x3] - 41.169 \cos [x5] \sin [x3] - 8.944 \cos [x6] \sin [x3] - 14.572 \sin [x3]^2 +$
$227.883 \sin [x4] - 8.815 \sin [x4]^2 + 0.949 \sin [x5] - 40.801 \sin [x5]^2 - 16.214 \sin [x6] -$
$8.216 \cos [x1] \sin [x6] - 5.323 \cos [x2] \sin [x6] + 10.023 \cos [x3] \sin [x6] + 8.718 \cos [x4] \sin [x6] -$
$12.411 \cos [x5] \sin [x6] - 0.746 \cos [x6] \sin [x6] - 37.612 \sin [x1] \sin [x6] - 45.843 \sin [x2] \sin [x6] +$
$31.088 \sin [x3] \sin [x6] + 55.416 \sin [x4] \sin [x6] + 1.067 \sin [x5] \sin [x6] - 10.296 \sin [x6]^2$

(Continued)

TABLE A7.5 *(Continued)*

Models Obtained by NRM and SNRM Using the Bootstrap Data Separation Method (80 Training–15 Testing–5 Validation*) (*Models Obtained Using the NRM and SNRM Methods*)(*Related to Table 7.6*)

$In[\bullet]:= (*2*)$

$Out[\bullet]//NumberForm=$

17.975 + 16.616 Cos [$x1$] + 325.814 Cos [$x1$]² + 1288.760 Cos [$x2$] − 30.370 Cos [$x2$]² + 537.979 Cos [$x3$] − 399.926 Cos [$x1$] Cos [$x3$] − 450.776 Cos [$x2$] Cos [$x3$] − 108.800 Cos [$x3$]² − 9.301 Cos [$x4$] − 27.862 Cos [$x2$] Cos [$x4$] − 24.170 Cos [$x4$]² + 283.655 Cos [$x5$] + 22.455 Cos [$x5$]² + 25.596 Cos [$x6$] + 26.161 Cos [$x6$]² + 28.456 Sin [$x1$] + 381.804 Cos [$x1$] Sin [$x1$] − 21.368 Cos [$x2$] Sin [$x1$] − 449.529 Cos [$x3$] Sin [$x1$] + 210.493 Cos [$x4$] Sin [$x1$] − 237.270 Cos [$x5$] Sin [$x1$] − 30.826 Cos [$x6$] Sin [$x1$] + 16.175 Sin [$x1$]² + 18.653 Sin [$x2$]² − 6.970 Sin [$x3$] − 369.935 Cos [$x1$] Sin [$x3$] − 1187.000 Cos [$x2$] Sin [$x3$] + 36.283 Cos [$x3$] Sin [$x3$] − 201.182 Cos [$x4$] Sin [$x3$] + 19.987 Cos [$x5$] Sin [$x3$] + 3.822 Cos [$x6$] Sin [$x3$] + 25.801 Sin [$x3$]² − 28.755 Sin [$x4$] + 19.931 Sin [$x4$]² + 0.344 Sin [$x5$] − 16.749 Sin [$x5$]² − 180.864 Sin [$x6$] + 10.653 Cos [$x1$] Sin [$x6$] − 13.268 Cos [$x2$] Sin [$x6$] − 2.745 Cos [$x3$] Sin [$x6$] + 4.663 Cos [$x4$] Sin [$x6$] − 4.323 Cos [$x5$] Sin [$x6$] − 0.164 Cos [$x6$] Sin [$x6$] − 4.525 Sin [$x1$] Sin [$x6$] + 162.488 Sin [$x2$] Sin [$x6$] + 21.369 Sin [$x3$] Sin [$x6$] − 3.802 Sin [$x4$] Sin [$x6$] + 4.802 Sin [$x5$] Sin [$x6$] + 23.892 Sin [$x6$]²

$In[\bullet]:= (*3*)$

$Out[\bullet]//NumberForm=$

13.259 − 418.010 Cos [$x1$] − 267.608 Cos [$x1$]² − 698.885 Cos [$x2$] + 1.074 Cos [$x2$]² − 719.207 Cos [$x3$] + 480.091 Cos [$x1$] Cos [$x3$] + 239.406 Cos [$x2$] Cos [$x3$] + 282.482 Cos [$x3$]² + 154.705 Cos [$x4$] + 18.157 Cos [$x2$] Cos [$x4$] + 27.164 Cos [$x4$]² − 213.445 Cos [$x5$] + 15.959 Cos [$x5$]² − 17.106 Cos [$x6$] + 18.668 Cos [$x6$]² − 18.213 Sin [$x1$] − 381.188 Cos [$x1$] Sin [$x1$] − 105.805 Cos [$x2$] Sin [$x1$] + 346.296 Cos [$x3$] Sin [$x1$] − 17.674 Cos [$x4$] Sin [$x1$] + 202.368 Cos [$x5$] Sin [$x1$] + 10.883 Cos [$x6$] Sin [$x1$] + 21.517 Sin [$x1$]² + 13.808 Sin [$x2$]² + 47.898 Sin [$x3$] + 770.905 Cos [$x1$] Sin [$x3$] + 765.765 Cos [$x2$] Sin [$x3$] + 173.832 Cos [$x3$] Sin [$x3$] − 143.057 Cos [$x4$] Sin [$x3$] + 38.353 Cos [$x5$] Sin [$x3$] + 6.570 Cos [$x6$] Sin [$x3$] + 10.777 Sin [$x3$]² + 9.154 Sin [$x4$] + 14.287 Sin [$x4$]² + 2.010 Sin [$x5$] + 9.088 Sin [$x5$]² + 176.279 Sin [$x6$] + 6.669 Cos [$x1$] Sin [$x6$] + 4.826 Cos [$x2$] Sin [$x6$] − 60.813 Cos [$x3$] Sin [$x6$] − 6.952 Cos [$x4$] Sin [$x6$] − 9.394 Cos [$x5$] Sin [$x6$] + 0.135 Cos [$x6$] Sin [$x6$] + 33.199 Sin [$x1$] Sin [$x6$] − 29.396 Sin [$x2$] Sin [$x6$] − 166.273 Sin [$x3$] Sin [$x6$] − 12. 807 Sin [$x4$] Sin [$x6$] + 1.384 Sin [$x5$] Sin [$x6$] + 18.628 Sin [$x6$]²)

$In[\bullet]:= (*4*)$

$Out[\bullet]//NumberForm=$

5.719 + 298.583 Cos [$x1$] − 68.278 Cos [$x1$]² − 791.866 Cos [$x2$] + 28.560 Cos [$x2$]² − 559.500 Cos [$x3$] − 100.829 Cos [$x1$] Cos [$x3$] + 239.394 Cos [$x2$] Cos [$x3$] + 170.826 Cos [$x3$]² + 22.118 Cos [$x4$] − 32.975 Cos [$x2$] Cos [$x4$] − 43.592 Cos [$x4$]² − 218.299 Cos [$x5$] + 6.956 Cos [$x5$]² + 8.449 Cos [$x6$] + 8.073 Cos [$x6$]² − 4.160 Sin [$x1$] + 54.847 Cos [$x1$] Sin [$x1$] + 150.334 Cos [$x2$] Sin [$x1$] + 230.359 Cos [$x3$] Sin [$x1$] − 57.896 Cos [$x4$] Sin [$x1$] + 253.554 Cos [$x5$] Sin [$x1$] − 5.421 Cos [$x6$] Sin [$x1$] + 8.561 Sin [$x1$]² + 5.956 Sin [$x2$]² + 62.701 Cos [$x3$] − 336.576 Cos [$x1$] Sin [$x3$] + 611.094 Cos [$x2$] Sin [$x3$] + 304.800 Cos [$x3$] Sin [$x3$] + 48.565 Cos [$x4$] Sin [$x3$] − 33.812 Cos [$x5$] Sin [$x3$] − 4.048 Cos [$x6$] Sin [$x3$] + 2.519 Sin [$x3$]² − 11.310 Sin [$x4$] + 6.305 Sin [$x4$]² + 2.108 Sin [$x5$] + 7.628 Sin [$x5$]² + 31.731 Sin [$x6$] + 16.600 Cos [$x1$] Sin [$x6$] − 13.225 Cos [$x2$] Sin [$x6$] − 0.003 Cos [$x3$] Sin [$x6$] − 4.511 Cos [$x4$] Sin [$x6$] − 17.619 Cos [$x5$] Sin [$x6$] + 0.451 Cos [$x6$] Sin [$x6$] + 41.346 Sin [$x1$] Sin [$x6$] − 77.982 Sin [$x2$] Sin [$x6$] + 2.901 Sin [$x3$] Sin [$x6$] − 17.110 Sin [$x4$] Sin [$x6$] + 3.352 Sin [$x5$] Sin [$x6$] + 7.608 Sin [$x6$]²

(Continued)

TABLE A7.5 *(Continued)*
Models Obtained by NRM and SNRM Using the Bootstrap Data Separation Method (80 Training–15 Testing–5 Validation*) (*Models Obtained Using the NRM and SNRM Methods*)(*Related to Table 7.6*)

In[•]:= (*5*)
Out[•]//NumberForm=

$7.480 + 46.483$ Cos $[x1] + 42.241$ Cos $[x1]^2 + 238.691$ Cos $[x2] + 5.115$ Cos $[x2]^2 + 65.042$ Cos $[x3] +$
71.689 Cos $[x1]$ Cos $[x3] - 86.656$ Cos $[x2]$ Cos $[x3] - 217.787$ Cos $[x3]^2 - 5.390$ Cos $[x4] -$
4.947 Cos $[x2]$ Cos $[x4] + 27.302$ Cos $[x4]^2 - 46.340$ Cos $[x5] + 8.433$ Cos $[x5]^2 + 3.680$ Cos $[x6] +$
10.301 Cos $[x6]^2 + 6.331$ Sin $[x1] - 25.945$ Cos $[x1]$ Sin $[x1] - 30.271$ Cos $[x2]$ Sin $[x1] +$
309.891 Cos $[x3]$ Sin $[x1] + 60.870$ Cos $[x4]$ Sin $[x1] - 4.427$ Cos $[x5]$ Sin $[x1] + 7.862$ Cos $[x6]$ Sin $[x1] +$
8.341 Sin $[x1]^2 + 7.830$ Sin $[x2]^2 - 11.295$ Sin $[x3] - 25.462$ Cos $[x1]$ Sin $[x3] -$
186.043 Cos $[x2]$ Sin $[x3] - 306.348$ Cos $[x3]$ Sin $[x3] - 59.890$ Cos $[x4]$ Sin $[x3] +$
27.491 Cos $[x5]$ Sin $[x3] - 11.694$ Cos $[x6]$ Sin $[x3] + 16.744$ Sin $[x3]^2 - 4.171$ Sin $[x4] +$
7.959 Sin $[x4]^2 - 0.193$ Sin $[x5] + 26.495$ Sin $[x5]^2 + 45.213$ Sin $[x6] + 6.836$ Cos $[x1]$ Sin $[x6] +$
1.271 Cos $[x2]$ Sin $[x6] - 13.323$ Cos $[x3]$ Sin $[x6] - 3.399$ Cos $[x4]$ Sin $[x6] - 0.878$ Cos $[x5]$ Sin $[x6] -$
0.485 Cos $[x6]$ Sin $[x6] + 20.785$ Sin $[x1]$ Sin $[x6] - 3.635$ Sin $[x2]$ Sin $[x6] - 46.607$ Sin $[x3]$ Sin $[x6] -$
15.492 Sin $[x4]$ Sin $[x6] - 1.041$ Sin $[x5]$ Sin $[x6] + 9.993$ Sin $[x6]^2)$

In[•]:= (*6*)
Out[•]//NumberForm=

$3.672 - 424.456$ Cos $[x1] - 168.717$ Cos $[x1]^2 + 710.419$ Cos $[x2] + 37.392$ Cos $[x2]^2 - 8.705$ Cos $[x3] +$
45.707 Cos $[x1]$ Cos $[x3] - 343.763$ Cos $[x2]$ Cos $[x3] + 115.065$ Cos $[x3]^2 + 68.934$ Cos $[x4] -$
35.854 Cos $[x2]$ Cos $[x4] + 35.121$ Cos $[x4]^2 - 173.692$ Cos $[x5] + 5.618$ Cos $[x5]^2 + 19.502$ Cos $[x6] +$
4.715 Cos $[x6]^2 + 33.870$ Sin $[x1] + 286.091$ Cos $[x1]$ Sin $[x1] + 166.688$ Cos $[x2]$ Sin $[x1] -$
65.558 Cos $[x3]$ Sin $[x1] + 132.488$ Cos $[x4]$ Sin $[x1] + 419.176$ Cos $[x5]$ Sin $[x1] -$
5.180 Cos $[x6]$ Sin $[x1] + 8.526$ Sin $[x1]^2 + 3.769$ Sin $[x2]^2 + 9.614$ Sin $[x3] + 177.213$ Cos $[x1]$ Sin $[x3] -$
798.528 Cos $[x2]$ Sin $[x3] + 55.178$ Cos $[x3]$ Sin $[x3] - 198.621$ Cos $[x4]$ Sin $[x3] -$
184.378 Cos $[x5]$ Sin $[x3] - 13.906$ Cos $[x6]$ Sin $[x3] + 2.246$ Sin $[x3]^2 + 69.570$ Sin $[x4] +$
3.870 Sin $[x4]^2 - 1.860$ Sin $[x5] - 19.485$ Sin $[x5]^2 + 18.159$ Sin $[x6] - 2.154$ Cos $[x1]$ Sin $[x6] -$
17.383 Cos $[x2]$ Sin $[x6] + 27.234$ Cos $[x3]$ Sin $[x6] + 12.141$ Cos $[x4]$ Sin $[x6] +$
36.289 Cos $[x5]$ Sin $[x6] + 1.684$ Cos $[x6]$ Sin $[x6] - 37.052$ Sin $[x1]$ Sin $[x6] - 77.425$ Sin $[x2]$ Sin $[x6] +$
55.057 Sin $[x3]$ Sin $[x6] + 70.555$ Sin $[x4]$ Sin $[x6] - 8.708$ Sin $[x5]$ Sin $[x6] + 4.808$ Sin $[x6]^2$

In[•]:= (*7*)
Out[•]//NumberForm=

$15.537 + 121.984$ Cos $[x1] - 34.431$ Cos $[x1]^2 - 25.762$ Cos $[x2] + 30.704$ Cos $[x2]^2 - 681.644$ Cos $[x3] +$
83.539 Cos $[x1]$ Cos $[x3] + 0.934$ Cos $[x2]$ Cos $[x3] + 124.506$ Cos $[x3]^2 + 104.147$ Cos $[x4] -$
21.717 Cos $[x2]$ Cos $[x4] - 54.979$ Cos $[x4]^2 - 200.905$ Cos $[x5] + 17.800$ Cos $[x5]^2 + 7.763$ Cos $[x6] +$
20.728 Cos $[x6]^2 + 6.174$ Sin $[x1] - 129.283$ Cos $[x1]$ Sin $[x1] + 66.553$ Cos $[x2]$ Sin $[x1] +$
312.881 Cos $[x3]$ Sin $[x1] - 30.116$ Cos $[x4]$ Sin $[x1] + 54.341$ Cos $[x5]$ Sin $[x1] -$
1.823 Cos $[x6]$ Sin $[x1] + 19.658$ Sin $[x1]^2 + 16.265$ Sin $[x]^2 + 62.717$ Sin $[x3] + 8.603$ Cos $[x1]$ Sin $[x3] -$
34.750 Cos $[x2]$ Sin $[x3] + 318.444$ Cos $[x3]$ Sin $[x3] - 73.309$ Cos $[x4]$ Sin $[x3] +$
135.139 Cos $[x5]$ Sin $[x3] - 6.722$ Cos $[x6]$ Sin $[x3] + 16.885$ Sin $[x3]^2 - 110.340$ Sin $[x4] +$
17.084 Sin $[x4]^2 + 1.233$ Sin $[x5] + 31.314$ Sin $[x5]^2 + 36.790$ Sin $[x6] - 15.173$ Cos $[x1]$ Sin $[x6] +$
0.961 Cos $[x2]$ Sin $[x6] + 21.004$ Cos $[x3]$ Sin $[x6] - 3.457$ Cos $[x4]$ Sin $[x6] - 7.537$ Cos $[x5]$ Sin $[x6] +$
0.141 Cos $[x6]$ Sin $[x6] - 55.088$ Sin $[x1]$ Sin $[x6] - 18.945$ Sin $[x2]$ Sin $[x6] + 56.648$ Sin $[x3]$ Sin $[x6] -$
29.254 Sin $[x4]$ Sin $[x6] + 3.071$ Sin $[x5]$ Sin $[x6] + 20.214$ Sin $[x6]^2$

(Continued)

TABLE A7.5 *(Continued)*

Models Obtained by NRM and SNRM Using the Bootstrap Data Separation Method (80 Training–15 Testing–5 Validation*) (*Models Obtained Using the NRM and SNRM Methods*)(*Related to Table 7.6*)

$In[\bullet]:= (*8*)$

$Out[\bullet]//NumberForm=$

8.921 − 14.679 Cos [$x1$] − 114.769 Cos [$x1$]^2 + 826.256 Cos [$x2$] + 16.041 Cos [$x2$]^2 − 74.463 Cos [$x3$] +
105.136 Cos [$x1$] Cos [$x3$] − 245.045 Cos [$x2$] Cos [$x3$] + 63.945 Cos [$x3$]^2 + 16.477 Cos [$x4$] −
2.766 Cos [$x2$] Cos [$x4$] − 5.054 Cos [$x4$]^2 − 75.525 Cos [$x5$] + 10.063 Cos [$x5$]^2 − 23.204 Cos [$x6$] +
12.475 Cos [$x6$]^2 − 17.913 Sin [$x1$] − 113.093 Cos [$x1$] Sin [$x1$] − 14.779 Cos [$x2$] Sin [$x1$] +
64.122 Cos [$x3$] Sin [$x1$] − 91.212 Cos [$x4$] Sin [$x1$] + 121.598 Cos [$x5$] Sin [$x1$] +
16.606 Cos [$x6$] Sin [$x1$] + 12.498 Sin [$x1$]^2 + 9.387 Sin [$x2$]^2 + 16.793 Sin [$x3$] +
123.727 Cos [$x1$] Sin [$x3$] − 776.435 Cos [$x2$] Sin [$x3$] − 7.487 Cos [$x3$] Sin [$x3$] +
82.313 Cos [$x4$] Sin [$x3$] − 63.816 Cos [$x5$] Sin [$x3$] + 7.293 Cos [$x6$] Sin [$x3$] + 9.762 Sin [$x3$]^2 +
2.637 Sin [$x4$] + 9.656 Sin [$x4$]^2 − 2.401 Sin [$x5$] + 27.384 Sin [$x5$]^2 + 36.016 Sin [$x6$] +
16.137 Cos [$x1$] Sin [$x6$] + 1.105 Cos [$x2$] Sin [$x6$] − 48.855 Cos [$x3$] Sin [$x6$] + 7.745 Cos [$x4$] Sin [$x6$] −
2.500 Cos [$x5$] Sin [$x6$] + 0.094 Cos [$x6$] Sin [$x6$] + 65.814 Sin [$x1$] Sin [$x6$] − 21.630 Sin [$x2$] Sin [$x6$] −
114.923 Sin [$x3$] Sin [$x6$] + 39.572 Sin [$x4$] Sin [$x6$] + 1.058 Sin [$x5$] Sin [$x6$] + 11.729 Sin [$x6$]^2

$In[\bullet]:= (*9*)$

$Out[\bullet]//NumberForm=$

20.043 − 46.200 Cos [$x1$] + 375.577 Cos [$x1$]^2 − 937.956 Cos [$x2$] + 84.249 Cos [$x2$]^2 −
296.317 Cos [$x3$] + 38.766 Cos [$x1$] Cos [$x3$] + 332.081 Cos [$x2$] Cos [$x3$] − 256.459 Cos [$x3$]^2 +
60.271 Cos [$x4$] + 18.887 Cos [$x2$] Cos [$x4$] − 56.722 Cos [$x4$]^2 − 10.624 Cos [$x5$] +
23.823 Cos [$x5$]^2 − 5.607 Cos [$x6$] + 26.405 Cos [$x6$]^2 + 14.477 Sin [$x1$] + 612.598 Cos [$x1$] Sin [$x1$] +
78.044 Cos [$x2$] Sin [$x1$] + 1124.210 Cos [$x3$] Sin [$x1$] + 200.331 Cos [$x4$] Sin [$x1$] +
230.532 Cos [$x5$] Sin [$x1$] − 0.041 Cos [$x6$] Sin [$x1$] + 17.357 Sin [$x1$]^2 + 20.807 Sin [$x2$]^2 −
13.538 Sin [$x3$] − 659.616 Cos [$x1$] Sin [$x3$] + 789.591 Cos [$x2$] Sin [$x3$] − 757.117 Cos [$x3$] Sin [$x3$] −
259.577 Cos [$x4$] Sin [$x3$] − 267.538 Cos [$x5$] Sin [$x3$] + 6.788 Cos [$x6$] Sin [$x3$] + 32.192 Sin [$x3$]^2 −
115.206 Sin [$x4$] + 22.075 Sin [$x4$]^2 − 0.565 Sin [$x5$] + 37.196 Sin [$x5$]^2 + 30.638 Sin [$x6$] −
12.260 Cos [$x1$] Sin [$x6$] − 20.360 Cos [$x2$] Sin [$x6$] − 2.837 Cos [$x3$] Sin [$x6$] + 4.217 Cos [$x4$] Sin [$x6$] +
6.886 Cos [$x5$] Sin [$x6$] − 2.056 Cos [$x6$] Sin [$x6$] − 41.429 Sin [$x1$] Sin [$x6$] − 31.069 Sin [$x2$] Sin [$x6$] −
13.010 Sin [$x3$] Sin [$x6$] + 65.357 Sin [$x4$] Sin [$x6$] + 4.796 Sin [$x5$] Sin [$x6$] + 26.986 Sin [$x6$]^2

$In[\bullet]:= (*10*)$

$Out[\bullet]//NumberForm=$

9.870 − 724.554 Cos [$x1$] + 349.814 Cos [$x1$]^2 − 210.898 Cos [$x2$] + 47.491 Cos [$x2$]^2 +
609.304 Cos [$x3$] − 8.592 Cos [$x1$] Cos [$x3$] + 45.142 Cos [$x2$] Cos [$x3$] − 374.653 Cos [$x3$]^2 +
145.445 Cos [$x4$] − 66.687 Cos [$x2$] Cos [$x4$] + 72.445 Cos [$x4$]^2 + 113.383 Cos [$x5$] +
14.087 Cos [$x5$]^2 − 18.369 Cos [$x6$] + 15.591 Cos [$x6$]^2 + 52.566 Sin [$x1$] + 441.185 Cos [$x1$] Sin [$x1$] +
29.943 Cos [$x2$] Sin [$x1$] + 49.930 Cos [$x3$] Sin [$x1$] + 43.255 Cos [$x4$] Sin [$x1$] −
171.618 Cos [$x5$] Sin [$x1$] + 43.528 Cos [$x6$] Sin [$x1$] + 4.216 Sin [$x1$]^2 + 10.301 Sin [$x2$]^2 −
42.018 Sin [$x3$] + 252.558 Cos [$x1$] Sin [$x3$] + 192.269 Cos [$x2$] Sin [$x3$] − 561.318 Cos [$x3$] Sin [$x3$] −
206.732 Cos [$x4$] Sin [$x3$] + 165.373 Cos [$x5$] Sin [$x3$] − 25.220 Cos [$x6$] Sin [$x3$] + 21.187 Sin [$x3$]^2 +
93.084 Sin [$x4$] + 10.438 Sin [$x4$]^2 + 8.998 Sin [$x5$] − 59.698 Sin [$x5$]^2 + 71.674 Sin [$x6$] −
6.018 Cos [$x1$] Sin [$x6$] − 6.035 Cos [$x2$] Sin [$x6$] + 2.719 Cos [$x3$] Sin [$x6$] +

(Continued)

TABLE A7.5   *(Continued)*
Models Obtained by NRM and SNRM Using the Bootstrap Data Separation Method (80 Training–15 Testing–5 Validation*) (*Models Obtained Using the NRM and SNRM Methods*)(*Related to Table 7.6*)

3.343 Cos [$x4$] Sin [$x6$] + 25.425 Cos [$x5$] Sin [$x6$] − 1.743 Cos [$x6$] Sin [$x6$] −
18.163 Sin [$x1$] Sin [$x6$] − 19.005 Sin [$x2$] Sin [$x6$] + 38.636 Sin [$x3$] Sin [$x6$] −
44.519 Sin [$x4$] Sin [$x6$] − 5.002 Sin [$x5$] Sin [$x6$] + 10.204 Sin [$x6$]^2

In[•]:= (*11*)
Out[•]//NumberForm=

10.645 + 82.251 Cos [$x1$] − 1.003 Cos [$x1$]^2 − 762.189 Cos [$x2$] + 17.334 Cos [$x2$]^2 − 228.710 Cos [$x3$] −
235.486 Cos [$x1$] Cos [$x3$] + 259.377 Cos [$x2$] Cos [$x3$] + 345.190 Cos [$x3$]^2 − 130.913 Cos [$x4$] −
5.204 Cos [$x2$] Cos [$x4$] + 136.743 Cos [$x4$]^2 + 359.175 Cos [$x5$] + 15.501 Cos [$x5$]^2 + 23.917 Cos [$x6$] +
14.083 Cos [$x6$]^2 + 17.857 Sin [$x1$] − 33.455 Cos [$x1$] Sin [$x1$] + 13.758 Cos [$x2$] Sin [$x1$] −
578.617 Cos [$x3$] Sin [$x1$] + 0.452 Cos [$x4$] Sin [$x1$] − 144.912 Cos [$x5$] Sin [$x1$] −
10.148 Cos [$x6$] Sin [$x1$] + 13.417 Sin [$x1$]^2 + 11.185 Sin [$x2$]^2 + 17.928 Sin [$x3$] +
16.606 Cos [$x1$] Sin [$x3$] + 708.111 Cos [$x2$] Sin [$x3$] + 741.939 Cos [$x3$] Sin [$x3$] +
145.540 Cos [$x4$] Sin [$x3$] + 86.785 Cos [$x5$] Sin [$x3$] − 15.604 Cos [$x6$] Sin [$x3$] + 4.126 Sin [$x3$]^2 +
273.379 Sin [$x4$] + 11.053 Sin [$x4$]^2 + 1.680 Sin [$x5$] − 147.257 Sin [$x5$]^2 + 142.742 Sin [$x6$] −
35.511 Cos [$x1$] Sin [$x6$] − 0.454 Cos [$x2$] Sin [$x6$] − 13.053 Cos [$x3$] Sin [$x6$] − 0.664 Cos [$x4$] Sin [$x6$] −
10.407 Cos [$x5$] Sin [$x6$] + 0.012 Cos [$x6$] Sin [$x6$] − 95.626 Sin [$x1$] Sin [$x6$] − 0.158 Sin [$x2$] Sin [$x6$] −
52.652 Sin [$x3$] Sin [$x6$] + 2.649 Sin [$x4$] Sin [$x6$] + 2.046 Sin [$x5$] Sin [$x6$] + 13.989 Sin [$x6$]^2

In[•]:= (*12*)
Out[•]//NumberForm=

−5.802 + 255.331 Cos [$x1$] − 867.039 Cos [$x1$]^2 + 24.503 Cos [$x2$] + 8.145 Cos [$x2$]^2 + 452.790 Cos [$x3$] −
116.096 Cos [$x1$] Cos [$x3$] − 73.686 Cos [$x2$] Cos [$x3$] + 687.568 Cos [$x3$]^2 + 111.608 Cos [$x4$] −
18.938 Cos [$x2$] Cos [$x4$] + 121.782 Cos [$x4$]^2 − 92.773 Cos [$x5$] − 5.919 Cos [$x5$]^2 − 27.861 Cos [$x6$] −
7.921 Cos [$x6$]^2 − 46.031 Sin [$x1$] − 924.899 Cos [$x1$] Sin [$x1$] + 124.334 Cos [$x2$] Sin [$x1$] −
1566.000 Cos [$x3$] Sin [$x1$] − 188.919 Cos [$x4$] Sin [$x1$] + 202.874 Cos [$x5$] Sin [$x1$] +
19.924 Cos [$x6$] Sin [$x1$] + 5.265 Sin [$x1$]^2 − 6.097 Sin [$x2$]^2 − 13.377 Sin [$x3$] +
894.821 Cos [$x1$] Sin [$x3$] − 125.690 Cos [$x2$] Sin [$x3$] + 883.645 Cos [$x3$] Sin [$x3$] +
78.335 Cos [$x4$] Sin [$x3$] − 65.413 Cos [$x5$] Sin [$x3$] + 8.518 Cos [$x6$] Sin [$x3$] − 25.415 Sin [$x3$]^2 +
253.472 Sin [$x4$] − 6.713 Sin [$x4$]^2 − 0.878 Sin [$x5$] − 32.459 Sin [$x5$]^2 − 230.065 Sin [$x6$] +
29.864 Cos [$x1$] Sin [$x6$] + 15.074 Cos [$x2$] Sin [$x6$] + 20.285 Cos [$x3$] Sin [$x6$] − 5.414 Cos [$x4$] Sin [$x6$] +
7.203 Cos [$x5$] Sin [$x6$] + 0.795 Cos [$x6$] Sin [$x6$] + 73.888 Sin [$x1$] Sin [$x6$] + 73.192 Sin [$x2$] Sin [$x6$] +
97.962 Sin [$x3$] Sin [$x6$] − 14.759 Sin [$x4$] Sin [$x6$] + 3.216 Sin [$x5$] Sin [$x6$] − 7.553 Sin [$x6$]^2)

In[•]:= (*13*)
Out[•]//NumberForm=

14.414 + 1117.360 Cos [$x1$] − 56.485 Cos [$x1$]^2 − 1108.310 Cos [$x2$] + 19.217 Cos [$x2$]^2 +
240.159 Cos [$x3$] − 82.853 Cos [$x1$] Cos [$x3$] + 341.826 Cos [$x2$] Cos [$x3$] − 322.891 Cos [$x3$]^2 +
0.301 Cos [$x4$] + 3.042 Cos [$x2$] Cos [$x4$] + 54.693 Cos [$x4$]^2 + 271.083 Cos [$x5$] + 19.204 Cos [$x5$]^2 +
4.570 Cos [$x6$] + 20.542 Cos [$x6$]^2 − 81.814 Sin [$x1$] − 420.741 Cos [$x1$] Sin [$x1$] +
18.677 Cos [$x2$] Sin [$x1$] + 297.797 Cos [$x3$] Sin [$x1$] − 48.432 Cos [$x4$] Sin [$x1$] −
231.402 Cos [$x5$] Sin [$x1$] + 3.599 Cos [$x6$] Sin [$x1$] + 18.721 Sin [$x1$]^2 + 14.967 Sin [$x2$]^2 −
10.539 Sin [$x3$] − 652.487 Cos [$x1$] Sin [$x3$] + 1043.130 Cos [$x2$] Sin [$x3$] − 429.022 Cos [$x3$] Sin [$x3$] +

(Continued)

TABLE A7.5 *(Continued)*

Models Obtained by NRM and SNRM Using the Bootstrap Data Separation Method (80 Training–15 Testing–5 Validation*) (*Models Obtained Using the NRM and SNRM Methods*)(*Related to Table 7.6*)

55.144 Cos [x4] Sin [x3] + 23.935 Cos [x5] Sin [x3] − 8.978 Cos [x6] Sin [x3] + 26.066 Sin [x3]² + 105.984 Sin [x4] + 15.213 Sin [x4]² + 4.192 Sin [x5] − 26.901 Sin [x5]² + 47.120 Sin [x6] + 0.499 Cos [x1] Sin [x6] + 5.610 Cos [x2] Sin [x6] − 30.843 Cos [x3] Sin [x6] + 0.595 Cos [x4] Sin [x6] + 4.298 Cos [x5] Sin [x6] − 0.829 Cos [x6] Sin [x6] + 24.710 Sin [x1] Sin [x6] + 9.603 Sin [x2] Sin [x6] − 77.288 Sin [x3] Sin [x6] + 7.522 Sin [x4] Sin [x6] − 1.494 Sin [x5] Sin [x6] + 19.426 Sin [x6]²

In[•]:= (***14***)

Out[•]//NumberForm=

−17.228 − 831.562 Cos [x1] − 110.232 Cos [x1]² − 851.313 Cos [x2] − 12.459 Cos [x2]² − 620.439 Cos [x3] + 213.053 Cos [x1] Cos [x3] + 265.490 Cos [x2] Cos [x3] + 386.273 Cos [x3]² − 8.004 Cos [x4] − 0.138 Cos [x2] Cos [x4] − 4.348 Cos [x4]² − 210.816 Cos [x5] − 24.924 Cos [x5]² − 9.001 Cos [x6] − 23.011 Cos [x6]² + 24.215 Sin [x1] + 86.568 Cos [x1] Sin [x1] + 67.932 Cos [x2] Sin [x1] + 49.181 Cos [x3] Sin [x1] + 30.593 Cos [x4] Sin [x1] + 161.758 Cos [x5] Sin [x1] + 3.410 Cos [x6] Sin [x1] − 18.838 Sin [x1]² − 18.269 Sin [x2]² + 2.217 Sin [x3] + 753.215 Cos [x1] Sin [x3] + 746.338 Cos [x2] Sin [x3] + 416.725 Cos [x3] Sin [x3] − 14.008 Cos [x4] Sin [x3] − 105.011 Cos [x5] Sin [x3] + 6.353 Cos [x6] Sin [x3] − 31.344 Sin [x3]² + 74.472 Sin [x4] − 18.577 Sin [x4]² + 1.349 Sin [x5] + 63.459 Sin [x5]² + 160.027 Sin [x6] − 26.627 Cos [x1] Sin [x6] − 3.350 Cos [x2] Sin [x6] − 17.373 Cos [x3] Sin [x6] − 12.622 Cos [x4] Sin [x6] + 3.004 Cos [x5] Sin [x6] − 0.231 Cos [x6] Sin [x6] − 94.359 Sin [x1] Sin [x6] − 43.964 Sin [x2] Sin [x6] − 3.036 Sin [x3] Sin [x6] − 6.567 Sin [x4] Sin [x6] − 0.284 Sin [x5] Sin [x6] − 23.529 Sin [x6]²

In[•]:= (***15***)

Out[•]//NumberForm=

−53.384 − 991.432 Cos [x1] + 509.566 Cos [x1]² + 340.309 Cos [x2] − 55.039 Cos [x2]² − 2104.950 Cos [x3] + 737.107 Cos [x1] Cos [x3] − 266.780 Cos [x2] Cos [x3] + 83.011 Cos [x3]² + 204.621 Cos [x4] + 33.833 Cos [x2] Cos [x4] + 12.947 Cos [x4]² + 97.861 Cos [x5] − 68.651 Cos [x5]² + 24.758 Cos [x6] − 71.435 Cos [x6]² + 0.924 Sin [x1] + 31.639 Cos [x1] Sin [x1] + 138.350 Cos [x2] Sin [x1] + 2342.610 Cos [x3] Sin [x1] − 501.781 Cos [x4] Sin [x1] − 250.308 Cos [x5] Sin [x1] + 5.690 Cos [x6] Sin [x1] − 71.436 Sin [x1]² − 55.597 Sin [x2]² + 96.694 Sin [x3] + 729.675 Cos [x1] Sin [x3] − 416.687 Cos [x2] Sin [x3] − 306.942 Cos [x3] Sin [x3] + 282.744 Cos [x4] Sin [x3] − 170.356 Cos [x5] Sin [x3] − 33.138 Cos [x6] Sin [x3] − 71.382 Sin [x3]² + 143.277 Sin [x4] − 59.524 Sin [x4]² − 12.095 Sin [x5] + 97.276 Sin [x5]² − 426.897 Sin [x6] + 20.816 Cos [x1] Sin [x6] + 50.880 Cos [x2] Sin [x6] + 157.324 Cos [x3] Sin [x6] − 46.929 Cos [x4] Sin [x6] − 19.687 Cos [x5] Sin [x6] − 3.549 Cos [x6] Sin [x6] + 115.277 Sin [x1] Sin [x6] + 156.820 Sin [x2] Sin [x6] + 491.460 Sin [x3] Sin [x6] − 384.028 Sin [x4] Sin [x6] + 17.522 Sin [x6] Sin [x6] − 70.711 Sin [x6]²))

In[•]:= (***16***)

Out[•]//NumberForm=

−4.144 − 295.510 Cos [x1] − 129.071 Cos [x1]² − 538.788 Cos [x2] − 5.394 Cos [x2]² − 519.582 Cos [x3] − 99.751 Cos [x1] Cos [x3] + 212.299 Cos [x2] Cos [x3] + 457.594 Cos [x3]² − 179.717 Cos [x4] + 22.999 Cos [x2] Cos [x4] − 49.192 Cos [x4]² − 323.941 Cos [x5] −

(Continued)

TABLE A7.5 *(Continued)*

Models Obtained by NRM and SNRM Using the Bootstrap Data Separation Method (80 Training–15 Testing–5 Validation*) (*Models Obtained Using the NRM and SNRM Methods*)(*Related to Table 7.6*)

7.573 Cos $[x5]^2$ − 4.595 Cos $[x6]$ − 5.320 Cos $[x6]^2$ − 13.948 Sin $[x1]$ + 206. 211 Cos $[x1]$ Sin $[x1]$ − 136.866 Cos $[x2]$ Sin $[x1]$ − 121.441 Cos $[x3]$ Sin $[x1]$ + 67.099 Cos $[x4]$ Sin $[x1]$ + 353.126 Cos $[x5]$ Sin $[x1]$ + 6.144 Cos $[x6]$ Sin $[x1]$ − 2.964 Sin $[x1]^2$ − 4.285 Sin $[x2]^2$ + 37.231 Sin $[x3]$ + 107.347 Cos $[x1]$ Sin $[x3]$ + 635.359 Cos $[x2]$ Sin $[x3]$ + 554.143 Cos $[x3]$ Sin $[x3]$ + 128.930 Cos $[x4]$ Sin $[x3]$ − 203.665 Cos $[x5]$ Sin $[x3]$ − 2.159 Cos $[x6]$ Sin $[x3]$ − 14.019 Sin $[x3]^2$ − 62.446 Sin $[x4]$ − 4.286 Sin $[x4]^2$ − 0.071 Sin $[x5]$ + 92.354 Sin $[x5]^2$ − 104.945 Sin $[x6]$ − 7.096 Cos $[x1]$ Sin $[x6]$ + 15.092 Cos $[x2]$ Sin $[x6]$ + 7.882 Cos $[x3]$ Sin $[x6]$ − 11.465 Cos $[x4]$ Sin $[x6]$ + 13.683 Cos $[x5]$ Sin $[x6]$ + 1.533 Cos $[x6]$ Sin $[x6]$ − 28.119 Sin $[x1]$ Sin $[x6]$ + 67.443 Sin $[x2]$ Sin $[x6]$ + 92.815 Sin $[x3]$ Sin $[x6]$ − 10.747 Sin $[x4]$ Sin $[x6]$ − 0.659 Sin $[x5]$ Sin $[x6]$ − 6.125 Sin $[x6]^2$)

In[•]:= (***17***)

Out[•]//NumberForm=

−32.350 − 800.532 Cos $[x1]$ + 247.017 Cos $[x1]^2$ + 988.507 Cos $[x2]$ − 58.999 Cos $[x2]^2$ + 3026.800 Cos $[x3]$ − 240.367 Cos $[x1]$ Cos $[x3]$ − 302.175 Cos $[x2]$ Cos $[x3]$ − 1115.940 Cos $[x3]^2$ + 118.604 Cos $[x4]$ + 26.737 Cos $[x2]$ Cos $[x4]$ + 440.095 Cos $[x4]^2$ + 607.439 Cos $[x5]$ − 32.573 Cos $[x5]^2$ − 45.675 Cos $[x6]$ − 43.653 Cos $[x6]^2$ − 40.827 Sin $[x1]$ + 1011.760 Cos $[x1]$ Sin $[x1]$ − 29.541 Cos $[x2]$ Sin $[x1]$ − 821.893 Cos $[x3]$ Sin $[x1]$ − 80.357 Cos $[x4]$ Sin $[x1]$ − 109.921 Cos $[x5]$ Sin $[x1]$ + 29.861 Cos $[x6]$ Sin $[x1]$ − 42.969 Sin $[x1]^2$ − 33.706 Sin $[x2]^2$ − 262.467 Sin $[x3]$ − 215.607 Cos $[x1]$ Sin $[x3]$ − 917.113 Cos $[x2]$ Sin $[x3]$ − 1831.450 Cos $[x3]$ Sin $[x3]$ − 46.931 Cos $[x4]$ Sin $[x3]$ − 238.021 Cos $[x5]$ Sin $[x3]$ + 17.782 Cos $[x6]$ Sin $[x3]$ − 18.615 Sin $[x3]^2$ + 855.894 Sin $[x4]$ − 38.407 Sin $[x4]^2$ + 1.348 Sin $[x5]$ − 180.856 Sin $[x5]^2$ − 147.214 Sin $[x6]$ − 6.148 Cos $[x1]$ Sin $[x6]$ − 5.356 Cos $[x2]$ Sin $[x6]$ + 67.729 Cos $[x3]$ Sin $[x6]$ − 4.388 Cos $[x4]$ Sin $[x6]$ + 25.501 Cos $[x5]$ Sin $[x6]$ − 2.843 Cos $[x6]$ Sin $[x6]$ − 73.471 Sin $[x1]$ Sin $[x6]$ − 15.888 Sin $[x2]$ Sin $[x6]$ + 218.417 Sin $[x3]$ Sin $[x6]$ + 35.384 Sin $[x4]$ Sin $[x6]$ − 13.924 Sin $[x5]$ Sin $[x6]$ − 42.796 Sin $[x6]^2$)

In[•]:= (***18***)

Out[•]//NumberForm=

14.207 − 1377.070 Cos $[x1]$ − 101.417 Cos $[x1]^2$ + 1743.760 Cos $[x2]$ + 69.100 Cos $[x2]^2$ − 1025.370 Cos $[x3]$ + 978.254 Cos $[x1]$ Cos $[x3]$ − 23.190 Cos $[x2]$ Cos $[x3]$ + 58.469 Cos $[x3]^2$ + 566.353 Cos $[x4]$ + 26.459 Cos $[x2]$ Cos $[x4]$ − 14.182 Cos $[x4]^2$ + 463.014 Cos $[x5]$ + 19.512 Cos $[x5]^2$ + 55.005 Cos $[x6]$ + 20.447 Cos $[x6]^2$ − 73.794 Sin $[x1]$ − 1087.040 Cos $[x1]$ Sin $[x1]$ − 1232.420 Cos $[x2]$ Sin $[x1]$ + 1211.830 Cos $[x3]$ Sin $[x1]$ − 339.031 Cos $[x4]$ Sin $[x1]$ − 434.061 Cos $[x5]$ Sin $[x1]$ − 16.890 Cos $[x6]$ Sin $[x1]$ + 18.108 Sin $[x1]^2$ + 14.673 Sin $[x2]^2$ + 138.444 Sin $[x3]$ + 2408.570 Cos $[x1]$ Sin $[x3]$ − 572.519 Cos $[x2]$ Sin $[x3]$ − 259.770 Cos $[x3]$ Sin $[x3]$ − 232.016 Cos $[x4]$ Sin $[x3]$ + 78.198 Cos $[x5]$ Sin $[x3]$ − 41.354 Cos $[x6]$ Sin $[x3]$ + 16.557 Sin $[x3]^2$ + 23.424 Sin $[x4]$ + 15.606 Sin $[x4]^2$ + 5.130 Sin $[x5]$ − 55.903 Sin $[x5]^2$ − 174.035 Sin $[x6]$ − 69.758 Cos $[x1]$ Sin $[x6]$ + 15.542 Cos $[x2]$ Sin $[x6]$ + 78.938 Cos $[x3]$ Sin $[x6]$ − 14.282 Cos $[x4]$ Sin $[x6]$ + 22.739 Cos $[x5]$ Sin $[x6]$ − 0.228 Cos $[x6]$ Sin $[x6]$ − 292.713 Sin $[x1]$ Sin $[x6]$ + 287.825 Sin $[x2]$ Sin $[x6]$ + 227.485 Sin $[x3]$ Sin $[x6]$ − 29.121 Sin $[x4]$ Sin $[x6]$ − 1.176 Sin $[x5]$ Sin $[x6]$ + 16.286 Sin $[x6]^2$

(Continued)

TABLE A7.5 *(Continued)*

Models Obtained by NRM and SNRM Using the Bootstrap Data Separation Method (80 Training–15 Testing–5 Validation*) (*Models Obtained Using the NRM and SNRM Methods*)(*Related to Table 7.6*)

In[•]:= (*19*)
Out[•]//NumberForm=

$-56.610 - 292.272$ Cos $[x1] + 186.688$ Cos $[x1]^2 - 1347.870$ Cos $[x2] - 46.132$ Cos $[x2]^2 -$
1417.650 Cos $[x3] + 56.936$ Cos $[x1]$ Cos $[x3] + 363.058$ Cos $[x2]$ Cos $[x3] + 124.767$ Cos $[x3]^2 +$
20.008 Cos $[x4] + 77.593$ Cos $[x2]$ Cos $[x4] + 280.772$ Cos $[x4]^2 - 166.232$ Cos $[x5] -$
69.430 Cos $[x5]^2 + 9.757$ Cos $[x6] - 74.096$ Cos $[x6]^2 - 20.660$ Sin $[x1] + 699.404$ Cos $[x1]$ Sin $[x1] +$
191.561 Cos $[x2]$ Sin $[x1] + 1033.380$ Cos $[x3]$ Sin $[x1] + 79.235$ Cos $[x4]$ Sin $[x1] +$
466.879 Cos $[x5]$ Sin $[x1] - 17.492$ Cos $[x6]$ Sin $[x1] - 71.171$ Sin $[x1]^2 - 59.319$ Sin $[x2]^2 -$
10.890 Sin $[x3] - 519.227$ Cos $[x1]$ Sin $[x3] + 1101.820$ Cos $[x2]$ Sin $[x3] + 368.879$ Cos $[x3]$ Sin $[x3] -$
118.333 Cos $[x4]$ Sin $[x3] - 301.344$ Cos $[x5]$ Sin $[x3] + 7.124$ Cos $[x6]$ Sin $[x3] - 75.108$ Sin $[x3]^2 +$
600.334 Sin $[x4] - 65.149$ Sin $[x4]^2 + 1.292$ Sin $[x5] - 58.090$ Sin $[x5]^2 - 92.392$ Sin $[x6] -$
75.555 Cos $[x1]$ Sin $[x6] - 17.847$ Cos $[x2]$ Sin $[x6] + 71.548$ Cos $[x3]$ Sin $[x6] +$
27.636 Cos $[x4]$ Sin $[x6] + 7.807$ Cos $[x5]$ Sin $[x6] + 0.122$ Cos $[x6]$ Sin $[x6] -$
237.101 Sin $[x1]$ Sin $[x6] - 25.142$ Sin $[x2]$ Sin $[x6] + 225.877$ Sin $[x3]$ Sin$[x6] +$
136.525 Sin $[x4]$ Sin $[x6] - 3.178$ Sin $[x5]$ Sin $[x6] - 78.970$ Sin $[x6]^2$

In[•]:= (*20*)
Out[•]//NumberForm=

$6.165 + 70.678$ Cos $[x1] - 112.136$ Cos $[x1]^2 - 737.188$ Cos $[x2] + 18.881$ Cos $[x2]^2 -$
128.061 Cos $[x3] - 24.106$ Cos $[x1]$ Cos $[x3] + 312.236$ Cos $[x2]$ Cos $[x3] + 54.911$ Cos $[x3]^2 +$
20.680 Cos $[x4] - 21.014$ Cos $[x2]$ Cos $[x4] + 107.625$ Cos $[x4]^2 - 276.676$ Cos $[x5] +$
7.926 Cos $[x5]^2 - 35.669$ Cos $[x6] + 8.227$ Cos $[x6]^2 - 51.868$ Sin $[x1] - 172.956$ Cos $[x1]$ Sin $[x1] -$
35.700 Cos $[x2]$ Sin $[x1] - 144.383$ Cos $[x3]$ Sin $[x1] + 12.919$ Cos $[x4]$ Sin $[x1] +$
2.307 Cos $[x5]$ Sin $[x1] - 6.577$ Cos $[x6]$ Sin $[x1] + 9.738$ Sin $[x1]^2 + 6.476$ Sin $[x2]^2 +$
7.672 Sin $[x3] + 142.484$ Cos $[x1]$ Sin $[x3] + 716.576$ Cos $[x2]$ Sin $[x3] + 153.449$ Cos $[x3]$ Sin $[x3] -$
36.058 Cos $[x4]$ Sin $[x3] + 355.706$ Cos $[x5]$ Sin $[x3] + 45.183$ Cos $[x6]$ Sin $[x3] + 6.614$ Sin $[x3]^2 +$
177.553 Sin $[x4] + 6.147$ Sin $[x4]^2 - 0.269$ Sin $[x5] - 22.091$ Sin $[x5]^2 - 39.501$ Sin $[x6] +$
10.163 Cos $[x1]$ Sin $[x6] - 5.858$ Cos $[x2]$ Sin $[x6] - 0.118$ Cos $[x3]$ Sin $[x6] + 0.457$ Cos $[x4]$ Sin $[x6] -$
14.316 Cos $[x5]$ Sin $[x6] - 1.768$ Cos $[x6]$ Sin $[x6] + 64.378$ Sin $[x1]$ Sin $[x6] + 0.683$ Sin $[x2]$ Sin $[x6] -$
55.954 Sin $[x3]$ Sin $[x6] + 14.042$ Sin $[x4]$ Sin $[x6] + 2.001$ Sin $[x5]$ Sin $[x6] + 8.621$ Sin $[x6]^2)$

(*Models proposed by SNRM*)
In[•]:= (*1*)
Out[•]//NumberForm=

$43.199 - 70.000$ Cos $[x1] - 90.196$ Cos $[x1]^2 - 55.131$ Cos $[x2] + 27.875$ Cos $[x2]^2 - 52.742$ Cos $[x3] +$
62.051 Cos $[x1]$ Cos $[x3] + 19.220$ Cos $[x2]$ Cos $[x3] + 3.577$ Cos $[x3]^2 + 9.389$ Cos $[x4] +$
0.824 Cos $[x2]$ Cos $[x4] + 45.357$ Cos $[x4]^2 - 78.860$ Cos $[x5] + 24.915$ Cos $[x5]^2 + 4.947$ Cos $[x6] +$
24.411 Cos $[x6]^2 - 400.666$ Sin $[x1] + 81.609$ Cos $[x1]$ Sin $[x1] + 6.877$ Cos $[x2]$ Sin $[x1] +$
218.528 Cos $[x3]$ Sin $[x1] - 7.102$ Cos $[x4]$ Sin $[x1] + 45.679$ Cos $[x5]$ Sin $[x1] +$
2.981 Cos $[x6]$ Sin $[x1] + 118.031$ Sin $[x1]^2 + 25.145$ Sin $[x2]^2 + 140.449$ Sin $[x3] -$
11.628 Cos $[x1]$ Sin $[x3] + 48.842$ Cos $[x2]$ Sin $[x3] - 149.721$ Cos $[x3]$ Sin $[x3] +$
0.598 Cos $[x4]$ Sin $[x3] + 45.554$ Cos $[x5]$ Sin $[x3] - 8.303$ Cos $[x6]$ Sin $[x3] + 40.369$ Sin $[x3]^2 +$

(Continued)

TABLE A7.5 *(Continued)*

Models Obtained by NRM and SNRM Using the Bootstrap Data Separation Method (80 Training–15 Testing–5 Validation*) (*Models Obtained Using the NRM and SNRM Methods*)(*Related to Table 7.6*)

76.815 Sin [$x4$] + 7.699 Sin [$x4$]^2 + 1.153 Sin [$x5$] + 21.395 Sin [$x5$]^2 + 37.892 Sin [$x6$] −
4.113 Cos [$x1$] Sin [$x6$] + 0.448 Cos [$x2$] Sin [$x6$] − 7.931 Cos [$x3$] Sin [$x6$] − 0.328 Cos [$x4$] Sin [$x6$] −
1.153 Cos [$x5$] Sin [$x6$] − 0.345 Cos [$x6$] Sin [$x6$] − 17.361 Sin [$x1$] Sin [$x6$] − 5.464 Sin [$x2$] Sin [$x6$] −
22.164 Sin [$x3$] Sin [$x6$] + 7.821 Sin [$x4$] Sin [$x6$] + 0.339 Sin [$x5$] Sin [$x6$] + 23.908 Sin [$x6$]^2

In[•]:= **(*2*)**

Out[•]//NumberForm=

43.199 − 70.000 Cos [$x1$] − 90.196 Cos [$x1$]^2 − 55.131 Cos [$x2$] + 27.875 Cos [$x2$]^2 − 52.742 Cos [$x3$] +
62.051 Cos [$x1$] Cos [$x3$] + 19.220 Cos [$x2$] Cos [$x3$] + 3.577 Cos [$x3$]^2 + 9.389 Cos [$x4$] +
0.824 Cos [$x2$] Cos [$x4$] + 45.357 Cos [$x4$]^2 − 78.860 Cos [$x5$] + 24.915 Cos [$x5$]^2 + 4.947 Cos [$x6$] +
24.411 Cos [$x6$]^2 − 400.666 Sin [$x1$] + 81.609 Cos [$x1$] Sin [$x1$] + 6.877 Cos [$x2$] Sin [$x1$] +
218.528 Cos [$x3$] Sin [$x1$] − 7.102 Cos [$x4$] Sin [$x1$] + 45.679 Cos [$x5$] Sin [$x1$] +
2.981 Cos [$x6$] Sin [$x1$] + 118.031 Sin [$x1$]^2 + 25.145 Sin [$x2$]^2 + 140.449 Sin [$x3$] −
11.628 Cos [$x1$] Sin [$x3$] + 48.842 Cos [$x2$] Sin [$x3$] − 149.721 Cos [$x3$] Sin [$x3$] +
0.598 Cos [$x4$] Sin [$x3$] + 45.554 Cos [$x5$] Sin [$x3$] − 8.303 Cos [$x6$] Sin [$x3$] + 40.369 Sin [$x3$]^2 +
76.815 Sin [$x4$] + 7.699 Sin [$x4$]^2 + 1.153 Sin [$x5$] + 21.395 Sin [$x5$]^2 + 37.892 Sin [$x6$] −
4.113 Cos [$x1$] Sin [$x6$] + 0.448 Cos [$x2$] Sin [$x6$] − 7.931 Cos [$x3$] Sin [$x6$] − 0.328 Cos [$x4$] Sin [$x6$] −
1.153 Cos [$x5$] Sin [$x6$] − 0.345 Cos [$x6$] Sin [$x6$] − 17.361 Sin [$x1$] Sin [$x6$] − 5.464 Sin [$x2$] Sin [$x6$] −
22.164 Sin [$x3$] Sin [$x6$] + 7.821 Sin [$x4$] Sin [$x6$] + 0.339 Sin [$x5$] Sin [$x6$] + 23.908 Sin [$x6$]^2

In[•]:= **(*3*)**

Out[•]//NumberForm=

47.442 − 71.611 Cos [$x1$] − 84.857 Cos [$x1$]^2 − 46.061 Cos [$x2$] + 25.895 Cos [$x2$]^2 − 56.170 Cos [$x3$] +
63.236 Cos [$x1$] Cos [$x3$] + 16.522 Cos [$x2$] Cos [$x3$] + 3.186 Cos [$x3$]^2 + 9.800 Cos [$x4$] +
0.923 Cos [$x2$] Cos [$x4$] + 44.084 Cos [$x4$]^2 − 69.122 Cos [$x5$] + 25.374 Cos [$x5$]^2 + 5.053 Cos [$x6$] +
21.298 Cos [$x6$]^2 − 395.248 Sin [$x1$] + 80.015 Cos [$x1$] Sin [$x1$] + 8.629 Cos [$x2$] Sin [$x1$] +
221.574 Cos [$x3$] Sin [$x1$] − 7.362 Cos [$x4$] Sin [$x1$] + 42.674 Cos [$x5$] Sin [$x1$] + 3.363 Cos [$x6$] Sin [$x1$] +
118.803 Sin [$x1$]^2 + 22.651 Sin [$x2$]^2 + 138.274 Sin [$x3$] − 8.598 Cos [$x1$] Sin [$x3$] +
38.175 Cos [$x2$] Sin [$x3$] − 149.554 Cos [$x3$] Sin [$x3$] + 0.211 Cos [$x4$] Sin [$x3$] +
42.158 Cos [$x5$] Sin [$x3$] − 8.804 Cos [$x6$] Sin [$x3$] + 38.937 Sin [$x3$]^2 + 84.241 Sin [$x4$] +
1.930 Sin [$x4$]^2 + 1.196 Sin [$x5$] + 19.789 Sin [$x5$]^2 + 37.038 Sin [$x6$] − 4.061 Cos [$x1$] Sin [$x6$] +
0.206 Cos [$x2$] Sin [$x6$] − 7.655 Cos [$x3$] Sin [$x6$] − 0.082 Cos [$x4$] Sin [$x6$] − 0.998 Cos [$x5$] Sin [$x6$] −
0.358 Cos [$x6$] Sin [$x6$] − 17.338 Sin [$x1$] Sin [$x6$] − 7.121 Sin [$x2$] Sin [$x6$] − 22.138 Sin [$x3$] Sin [$x6$] +
10.386 Sin [$x4$] Sin [$x6$] + 0.297 Sin [$x5$] Sin [$x6$] + 20.790 Sin [$x6$]^2

In[•]:= **(*4*)**

Out[•]//NumberForm=

65.332 − 116.273 Cos [$x1$] − 133.476 Cos [$x1$]^2 − 53.652 Cos [$x2$] + 40.281 Cos [$x2$]^2 +
4.324 Cos [$x3$] + 54.587 Cos [$x1$] Cos [$x3$] + 20.824 Cos [$x2$] Cos [$x3$] + 28.341 Cos [$x3$]^2 +
6.071 Cos [$x4$] + 1.753 Cos [$x2$] Cos [$x4$] + 64.982 Cos [$x4$]^2 − 53.289 Cos [$x5$] + 42.551 Cos [$x5$]^2 +
5.364 Cos [$x6$] + 35.937 Cos [$x6$]^2 − 673.802 Sin [$x1$] + 147.727 Cos [$x1$] Sin [$x1$] +
3.592 Cos [$x2$] Sin [$x1$] + 215.771 Cos [$x3$] Sin [$x1$] − 4.346 Cos [$x4$] Sin [$x1$] +
29.979 Cos [$x5$] Sin [$x1$] + 3.162 Cos [$x6$] Sin [$x1$] + 193.294 Sin [$x1$]^2 + 37.786 Sin [$x2$]^2 +

(Continued)

TABLE A7.5 *(Continued)*

Models Obtained by NRM and SNRM Using the Bootstrap Data Separation Method (80 Training–15 Testing–5 Validation*) (*Models Obtained Using the NRM and SNRM Methods*)(*Related to Table 7.6*)

256.999 Sin [$x3$] − 31.423 Cos [$x1$] Sin [$x3$] + 50.015 Cos [$x2$] Sin [$x3$] − 198.774 Cos [$x3$] Sin [$x3$] + 1.235 Cos [$x4$] Sin [$x3$] + 38.378 Cos [$x5$] Sin [$x3$] − 8.926 Cos [$x6$] Sin [$x3$] + 20.719 Sin [$x3$]^2 + 119.024 Sin [$x4$] + 4.802 Sin [$x4$]^2 + 1.249 Sin [$x5$] + 36.680 Sin [$x5$]^2 + 34.793 Sin [$x6$] − 2.627 Cos [$x1$] Sin [$x6$] + 0.516 Cos [$x2$] Sin [$x6$] − 8.691 Cos [$x3$] Sin [$x6$] − 0.360 Cos [$x4$] Sin [$x6$] − 1.081 Cos [$x5$] Sin [$x6$] − 0.349 Cos [$x6$] Sin [$x6$] − 11.796 Sin [$x1$] Sin [$x6$] − 5.624 Sin [$x2$] Sin [$x6$] − 24.094 Sin [$x3$] Sin [$x6$] + 7.544 Sin [$x4$] Sin [$x6$] + 0.141 Sin [$x5$] Sin [$x6$] + 35.326 Sin [$x6$]^2

In[•]:= (*5*)

Out[•]//NumberForm=

54.800 − 80.256 Cos [$x1$] − 110.107 Cos [$x1$]^2 − 41.365 Cos [$x2$] + 37.484 Cos [$x2$]^2 − 39.642 Cos [$x3$] + 57.563 Cos [$x1$] Cos [$x3$] + 16.770 Cos [$x2$] Cos [$x3$] + 34.100 Cos [$x3$]^2 + 5.389 Cos [$x4$] − 0.163 Cos [$x2$] Cos [$x4$] + 41.868 Cos [$x4$]^2 − 70.927 Cos [$x5$] + 35.620 Cos [$x5$]^2 + 4.602 Cos [$x6$] + 29.196 Cos [$x6$]^2 − 519.714 Sin [$x1$] + 106.992 Cos [$x1$] Sin [$x1$] + 6.740 Cos [$x2$] Sin [$x1$] + 223.600 Cos [$x3$] Sin [$x1$] − 10.948 Cos [$x4$] Sin [$x1$] + 45.921 Cos [$x5$] Sin [$x1$] + 2.822 Cos [$x6$] Sin [$x1$] + 154.021 Sin [$x1$]^2 + 34.853 Sin [$x2$]^2 + 238.357 Sin [$x3$] − 26.445 Cos [$x1$] Sin [$x3$] + 35.327 Cos [$x2$] Sin [$x3$] − 167.127 Cos [$x3$] Sin [$x3$] + 9.184 Cos [$x4$] Sin [$x3$] + 44.467 Cos [$x5$] Sin [$x3$] − 7.813 Cos [$x6$] Sin [$x3$] + 23.042 Sin [$x3$]^2 + 23.043 Sin [$x4$] + 34.165 Sin [$x4$]^2 + 1.335 Sin [$x5$] + 27.541 Sin [$x5$]^2 + 26.341 Sin [$x6$] − 3.609 Cos [$x1$] Sin [$x6$] + 0.214 Cos [$x2$] Sin [$x6$] − 5.507 Cos [$x3$] Sin [$x6$] + 0.171 Cos [$x4$] Sin [$x6$] − 1.883 Cos [$x5$] Sin [$x6$] − 0.282 Cos [$x6$] Sin [$x6$] − 14.695 Sin [$x1$] Sin [$x6$] − 7.150 Sin [$x2$] Sin [$x6$] − 14.508 Sin [$x3$] Sin [$x6$] + 9.628 Sin [$x4$] Sin [$x6$] + 0.540 Sin [$x5$] Sin [$x6$] + 28.682 Sin [$x6$]^2

In[•]:= (*6*)

Out[•]//NumberForm=

43.199 − 70.000 Cos [$x1$] − 90.196 Cos [$x1$]^2 − 55.131 Cos [$x2$] + 27.875 Cos [$x2$]^2 − 52.742 Cos [$x3$] + 62.051 Cos [$x1$] Cos [$x3$] + 19.220 Cos [$x2$] Cos [$x3$] + 3.577 Cos [$x3$]^2 + 9.389 Cos [$x4$] + 0.824 Cos [$x2$] Cos [$x4$] + 45.357 Cos [$x4$]^2 − 78.860 Cos [$x5$] + 24.915 Cos [$x5$]^2 + 4.947 Cos [$x6$] + 24.411 Cos [$x6$]^2 − 400.666 Sin [$x1$] + 81.609 Cos [$x1$] Sin [$x1$] + 6.877 Cos [$x2$] Sin [$x1$] + 218.528 Cos [$x3$] Sin [$x1$] − 7.102 Cos [$x4$] Sin [$x1$] + 45.679 Cos [$x5$] Sin [$x1$] + 2.981 Cos [$x6$] Sin [$x1$] + 118.031 Sin [$x1$]^2 + 25.145 Sin [$x2$]^2 + 140.449 Sin [$x3$] − 11.628 Cos [$x1$] Sin [$x3$] + 48.842 Cos [$x2$] Sin [$x3$] − 149.721 Cos [$x3$] Sin [$x3$] + 0.598 Cos [$x4$] Sin [$x3$] + 45.554 Cos [$x5$] Sin [$x3$] − 8.303 Cos [$x6$] Sin [$x3$] + 40.369 Sin [$x3$]^2 + 76.815 Sin [$x4$] + 7.699 Sin [$x4$]^2 + 1.153 Sin [$x5$] + 21.395 Sin [$x5$]^2 + 37.892 Sin [$x6$] − 4.113 Cos [$x1$] Sin [$x6$] + 0.448 Cos [$x2$] Sin [$x6$] − 7.931 Cos [$x3$] Sin [$x6$] − 0.328 Cos [$x4$] Sin [$x6$] − 1.153 Cos [$x5$] Sin [$x6$] − 0.345 Cos [$x6$] Sin [$x6$] − 17.361 Sin [$x1$] Sin [$x6$] − 5.464 Sin [$x2$] Sin [$x6$] − 22.164 Sin [$x3$] Sin [$x6$] + 7.821 Sin [$x4$] Sin [$x6$] + 0.339 Sin [$x5$] Sin [$x6$] + 23.908 Sin [$x6$]^2

In[•]:= (*7*)

Out[•]//NumberForm=

43.199 − 70.000 Cos [$x1$] − 90.196 Cos [$x1$]^2 − 55.131 Cos [$x2$] + 27.875 Cos [$x2$]^2 − 52.742 Cos [$x3$] + 62.051 Cos [$x1$] Cos [$x3$] + 19.220 Cos [$x2$] Cos [$x3$] + 3.577 Cos [$x3$]^2 + 9.389 Cos [$x4$] + 0.824 Cos [$x2$] Cos [$x4$] + 45.357 Cos [$x4$]^2 − 78.860 Cos [$x5$] + 24.915 Cos [$x5$]^2 + 4.947 Cos [$x6$] + 24.411 Cos [$x6$]^2 − 400.666 Sin [$x1$] + 81.609 Cos [$x1$] Sin [$x1$] +

(Continued)

TABLE A7.5 *(Continued)*
Models Obtained by NRM and SNRM Using the Bootstrap Data Separation Method (80 Training–15 Testing–5 Validation*) (*Models Obtained Using the NRM and SNRM Methods*)(*Related to Table 7.6*)

6.877 Cos [$x2$] Sin [$x1$] + 218.528 Cos [$x3$] Sin [$x1$] − 7.102 Cos [$x4$] Sin [$x1$] +
45.679 Cos [$x5$] Sin [$x1$] + 2.981 Cos [$x6$] Sin [$x1$] + 118.031 Sin [$x1$]^2 + 25.145 Sin [$x2$]^2 +
140.449 Sin [$x3$] − 11.628 Cos [$x1$] Sin [$x3$] + 48.842 Cos [$x2$] Sin [$x3$] − 149.721 Cos [$x3$] Sin [$x3$] +
0.598 Cos [$x4$] Sin [$x3$] + 45.554 Cos [$x5$] Sin [$x3$] − 8.303 Cos [$x6$] Sin [$x3$] + 40.369 Sin [$x3$]^2 +
76.815 Sin [$x4$] + 7.699 Sin [$x4$]^2 + 1.153 Sin [$x5$] + 21.395 Sin [$x5$]^2 + 37.892 Sin [$x6$] −
4.113 Cos [$x1$] Sin [$x6$] + 0.448 Cos [$x2$] Sin [$x6$] − 7.931 Cos [$x3$] Sin [$x6$] − 0.328 Cos [$x4$] Sin [$x6$] −
1.153 Cos [$x5$] Sin [$x6$] − 0.345 Cos [$x6$] Sin [$x6$] − 17.361 Sin [$x1$] Sin [$x6$] − 5.464 Sin [$x2$] Sin [$x6$] −
22.164 Sin [$x3$] Sin [$x6$] + 7.821 Sin [$x4$] Sin [$x6$] + 0.339 Sin [$x5$] Sin [$x6$] + 23.908 Sin [$x6$]^2

In[•]:= **(*8*)**
Out[•]//NumberForm=

47.442 − 71.611 Cos [$x1$] − 84.857 Cos [$x1$]^2 − 46.061 Cos [$x2$] + 25.895 Cos [$x2$]^2 − 56.170 Cos [$x3$] +
63.236 Cos [$x1$] Cos [$x3$] + 16.522 Cos [$x2$] Cos [$x3$] + 3.186 Cos [$x3$]^2 + 9.800 Cos [$x4$] +
0.923 Cos [$x2$] Cos [$x4$] + 44.084 Cos [$x4$]^2 − 69.122 Cos [$x5$] + 25.374 Cos [$x5$]^2 + 5.053 Cos [$x6$] +
21.298 Cos [$x6$]^2 − 395.248 Sin [$x1$] + 80.015 Cos [$x1$] Sin [$x1$] + 8.629 Cos [$x2$] Sin [$x1$] +
221.574 Cos [$x3$] Sin [$x1$] − 7.362 Cos [$x4$] Sin [$x1$] + 42.674 Cos [$x5$] Sin [$x1$] +
3.363 Cos [$x6$] Sin [$x1$] + 118.803 Sin [$x1$]^2 + 22.651 Sin [$x2$]^2 + 138.274 Sin [$x3$] −
8.598 Cos [$x1$] Sin [$x3$] + 38.175 Cos [$x2$] Sin [$x3$] − 149.554 Cos [$x3$] Sin [$x3$] +
0.211 Cos [$x4$] Sin [$x3$] + 42.158 Cos [$x5$] Sin [$x3$] − 8.804 Cos [$x6$] Sin [$x3$] + 38.937 Sin [$x3$]^2 +
84.241 Sin [$x4$] + 1.930 Sin [$x4$]^2 + 1.196 Sin [$x5$] + 19.789 Sin [$x5$]^2 + 37.038 Sin [$x6$] −
4.061 Cos [$x1$] Sin [$x6$] + 0.206 Cos [$x2$] Sin [$x6$] − 7.655 Cos [$x3$] Sin [$x6$] − 0.082 Cos [$x4$] Sin [$x6$] −
0.998 Cos [$x5$] Sin [$x6$] − 0.358 Cos [$x6$] Sin [$x6$] − 17.338 Sin [$x1$] Sin [$x6$] − 7.121 Sin [$x2$] Sin [$x6$] −
22.138 Sin [$x3$] Sin [$x6$] + 10.386 Sin [$x4$] Sin [$x6$] + 0.297 Sin [$x5$] Sin [$x6$] + 20.790 Sin [$x6$]^2

In[•]:= **(*9*)**
Out[•]//NumberForm=

52.535 − 107.453 Cos [$x1$] − 123.094 Cos [$x1$]^2 − 32.799 Cos [$x2$] + 39.313 Cos [$x2$]^2 +
42.783 Cos [$x3$] + 51.477 Cos [$x1$] Cos [$x3$] + 13.619 Cos [$x2$] Cos [$x3$] + 16.949 Cos [$x3$]^2 +
11.247 Cos [$x4$] − 1.214 Cos [$x2$] Cos [$x4$] + 49.842 Cos [$x4$]^2 − 87.906 Cos [$x5$] + 23.970 Cos [$x5$]^2 +
4.521 Cos [$x6$] + 23.038 Cos [$x6$]^2 − 547.260 Sin [$x1$] + 123.731 Cos [$x1$] Sin [$x1$] +
5.492 Cos [$x2$] Sin [$x1$] + 179.068 Cos [$x3$] Sin [$x1$] − 5.180 Cos [$x4$] Sin [$x1$] + 53.702 Cos [$x5$] Sin [$x1$] +
3.904 Cos [$x6$] Sin [$x1$] + 162.769 Sin [$x1$]^2 + 36.226 Sin [$x2$]^2 + 229.583 Sin [$x3$] −
13.099 Cos [$x1$] Sin [$x3$] + 28.576 Cos [$x2$] Sin [$x3$] − 200.428 Cos [$x3$] Sin [$x3$] −
3.736 Cos [$x4$] Sin [$x3$] + 38.213 Cos [$x5$] Sin [$x3$] − 8.796 Cos [$x6$] Sin [$x3$] + 23.898 Sin [$x3$]^2 +
74.043 Sin [$x4$] + 12.494 Sin [$x4$]^2 + 1.126 Sin [$x5$] + 25.415 Sin [$x5$]^2 + 46.802 Sin [$x6$] −
4.493 Cos [$x1$] Sin [$x6$] + 0.544 Cos [$x2$] Sin [$x6$] − 11.292 Cos [$x3$] Sin [$x6$] + 0.172 Cos [$x4$] Sin [$x6$] −
0.622 Cos [$x5$] Sin [$x6$] − 0.215 Cos [$x6$] Sin [$x6$] − 17.189 Sin [$x1$] Sin [$x6$] − 6.220 Sin [$x2$] Sin [$x6$] −
31.289 Sin [$x3$] Sin [$x6$] + 9.792 Sin [$x4$] Sin [$x6$] + 0.055 Sin [$x5$] Sin [$x6$] + 22.631 Sin [$x6$]^2

In[•]:= **(*10*)**
Out[•]//NumberForm=

51.903 − 71.330 Cos [$x1$] − 70.910 Cos [$x1$]^2 − 82.830 Cos [$x2$] + 28.824 Cos [$x2$]^2 − 117.942 Cos [$x3$] +
67.388 Cos [$x1$] Cos [$x3$] + 27.687 Cos [$x2$] Cos [$x3$] + 8.087 Cos [$x3$]^2 + 15.651 Cos [$x4$] −

(Continued)

TABLE A7.5 *(Continued)*

Models Obtained by NRM and SNRM Using the Bootstrap Data Separation Method (80 Training–15 Testing–5 Validation*) (*Models Obtained Using the NRM and SNRM Methods*)(*Related to Table 7.6*)

0.039 Cos [$x2$] Cos [$x4$] + 38.323 Cos [$x4$]^2 − 87.572 Cos [$x5$] + 22.188 Cos [$x5$]^2 + 4.701 Cos [$x6$] + 30.081 Cos [$x6$]^2 − 367.270 Sin [$x1$] + 73.478 Cos [$x1$] Sin [$x1$] + 9.706 Cos [$x2$] Sin [$x1$] + 235.919 Cos [$x3$] Sin [$x1$] − 10.223 Cos [$x4$] Sin [$x1$] + 47.939 Cos [$x5$] Sin [$x1$] + 3.177 Cos [$x6$] Sin [$x1$] + 119.108 Sin [$x1$]^2 + 24.041 Sin [$x2$]^2 + 103.659 Sin [$x3$] − 3.913 Cos [$x1$] Sin [$x3$] + 72.311 Cos [$x2$] Sin [$x3$] − 109.036 Cos [$x3$] Sin [$x3$] − 2.674 Cos [$x4$] Sin [$x3$] + 43.539 Cos [$x5$] Sin [$x3$] − 8.211 Cos [$x6$] Sin [$x3$] + 41.472 Sin [$x3$]^2 + 55.009 Sin [$x4$] + 12.492 Sin [$x4$]^2 + 1.047 Sin [$x5$] + 23.508 Sin [$x5$]^2 + 26.419 Sin [$x6$] − 4.095 Cos [$x1$] Sin [$x6$] + 0.380 Cos [$x2$] Sin [$x6$] − 5.427 Cos [$x3$] Sin [$x6$] + 0.046 Cos [$x4$] Sin [$x6$] − 1.527 Cos [$x5$] Sin [$x6$] − 0.424 Cos [$x6$] Sin [$x6$] − 17.187 Sin [$x1$] Sin [$x6$] − 4.579 Sin [$x2$] Sin [$x6$] − 16.310 Sin [$x3$] Sin [$x6$] + 11.495 Sin [$x4$] Sin [$x6$] + 0.713 Sin [$x5$] Sin [$x6$] + 29.594 Sin [$x6$]^2

In[•]:= (*11*)

Out[•]//NumberForm=

85.236 − 168.566 Cos [$x1$] − 113.275 Cos [$x1$]^2 − 81.518 Cos [$x2$] + 46.099 Cos [$x2$]^2 − 69.875 Cos [$x3$] + 72.017 Cos [$x1$] Cos [$x3$] + 24.867 Cos [$x2$] Cos [$x3$] + 26.924 Cos [$x3$]^2 + 18.284 Cos [$x4$] + 3.165 Cos [$x2$] Cos [$x4$] + 62.974 Cos [$x4$]^2 − 96.863 Cos [$x5$] + 40.537 Cos [$x5$]^2 + 4.722 Cos [$x6$] + 43.321 Cos [$x6$]^2 − 626.220 Sin [$x1$] + 167.204 Cos [$x1$] Sin [$x1$] + 9.461 Cos [$x2$] Sin [$x1$] + 249.552 Cos [$x3$] Sin [$x1$] − 10.390 Cos [$x4$] Sin [$x1$] + 63.209 Cos [$x5$] Sin [$x1$] + 2.415 Cos [$x6$] Sin [$x1$] + 191.230 Sin [$x1$]^2 + 43.066 Sin [$x2$]^2 + 168.081 Sin [$x3$] − 4.057 Cos [$x1$] Sin [$x3$] + 71.825 Cos [$x2$] Sin [$x3$] − 164.079 Cos [$x3$] Sin [$x3$] − 5.839 Cos [$x4$] Sin [$x3$] + 44.426 Cos [$x5$] Sin [$x3$] − 7.515 Cos [$x6$] Sin [$x3$] + 52.714 Sin [$x3$]^2 + 75.016 Sin [$x4$] + 26.086 Sin [$x4$]^2 + 1.112 Sin [$x5$] + 38.402 Sin [$x5$]^2 + 39.148 Sin [$x6$] − 3.400 Cos [$x1$] Sin [$x6$] + 0.447 Cos [$x2$] Sin [$x6$] − 8.686 Cos [$x3$] Sin [$x6$] − 0.923 Cos [$x4$] Sin [$x6$] − 0.589 Cos [$x5$] Sin [$x6$] − 0.319 Cos [$x6$] Sin [$x6$] − 15.259 Sin [$x1$] Sin [$x6$] − 4.176 Sin [$x2$] Sin [$x6$] − 25.220 Sin [$x3$] Sin [$x6$] + 6.843 Sin [$x4$] Sin [$x6$] + 0.272 Sin [$x5$] Sin [$x6$] + 42.753 Sin [$x6$]^2

In[•]:= (*12*)

Out[•]//NumberForm=

80.651 − 120.683 Cos [$x1$] − 113.071 Cos [$x1$]^2 − 32.430 Cos [$x2$] + 47.253 Cos [$x2$]^2 − 38.624 Cos [$x3$] + 60.092 Cos [$x1$] Cos [$x3$] + 12.980 Cos [$x2$] Cos [$x3$] + 38.241 Cos [$x3$]^2 + 14.423 Cos [$x4$] − 0.520 Cos [$x2$] Cos [$x4$] + 51.610 Cos [$x4$]^2 − 57.994 Cos [$x5$] + 48.216 Cos [$x5$]^2 + 6.222 Cos [$x6$] + 45.451 Cos [$x6$]^2 − 608.771 Sin [$x1$] + 141.022 Cos [$x1$] Sin [$x1$] + 0.991 Cos [$x2$] Sin [$x1$] + 234.064 Cos [$x3$] Sin [$x1$] − 6.857 Cos [$x4$] Sin [$x1$] + 36.545 Cos [$x5$] Sin [$x1$] + 2.681 Cos [$x6$] Sin [$x1$] + 178.458 Sin [$x1$]^2 + 44.725 Sin [$x2$]^2 + 205.525 Sin [$x3$] − 22.994 Cos [$x1$] Sin [$x3$] + 33.249 Cos [$x2$] Sin [$x3$] − 176.323 Cos [$x3$] Sin [$x3$] − 4.752 Cos [$x4$] Sin [$x3$] + 38.176 Cos [$x5$] Sin [$x3$] −)9.324 Cos [$x6$] Sin [$x3$] + 48.584 Sin [$x3$]^2 + 13.077 Sin [$x4$] + 48.765 Sin [$x4$]^2 + 1.146 Sin [$x5$] + 41.672 Sin [$x5$]^2 + 26.019 Sin [$x6$] − 3.012 Cos [$x1$] Sin [$x6$] + 0.443 Cos [$x2$] Sin [$x6$] − 6.163 Cos [$x3$] Sin [$x6$] − 0.124 Cos [$x4$] Sin [$x6$] − 0.259 Cos [$x5$] Sin [$x6$] − 0.343 Cos [$x6$] Sin [$x6$] − 12.675 Sin [$x1$] Sin [$x6$] − 3.923 Sin [$x2$] Sin [$x6$] − 18.748 Sin [$x3$] Sin [$x6$] + 10.483 Sin [$x4$] Sin [$x6$] + 0.249 Sin [$x5$] Sin [$x6$] + 44.845 Sin [$x6$]^2

(Continued)

TABLE A7.5 *(Continued)*

Models Obtained by NRM and SNRM Using the Bootstrap Data Separation Method (80 Training–15 Testing–5 Validation*) (*Models Obtained Using the NRM and SNRM Methods*)(*Related to Table 7.6*)

In[•]:= (*13*)

Out[•]///NumberForm=

63.561 − 135.463 Cos [x1] − 125.039 Cos [x1]² − 47.397 Cos [x2] + 35.967 Cos [x2]² −
21.024 Cos [x3] + 57.829 Cos [x1] Cos [x3] + 15.191 Cos [x2] Cos [x3] + 45.872 Cos [x3]² +
9.323 Cos [x4] − 0.373 Cos [x2] Cos [x4] + 48.968 Cos [x4]² − 84.883 Cos [x5] + 38.245 Cos [x5]² +
3.910 Cos [x6] + 37.305 Cos [x6]² − 594.317 Sin [x1] + 143.876 Cos [x1] Sin [x1] +
13.525 Cos [x2] Sin [x1] + 205.008 Cos [x3] Sin [x1] − 13.407 Cos [x4] Sin [x1] +
49.653 Cos [x5] Sin [x1] + 3.534 Cos [x6] Sin [x1] + 171.301 Sin [x1]² + 32.603 Sin [x2]² +
274.738 Sin [x3] − 9.154 Cos [x1] Sin [x3] + 35.477 Cos [x2] Sin [x3] − 169.462 Cos [x3] Sin [x3] +
7.297 Cos [x4] Sin [x3] + 50.012 Cos [x5] Sin [x3] − 7.782 Cos [x6] Sin [x3] + 13.639 Sin [x3]² +
28.594 Sin [x4] + 37.133 Sin [x4]² + 1.201 Sin [x5] + 33.485 Sin [x5]² + 33.958 Sin [x6] −
5.050 Cos [x1] Sin [x6] + 0.618 Cos [x2] Sin [x6] − 5.905 Cos [x3] Sin [x6] − 0.017 Cos [x4] Sin [x6] −
1.457 Cos [x5] Sin [x6] − 0.313 Cos [x6] Sin [x6] − 20.410 Sin [x1] Sin [x6] − 7.261 Sin [x2] Sin [x6] −
16.044 Sin [x3] Sin [x6] + 9.900 Sin [x4] Sin [x6] + 0.586 Sin [x5] Sin [x6] + 36.835 Sin [x6]²

In[•]:= (*14*)

Out[•]///NumberForm=

57.114 − 92.715 Cos [x1] − 113.077 Cos [x1]² − 51.182 Cos [x2] + 32.283 Cos [x2]² +
0.517 Cos [x3] + 54.280 Cos [x1] Cos [x3] + 17.552 Cos [x2] Cos [x3] + 26.798 Cos [x3]² +
11.025 Cos [x4] + 0.366 Cos [x2] Cos [x4] + 51.825 Cos [x4]² − 56.318 Cos [x5] + 33.017 Cos [x5]² +
4.732 Cos [x6] + 28.321 Cos [x6]² − 545.266 Sin [x1] + 119.400 Cos [x1] Sin [x1] +
6.985 Cos [x2] Sin [x1] + 204.591 Cos [x3] Sin [x1] − 7.781 Cos [x4] Sin [x1] +
37.752 Cos [x5] Sin [x1] + 3.402 Cos [x6] Sin [x1] + 157.536 Sin [x1]² + 29.733 Sin [x2]² +
235.489 Sin [x3] − 26.718 Cos [x1] Sin [x3] + 45.327 Cos [x2] Sin [x3] − 186.827 Cos [x3] Sin [x3] −
0.450 Cos [x4] Sin [x3] + 36.560 Cos [x5] Sin [x3] − 8.463 Cos [x6] Sin [x3] + 21.675 Sin [x3]² +
71.771 Sin [x4] + 16.553 Sin [x4]² + 1.151 Sin [x5] + 25.723 Sin [x5]² + 36.431 Sin [x6] −
2.932 Cos [x1] Sin [x6] + 0.422 Cos [x2] Sin [x6] − 8.330 Cos [x3] Sin [x6] − 0.216 Cos [x4] Sin [x6] −
0.488 Cos [x5] Sin [x6] − 0.292 Cos [x6] Sin [x6] − 12.526 Sin [x1] Sin [x6] − 5.820 Sin [x2] Sin [x6] −
24.089 Sin [x3] Sin [x6] + 7.308 Sin [x4] Sin [x6] + 0.123 Sin [x5] Sin [x6] + 27.814 Sin [x6]²

In[•]:= (*15*)

Out[•]///NumberForm=

78.204 − 138.507 Cos [x1] − 93.546 Cos [x1]² − 86.352 Cos [x2] + 43.833 Cos [x2]² −
117.933 Cos [x3] + 76.449 Cos [x1] Cos [x3] + 28.428 Cos [x2] Cos [x3] + 15.900 Cos [x3]² +
13.908 Cos [x4] + 0.867 Cos [x2] Cos [x4] + 71.675 Cos [x4]² − 72.091 Cos [x5] + 40.261 Cos [x5]² +
4.615 Cos [x6] + 37.998 Cos [x6]² − 508.714 Sin [x1] + 115.438 Cos [x1] Sin [x1] +
10.290 Cos [x2] Sin [x1] + 233.945 Cos [x3] Sin [x1] − 9.805 Cos [x4] Sin [x1] +
38.996 Cos [x5] Sin [x1] + 3.303 Cos [x6] Sin [x1] + 156.946 Sin [x1]² + 41.013 Sin [x2]² +
84.557 Sin [x3] + 20.005 Cos [x1] Sin [x3] + 75.585 Cos [x2] Sin [x3] − 108.990 Cos [x3] Sin [x3] −
1.810 Cos [x4] Sin [x3] + 44.178 Cos [x5] Sin [x3] − 8.223 Cos [x6] Sin [x3] + 57.160 Sin [x3]² +
82.434 Sin [x4] + 30.220 Sin [x4]² + 1.000 Sin [x5] + 37.577 Sin [x5]² + 40.279 Sin [x6] −
4.726 Cos [x1] Sin [x6] + 0.635 Cos [x2] Sin [x6] − 8.315 Cos [x3] Sin [x6] − 0.473 Cos [x4] Sin [x6] −

(Continued)

TABLE A7.5 *(Continued)*

Models Obtained by NRM and SNRM Using the Bootstrap Data Separation Method (80 Training–15 Testing–5 Validation*) (*Models Obtained Using the NRM and SNRM Methods*)(*Related to Table 7.6*)

1.003 Cos [$x5$] Sin [$x6$] − 0.344 Cos [$x6$] Sin [$x6$] − 19.673 Sin [$x1$] Sin [$x6$] − 5.644 Sin [$x2$] Sin [$x6$] − 23.633 Sin [$x3$] Sin [$x6$] + 9.661 Sin [$x4$] Sin [$x6$] + 0.279 Sin [$x5$] Sin [$x6$] + 37.515 Sin [$x6$]^2

In[•]:= (*16*)

Out[•]//NumberForm=

43.199 − 70.000 Cos [$x1$] − 90.196 Cos [$x1$]^2 − 55.131 Cos [$x2$] + 27.875 Cos [$x2$]^2 − 52.742 Cos [$x3$] + 62.051 Cos [$x1$] Cos [$x3$] + 19.220 Cos [$x2$] Cos [$x3$] + 3.577 Cos [$x3$]^2 + 9.389 Cos [$x4$] + 0.824 Cos [$x2$] Cos [$x4$] + 45.357 Cos [$x4$]^2 − 78.860 Cos [$x5$] + 24.915 Cos [$x5$]^2 + 4.947 Cos [$x6$] + 24.411 Cos [$x6$]^2 − 400.666 Sin [$x1$] + 81.609 Cos [$x1$] Sin [$x1$] + 6.877 Cos [$x2$] Sin [$x1$] + 218.528 Cos [$x3$] Sin [$x1$] − 7.102 Cos [$x4$] Sin [$x1$] + 45.679 Cos [$x5$] Sin [$x1$] + 2.981 Cos [$x6$] Sin [$x1$] + 118.031 Sin [$x1$]^2 + 25.145 Sin [$x2$]^2 + 140.449 Sin [$x3$] − 11.628 Cos [$x1$] Sin [$x3$] + 48.842 Cos [$x2$] Sin [$x3$] − 149.721 Cos [$x3$] Sin [$x3$] + 0.598 Cos [$x4$] Sin [$x3$] + 45.554 Cos [$x5$] Sin [$x3$] − 8.303 Cos [$x6$] Sin [$x3$] + 40.369 Sin [$x3$]^2 + 76.815 Sin [$x4$] + 7.699 Sin [$x4$]^2 + 1.153 Sin [$x5$] + 21.395 Sin [$x5$]^2 + 37.892 Sin [$x6$] − 4.113 Cos [$x1$] Sin [$x6$] + 0.448 Cos [$x2$] Sin [$x6$] − 7.931 Cos [$x3$] Sin [$x6$] − 0.328 Cos [$x4$] Sin [$x6$] − 1.153 Cos [$x5$] Sin [$x6$] − 0.345 Cos [$x6$] Sin [$x6$] − 17.361 Sin [$x1$] Sin [$x6$] − 5.464 Sin [$x2$] Sin [$x6$] − 22.164 Sin [$x3$] Sin [$x6$] + 7.821 Sin [$x4$] Sin [$x6$] + 0.339 Sin [$x5$] Sin [$x6$] + 23.908 Sin [$x6$]^2

In[•]:= (*17*)

Out[•]//NumberForm=

62.655 − 106.002 Cos [$x1$] − 92.268 Cos [$x1$]^2 − 93.019 Cos [$x2$] + 33.278 Cos [$x2$]^2 − 121.512 Cos [$x3$] + 76.381 Cos [$x1$] Cos [$x3$] + 28.733 Cos [$x2$] Cos [$x3$] + 3.842 Cos [$x3$]^2 + 16.897 Cos [$x4$] + 1.146 Cos [$x2$] Cos [$x4$] + 66.260 Cos [$x4$]^2 − 69.141 Cos [$x5$] + 44.113 Cos [$x5$]^2 + 4.542 Cos [$x6$] + 38.873 Cos [$x6$]^2 − 483.003 Sin [$x1$] + 103.348 Cos [$x1$] Sin [$x1$] + 9.428 Cos [$x2$] Sin [$x1$] + 267.060 Cos [$x3$] Sin [$x1$] − 12.112 Cos [$x4$] Sin [$x1$] + 38.248 Cos [$x5$] Sin [$x1$] + 2.919 Cos [$x6$] Sin [$x1$] + 140.609 Sin [$x1$]^2 + 29.716 Sin [$x2$]^2 + 123.674 Sin [$x3$] − 2.501 Cos [$x1$] Sin [$x3$] + 83.124 Cos [$x2$] Sin [$x3$] − 133.852 Cos [$x3$] Sin [$x3$] − 2.754 Cos [$x4$] Sin [$x3$] + 47.568 Cos [$x5$] Sin [$x3$] − 7.837 Cos [$x6$] Sin [$x3$] + 40.721 Sin [$x3$]^2 + 83.550 Sin [$x4$] + 24.572 Sin [$x4$]^2 + 1.210 Sin [$x5$] + 37.877 Sin [$x5$]^2 + 31.806 Sin [$x6$] − 4.099 Cos [$x1$] Sin [$x6$] + 0.334 Cos [$x2$] Sin [$x6$] − 5.720 Cos [$x3$] Sin [$x6$] − 0.276 Cos [$x4$] Sin [$x6$] − 0.997 Cos [$x5$] Sin [$x6$] − 0.390 Cos [$x6$] Sin [$x6$] − 18.075 Sin [$x1$] Sin [$x6$] − 5.861 Sin [$x2$] Sin [$x6$] − 17.289 Sin [$x3$] Sin [$x6$] + 9.771 Sin [$x4$] Sin [$x6$] + 0.709 Sin [$x5$] Sin [$x6$] + 38.345 Sin [$x6$]^2

In[•]:= (*18*)

Out[•]//NumberForm=

60.237 − 96.162 Cos [$x1$] − 99.841 Cos [$x1$]^2 − 32.201 Cos [$x2$] + 40.418 Cos [$x2$]^2 − 37.520 Cos [$x3$] + 54.758 Cos [$x1$] Cos [$x3$] + 7.567 Cos [$x2$] Cos [$x3$] + 4.897 Cos [$x3$]^2 + 6.017 Cos [$x4$] + 1.016 Cos [$x2$] Cos [$x4$] + 59.121 Cos [$x4$]^2 − 79.038 Cos [$x5$] + 35.177 Cos [$x5$]^2 + 5.660 Cos [$x6$] + 29.160 Cos [$x6$]^2 − 507.135 Sin [$x1$] + 124.402 Cos [$x1$] Sin [$x1$] + 11.851 Cos [$x2$] Sin [$x1$] + 226.907 Cos [$x3$] Sin [$x1$] − 3.375 Cos [$x4$] Sin [$x1$] + 50.133 Cos [$x5$] Sin [$x1$] + 2.609 Cos [$x6$] Sin [$x1$] + 149.586 Sin [$x1$]^2 + 37.513 Sin [$x2$]^2 + 150.979 Sin [$x3$] − 29.884 Cos [$x1$] Sin [$x3$] + 23.566 Cos [$x2$] Sin [$x3$] − 165.634 Cos [$x3$] Sin [$x3$] + 0.437 Cos [$x4$] Sin [$x3$] + 48.385 Cos [$x5$] Sin [$x3$] − 8.744 Cos [$x6$] Sin [$x3$] + 51.962 Sin [$x3$]^2 +

(Continued)

TABLE A7.5 *(Continued)*
Models Obtained by NRM and SNRM Using the Bootstrap Data Separation Method (80 Training–15 Testing–5 Validation*) (*Models Obtained Using the NRM and SNRM Methods*)(*Related to Table 7.6*)

76.840 Sin [x4] + 22.173 Sin [x4]^2 + 1.306 Sin [x5] + 27.717 Sin [x5]^2 + 42.566 Sin [x6] −
5.230 Cos [x1] Sin [x6] + 0.813 Cos [x2] Sin [x6] − 6.889 Cos [x3] Sin [x6] − 0.455 Cos [x4] Sin [x6] +
0.069 Cos [x5] Sin [x6] − 0.237 Cos [x6] Sin [x6] − 20.919 Sin [x1] Sin [x6] − 5.624 Sin [x2] Sin [x6] −
19.734 Sin [x3] Sin [x6] + 5.470 Sin [x4] Sin [x6] + 0.313 Sin [x5] Sin [x6] + 28.601 Sin [x6]^2

In[•]: = (***19***)
Out[•]//NumberForm=

60.840 − 65.000 Cos [x1] − 113.681 Cos [x1]^2 − 45.752 Cos [x2] + 39.087 Cos [x2]^2 − 61.564 Cos [x3] +
56.666 Cos [x1] Cos [x3] + 17.839 Cos [x2] Cos [x3] − 0.292 Cos [x3]^2 + 7.318 Cos [x4] +
1.897 Cos [x2] Cos [x4] + 66.280 Cos [x4]^2 − 85.763 Cos [x5] + 31.867 Cos [x5]^2 + 5.205 Cos [x6] +
32.195 Cos [x6]^2 − 459.425 Sin [x1] + 91.912 Cos [x1] Sin [x1] + 6.009 Cos [x2] Sin [x1] +
213.925 Cos [x3] Sin [x1] − 2.665 Cos [x4] Sin [x1] + 60.109 Cos [x5] Sin [x1] +
3.240 Cos [x6] Sin [x1] + 129.611 Sin [x1]^2 + 36.486 Sin [x2]^2 + 104.653 Sin [x3] −
26.825 Cos [x1] Sin [x3] + 40.076 Cos [x2] Sin [x3] − 136.090 Cos [x3] Sin [x3] −
1.638 Cos [x4] Sin [x3] + 38.423 Cos [x5] Sin [x3] − 8.847 Cos [x6] Sin [x3] + 49.237 Sin [x3]^2 +
97.733 Sin [x4] + 17.588 Sin [x4]^2 + 1.236 Sin [x5] + 28.199 Sin [x5]^2 + 32.495 Sin [x6] −
2.463 Cos [x1] Sin [x6] + 0.522 Cos [x2] Sin [x6] − 7.851 Cos [x3] Sin [x6] − 0.336 Cos [x4] Sin [x6] −
0.746 Cos [x5] Sin [x6] − 0.325 Cos [x6] Sin [x6] − 10.355 Sin [x1] Sin [x6] − 5.858 Sin [x2] Sin [x6] −
22.922 Sin [x3] Sin [x6] + 7.598 Sin [x4] Sin [x6] + 0.225 Sin [x5] Sin [x6] + 31.678 Sin [x6]^2

In[•]: = (***20***)
Out[•]//NumberForm=

44.309 − 44.951 Cos [x1] − 84.526 Cos [x1]^2 − 55.762 Cos [x2] + 21.943 Cos [x2]^2 − 72.022 Cos
[x3] + 65.207 Cos [x1] Cos [x3] + 18.440 Cos [x2] Cos [x3] − 12.617 Cos [x3]^2 + 5.123 Cos [x4] +
1.394 Cos [x2] Cos [x4] + 48.043 Cos [x4]^2 − 78.544 Cos [x5] + 25.113 Cos [x5]^2 + 6.131 Cos [x6] +
20.073 Cos [x6]^2 − 367.062 Sin [x1] + 66.221 Cos [x1] Sin [x1] + 8.807 Cos [x2] Sin [x1] + 239.394
Cos [x3] Sin [x1] − 3.402 Cos [x4] Sin [x1] + 54.347 Cos [x5] Sin [x1] + 3.158 Cos [x6] Sin [x1] +
112.364 Sin [x1]^2 + 19.583 Sin [x2]^2 + 104.473 Sin [x3] − 21.586 Cos [x1] Sin [x3] +
47.715 Cos [x2] Sin [x3] − 149.787 Cos [x3] Sin [x3] + 1.331 Cos [x4] Sin [x3] +
41.870 Cos [x5] Sin [x3] − 9.746 Cos [x6] Sin [x3] + 45.276 Sin [x3]^2 + 107.881 Sin [x4] −
6.548 Sin [x4]^2 + 1.221 Sin [x5] + 18.689 Sin [x5]^2 + 43.691 Sin [x6] − 4.169 Cos [x1] Sin [x6] +
0.467 Cos [x2] Sin [x6] − 8.441 Cos [x3] Sin [x6] − 0.514 Cos [x4] Sin [x6] − 0.043 Cos [x5] Sin [x6] −
0.286 Cos [x6] Sin [x6] − 16.633 Sin [x1] Sin [x6] − 6.755 Sin [x2] Sin [x6] − 24.822 Sin [x3] Sin [x6] +
6.433 Sin [x4] Sin [x6] + 0.015 Sin [x5] Sin [x6] + 19.569 Sin [x6]^2

TABLE A7.6

Models Obtained by NRM and SNRM Using the Cross-Validation Data Separation Method (*(10 K-fold Cross-Validation: 80 Training–20 Testing)*) (*Training Was Divided into 10 Parts: 9 Training–1 Validation*) (*Models Obtained Using the NRM and SNRM Methods*)(*(Related to Table 7.7)*)

(*Models proposed by NRM*)

In[•]:= (*1*)

Out[•]//NumberForm=

$-268.773 - 455.725$ Cos $[x1] - 1017.269$ Cos $[x1]^2 + 39.202$ Cos $[x2] - 283.983$ Cos $[x2]^2 +$
1084.608 Cos $[x3] + 38.442$ Cos $[x1]$ Cos $[x3] - 0.369$ Cos $[x2]$ Cos $[x3] + 1152.645$ Cos $[x3]^2 +$
63.197 Cos $[x4] - 2.517$ Cos $[x2]$ Cos $[x4] - 207.228$ Cos $[x4]^2 - 79.818$ Cos $[x5] - 315.775$ Cos $[x5]^2 +$
4.603 Cos $[x6] - 355.948$ Cos $[x6]^2 - 1522.423$ Sin $[x1] + 431.629$ Cos $[x1]$ Sin $[x1] -$
20.976 Cos $[x2]$ Sin $[x1] + 122.079$ Cos $[x3]$ Sin $[x1] - 20.783$ Cos $[x4]$ Sin $[x1] +$
37.894 Cos $[x5]$ Sin $[x1] + 2.291$ Cos $[x6]$ Sin $[x1] - 309.605$ Sin $[x1]^2 - 282.461$ Sin $[x2]^2 +$
3628.706 Sin $[x3] + 21.447$ Cos $[x1]$ Sin $[x3] - 16.112$ Cos $[x2]$ Sin $[x3] - 1162.890$ Cos $[x3]$ Sin $[x3] -$
42.965 Cos $[x4]$ Sin $[x3] + 32.383$ Cos $[x5]$ Sin $[x3] - 7.189$ Cos $[x6]$ Sin $[x3] - 368.222$ Sin $[x3]^2 +$
164.083 Sin $[x4] - 292.028$ Sin $[x4]^2 + 0.742$ Sin $[x5] - 306.737$ Sin $[x5]^2 + 46.550$ Sin $[x6] +$
1.687 Cos $[x1]$ Sin $[x6] - 0.417$ Cos $[x2]$ Sin $[x6] - 10.663$ Cos $[x3]$ Sin $[x6] - 2.253$ Cos $[x4]$ Sin $[x6] -$
0.274 Cos $[x5]$ Sin $[x6] - 0.334$ Cos $[x6]$ Sin $[x6] - 0.891$ Sin $[x1]$ Sin $[x6] - 9.507$ Sin $[x2]$ Sin $[x6] -$
29.871 Sin $[x3]$ Sin $[x6] - 5.211$ Sin $[x4]$ Sin $[x6] - 0.163$ Sin $[x5]$ Sin $[x6] - 356.024$ Sin $[x6]^2$

In[•]:= (*2*)

Out[•]//NumberForm=

$244.412 - 939.794$ Cos $[x1] - 1142.451$ Cos $[x1]^2 + 46.588$ Cos $[x2] +$
256.410 Cos $[x2]^2 132.615$ Cos $[x3] + 45.376$ Cos $[x1]$ Cos $[x3] + 2.715$ Cos $[x2]$ Cos $[x3] +$
714.732 Cos $[x3]^2 + 77.101$ Cos $[x4] + 4.179$ Cos $[x2]$ Cos $[x4] + 433.444$ Cos $[x4]^2 - 82.182$ Cos $[x5] +$
290.095 Cos $[x5]^2 + 5.364$ Cos $[x6] + 327.300$ Cos $[x6]^2 - 3192.139$ Sin $[x1] +$
890.875 Cos $[x1]$ Sin $[x1] - 50.946$ Cos $[x2]$ Sin $[x1] + 109.683$ Cos $[x3]$ Sin $[x1] -$
44.342 Cos $[x4]$ Sin $[x1] + 42.764$ Cos $[x5]$ Sin $[x1] + 0.818$ Cos $[x6]$ Sin $[x1] + 318.957$ Sin $[x1]^2 +$
256.608 Sin $[x2]^2 + 968.020$ Sin $[x3] + 30.872$ Cos $[x1]$ Sin $[x3] + 3.810$ Cos $[x2]$ Sin $[x3] -$
244.652 Cos $[x3]$ Sin $[x3] - 35.894$ Cos $[x4]$ Sin $[x3] + 38.758$ Cos $[x5]$ Sin $[x3] -$
6.699 Cos $[x6]$ Sin $[x3] + 295.885$ Sin $[x3]^2 + 317.747$ Sin $[x4] + 263.561$ Sin $[x4]^2 + 1.137$ Sin $[x5] +$
293.067 Sin $[x5]^2 + 34.461$ Sin $[x6] + 3.929$ Cos $[x1]$ Sin $[x6] + 0.467$ Cos $[x2]$ Sin $[x6] -$
5.751 Cos $[x3]$ Sin $[x6] - 1.760$ Cos $[x4]$ Sin $[x6] - 1.068$ Cos $[x5]$ Sin $[x6] - 0.301$ Cos $[x6]$ Sin $[x6] +$
10.797 Sin $[x1]$ Sin $[x6] - 16.003$ Sin $[x2]$ Sin $[x6] - 22.056$ Sin $[x3]$ Sin $[x6] - 8.646$ Sin $[x4]$ Sin $[x6] +$
0.073 Sin $[x5]$ Sin $[x6] + 327.033$ Sin $[x6]^2$

In[•]:= (*3*)

Out[•]//NumberForm=

$1197.420 - 846.357$ Cos $[x1] - 229.101$ Cos $[x1]^2 - 150.690$ Cos $[x2] + 1254.020$ Cos $[x2]^2 -$
2172.700 Cos $[x3] - 78.832$ Cos $[x1]$ Cos $[x3] + 65.035$ Cos $[x2]$ Cos $[x3] - 965.240$ Cos $[x3]^2 +$
53.376 Cos $[x4] + 9.084$ Cos $[x2]$ Cos $[x4] + 1460.930$ Cos $[x4]^2 + 205.897$ Cos $[x5] +$
1407.500 Cos $[x5]^2 + 11.214$ Cos $[x6] + 1584.600$ Cos $[x6]^2 - 3769.150$ Sin $[x1] +$
1078.960 Cos $[x1]$ Sin $[x1] - 16.545$ Cos $[x2]$ Sin $[x1] + 12.010$ Cos $[x3]$ Sin $[x1] -$
65.113 Cos $[x4]$ Sin $[x1] - 18.365$ Cos $[x5]$ Sin $[x1] - 2.433$ Cos $[x6]$ Sin $[x1] + 1464.490$ Sin $[x1]^2 +$
1251.400 Sin $[x2]^2 - 6156.220$ Sin $[x3] - 231.695$ Cos $[x1]$ Sin $[x3] + 157.450$ Cos $[x2]$ Sin $[x3] +$

(Continued)

TABLE A7.6 *(Continued)*

Models Obtained by NRM and SNRM Using the Cross-Validation Data Separation Method (*(10 K-fold Cross-Validation: 80 Training–20 Testing)*) (*Training Was Divided into 10 Parts: 9 Training–1 Validation*) (*Models Obtained Using the NRM and SNRM Methods*)(*(Related to Table 7.7)*)

2068.660 Cos [$x3$] Sin [$x3$] + 11.242 Cos [$x4$] Sin [$x3$] − 146.223 Cos [$x5$] Sin [$x3$] −
9.604 Cos [$x6$] Sin [$x3$] + 1549.240 Sin [$x3$]^2 + 313.300 Sin [$x4$] + 1292.920 Sin [$x4$]^2 +
0.562 Sin [$x5$] + 1381.620 Sin [$x5$]^2 + 59.382 Sin [$x6$] + 7.579 Cos [$x1$] Sin [$x6$] −
0.060 Cos [$x2$] Sin [$x6$] − 20.371 Cos [$x3$] Sin [$x6$] − 3.906 Cos [$x4$] Sin [$x6$] − 1.206 Cos [$x5$] Sin [$x6$] −
0.466 Cos [$x6$] Sin [$x6$] + 20.125 Sin [$x1$] Sin [$x6$] + 4.142 Sin [$x2$] Sin [$x6$] − 61.861 Sin [$x3$] Sin [$x6$] −
21.097 Sin [$x4$] Sin [$x6$] + 0.218 Sin [$x5$] Sin [$x6$] + 1584.830 Sin [$x6$]^2

In[•]:= (*4*)

Out[•]///NumberForm=

2.484 − 525.111 Cos [$x1$] − 551.118 Cos [$x1$]^2 − 88.306 Cos [$x2$] + 3.134 Cos [$x2$]^2 + 119.974 Cos [$x3$] +
143.024 Cos [$x1$] Cos [$x3$] + 37.525 Cos [$x2$] Cos [$x3$] + 414.182 Cos [$x3$]^2 + 19.812 Cos [$x4$] +
6.344 Cos [$x2$] Cos [$x4$] + 152.284 Cos [$x4$]^2 − 49.890 Cos [$x5$] + 2.801 Cos [$x5$]^2 + 5.413 Cos [$x6$] +
3.368 Cos [$x6$]^2 − 1260.770 Sin [$x1$] + 326.684 Cos [$x1$] Sin [$x1$] − 23.204 Cos [$x2$] Sin [$x1$] +
303.214 Cos [$x3$] Sin [$x1$] − 23.515 Cos [$x4$] Sin [$x1$] + 24.237 Cos [$x5$] Sin [$x1$] +
5.799 Cos [$x6$] Sin [$x1$] + 12.125 Sin [$x1$]^2 + 2.614 Sin [$x2$]^2 + 1053.322 Sin [$x3$] +
174.952 Cos [$x1$] Sin [$x3$] + 107.418 Cos [$x2$] Sin [$x3$] − 428.235 Cos [$x3$] Sin [$x3$] +
3.548 Cos [$x4$] Sin [$x3$] + 22.008 Cos [$x5$] Sin [$x3$] − 11.631 Cos [$x6$] Sin [$x3$] − 7.259 Sin [$x3$]^2 +
278.082 Sin [$x4$] + 1.935 Sin [$x4$]^2 + 1.000 Sin [$x5$] + 6.535 Sin [$x5$]^2 + 37.266 Sin [$x6$] +
8.781 Cos [$x1$] Sin [$x6$] − 0.308 Cos [$x2$] Sin [$x6$] − 11.509 Cos [$x3$] Sin [$x6$] − 3.462 Cos [$x4$] Sin [$x6$] +
0.221 Cos [$x5$] Sin [$x6$] − 0.354 Cos [$x6$] Sin [$x6$] + 23.137 Sin [$x1$] Sin [$x6$] − 7.935 Sin [$x2$] Sin [$x6$] −
31.046 Sin [$x3$] Sin [$x6$] − 20.822 Sin [$x4$] Sin [$x6$] − 0.517 Sin [$x5$] Sin [$x6$] + 3.230 Sin [$x6$]^2

In[•]:= (*5*)

Out[•]///NumberForm=

−162.599 − 399.975 Cos [$x1$] − 801.211 Cos [$x1$]^2 − 59.274 Cos [$x2$] − 171.642 Cos [$x2$]^2 +
551.252 Cos [$x3$] + 68.223 Cos [$x1$] Cos [$x3$] + 31.716 Cos [$x2$] Cos [$x3$] + 849.282 Cos [$x3$]^2 +
61.889 Cos [$x4$] − 1.030 Cos [$x2$] Cos [$x4$] − 69.834 Cos [$x4$]^2 − 133.901 Cos [$x5$] − 191.935 Cos [$x5$]^2 +
1.320 Cos [$x6$] − 212.713 Cos [$x6$]^2 − 1250.430 Sin [$x1$] + 327.165 Cos [$x1$] Sin [$x1$] −
11.270 Cos [$x2$] Sin [$x1$] + 169.993 Cos [$x3$] Sin [$x1$] − 30.299 Cos [$x4$] Sin [$x1$] +
93.607 Cos [$x5$] Sin [$x1$] + 4.902 Cos [$x6$] Sin [$x1$] − 184.703 Sin [$x1$]^2 − 170.506 Sin [$x2$]^2 +
2455.600 Sin [$x3$] + 66.463 Cos [$x1$] Sin [$x3$] + 67.394 Cos [$x2$] Sin [$x3$] − 718.629 Cos [$x3$] Sin [$x3$] −
32.133 Cos [$x4$] Sin [$x3$] + 27.696 Cos [$x5$] Sin [$x3$] − 6.361 Cos [$x6$] Sin [$x3$] − 231.192 Sin [$x3$]^2 +
203.408 Sin [$x4$] − 176.405 Sin [$x4$]^2 + 1.013 Sin [$x5$] − 180.989 Sin [$x5$]^2 + 39.189 Sin [$x6$] +
2.199 Cos [$x1$] Sin [$x6$] − 0.531 Cos [$x2$] Sin [$x6$] − 3.313 Cos [$x3$] Sin [$x6$] − 3.559 Cos [$x4$] Sin [$x6$] −
1.588 Cos [$x5$] Sin [$x6$] − 0.281 Cos [$x6$] Sin [$x6$] + 4.613 Sin [$x1$] Sin [$x6$] − 13.444 Sin [$x2$] Sin [$x6$] −
10.526 Sin [$x3$] Sin [$x6$] − 21.394 Sin [$x4$] Sin [$x6$] + 0.147 Sin [$x5$] Sin [$x6$] − 212.681 Sin [$x6$]^2

In[•]:= (*6*)

Out[•]///NumberForm=

−208.500 − 234.666 Cos [$x1$] − 581.085 Cos [$x1$]^2 − 12.954 Cos [$x2$] − 219.370 Cos [$x2$]^2 +
530.672 Cos [$x3$] + 85.607 Cos [$x1$] Cos [$x3$] + 17.254 Cos [$x2$] Cos [$x3$] + 602.094 Cos [$x3$]^2 +
28.210 Cos [$x4$] + 1.876 Cos [$x2$] Cos [$x4$] − 93.957 Cos [$x4$]^2 − 173.635 Cos [$x5$] − 247.918 Cos [$x5$]^2 +

(Continued)

TABLE A7.6 *(Continued)*

Models Obtained by NRM and SNRM Using the Cross-Validation Data Separation Method (*(10 K-fold Cross-Validation: 80 Training–20 Testing)*) (*Training Was Divided into 10 Parts: 9 Training–1 Validation*) (*Models Obtained Using the NRM and SNRM Methods*)(*(Related to Table 7.7)*)

1.819 Cos [$x6$] − 275.992 Cos [$x6$]^2 − 601.465 Sin [$x1$] + 128.133 Cos [$x1$] Sin [$x1$] −
18.704 Cos [$x2$] Sin [$x1$] + 180.829 Cos [$x3$] Sin [$x1$] − 14.824 Cos [$x4$] Sin [$x1$] +
104.543 Cos [$x5$] Sin [$x1$] + 3.468 Cos [$x6$] Sin [$x1$] − 243.853 Sin [$x1$]^2 − 219.168 Sin [$x2$]^2 +
2121.370 Sin [$x3$] + 103.095 Cos [$x1$] Sin [$x3$] + 29.070 Cos [$x2$] Sin [$x3$] − 700.340 Cos [$x3$] Sin [$x3$] −
12.783 Cos [$x4$] Sin [$x3$] + 46.590 Cos [$x5$] Sin [$x3$] − 5.602 Cos [$x6$] Sin [$x3$] − 280.996 Sin [$x3$]^2 +
251.655 Sin [$x4$] − 227.558 Sin [$x4$]^2 + 1.456 Sin [$x5$] − 230.687 Sin [$x5$]^2 + 61.185 Sin [$x6$] +
0.446 Cos [$x1$] Sin [$x6$] − 0.919 Cos [$x2$] Sin [$x6$] − 9.844 Cos [$x3$] Sin [$x6$] − 3.550 Cos [$x4$] Sin [$x6$] −
2.421 Cos [$x5$] Sin [$x6$] − 0.270 Cos [$x6$] Sin [$x6$] − 4.950 Sin [$x1$] Sin [$x6$] − 10.832 Sin [$x2$] Sin [$x6$] −
25.906 Sin [$x3$] Sin [$x6$] − 20.675 Sin [$x4$] Sin [$x6$] + 0.246 Sin [$x5$] Sin [$x6$] − 276.165 Sin [$x6$]^2

In[•]:= (**7***)

Out[•]//NumberForm=

−96.357 − 55.670 Cos [$x1$] − 79.697 Cos [$x1$]^2 − 30.670 Cos [$x2$] − 103.229 Cos [$x2$]^2 − 0.162 Cos [$x3$] +
94.938 Cos [$x1$] Cos [$x3$] + 12.712 Cos [$x2$] Cos [$x3$] + 6.214 Cos [$x3$]^2 + 6.379 Cos [$x4$] +
1.645 Cos [$x2$] Cos [$x4$] + 23.090 Cos [$x4$]^2 − 179.798 Cos [$x5$] − 113.620 Cos [$x5$]^2 +
4.571 Cos [$x6$] − 130.710 Cos [$x6$]^2 + 272.683 Sin [$x1$] − 80.208 Cos [$x1$] Sin [$x1$] −
6.745 Cos [$x2$] Sin [$x1$] + 186.377 Cos [$x3$] Sin [$x1$] − 4.173 Cos [$x4$] Sin [$x1$] +
174.808 Cos [$x5$] Sin [$x1$] + 0.832 Cos [$x6$] Sin [$x1$] − 115.880 Sin [$x1$]^2 − 101.200 Sin [$x2$]^2 +
302.035 Sin [$x3$] + 133.519 Cos [$x1$] Sin [$x3$] + 37.866 Cos [$x2$] Sin [$x3$] − 190.621 Cos [$x3$] Sin [$x3$] −
0.601 Cos [$x4$] Sin [$x3$] − 56.089 Cos [$x5$] Sin [$x3$] − 5.678 Cos [$x6$] Sin [$x3$] − 122.774 Sin [$x3$]^2 −
239.612 Sin [$x4$] − 104.404 Sin [$x4$]^2 + 0.963 Sin [$x5$] − 76.864 Sin [$x5$]^2 + 48.660 Sin [$x6$] +
0.447 Cos [$x1$] Sin [$x6$] + 0.240 Cos [$x2$] Sin [$x6$] − 8.407 Cos [$x3$] Sin [$x6$] − 2.530 Cos [$x4$] Sin [$x6$] +
3.587 Cos [$x5$] Sin [$x6$] − 0.269 Cos [$x6$] Sin [$x6$] − 3.619 Sin [$x1$] Sin [$x6$] − 5.567 Sin [$x2$] Sin [$x6$] −
20.676 Sin [$x3$] Sin [$x6$] − 13.991 Sin [$x4$] Sin [$x6$] − 1.234 Sin [$x5$] Sin [$x6$] − 130.668 Sin [$x6$]^2

In[•]:= (**8***)

Out[•]//NumberForm=

65.345 − 462.100 Cos [$x1$] − 582.866 Cos [$x1$]^2 − 25.782 Cos [$x2$] + 66.677 Cos [$x2$]^2 +
195.749 Cos [$x3$] + 46.336 Cos [$x1$] Cos [$x3$] + 17.234 Cos [$x2$] Cos [$x3$] + 430.772 Cos [$x3$]^2 +
24.186 Cos [$x4$] − 2.330 Cos [$x2$] Cos [$x4$] + 124.713 Cos [$x4$]^2 − 82.806 Cos [$x5$] + 77.203 Cos [$x5$]^2 +
0.304 Cos [$x6$] + 87.444 Cos [$x6$]^2 − 1411.650 Sin [$x1$] + 360.813 Cos [$x1$] Sin [$x1$] −
26.221 Cos [$x2$] Sin [$x1$] + 61.941 Cos [$x3$] Sin [$x1$] − 16.859 Cos [$x4$] Sin [$x1$] +
25.056 Cos [$x5$] Sin [$x1$] + 2.430 Cos [$x6$] Sin [$x1$] + 91.440 Sin [$x1$]^2 + 68.883 Sin [$x2$]^2 +
852.755 Sin [$x3$] + 102.020 Cos [$x1$] Sin [$x3$] + 51.898 Cos [$x2$] Sin [$x3$] − 261.846 Cos [$x3$] Sin [$x3$] −
4.856 Cos [$x4$] Sin [$x3$] + 51.429 Cos [$x5$] Sin [$x3$] − 2.800 Cos [$x6$] Sin [$x3$] + 73.125 Sin [$x3$]^2 +
108.444 Sin [$x4$] + 70.703 Sin [$x4$]^2 + 1.301 Sin [$x5$] + 82.760 Sin [$x5$]^2 + 37.112 Sin [$x6$] −
1.735 Cos [$x1$] Sin [$x6$] − 0.065 Cos [$x2$] Sin [$x6$] − 7.853 Cos [$x3$] Sin [$x6$] − 1.082 Cos [$x4$] Sin [$x6$] −
2.897 Cos [$x5$] Sin [$x6$] − 0.228 Cos [$x6$] Sin [$x6$] − 5.450 Sin [$x1$] Sin [$x6$] − 6.308 Sin [$x2$] Sin [$x6$] −
25.759 Sin [$x3$] Sin [$x6$] − 1.121 Sin [$x4$] Sin [$x6$] + 0.388 Sin [$x5$] Sin [$x6$] + 87.454 Sin [$x6$]^2

(Continued)

TABLE A7.6 *(Continued)*

Models Obtained by NRM and SNRM Using the Cross-Validation Data Separation Method (*(10 K-fold Cross-Validation: 80 Training–20 Testing)*) (*Training Was Divided into 10 Parts: 9 Training–1 Validation*) (*Models Obtained Using the NRM and SNRM Methods*)(*(Related to Table 7.7)*)

In[•]:= (*9*)

Out[•]///NumberForm=

57.895 − 402.912 Cos [$x1$] − 531.716 Cos [$x1$]^2 − 17.347 Cos [$x2$] + 60.749 Cos [$x2$]^2 +
47.413 Cos [$x3$] + 78.286 Cos [$x1$] Cos [$x3$] + 17.160 Cos [$x2$] Cos [$x3$] + 343.546 Cos [$x3$]^2 +
34.672 Cos [$x4$] + 4.805 Cos [$x2$] Cos [$x4$] + 165.189 Cos [$x4$]^2 − 103.727 Cos [$x5$] + 68.350 Cos [$x5$]^2 +
3.530 Cos [$x6$] + 77.635 Cos [$x6$]^2 − 1276.220 Sin [$x1$] + 320.261 Cos [$x1$] Sin [$x1$] −
19.804 Cos [$x2$] Sin [$x1$] + 178.858 Cos [$x3$] Sin [$x1$] − 19.353 Cos [$x4$] Sin [$x1$] +
63.625 Cos [$x5$] Sin [$x1$] + 3.018 Cos [$x6$] Sin [$x1$] + 79.987 Sin [$x1$]^2 + 60.941 Sin [$x2$]^2 +
686.167 Sin [$x3$] + 71.751 Cos [$x1$] Sin [$x3$] + 34.816 Cos [$x2$] Sin [$x3$] − 231. 285 Cos [$x3$] Sin [$x3$] −
14.764 Cos [$x4$] Sin [$x3$] + 26.947 Cos [$x5$] Sin [$x3$] − 6.931 Cos [$x6$] Sin [$x3$] + 65.496 Sin [$x3$]^2 +
195.316 Sin [$x4$] + 62.698 Sin [$x4$]^2 + 1.011 Sin [$x5$] + 78.933 Sin [$x5$]^2 + 28.643 Sin [$x6$] +
5.985 Cos [$x1$] Sin [$x6$] + 0.315 Cos [$x2$] Sin [$x6$] − 7.298 Cos [$x3$] Sin [$x6$] − 3.453 Cos [$x4$] Sin [$x6$] −
1.371 Cos [$x5$] Sin [$x6$] − 0.158 Cos [$x6$] Sin [$x6$] + 15.891 Sin [$x1$] Sin [$x6$] − 8.038 Sin [$x2$] Sin [$x6$] −
21.154 Sin [$x3$] Sin [$x6$] − 16.406 Sin [$x4$] Sin [$x6$] + 0.040 Sin [$x5$] Sin [$x6$] + 77.412 Sin [$x6$]^2

In[•]:= (*10*)

Out[•]///NumberForm=

208.275 − 153.726 Cos [$x1$] − 221.703 Cos [$x1$]^2 − 77.215 Cos [$x2$] + 226.740 Cos [$x2$]^2 −
264.441 Cos [$x3$] + 24.544 Cos [$x1$] Cos [$x3$] + 33.785 Cos [$x2$] Cos [$x3$] − 156.538 Cos [$x3$]^2 +
45.231 Cos [$x4$] − 0.572 Cos [$x2$] Cos [$x4$] + 231.003 Cos [$x4$]^2 − 171.956 Cos [$x5$] +
245.178 Cos [$x5$]^2 + 4.589 Cos [$x6$] + 275.210 Cos [$x6$]^2 − 913.924 Sin [$x1$] + 248.065 Cos [$x1$] Sin [$x1$] +
4.770 Cos [$x2$] Sin [$x1$] + 160.348 Cos [$x3$] Sin [$x1$] − 24.226 Cos [$x4$] Sin [$x1$] +
123.422 Cos [$x5$] Sin [$x1$] + 1.936 Cos [$x6$] Sin [$x1$] + 261.916 Sin [$x1$]^2 + 217.232 Sin [$x2$]^2 −
716.623 Sin [$x3$] − 89.945 Cos [$x1$] Sin [$x3$] + 70.225 Cos [$x2$] Sin [$x3$] + 118.786 Cos [$x3$] Sin [$x3$] −
18.647 Cos [$x4$] Sin [$x3$] + 53.212 Cos [$x5$] Sin [$x3$] − 6.919 Cos [$x6$] Sin [$x3$] + 269.803 Sin [$x3$]^2 +
17.845 Sin [$x4$] + 227.051 Sin [$x4$]^2 + 1.616 Sin [$x5$] + 247.324 Sin [$x5$]^2 + 65.809 Sin [$x6$] −
5.206 Cos [$x1$] Sin [$x6$] − 0.670 Cos [$x2$] Sin [$x6$] − 13.567 Cos [$x3$] Sin [$x6$] − 2.614 Cos [$x4$] Sin [$x6$] −
0.284 Cos [$x5$] Sin [$x6$] − 0.047 Cos [$x6$] Sin [$x6$] − 20.430 Sin [$x1$] Sin [$x6$] + 9.909 Sin [$x2$] Sin [$x6$] −
43.635 Sin [$x3$] Sin [$x6$] − 9.254 Sin [$x4$] Sin [$x6$] − 0.277 Sin [$x5$] Sin [$x6$] + 274.772 Sin [$x6$]^2

(*Models proposed by SNRM*)

In[•]:= (*1*)

Out[•]///NumberForm=

115.988 − 377.937 Cos [$x1$] − 281.001 Cos [$x1$]^2 − 0.595 Cos [$x2$] + 50.120 Cos [$x2$]^2 +
95.405 Cos [$x3$] + 47.577 Cos [$x1$] Cos [$x3$] + 10.090 Cos [$x2$] Cos [$x3$] + 191.778 Cos [$x3$]^2 +
26.491 Cos [$x4$] + 1.677 Cos [$x2$] Cos [$x4$] + 86.871 Cos [$x4$]^2 − 89.892 Cos [$x5$] + 62.334 Cos [$x5$]^2 +
2.354 Cos [$x6$] + 56.640 Cos [$x6$]^2 − 1300.753 Sin [$x1$] + 339.589 Cos [$x1$] Sin [$x1$] −
12.224 Cos [$x2$] Sin [$x1$] + 118.697 Cos [$x3$] Sin [$x1$] − 15.407 Cos [$x4$] Sin [$x1$] +
63.547 Cos [$x5$] Sin [$x1$] + 2.788 Cos [$x6$] Sin [$x1$] + 347.711 Sin [$x1$]^2 + 50.137 Sin [$x2$]^2 +
689.848 Sin [$x3$] + 34.696 Cos [$x1$] Sin [$x3$] + 12.573 Cos [$x2$] Sin [$x3$] − 215.010 Cos [$x3$] Sin [$x3$] −

(Continued)

TABLE A7.6 *(Continued)*

Models Obtained by NRM and SNRM Using the Cross-Validation Data Separation Method (*(10 K-fold Cross-Validation: 80 Training–20 Testing)*) (*Training Was Divided into 10 Parts: 9 Training–1 Validation*) (*Models Obtained Using the NRM and SNRM Methods*)(*(Related to Table 7.7)*)

10.098 Cos $[x4]$ Sin $[x3]$ + 30.030 Cos $[x5]$ Sin $[x3]$ – 5.370 Cos $[x6]$ Sin $[x3]$ – 88.411 Sin $[x3]^2$ + 154.709 Sin $[x4]$ + 5.704 Sin $[x4]^2$ + 1.250 Sin $[x5]$ + 63.125 Sin $[x5]^2$ + 36.390 Sin $[x6]$ + 2.044 Cos $[x1]$ Sin $[x6]$ – 0.485 Cos $[x2]$ Sin $[x6]$ – 6.114 Cos $[x3]$ Sin $[x6]$ – 2.130 Cos $[x4]$ Sin $[x6]$ – 1.428 Cos $[x5]$ Sin $[x6]$ – 0.188 Cos $[x6]$ Sin $[x6]$ + 3.940 Sin $[x1]$ Sin $[x6]$ – 12.469 Sin $[x2]$ Sin $[x6]$ – 18.064 Sin $[x3]$ Sin $[x6]$ – 10.737 Sin $[x4]$ Sin $[x6]$ + 0.113 Sin $[x5]$ Sin $[x6]$ + 56.481 Sin $[x6]^2$

In[•]:= (*2*)

Out[•]//NumberForm=

51.944 – 218.735 Cos $[x1]$ – 168.406 Cos $[x1]^2$ – 10.245 Cos $[x2]$ + 29.394 Cos $[x2]^2$ + 10.766 Cos $[x3]$ + 48.199 Cos $[x1]$ Cos $[x3]$ + 13.899 Cos $[x2]$ Cos $[x3]$ + 95.527 Cos $[x3]^2$ + 18.702 Cos $[x4]$ + 0.095 Cos $[x2]$ Cos $[x4]$ + 60.038 Cos $[x4]^2$ – 94.476 Cos $[x5]$ + 28.075 Cos $[x5]^2$ + 2.438 Cos $[x6]$ + 26.895 Cos $[x6]^2$ – 705.364 Sin $[x1]$ + 174.940 Cos $[x1]$ Sin $[x1]$ – 9.730 Cos $[x2]$ Sin $[x1]$ + 108.146 Cos $[x3]$ Sin $[x1]$ – 6.308 Cos $[x4]$ Sin $[x1]$ + 63.269 Cos $[x5]$ Sin $[x1]$ + 2.345 Cos $[x6]$ Sin $[x1]$ + 193.842 Sin $[x1]^2$ + 29.119 Sin $[x2]^2$ + 390.621 Sin $[x3]$ + 44.528 Cos $[x1]$ Sin $[x3]$ + 19.210 Cos $[x2]$ Sin $[x3]$ – 124.058 Cos $[x3]$ Sin $[x3]$ – 10.645 Cos $[x4]$ Sin $[x3]$ + 30.934 Cos $[x5]$ Sin $[x3]$ – 4.979 Cos $[x6]$ Sin $[x3]$ – 55.121 Sin $[x3]^2$ + 145.677 Sin $[x4]$ – 15.636 Sin $[x4]^2$ + 1.108 Sin $[x5]$ + 31.613 Sin $[x5]^2$ + 35.596 Sin $[x6]$ + 1.207 Cos $[x1]$ Sin $[x6]$ – 0.712 Cos $[x2]$ Sin $[x6]$ – 5.950 Cos $[x3]$ Sin $[x6]$ – 1.895 Cos $[x4]$ Sin $[x6]$ – 1.078 Cos $[x5]$ Sin $[x6]$ – 0.218 Cos $[x6]$ Sin $[x6]$ + 0.432 Sin $[x1]$ Sin $[x6]$ – 10.820 Sin $[x2]$ Sin $[x6]$ – 16.374 Sin $[x3]$ Sin $[x6]$ – 9.205 Sin $[x4]$ Sin $[x6]$ + 0.050 Sin $[x5]$ Sin $[x6]$ + 26.785 Sin $[x6]^2$

In[•]:= (*3*)

Out[•]//NumberForm=

82.160 – 328.630 Cos $[x1]$ – 229.571 Cos $[x1]^2$ + 0.469 Cos $[x2]$ + 35.983 Cos $[x2]^2$ + 68.475 Cos $[x3]$ + 50.603 Cos $[x1]$ Cos $[x3]$ + 10.819 Cos $[x2]$ Cos $[x3]$ + 168.047 Cos $[x3]^2$ + 26.639 Cos $[x4]$ – 0.110 Cos $[x2]$ Cos $[x4]$ + 68.508 Cos $[x4]^2$ – 99.220 Cos $[x5]$ + 41.898 Cos $[x5]^2$ + 2.247 Cos $[x6]$ + 30.895 Cos $[x6]^2$ – 1039.240 Sin $[x1]$ + 276.025 Cos $[x1]$ Sin $[x1]$ – 16.353 Cos $[x2]$ Sin $[x1]$ + 113.384 Cos $[x3]$ Sin $[x1]$ – 10.059 Cos $[x4]$ Sin $[x1]$ + 63.231 Cos $[x5]$ Sin $[x1]$ + 3.406 Cos $[x6]$ Sin $[x1]$ + 278.868 Sin $[x1]^2$ + 36.390 Sin $[x2]^2$ + 574.320 Sin $[x3]$ + 50.104 Cos $[x1]$ Sin $[x3]$ + 15.778 Cos $[x2]$ Sin $[x3]$ – 184.187 Cos $[x3]$ Sin $[x3]$ – 15.331 Cos $[x4]$ Sin $[x3]$ + 35.173 Cos $[x5]$ Sin $[x3]$ – 5.864 Cos $[x6]$ Sin $[x3]$ – 58.042 Sin $[x3]^2$ + 152.967 Sin $[x4]$ – 10.734 Sin $[x4]^2$ + 1.127 Sin $[x5]$ + 45.639 Sin $[x5]^2$ + 32.091 Sin $[x6]$ + 1.566 Cos $[x1]$ Sin $[x6]$ – 0.334 Cos $[x2]$ Sin $[x6]$ – 5.500 Cos $[x3]$ Sin $[x6]$ – 1.819 Cos $[x4]$ Sin $[x6]$ – 1.009 Cos $[x5]$ Sin $[x6]$ – 0.261 Cos $[x6]$ Sin $[x6]$ + 2.108 Sin $[x1]$ Sin $[x6]$ – 10.285 Sin $[x2]$ Sin $[x6]$ – 16.310 Sin $[x3]$ Sin$[x6]$ – 8.140 Sin $[x4]$ Sin $[x6]$ + 0.055 Sin $[x5]$ Sin $[x6]$ + 30.715 Sin $[x6]^2$

In[•]:= (*4*)

Out[•]//NumberForm=

74.249 – 228.526 Cos $[x1]$ – 167.394 Cos $[x1]^2$ – 11.188 Cos $[x2]$ + 30.061 Cos $[x2]^2$ – 28.656 Cos $[x3]$ + 44.288 Cos $[x1]$ Cos $[x3]$ + 13.534 Cos $[x2]$ Cos $[x3]$ + 97.041 Cos $[x3]^2$ + 27.887 Cos $[x4]$ + 0.657 Cos $[x2]$ Cos $[x4]$ + 69.802 Cos $[x4]^2$ – 103.379 Cos $[x5]$ + 34.572 Cos $[x5]^2$ +

(Continued)

TABLE A7.6 *(Continued)*

Models Obtained by NRM and SNRM Using the Cross-Validation Data Separation Method (*(10 K-fold Cross-Validation: 80 Training–20 Testing)*) (*Training Was Divided into 10 Parts: 9 Training–1 Validation*) (*Models Obtained Using the NRM and SNRM Methods*)(*(Related to Table 7.7)*)

2.422 Cos [x6] + 37.938 Cos [x6]^2 – 716.950 Sin [x1] + 188.937 Cos [x1] Sin [x1] –
8.058 Cos [x2] Sin [x1] + 102.464 Cos [x3] Sin [x1] – 11.879 Cos [x4] Sin [x1] +
73.211 Cos [x5] Sin [x1] + 1.946 Cos [x6] Sin [x1] + 202.218 Sin [x1]^2 + 30.280 Sin [x2]^2 +
317.730 Sin [x3] + 40.197 Cos [x1] Sin [x3] + 18.459 Cos [x2] Sin [x3] – 81.542 Cos [x3] Sin [x3] –
14.696 Cos [x4] Sin [x3] + 27.850 Cos [x5] Sin [x3] – 4.598 Cos [x6] Sin [x3] – 32.467 Sin [x3]^2 +
146.940 Sin [x4] – 6.295 Sin [x4]^2 + 1.127 Sin [x5] + 39.350 Sin [x5]^2 + 36.665 Sin [x6] +
0.112 Cos [x1] Sin [x6] – 0.480 Cos [x2] Sin [x6] – 4.857 Cos [x3] Sin [x6] – 1.755 Cos [x4] Sin [x6] –
1.093 Cos [x5] Sin [x6] – 0.244 Cos [x6] Sin [x6] – 3.692 Sin [x1] Sin [x6] – 11.724 Sin [x2] Sin [x6] –
14.870 Sin [x3] Sin [x6] – 6.951 Sin [x4] Sin [x6] + 0.088 Sin [x5] Sin [x6] + 37.779 Sin [x6]^2

*In[•]:= (*5*)*

Out[•]//NumberForm=

68.598 – 306.929 Cos [x1] – 217.964 Cos [x1]^2 – 0.932 Cos [x2] + 37.398 Cos [x2]^2 + 77.842 Cos [x3] +
51.430 Cos [x1] Cos [x3] + 8.705 Cos [x2] Cos [x3] + 130.467 Cos [x3]^2 + 22.552 Cos [x4] +
2.216 Cos [x2] Cos [x4] + 69.492 Cos [x4]^2 – 78.827 Cos [x5] + 35.087 Cos [x5]^2 + 3.247 Cos [x6] +
45.507 Cos [x6]^2 – 972.454 Sin [x1] + 259.593 Cos [x1] Sin [x1] – 10.221 Cos [x2] Sin [x1] +
123.195 Cos [x3] Sin [x1] – 10.026 Cos [x4] Sin [x1] + 50.717 Cos [x5] Sin [x1] +
2.225 Cos [x6] Sin [x1] + 252.199 Sin [x1]^2 + 37.182 Sin [x2]^2 + 545.357 Sin [x3] +
44.853 Cos [x1] Sin [x3] + 11.162 Cos [x2] Sin [x3] – 197.960 Cos [x3] Sin [x3] –
11.333 Cos [x4] Sin [x3] + 31.486 Cos [x5] Sin [x3] – 5.759 Cos [x6] Sin [x3] – 69.839 Sin [x3]^2 +
171.323 Sin [x4] – 19.893 Sin [x4]^2 + 1.173 Sin [x5] + 36.071 Sin [x5]^2 + 44.156 Sin [x6] +
1.094 Cos [x1] Sin [x6] – 0.399 Cos [x2] Sin [x6] – 8.291 Cos [x3] Sin [x6] – 2.060 Cos [x4] Sin [x6] –
0.655 Cos [x5] Sin [x6] – 0.218 Cos [x6] Sin [x6] – 1.606 Sin [x1] Sin [x6] – 10.950 Sin [x2] Sin [x6] –
23.158 Sin [x3] Sin [x6] – 8.114 Sin [x4] Sin [x6] – 0.076 Sin [x5] Sin [x6] + 45.282 Sin [x6]^2

*In[•]:= (*6*)*

Out[•]//NumberForm=

71.208 – 233.848 Cos [x1] – 165.288 Cos [x1]^2 – 16.018 Cos [x2] + 37.694 Cos [x2]^2 + 3.638 Cos [x3] +
43.176 Cos [x1] Cos [x3] + 13.499 Cos [x2] Cos [x3] + 107.028 Cos [x3]^2 + 25.173 Cos [x4] +
0.829 Cos [x2] Cos [x4] + 71.898 Cos [x4]^2 – 72.007 Cos [x5] + 31.174 Cos [x5]^2 + 2.346 Cos [x6] +
37.664 Cos [x6]^2 – 766.672 Sin [x1] + 200.736 Cos [x1] Sin [x1] – 5.840 Cos [x2] Sin [x1] +
107.580 Cos [x3] Sin [x1] – 10.444 Cos [x4] Sin [x1] + 52.597 Cos [x5] Sin [x1] +
2.463 Cos [x6] Sin [x1] + 215.071 Sin [x1]^2 + 37.460 Sin [x2]^2 + 380.461 Sin [x3] +
33.280 Cos [x1] Sin [x3] + 21.560 Cos [x2] Sin [x3] – 115.112 Cos [x3] Sin [x3] –
13.406 Cos [x4] Sin [x3] + 26.205 Cos [x5] Sin [x3] – 4.998 Cos [x6] Sin [x3] – 41.794 Sin [x3]^2 +
124.982 Sin [x4] + 7.439 Sin [x4]^2 + 1.054 Sin [x5] + 30.214 Sin [x5]^2 + 37.618 Sin [x6] +
0.764 Cos [x1] Sin [x6] – 0.648 Cos [x2] Sin [x6] – 5.385 Cos [x3] Sin [x6] – 1.919 Cos [x4] Sin [x6] –
0.746 Cos [x5] Sin [x6] – 0.258 Cos [x6] Sin [x6] – 1.179 Sin [x1] Sin [x6] – 13.266 Sin [x2] Sin [x6] –
15.453 Sin [x3] Sin [x6] – 7.879 Sin [x4] Sin [x6] + 0.008 Sin [x5] Sin [x6] + 37.525 Sin [x6]^2

(Continued)

TABLE A7.6 *(Continued)*

Models Obtained by NRM and SNRM Using the Cross-Validation Data Separation Method (*(10 K-fold Cross-Validation: 80 Training–20 Testing)*) (*Training Was Divided into 10 Parts: 9 Training–1 Validation*) (*Models Obtained Using the NRM and SNRM Methods*)(*(Related to Table 7.7)*)

In[•]:= (*7*)

Out[•]//NumberForm=

120.494 – 447.944 Cos [x1] – 296.012 Cos [x1]² – 27.452 Cos [x2] + 56.927 Cos [x2]² +
120.692 Cos [x3] + 56.046 Cos [x1] Cos [x3] + 18.490 Cos [x2] Cos [x3] + 209.924 Cos [x3]² +
35.077 Cos [x4] + 2.783 Cos [x2] Cos [x4] + 97.311 Cos [x4]² – 88.363 Cos [x5] + 51.140 Cos [x5]² +
2.805 Cos [x6] + 46.447 Cos [x6]² – 1428.020 Sin [x1] + 388.654 Cos [x1] Sin [x1] –
11.313 Cos [x2] Sin [x1] + 129.903 Cos [x3] Sin [x1] – 19.892 Cos [x4] Sin [x1] +
59.166 Cos [x5] Sin [x1] + 2.769 Cos [x6] Sin [x1] + 382.465 Sin [x1]² + 56.949 Sin [x2]² +
794.129 Sin [x3] + 53.250 Cos [x1] Sin [x3] + 37.378 Cos [x2] Sin [x3] – 251.939 Cos [x3] Sin [x3] –
14.769 Cos [x4] Sin [x3] + 32.231 Cos [x5] Sin [x3] – 5.833 Cos [x6] Sin [x3] – 114.922 Sin [x3]² +
184.484 Sin [x4] + 0.332 Sin [x4]² + 1.090 Sin [x5] + 52.493 Sin [x5]² + 39.330 Sin [x6] +
2.110 Cos [x1] Sin [x6] – 0.571 Cos [x2] Sin [x6] – 7.364 Cos [x3] Sin [x6] – 2.356 Cos [x4] Sin [x6] –
1.938 Cos [x5] Sin [x6] – 0.270 Cos [x6] Sin [x6] + 2.679 Sin [x1] Sin [x6] – 11.397 Sin [x2] Sin [x6] –
21.578 Sin [x3] Sin [x6] – 10.300 Sin [x4] Sin [x6] + 0.270 Sin [x5] Sin [x6] + 46.247 Sin [x6]²

In[•]:= (*8*)

Out[•]//NumberForm=

38.359 – 226.249 Cos [x1] – 177.007 Cos [x1]² + 3.073 Cos [x2] + 40.714 Cos [x2]² + 4.374 Cos [x3] +
47.646 Cos [x1] Cos [x3] + 9.885 Cos [x2] Cos [x3] + 103.049 Cos [x3]² + 24.215 Cos [x4] +
0.888 Cos [x2] Cos [x4] + 76.139 Cos [x4]² – 97.758 Cos [x5] + 25.868 Cos [x5]² + 2.930 Cos [x6] +
28.356 Cos [x6]² – 756.919 Sin [x1] + 194.224 Cos [x1] Sin [x1] – 9.602 Cos [x2] Sin [x1] +
121.209 Cos [x3] Sin [x1] – 10.688 Cos [x4] Sin [x1] + 69.949 Cos [x5] Sin [x1] +
2.363 Cos [x6] Sin [x1] + 208.512 Sin [x1]² + 40.457 Sin [x2]² + 395.262 Sin [x3] +
31.730 Cos [x1] Sin [x3] + 6.095 Cos [x2] Sin [x3] – 128.397 Cos [x3] Sin [x3] –
12.315 Cos [x4] Sin [x3] + 29.913 Cos [x5] Sin [x3] – 5.560 Cos [x6] Sin [x3] – 45.084 Sin [x3]² +
169.964 Sin [x4] – 12.484 Sin [x4]² + 1.037 Sin [x5] + 28.243 Sin [x5]² + 37.739 Sin [x6] +
1.538 Cos [x1] Sin [x6] – 0.734 Cos [x2] Sin [x6] – 5.907 Cos [x3] Sin [x6] – 1.949 Cos [x4] Sin [x6] –
0.930 Cos [x5] Sin [x6] – 0.232 Cos [x6] Sin [x6] + 0.650 Sin [x1] Sin [x6] – 12.725 Sin [x2] Sin [x6] –
16.780 Sin [x3] Sin [x6] – 9.270 Sin [x4] Sin [x6] – 0.053 Sin [x5] Sin [x6] + 28.172 Sin [x6]²

In[•]:= (*9*)

Out[•]//NumberForm=

65.667 – 284.963 Cos [x1] – 189.808 Cos [x1]² – 13.135 Cos [x2] + 38.082 Cos [x2]² + 15.788 Cos [x3] +
51.335 Cos [x1] Cos [x3] + 12.935 Cos [x2] Cos [x3] + 123.628 Cos [x3]² + 23.724 Cos [x4] +
0.024 Cos [x2] Cos [x4] + 82.428 Cos [x4]² – 80.055 Cos [x5] + 35.585 Cos [x5]² + 2.689 Cos [x6] +
35.015 Cos [x6]² – 877.221 Sin [x1] + 232.906 Cos [x1] Sin [x1] – 8.942 Cos [x2] Sin [x1] +
118.820 Cos [x3] Sin [x1] – 10.935 Cos [x4] Sin [x1] + 53.798 Cos [x5] Sin [x1] +
2.692 Cos [x6] Sin [x1] + 240.603 Sin [x1]² + 38.476 Sin [x2]² + 458.731 Sin [x3] +
50.910 Cos [x1] Sin [x3] + 22.066 Cos [x2] Sin [x3] – 139.512 Cos [x3] Sin [x3] –
11.366 Cos [x4] Sin [x3] + 30.974 Cos [x5] Sin [x3] – 5.579 Cos [x6] Sin [x3] – 59.798 Sin [x3]² +
157.343 Sin [x4] + 0.598 Sin [x4]² + 1.159 Sin [x5] + 35.973 Sin [x5]² + 44.159 Sin [x6] –

(Continued)

TABLE A7.6 *(Continued)*

Models Obtained by NRM and SNRM Using the Cross-Validation Data Separation Method (*(10 K-fold Cross-Validation: 80 Training–20 Testing)*) (*Training Was Divided into 10 Parts: 9 Training–1 Validation*) (*Models Obtained Using the NRM and SNRM Methods*)(*(Related to Table 7.7)*)

0.204 Cos [x1] Sin [x6] – 0.777 Cos [x2] Sin [x6] – 5.723 Cos [x3] Sin [x6] – 2.158 Cos [x4] Sin [x6] – 0.637 Cos [x5] Sin [x6] – 0.264 Cos [x6] Sin [x6] – 5.119 Sin [x1] Sin [x6] – 12.610 Sin [x2] Sin [x6] – 16.743 Sin [x3] Sin [x6] – 9.648 Sin [x4] Sin [x6] – 0.049 Sin [x5] Sin [x6] + 34.869 Sin [x6]^2

In[•]:= (*10*)

Out[•]//NumberForm=

68.660 – 297.282 Cos [x1] – 199.968 Cos [x1]^2 – 4.297 Cos [x2] + 44.250 Cos [x2]^2 + 2.911 Cos [x3] + 57.593 Cos [x1] Cos [x3] + 13.883 Cos [x2] Cos [x3] + 140.659 Cos [x3]^2 + 29.250 Cos [x4] + 0.876 Cos [x2] Cos [x4] + 78.392 Cos [x4]^2 – 78.210 Cos [x5] + 31.167 Cos [x5]^2 + 1.803 Cos [x6] + 36.102 Cos [x6]^2 – 900.757 Sin [x1] + 226.788 Cos [x1] Sin [x1] – 17.213 Cos [x2] Sin [x1] + 110.943 Cos [x3] Sin [x1] – 12.338 Cos [x4] Sin [x1] + 51.169 Cos [x5] Sin [x1] + 2.813 Cos [x6] Sin [x1] + 244.536 Sin [x1]^2 + 44.850 Sin [x2]^2 + 475.688 Sin [x3] + 68.305 Cos [x1] Sin [x3] + 20.190 Cos [x2] Sin [x3] – 125.629 Cos [x3] Sin [x3] – 16.229 Cos [x4] Sin [x3] + 26.495 Cos [x5] Sin [x3] – 4.769 Cos [x6] Sin [x3] – 67.604 Sin [x3]^2 + 168.317 Sin [x4] – 10.158 Sin [x4]^2 + 0.996 Sin [x5] + 34.428 Sin [x5]^2 + 35.176 Sin [x6] + 2.040 Cos [x1] Sin [x6] – 0.922 Cos [x2] Sin [x6] – 6.089 Cos [x3] Sin [x6] – 1.946 Cos [x4] Sin [x6] – 1.524 Cos [x5] Sin [x6] – 0.240 Cos [x6] Sin [x6] + 2.869 Sin [x1] Sin [x6] – 12.290 Sin [x2] Sin [x6] – 17.524 Sin [x3] Sin [x6] – 9.247 Sin [x4] Sin [x6] + 0.185 Sin [x5] Sin [x6] + 36.001 Sin [x6]^2

TABLE A7.7

Models Obtained by NRM and SNRM Using the Cross-Validation Data Separation Method(*(5 K-fold Cross-Validation: 80 Training–20 Testing)*) (*The Training Was Divided into 5 Parts: 4 Training–1 Validation*) (*Models Obtained Using the NRM and SNRM Methods*)(*(Related to Table 7.8)*)

(*Models Proposed by NRM*)

In[•]:= (*1*)

Out[•]//NumberForm=

94.402 + 191.922 Cos [x1] + 23.656 Cos [x1]^2 – 362.410 Cos [x2] + 108.733 Cos [x2]^2 – 312.831 Cos [x3] + 47.948 Cos [x1] Cos [x3] + 107.612 Cos [x2] Cos [x3] – 259.274 Cos [x3]^2 + 19.804 Cos [x4] – 7.455 Cos [x2] Cos [x4] + 40.946 Cos [x4]^2 – 222.535 Cos [x5] + 109.998 Cos [x5]^2 – 2.053 Cos [x6] + 124.161 Cos [x6]^2 – 24.498 Sin [x1] – 45.804 Cos [x1] Sin [x1] + 28.113 Cos [x2] Sin [x1] + 273.936 Cos [x3] Sin [x1] – 10.385 Cos [x4] Sin [x1] + 125.501 Cos [x5] Sin [x1] + 1.883 Cos [x6] Sin [x1] + 115.936 Sin [x1]^2 + 98.487 Sin [x2]^2 – 576.402 Sin [x3] – 137.789 Cos [x1] Sin [x3] + 324.506 Cos [x2] Sin [x3] + 49.385 Cos [x3] Sin [x3] – 4.409 Cos [x4] Sin [x3] + 79.168 Cos [x5] Sin [x3] + 0.199 Cos [x6] Sin [x3] + 125.731 Sin [x3]^2 –

(Continued)

TABLE A7.7 *(Continued)*

Models Obtained by NRM and SNRM Using the Cross-Validation Data Separation Method(*(5 K-fold Cross-Validation: 80 Training–20 Testing)*) (*The Training Was Divided into 5 Parts: 4 Training–1 Validation*) (*Models Obtained Using the NRM and SNRM Methods*)(*(Related to Table 7.8)*)

100.671 Sin [$x4$] + 103.275 Sin [$x4$]^2 +1.512 Sin [$x5$] + 126.082 Sin [$x5$]^2 + 79.602 Sin [$x6$] – 9.342 Cos [$x1$] Sin [$x6$] – 0.264 Cos [$x2$] Sin [$x6$] – 14.607 Cos [$x3$] Sin [$x6$] – 2.022 Cos [$x4$] Sin [$x6$] – 0.077 Cos [$x5$] Sin [$x6$] + 0.098 Cos [$x6$] Sin [$x6$] – 36.340 Sin [$x1$] Sin [$x6$] + 6.149 Sin [$x2$] Sin [$x6$] – 43.533 Sin [$x3$] Sin [$x6$] – 2.489 Sin [$x4$] Sin [$x6$] – 0.253 Sin [$x5$] Sin [$x6$] + 123.740 Sin [$x6$]^2

In[•]:= (*2*)

Out[•]//NumberForm=

1510.100 – 1451.450 Cos [$x1$] – 746.795 Cos [$x1$]^2 – 168.108 Cos [$x2$] + 1577.820 Cos [$x2$]^2 – 2464.460 Cos [$x3$] – 95.777 Cos [$x1$] Cos [$x3$] + 70.516 Cos [$x2$] Cos [$x3$] – 819.710 Cos [$x3$]^2 + 89.633 Cos [$x4$] + 14.292 Cos [$x2$] Cos [$x4$] + 1924.880 Cos [$x4$]^2 + 276.830 Cos [$x5$] + 1787.230 Cos [$x5$]^2 + 14.284 Cos [$x6$] + 2011.480 Cos [$x6$]^2 – 5910.320 Sin [$x1$] + 1738.220 Cos[$x1$] Sin [$x1$] – 23.035 Cos[$x2$] Sin [$x1$] + 28.949 Cos [$x3$] Sin [$x1$] – 91.551 Cos [$x4$] Sin [$x1$] – 67.464 Cos [$x5$] Sin [$x1$] – 4.486 Cos [$x6$] Sin [$x1$] + 1853.050 Sin [$x1$]^2 + 1577.350 Sin [$x2$]^2 – 6827.020 Sin [$x3$] – 306.197 Cos [$x1$] Sin [$x3$] + 180.072 Cos [$x2$] Sin [$x3$] + 2333.250 Cos [$x3$] Sin [$x3$] – 2.459 Cos [$x4$] Sin [$x3$] – 163.376 Cos [$x5$] Sin [$x3$] – 10.896 Cos [$x6$] Sin [$x3$] + 1958.090 Sin [$x3$]^2 + 558.596 Sin [$x4$] + 1623.220 Sin [$x4$]^2 + 0.500 Sin [$x5$] + 1756.750 Sin [$x5$]^2 + 79.471 Sin [$x6$] + 8.420 Cos [$x1$] Sin [$x6$] + 0.564 Cos [$x2$] Sin [$x6$] – 26.185 Cos [$x3$] Sin [$x6$] – 3.105 Cos [$x4$] Sin [$x6$] – 0.592 Cos [$x5$] Sin [$x6$] – 0.616 Cos [$x6$] Sin [$x6$] + 20.932 Sin [$x1$] Sin [$x6$] + 2.916 Sin [$x2$] Sin [$x6$] – 83.221 Sin [$x3$] Sin [$x6$] – 18.631 Sin [$x4$] Sin [$x6$] + 0.288 Sin [$x5$] Sin [$x6$] + 2011.780 Sin [$x6$]^2)

In[•]:= (*3*)

Out[•]//NumberForm=

–369.810 – 187.783 Cos [$x1$] – 449.394 Cos [$x1$]^2 – 64.455 Cos [$x2$] – 386.267 Cos [$x2$]^2 + 604.393 Cos [$x3$] + 158.291 Cos [$x1$] Cos [$x3$] + 39.891 Cos [$x2$] Cos [$x3$] + 733.620 Cos [$x3$]^2 + 36.268 Cos [$x4$] + 10.619 Cos [$x2$] Cos [$x4$] – 277.971 Cos [$x4$]^2 – 112.550 Cos [$x5$] 435.695 Cos [$x5$]^2 + 1.951 Cos [$x6$] – 482.459 Cos [$x6$]^2 + 6.355 Sin [$x1$] – 16.425 Cos [$x1$] Sin [$x1$] – 49.233 Cos [$x2$] Sin [$x1$] + 336.144 Cos [$x3$] Sin [$x1$] – 37.004 Cos [$x4$] Sin [$x1$] + 64.514 Cos [$x5$] Sin [$x1$] + 8.826 Cos [$x6$] Sin [$x1$] – 444.185 Sin [$x1$]^2 – 389.029 Sin [$x2$]^2 + 2855.130 Sin [$x3$] + 183.111 Cos [$x1$] Sin [$x3$] + 107.665 Cos [$x2$] Sin [$x3$] – 939.322 Cos [$x3$] Sin [$x3$] + 0.479 Cos [$x4$] Sin [$x3$] + 25.904 Cos [$x5$] Sin [$x3$] – 10.913 Cos [$x6$] Sin [$x3$] – 496.964 Sin [$x3$]^2 + 229.655 Sin [$x4$] – 402.141 Sin [$x4$]^2 + 0.874 Sin [$x5$] – 420.195 Sin [$x5$]^2 + 23.795 Sin [$x6$] + 11.170 Cos [$x1$] Sin [$x6$] – 1.036 Cos [$x2$] Sin [$x6$] – 9.671 Cos [$x3$] Sin [$x6$] – 5.049 Cos [$x4$] Sin [$x6$] – 0.715 Cos [$x5$] Sin [$x6$] – 0.593 Cos [$x6$] Sin [$x6$] + 39.170 Sin [$x1$] Sin [$x6$] – 3.149 Sin [$x2$] Sin [$x6$] – 25.638 Sin [$x3$] Sin [$x6$] – 34.683 Sin [$x4$] Sin [$x6$] – 0.681 Sin [$x5$] Sin [$x6$] – 482.328 Sin [$x6$]^2

In[●]:= (*4*)

Out[●]//NumberForm=

–396.845 + 546.044 Cos [$x1$] + 608.834 Cos [$x1$]^2 – 17.071 Cos [$x2$] – 418.526 Cos [$x2$]^2 – 126.847 Cos [$x3$] + 166.447 Cos [$x1$] Cos [$x3$] + 6.728 Cos [$x2$] Cos [$x3$] – 553.223 Cos [$x3$]^2 – 16.232 Cos [$x4$] + 9.528 Cos [$x2$] Cos [$x4$] – 172.982 Cos [$x4$]^2 – 380.069 Cos [$x5$] –

(Continued)

TABLE A7.7 *(Continued)*
Models Obtained by NRM and SNRM Using the Cross-Validation Data Separation Method(*(5 K-fold Cross-Validation: 80 Training–20 Testing)*) (*The Training Was Divided into 5 Parts: 4 Training–1 Validation*) (*Models Obtained Using the NRM and SNRM Methods*)(*(Related to Table 7.8)*)

468.412 Cos [x5]² + 5.511 Cos [x6] − 536.793 Cos [x6]² + 2846.520 Sin [x1] −
808.323 Cos [x1] Sin [x1] − 13.387 Cos [x2] Sin [x1] + 285.782 Cos [x3] Sin [x1] −
8.817 Cos [x4] Sin [x1] + 363.993 Cos [x5] Sin [x1] − 0.772 Cos [x6] Sin [x1] − 491.914 Sin [x1]² −
417.413 Sin [x2]² − 132.531 Sin [x3] + 264.916 Cos [x1] Sin [x3] + 30.229 Cos [x2] Sin [x3] −
168.230 Cos [x3] Sin [x3] + 26.914 Cos [x4] Sin [x3] − 114.385 Cos [x5] Sin [x3] −
5.225 Cos [x6] Sin [x3] − 490.926 Sin [x3]² + 478.743 Sin [x4] − 429.817 Sin [x4]² + 0.811 Sin [x5] −
390.655 Sin [x5]² + 67.373 Sin [x6] − 2.417 Cos [x1] Sin [x6] − 0.675 Cos [x2] Sin [x6] −
12.738 Cos [x3] Sin [x6] − 3.088 Cos [x4] Sin [x6] + 5.201 Cos [x5] Sin [x6] − 0.452 Cos [x6] Sin [x6] −
15.188 Sin [x1] Sin [x6] + 7.951 Sin [x2] Sin [x6] − 28.414 Sin [x3] Sin [x6] − 24.083 Sin [x4] Sin [x6] −
1.877 Sin [x5] Sin [x6] − 536.801 Sin [x6]²

In[•]:= (*5*)
Out[•]///NumberForm=

214.240 − 478.079 Cos [x1] − 498.107 Cos [x1]² + 37.279 Cos [x2] + 225.341 Cos [x2]² −
145.883 Cos [x3] + 58.532 Cos [x1] Cos [x3] − 0.792 Cos [x2] Cos [x3] + 165.409 Cos [x3]² +
13.719 Cos [x4] + 0.296 Cos [x2] Cos [x4] + 299.630 Cos [x4]² − 68.563 Cos [x5] + 255.415 Cos [x5]² +
1.275 Cos [x6] + 289.166 Cos [x6]² − 1636.870 Sin [x1] + 389.441 Cos [x1] Sin [x1] −
35.630 Cos [x2] Sin [x1] + 100.356 Cos [x3] Sin [x1] − 12.869 Cos [x4] Sin [x1] +
9.479 Cos [x5] Sin [x1] + 2.151 Cos [x6] Sin [x1] + 271.985 Sin [x1]² + 226.606 Sin [x2]² −
174.805 Sin [x3] + 83.293 Cos [x1] Sin [x3] − 0.057 Cos [x2] Sin [x3] + 34.866 Cos [x3] Sin [x3] +
2.481 Cos [x4] Sin [x3] + 49.642 Cos [x5] Sin [x3] − 3.647 Cos [x6] Sin [x3] + 269.690 Sin [x3]² +
131.746 Sin [x4] + 234.102 Sin [x4]² + 1.209 Sin [x5] + 262.102 Sin [x5]² + 7.742 Sin [x6] +
4.371 Cos [x1] Sin [x6] + 0.275 Cos [x2] Sin [x6] − 6.853 Cos [x3] Sin [x6] − 0.945 Cos [x4] Sin [x6] −
4.683 Cos [x5] Sin [x6] − 0.106 Cos [x6] Sin [x6] + 15.806 Sin [x1] Sin [x6] − 2.477 Sin [x2] Sin [x6] −
23.206 Sin [x3] Sin [x6] − 1.936 Sin [x4] Sin [x6] + 0.643 Sin [x5] Sin [x6] + 289.122 Sin [x6]²

In[•]:= (*1*)
Out[•]///NumberForm=

46.938 − 242.620 Cos [x1] − 181.614 Cos [x1]² + 3.932 Cos [x2] + 19.086 Cos [x2]² −
1.906 Cos [x3] + 56.901 Cos [x1] Cos [x3] + 10.931 Cos [x2] Cos [x3] + 130.340 Cos [x3]² +
25.333 Cos [x4] + 1.214 Cos [x2] Cos [x4] + 67.070 Cos [x4]² − 86.315 Cos [x5] + 17.427 Cos [x5]² +
1.497 Cos [x6] + 24.729 Cos [x6]² − 778.020 Sin [x1] + 182.070 Cos [x1] Sin [x1] −
16.289 Cos [x2] Sin [x1] + 118.087 Cos [x3] Sin [x1] − 11.930 Cos [x4] Sin [x1] +
59.378 Cos [x5] Sin [x1] + 2.733 Cos [x6] Sin [x1] + 213.996 Sin [x1]² + 19.481 Sin [x2]² +
495.563 Sin [x3] + 59.074 Cos [x1] Sin [x3] + 11.352 Cos [x2] Sin [x3] − 127.724 Cos [x3] Sin [x3] −
12.445 Cos [x4] Sin [x3] + 27.512 Cos [x5] Sin [x3] − 4.391 Cos [x6] Sin [x3] − 87.078 Sin [x3]² +
151.322 Sin [x4] − 12.451 Sin [x4]² + 1.130 Sin [x5] + 19.975 Sin [x5]² + 34.947 Sin [x6] +
1.867 Cos [x1] Sin [x6] − 0.949 Cos [x2] Sin [x6] − 5.206 Cos [x3] Sin [x6] − 1.978 Cos [x4] Sin [x6] −
1.805 Cos [x5] Sin [x6] − 0.188 Cos [x6] Sin [x6] + 3.035 Sin [x1] Sin [x6] − 13.289 Sin [x2] Sin [x6] −
15.055 Sin [x3] Sin [x6] − 11.072 Sin [x4] Sin [x6] + 0.212 Sin [x5] Sin [x6] + 24.626 Sin [x6]²

(Continued)

TABLE A7.7 *(Continued)*

Models Obtained by NRM and SNRM Using the Cross-Validation Data Separation Method(*(5 K-fold Cross-Validation: 80 Training–20 Testing)*) (*The Training Was Divided into 5 Parts: 4 Training–1 Validation*) (*Models Obtained Using the NRM and SNRM Methods*)(*(Related to Table 7.8)*)

$In[\bullet]:=$ (*2*)

$Out[\bullet]//NumberForm=$

$88.622 - 271.092$ Cos $[x1] - 177.968$ Cos $[x1]^2 + 8.121$ Cos $[x2] + 49.910$ Cos $[x2]^2 + 37.755$ Cos $[x3] + 38.462$ Cos $[x1]$ Cos $[x3] + 6.441$ Cos $[x2]$ Cos $[x3] + 115.175$ Cos $[x3]^2 + 20.526$ Cos $[x4] + 0.817$ Cos $[x2]$ Cos $[x4] + 82.785$ Cos $[x4]^2 - 64.130$ Cos $[x5] + 53.765$ Cos $[x5]^2 + 1.751$ Cos $[x6] + 45.555$ Cos $[x6]^2 - 913.188$ Sin $[x1] + 244.014$ Cos $[x1]$ Sin $[x1] - 7.250$ Cos $[x2]$ Sin $[x1] + 94.997$ Cos $[x3]$ Sin $[x1] - 9.984$ Cos $[x4]$ Sin $[x1] + 46.269$ Cos $[x5]$ Sin $[x1] + 3.227$ Cos $[x6]$ Sin $[x1] + 265.712$ Sin $[x1]^2 + 49.385$ Sin $[x2]^2 + 401.772$ Sin $[x3] + 27.128$ Cos $[x1]$ Sin $[x3] - 0.301$ Cos $[x2]$ Sin $[x3] - 133.156$ Cos $[x3]$ Sin $[x3] - 9.053$ Cos $[x4]$ Sin $[x3] + 28.583$ Cos $[x5]$ Sin $[x3] - 5.141$ Cos $[x6]$ Sin $[x3] - 33.847$ Sin $[x3]^2 + 124.646$ Sin $[x4] + 18.119$ Sin $[x4]^2 + 1.125$ Sin $[x5] + 50.459$ Sin $[x5]^2 + 33.842$ Sin $[x6] + 0.826$ Cos $[x1]$ Sin $[x6] - 0.787$ Cos $[x2]$ Sin $[x6] - 5.850$ Cos $[x3]$ Sin $[x6] - 1.315$ Cos $[x4]$ Sin $[x6] - 1.367$ Cos $[x5]$ Sin $[x6] - 0.248$ Cos $[x6]$ Sin $[x6] - 1.081$ Sin $[x1]$ Sin $[x6] - 12.720$ Sin $[x2]$ Sin $[x6] - 16.712$ Sin $[x3]$ Sin $[x6] - 4.019$ Sin $[x4]$ Sin $[x6] + 0.223$ Sin $[x5]$ Sin $[x6] + 45.356$ Sin $[x6]^2$

$In[\bullet]:=$ (*3*)

$Out[\bullet]//NumberForm=$

$58.189 - 234.985$ Cos $[x1] - 164.841$ Cos $[x1]^2 - 0.493$ Cos $[x2] + 35.030$ Cos $[x2]^2 + 19.987$ Cos $[x3] + 42.495$ Cos $[x1]$ Cos $[x3] + 8.681$ Cos $[x2]$ Cos $[x3] + 103.921$ Cos $[x3]^2 + 23.634$ Cos $[x4] + 1.732$ Cos $[x2]$ Cos $[x4] + 63.140$ Cos $[x4]^2 - 93.478$ Cos $[x5] + 27.716$ Cos $[x5]^2 + 3.091$ Cos $[x6] + 29.824$ Cos $[x6]^2 - 740.689$ Sin $[x1] + 199.648$ Cos $[x1]$ Sin $[x1] - 8.005$ Cos $[x2]$ Sin $[x1] + 102.739$ Cos $[x3]$ Sin $[x1] - 9.448$ Cos $[x4]$ Sin $[x1] + 63.903$ Cos $[x5]$ Sin $[x1] + 1.512$ Cos $[x6]$ Sin $[x1] + 209.339$ Sin $[x1]^2 + 35.045$ Sin $[x2]^2 + 361.410$ Sin $[x3] + 36.039$ Cos $[x1]$ Sin $[x3] + 8.414$ Cos $[x2]$ Sin $[x3] - 122.929$ Cos $[x3]$ Sin $[x3] - 12.678$ Cos $[x4]$ Sin $[x3] + 28.829$ Cos $[x5]$ Sin $[x3] - 4.894$ Cos $[x6]$ Sin $[x3] - 25.783$ Sin $[x3]^2 + 146.207$ Sin $[x4] - 12.277$ Sin $[x4]^2 + 1.190$ Sin $[x5] + 31.351$ Sin $[x5]^2 + 43.501$ Sin $[x6] - 0.221$ Cos $[x1]$ Sin $[x6] - 0.324$ Cos $[x2]$ Sin $[x6] - 6.978$ Cos $[x3]$ Sin $[x6] - 1.834$ Cos $[x4]$ Sin $[x6] - 0.727$ Cos $[x5]$ Sin $[x6] - 0.225$ Cos $[x6]$ Sin $[x6] - 5.808$ Sin $[x1]$ Sin $[x6] - 11.057$ Sin $[x2]$ Sin $[x6] - 20.032$ Sin $[x3]$ Sin $[x6] - 6.407$ Sin $[x4]$ Sin $[x6] - 0.034$ Sin $[x5]$ Sin $[x6] + 29.599$ Sin $[x6]^2)$

$In[\bullet]:=$ (*4*)

$Out[\bullet]//NumberForm=$

$93.258 - 310.805$ Cos $[x1] - 196.282$ Cos $[x1]^2 - 28.141$ Cos $[x2] + 55.323$ Cos $[x2]^2 - 13.894$ Cos $[x3] + 50.601$ Cos $[x1]$ Cos $[x3] + 17.002$ Cos $[x2]$ Cos $[x3] + 129.196$ Cos $[x3]^2 + 29.912$ Cos $[x4] + 2.225$ Cos $[x2]$ Cos $[x4] + 97.166$ Cos $[x4]^2 - 78.161$ Cos $[x5] + 56.089$ Cos $[x5]^2 + 2.668$ Cos $[x6] + 62.086$ Cos $[x6]^2 - 977.280$ Sin $[x1] + 264.948$ Cos $[x1]$ Sin $[x1] - 5.296$ Cos $[x2]$ Sin $[x1] + 123.690$ Cos $[x3]$ Sin $[x1] - 15.214$ Cos $[x4]$ Sin $[x1] + 56.189$ Cos $[x5]$ Sin $[x1] + 2.255$ Cos $[x6]$ Sin $[x1] + 276.832$ Sin $[x1]^2 + 54.932$ Sin $[x2]^2 + 391.349$ Sin $[x3] + 42.978$ Cos $[x1]$ Sin $[x3] + 32.656$ Cos $[x2]$ Sin $[x3] - 115.398$ Cos $[x3]$ Sin $[x3] - 13.677$ Cos $[x4]$ Sin $[x3] + 30.119$ Cos $[x5]$ Sin $[x3] - 5.154$ Cos $[x6]$ Sin $[x3] - 25.632$ Sin $[x3]^2 + 144.871$ Sin $[x4] + 22.157$ Sin $[x4]^2 + 1.012$ Sin $[x5] + 54.523$ Sin $[x5]^2 + 38.669$ Sin $[x6] +$

(Continued)

TABLE A7.7 *(Continued)*

Models Obtained by NRM and SNRM Using the Cross-Validation Data Separation Method(*(5 K-fold Cross-Validation: 80 Training–20 Testing)*) (*The Training Was Divided into 5 Parts: 4 Training–1 Validation*) (*Models Obtained Using the NRM and SNRM Methods*)(*(Related to Table 7.8)*)

0.970 Cos $[x1]$ Sin $[x6] - 0.722$ Cos $[x2]$ Sin $[x6] - 5.643$ Cos $[x3]$ Sin $[x6] - 2.205$ Cos $[x4]$ Sin $[x6] - 1.572$ Cos $[x5]$ Sin $[x6] - 0.278$ Cos $[x6]$ Sin $[x6] - 0.969$ Sin $[x1]$ Sin $[x6] - 12.835$ Sin $[x2]$ Sin $[x6] - 16.658$ Sin $[x3]$ Sin $[x6] - 9.222$ Sin $[x4]$ Sin $[x6] + 0.225$ Sin $[x5]$ Sin $[x6] + 61.894$ Sin $[x6]^2$

In[•]:= (*5*)

Out[•]//NumberForm=

$53.756 - 262.784$ Cos $[x1] - 194.484$ Cos $[x1]^2 - 4.026$ Cos $[x2] + 25.986$ Cos $[x2]^2 + 31.525$ Cos $[x3] + 49.218$ Cos $[x1]$ Cos $[x3] + 11.185$ Cos $[x2]$ Cos $[x3] + 116.966$ Cos $[x3]^2 + 23.837$ Cos $[x4] + 0.101$ Cos $[x2]$ Cos $[x4] + 77.493$ Cos $[x4]^2 - 88.388$ Cos $[x5] + 27.970$ Cos $[x5]^2 + 3.157$ Cos $[x6] + 36.819$ Cos $[x6]^2 - 859.538$ Sin $[x1] + 226.093$ Cos $[x1]$ Sin $[x1] - 8.603$ Cos $[x2]$ Sin $[x1] + 126.042$ Cos $[x3]$ Sin $[x1] - 11.088$ Cos $[x4]$ Sin $[x1] + 62.782$ Cos $[x5]$ Sin $[x1] + 2.570$ Cos $[x6]$ Sin $[x1] + 232.800$ Sin $[x1]^2 + 26.079$ Sin $[x2]^2 + 489.320$ Sin $[x3] + 36.017$ Cos $[x1]$ Sin $[x3] + 12.535$ Cos $[x2]$ Sin $[x3] - 159.473$ Cos $[x3]$ Sin $[x3] - 11.403$ Cos $[x4]$ Sin $[x3] + 29.187$ Cos $[x5]$ Sin $[x3] - 5.977$ Cos $[x6]$ Sin $[x3] - 72.486$ Sin $[x3]^2 + 169.964$ Sin $[x4] - 11.066$ Sin $[x4]^2 + 1.086$ Sin $[x5] + 29.309$ Sin $[x5]^2 + 43.634$ Sin $[x6] + 0.656$ Cos $[x1]$ Sin $[x6] - 0.866$ Cos $[x2]$ Sin $[x6] - 5.966$ Cos $[x3]$ Sin $[x6] - 2.156$ Cos $[x4]$ Sin $[x6] - 0.528$ Cos $[x5]$ Sin $[x6] - 0.260$ Cos $[x6]$ Sin $[x6] - 2.770$ Sin $[x1]$ Sin $[x6] - 13.291$ Sin $[x2]$ Sin $[x6] - 17.186$ Sin $[x3]$ Sin $[x6] - 10.314$ Sin $[x4]$ Sin $[x6] - 0.166$ Sin $[x5]$ Sin $[x6] + 36.657$ Sin $[x6]^2$

8 Comparison of Different Model Assessment Criteria Used in Mathematical Modeling

This chapter compares various fit indices employed as model assessment criteria. To accomplish this, we examined the most prevalent model assessment criteria found in the literature and utilized them to evaluate the performance of mathematical models proposed for a particular problem. Table 8.1 enumerates the most frequently utilized model assessment criteria.

When assessing model fit indices, the proximity of values for criteria such as R-squared (R^2), adjusted R^2, and Kling Gupta Efficiency (KGE) to 1 indicates a more successful model. Conversely, criteria such as AIC, BIC, and AICC deem a model successful based on the smallness of the obtained value. Additionally, other assessment criteria evaluate the success of the model based on its ability to closely estimate actual values. A smaller error value signifies a more successful model.

CASE STUDY 8.1

In a study [3], a laminated composite problem was taken as a reference for mathematical modeling. It has an eight-layered symmetric hybrid structure containing glass/epoxy and flax/epoxy materials. The properties of these materials are given in Table 8.2.

The hybrid laminated composite considered within the scope of the study is shown in Figure 8.1.

The design parameters selected for a laminated composite material are the fiber angle and the number of glass/epoxy layers, with the objective of maximizing the difference between natural frequency modes. This choice of a laminated composite-based problem is motivated by several factors. First, the field of laminated composites boasts diverse applications and attracts interest from various sectors. Second, it enables the utilization of analytical formulas to generate data, facilitating the creation of datasets for testing mathematical models. This approach allows for the accumulation of sufficient data to assess the efficacy of the proposed models effectively.

The task of determining the natural frequency gap for a glass-flax/epoxy hybrid composite has been addressed by employing the Classical Lamination Theory. The reference study offers comprehensive information and assumptions regarding this theory. To compute the natural frequency gap using the analytical formula

DOI: 10.1201/9781003494843-8

TABLE 8.1
Model Assessment Criteria (Goodness of Fit Index) [1, 2]

Symbol	Criteria
R^2	$1 - \dfrac{SSE}{SST} \ldots\ldots\ldots\ldots 1 - \dfrac{\sum_{i=1}^{n}(O_i - P_i)^2}{\sum_{i=1}^{n}(O_i - \bar{O})^2}$
R^2 adj.	$1 - \dfrac{(n-1)(1-R^2)}{n-p}$
SE	$\left[\dfrac{1}{n}\sum_{i=1}^{n}\lvert O_i - P_i\rvert^2\right]^{1/2}$
MSE	$\dfrac{1}{n}\sum_{i=1}^{n}(O_i - P_i)^2$
RMSE	$\left[\dfrac{1}{n}\sum_{i=1}^{n}(O_i - P_i)^2\right]^{0.5}$
RMSPE	$\left[\dfrac{1}{n}\sum_{i=1}^{n}\left(\dfrac{P_i - O_i}{P_i}100\right)^2\right]^{0.5}$
RRMS	$\dfrac{RMSE}{\bar{O}}100$
SAE	$\sum_{i=1}^{n}\lvert P_i - O_i\rvert$
MAE	$\dfrac{1}{n}\sum_{i=1}^{n}\lvert P_i - O_i\rvert$
SRMSE	$\dfrac{1}{\bar{O}}\sqrt{\dfrac{1}{n}\sum_{i=1}^{n}(P_i - O_i)^2}$
MAPE	$\dfrac{1}{n}\sum_{i=1}^{n}\dfrac{\lvert P_i - O_i\rvert}{P_i}100$
SEP	$\dfrac{1}{\bar{O}}\sqrt{\dfrac{1}{n}\sum_{i=1}^{n}(O_i - P_i)^2}$
KGE	$1 - \sqrt{(R-1)^2 + (BJ-1)^2 + (AJ-1)^2}$ $R = \sqrt{1 - \dfrac{\sum_{i=1}^{n}(O_i - P_i)^2}{\sum_{i=1}^{n}(P_i)^2}}$

(Continued)

TABLE 8.1 *(Continued)*

Model Assessment Criteria (Goodness of Fit Index) [1, 2]

Symbol	Criteria
	$$BJ = \dfrac{\sum_{i=1}^{n}(P_i)}{\sum_{i=1}^{n}(O_i)}$$
	$$AJ = \dfrac{\sqrt{\sum_{i=1}^{n}\left(P_i - \bar{P}\right)^2 \frac{1}{n}}}{\sqrt{\sum_{i=1}^{n}\left(O_i - \bar{O}\right)^2 \frac{1}{n}}}$$
U2theil	$$\sqrt{\dfrac{\sum_{i=1}^{n}(O_i - P_i)^2}{\sum_{i=1}^{n}O_i^2}}$$
MSRE	$$\left| \frac{1}{n}\sum_{i=1}^{n}\frac{(O_i - P_i)^2}{O_i^2} \right|$$
X2	$$tr\left(E^{-1}S - I\right) - \log\left|E^{-1}S\right| \quad\ldots\ldots\ldots\ldots\ldots\quad \sum_{i=1}^{n}\frac{(O_i - P_i)^2}{P_i}$$
AIC	$$n\ln\left(\frac{SSE}{n}\right) + 2\,p$$ $$SSE = \sum_{i=1}^{n}(O_i - P_i)^2$$
BIC	$$n\ln\left(\frac{SSE}{n}\right) + \frac{2\,(p+2)\,n\,s^2}{SSE} - \frac{2\,n^2 s^4}{SSE^2} \quad\ldots\ldots\ldots\ldots\ldots\quad n\ln\left(\frac{SSE}{n}\right) + p\,\ln(n)$$
AICC	$$AIC + \left(\frac{2p^2 + 2p}{n - p - 1}\right)$$

O: Observed, P: Predicted, n: number of data rows, p: number of model terms, SE: Standard Error, MSE: Mean Square Error, RMSE: Root Mean Square Error, RMSEP: Root Mean Square Percentage Error, RRMS: Relative Root Mean Square, SAE: Sum of Absolute Error, MAE: Mean Absolute Error, SRMSE: Scaled Root Mean Square Error, MAPE: Mean Absolute Percentage Error, KGE: Kling Gupta Efficiency, MSRE: Mean Square Relative Error, X2: Chi Square, AIC: Akaike Information Criterion, AICC: Corrected Akaike Information Criterion, BIC: Bayesian information Criterion.

grounded in composite material mechanics, it is imperative to establish the angle-dependent properties of the laminated composite. Subsequently, the transformed reduced stiffness matrix and engineering constants are determined accordingly. The formulas utilized for calculating the natural frequency gap and the elements of the [D] matrix are delineated in Chapter 6, specifically Equations 6.1–6.5.

Table 8.3 lists the different levels that the fiber angles and the number of glass/epoxy layers can take when creating a test set. The fiber angles were chosen as

TABLE 8.2
Glass/Epoxy and Flax/Epoxy Composite Material Properties

Parameter	Glass/Epoxy [4]	Flax/Epoxy [4]
E1 (GPa)	34	22.8
E2 (GPa)	8.2	4.52
G12 (GPa)	4.5	1.96
V12	0.29	0.43
ρ (kg/m^3)	1910	1310

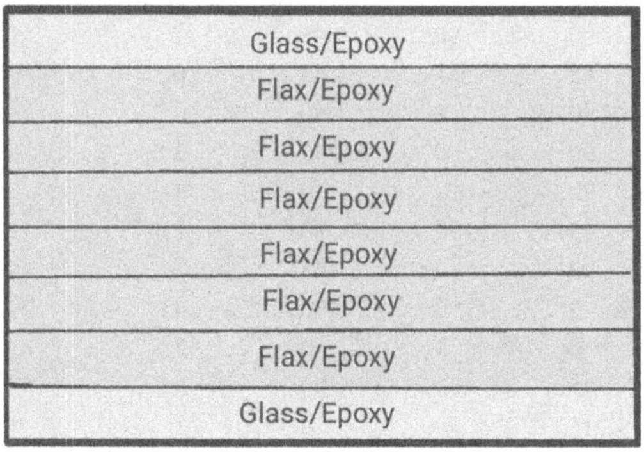

FIGURE 8.1   Hybrid laminated composite structure

TABLE 8.3
Design Parameters and Their Levels

	Level 1	Level 2	Level 3	Level 4	Level 5
x_1	0	45	90	–	–
x_2	0	45	90	–	–
x_3	0	45	90	–	–
x_4	0	45	90	–	–
x_5	0	1	2	3	4

0, 45, and 90 degrees, and the number of glass/epoxy layers was selected as 0, 1, 2, 3, and 4.

The experimental set was created using an optimization-based design method D-Optimal. As a result, the number of experiments required was reduced from 405 to 74 when D-optimal instead of Full Factorial design (FFD) was used. Table 8.4 presents the natural frequency gap results obtained using the experimental set and analytical formulas.

TABLE 8.4
Experiment Set Formed Using the D-Optimal Method

	x_1	x_2	x_3	x_4	x_5	Natural Frequency Gap
1	0	90	90	0	3	590.261
2	90	45	45	45	2	495.325
3	90	0	0	0	3	595.434
4	45	90	90	45	4	509.883
5	0	0	90	90	3	448.393
6	0	90	45	90	1	541.669
7	90	0	90	90	1	513.395
8	45	0	45	0	4	529.718
9	90	0	45	0	2	579.859
10	0	90	90	45	2	598.1
11	0	45	45	45	0	437.594
12	45	90	0	90	0	529.312
13	45	90	0	45	2	584.434
14	0	90	0	0	4	517.895
15	45	45	90	90	2	610.578
16	90	90	0	45	3	445.212
17	45	0	90	90	4	561.771
18	0	0	0	0	2	387.162
19	45	45	90	90	0	542.76
20	0	45	90	0	2	514.631
21	0	90	45	0	0	515.043
22	0	45	90	45	4	498.96
23	0	90	0	45	1	516.547
24	90	90	45	90	2	410.766
25	90	90	45	90	0	351.822
26	90	45	45	0	4	475.745
27	45	45	45	0	3	612.629
28	90	90	45	45	1	404.441
29	45	45	0	0	2	610.578
30	90	0	0	90	0	544.288
31	45	90	90	0	0	483.785
32	90	45	0	45	1	495.8
33	90	0	90	0	0	497.455

(Continued)

TABLE 8.4 *(Continued)*
Experiment Set Formed Using the D-Optimal Method

	x_1	x_2	x_3	x_4	x_5	Natural Frequency Gap
34	0	90	45	45	3	565.33
35	45	90	45	90	4	529.718
36	0	90	45	45	4	544.524
37	0	45	90	90	1	499.364
38	45	90	90	90	1	547.423
39	45	0	45	45	0	504.97
40	0	90	90	90	0	551.858
41	90	90	0	90	1	428.173
42	45	0	0	90	1	553.661
43	45	45	45	90	1	619.842
44	90	45	0	90	4	495.503
45	0	0	45	90	0	361.335
46	45	0	90	0	3	577.266
47	0	0	0	90	4	375.491
48	45	45	0	45	4	570.775
49	0	0	90	0	4	425.789
50	90	0	45	90	4	541.256
51	45	90	0	0	1	596.264
52	90	0	45	45	3	565.33
53	0	45	0	0	1	443.692
54	0	45	0	90	3	465.75
55	90	90	0	0	2	441.866
56	0	0	45	0	1	400.735
57	45	90	45	90	3	550.831
58	0	45	45	90	4	475.745
59	45	0	45	90	2	566.87
60	90	0	0	45	4	570.629
61	45	0	0	45	3	529.126
62	90	45	0	0	0	472.108
63	45	45	90	45	3	592.769
64	0	0	45	45	2	414.059
65	45	90	45	0	2	566.87
66	0	0	0	45	0	323.036
67	90	90	90	0	1	382.037
68	90	45	90	90	3	459.904
69	45	45	90	0	4	574.246
70	45	0	90	45	1	593.061
71	90	90	90	90	4	366.775
72	0	45	0	90	2	478.126
73	90	90	90	45	0	323.036
74	90	0	90	45	2	558.317

TABLE 8.5

Measuring the Prediction Performance of Mathematical Models Regarding Different Model Assessment Criteria

	SON*	SONR*	TON*	FOTN*	FOTNR*	SOTN*	BesselJ*	HM*
R^2	0.953	0.966	0.967	0.318	0.775	0.988	0.871	0.993
R^2 adj	0.966	0.992	0.968	0.378	0.895	0.991	0.821	0.994
SE	16.115	13.037	13.019	61.622	33.326	8.234	26.318	6.098
MSE	266.587	202.373	197.947	3809.293	1199.374	72.901	706.721	39.085
RMSE	16.115	13.037	13.019	61.622	33.326	8.234	26.318	6.098
RMSPE	3.053	2.557	2.387	12.711	7.820	1.608	5.380	1.157
RRMS	3.141	2.533	2.545	12.020	6.480	1.606	5.145	1.190
SAE	265.598	160.354	245.729	1167.011	548.926	138.768	594.269	106.705
MAE	13.411	11.824	11.126	51.071	27.358	7.111	21.842	4.872
SRMSE	0.031	0.025	0.025	0.120	0.065	0.016	0.051	0.012
MAPE	2.563	2.314	2.098	10.355	6.042	1.396	4.434	0.934
SEP	0.031	0.025	0.025	0.120	0.065	0.016	0.051	0.012
KGE	0.979	0.992	0.987	0.617	0.902	0.991	0.911	0.987
U2theil	0.031	0.025	0.025	0.119	0.064	0.016	0.051	0.012
MSRE	0.001	0.001	0.001	0.019	0.006	0.000	0.003	0.000
X^2	9.256	3.646	7.418	177.387	41.764	2.386	43.872	1.498
AIC	241.637	253.585	304.282	292.415	239.050	297.551	278.769	426.175
BIC	254.669	279.649	339.034	299.242	241.533	338.509	291.801	510.573
AICC	204.850	260.338	1281.870	195.288	228.186	−170.38	241.982	74.952

* Full form of models are given in appendix (Table A8.1).

Table 8.5 presents the assessment outcomes of mathematical models utilized in modeling the natural frequency gap across 19 distinct fit indices. This evaluation process involves partitioning the data into three distinct groups: training, testing, and validation. The results depicted in Table 8.5 represent the average prediction performance of the models across the training, testing, and validation phases. For instance, the R^2 value of 0.953 associated with the SON model signifies the average R^2 values obtained throughout the training, testing, and validation stages.

The models assessed in Table 8.5 encompass hybrid formulations comprising polynomial, rational, trigonometric, Bessel-type special functions, and polynomial-trigonometric expressions. Within the realm of model fit indices, metrics such as R^2, R^2 correction, and KGE are deemed more successful as their values approach 1. Conversely, criteria including AIC, BIC, and AICC consider smaller values indicative of model success. Beyond these delineated groups, additional model evaluation criteria gauge success based on the proximity of errors in estimating actual values to 0.

In this context, it can be inferred that results provided by metrics such as R^2, R^2 correction, and KGE are more readily interpretable and universally acceptable, irrespective of the dataset.

To illustrate with an example, certain assessment criteria like Mean Squared Error (MSE) and Root Mean Squared Error (RMSE) may consider a proposed model successful if the error approaches 0 in a dataset where the output parameter varies within a narrow range, such as 0–1. However, it's crucial to recognize that even seemingly low errors, like 0.3 or 0.5, could be misleading in such datasets.

If a similar evaluation were conducted on a dataset where the output parameter ranged from 0 to 1000, an error of 0.5 might be deemed acceptable. In essence, the interpretability of model success in assessment criteria based on minimizing error values may vary depending on the value range of the output parameter. What appears as a low error in one context may be considered high in another.

Contrastingly, R^2-based model assessment criteria provide a more reliable and realistic evaluation. These criteria factor in both the actual value and the mean of the actual value (SST), along with the error (SSE), offering a more comprehensive analysis compared to metrics like MSE, RMSE, and SE, which solely consider prediction error. It's important to note that R^2-based criteria assume a normal or close to normal distribution of the output parameter. If the output parameter is heavily clustered in a particular region, the R^2 result may be biased due to its effect on the SST calculation.

Criteria such as AIC, BIC, and AICC, discussed under alternative model assessment criteria, take into account both the number of terms incorporated in the model and the model's prediction performance. These criteria aim to achieve success by minimizing both estimation error and the number of model terms. Consequently, it is more meaningful to employ a combination of diverse model assessment criteria to gauge the performance and usability of a model effectively.

Table 8.6 presents the criteria utilized to evaluate different models and identifies the top 5 most successful models recommended based on each criterion. Among the recommended models, HM emerges as the most successful across most criteria. Another notable recommendation is the SOTN model. However, it is imperative to interpret the results of Table 8.6 in conjunction with the numerical values provided in Table 8.5. For instance, while HM ranks highest in Table 8.6 based on the R^2 criterion, the SOTN, TON, SONR, and SON models also exhibit commendable performance.

It is essential to acknowledge that relying solely on R^2 as a criterion may not suffice when selecting among the aforementioned models. The disparities in R^2 values among the models may not be substantial enough to decisively influence the decision-making process. Similarly, when considering other criteria, the results indicate that the HM model stands out as the most successful. Nevertheless, the SOTN, TON, SONR, and SON models also demonstrate comparable levels of success to the HM model across various assessment criteria.

Table 8.7 presents the evaluation of models, considering the boundedness control criterion proposed as an original model assessment criterion within the scope of this book. This criterion examines whether the maximum and minimum values that the models can produce while the design parameters vary continuously align with engineering reality. In this study, based on the results presented in Table 8.4, the maximum and minimum natural frequency gaps were found to be 619.842 rad/sec and 323.036 rad/sec, respectively. Table 8.7 displays the maximum

TABLE 8.6

The Top Five Most Successful Models Based on Different Model Assessment Criteria

Criteria	1	2	3	4	5
R^2	HM	SOTN	TON	SONR	SON
R^2 adj	HM	SONR	SOTN	TON	SON
SE	HM	SOTN	TON	SONR	SON
MSE	HM	SOTN	TON	SONR	SON
RMSE	HM	SOTN	TON	SONR	SON
RMSPE	HM	SOTN	TON	SONR	SON
RRMS	HM	SOTN	SONR	TON	SON
SAE	HM	SOTN	SONR	TON	SON
MAE	HM	SOTN	TON	SONR	SON
SRMSE	HM	SOTN	TON	SONR	SON
MAPE	HM	SOTN	TON	SONR	SON
SEP	HM	SOTN	TON	SONR	SON
KGE	SOTN	SONR	HM	TON	SON
U2theil	HM	SOTN	TON	SONR	SON
MSRE	HM	SOTN	TON	SONR	SON
X^2	HM	SOTN	SONR	TON	SON
AIC	FOTNR	SON	SONR	BesselJ	FOTN
BIC	FOTNR	SON	SONR	BesselJ	FOTN
AICC	SOTN	HM	FOTN	SON	FOTNR

and minimum natural frequency gaps obtained from the proposed mathematical models. A comparison between the results obtained using mathematical models and the actual results derived from the analytical formula in Table 8.4 reveals that only the SON and SONR models yield results consistent with the actual values in terms of both maximum and minimum natural frequency gaps. This outcome is

TABLE 8.7

Boundedness Check Assessment of Models

Model	Maximum	Minimum	Model Term Number
SON	606.85	317.72	21
SONR	629.84	297.89	42
TON	9.94E+04	−9.84E+04	56
FOTN	604.3	100.06	11
FOTNR	1.15E+10	−2.35E+07	22
SOTN	652.67	−629.4	66
BesselJ	799	−21535.2	21
HM	3931.95	−11596	136

particularly noteworthy. The HM and SOTN models, which emerged as the most successful models based on 19 different model assessment criteria in Table 8.5 and Table 8.6, have been identified as models that are not suitable according to the results of the boundedness control criterion in Table 8.7.

For this reason, it can be asserted that boundedness control serves as a pivotal criterion directly influencing model selection. The SON and SONR models yielded results congruent with reality and closely aligned with each other in terms of both maximum and minimum natural frequency gap. Upon comparing these two models with regard to their respective term counts, the SON model emerges as more streamlined than the SONR model.

The evaluations presented in Tables 8.5, 8.6, and 8.7 have demonstrated that the SON and SONR models stand out as the two most suitable options for modeling the natural frequency range output.

To further assess the performance of these two models (SON and SONR), test data comprising 50, 100, and 250 lines was randomly generated. The accuracy of each model's predictions regarding the natural frequency gap values for the test data was evaluated using various criteria, as outlined in Table 8.8.

TABLE 8.8

Measuring the Prediction Performance of SON and SONR Models Based on Randomly Generated Datasets and Different Assessment Criteria

Model Assessment Criteria	50-Row Testing Data		100-Row Testing Data		250-Row Testing Data	
	SON	SONR	SON	SONR	SON	SONR
R^2test	0.81	0.88	0.72	0.86	0.32	0.22
R^2adj	0.79	0.87	0.71	0.85	0.30	0.20
SE	29.52	22.94	35.19	25.20	52.72	56.47
MSE	871.57	526.25	1238.56	634.94	2779.48	3188.35
RMSE	29.52	22.94	35.19	25.20	52.72	56.47
RMSPE	5.61	4.35	6.80	4.76	10.42	11.30
RRMS	5.59	4.35	6.72	4.81	9.79	10.49
SAE	1222.04	967.88	2968.91	2136.61	9883.96	9767.99
MAE	24.44	19.36	29.69	21.37	39.54	39.07
SRMSE	0.06	0.04	0.07	0.05	0.10	0.10
MAPE	4.62	3.64	5.64	4.04	7.60	7.56
SEP	0.06	0.04	0.07	0.05	0.10	0.10
KGE	0.78	0.89	0.68	0.88	0.70	0.87
U2theil	0.06	0.04	0.07	0.05	0.10	0.10
MSRE	0.00	0.00	0.01	0.00	0.01	0.01
X^2	82.47	49.64	238.04	119.36	1364.36	1581.94
AIC	522.41	539.18	1037.96	1013.14	2733.97	2810.28
BIC	562.56	619.49	1092.67	1122.56	2807.92	2958.19
AICC	555.41	1055.18	1049.80	1076.51	2738.03	2827.73

The results revealed that when comparing the accuracy of the two models using 50 and 100 lines of test data, the SONR model surpassed the SON model across all model assessment criteria except for AIC, BIC, and AICC.

However, with the utilization of 250 lines of test data for comparison, the SON model demonstrated superior performance compared to the SONR model, as indicated by the R^2, R^2 adjusted, SE, MSE, RMSE, RMSPE, RRMS, X^2, AIC, BIC, and AICC criteria.

When assessing the performance of prediction models, it is imperative to employ diverse criteria to avoid potential misinterpretations. The efficacy of a model can fluctuate depending on the evaluation criteria utilized, posing challenges in determining its overall success. For instance, while the R^2 and R^2 adjusted criteria showcased high success rates for models with 50 and 100-row datasets, the success rate diminished as the dataset size increased to 250 rows. Nevertheless, upon evaluating the models considering success criteria such as MAPE, KGE, and MSRE, it becomes evident that augmenting the test data volume does not significantly alter model success status. These findings underscore the importance of discerning the arbitrary application of criteria, stemming from differing approaches in model success evaluation, which may potentially lead to misinterpretations.

Therefore, when evaluating the predictive performance of models, it may be more meaningful to utilize the diverse criteria provided in the tables above collectively. However, in certain instances, employing various criteria concurrently may prove insufficient for decision-making or could yield misleading conclusions.

As illustrated in the example provided, the R^2 value of 0.22 obtained for the SONR model using 250 lines of test data suggests relatively poor performance in predicting actual values, whereas the MAPE of 7.56 for SONR indicates superior model performance. While this may not pose a significant issue when selecting the most successful model among several options, it can complicate assessments aimed at determining the overall success of a model. In such cases, deciding how successful a model truly is becomes more challenging, as there arises uncertainty regarding which assessment criterion should be given precedence.

When confronted with the need to identify the most successful model among alternatives, it may be meaningful to prioritize the model that excels across various assessment criteria. However, gauging the success or failure of a model becomes more intricate when considering multiple evaluation metrics. To address this challenge effectively, it becomes essential to understand the methodology underlying the calculation of assessment criteria and assess whether the dataset being analyzed aligns with this approach.

In essence, rather than solely focusing on the predictive performance of actual values resulting from model evaluations, attention should be directed toward factors that could influence this assessment. Crucial considerations include accurately determining design parameters and their levels, employing an appropriate experimental design methodology, ensuring an adequate number of experiments for modeling, and identifying the distribution type of the output parameter. Even if a model demonstrates apparent success without accounting for these factors, it may still yield misleading results.

In conclusion, while selecting the most successful model from a pool of alternatives is pivotal, it is equally imperative to take into account the diverse criteria used for evaluation and the underlying factors that may influence these assessments.

In light of these evaluations, several significant findings emerge regarding the use of model types and assessment criteria discussed within the scope of this study:

1. The proposed boundedness check model assessment criterion serves to evaluate the compatibility of model results with reality, rather than simply assessing the model's success. This criterion directly informs whether a model is usable. The importance of the boundedness control criterion becomes evident in assessing the success of mathematical models applied to model the natural frequency gap output. While many other assessment criteria highlight HM as the most successful model, the boundedness control criterion reveals that the maximum and minimum natural frequency gap values produced by HM are unattainable, rendering the model unusable. Conversely, it is determined that only the maximum and minimum natural frequency gap values produced by the SON and SONR models align with reality.

2. When evaluating model goodness-of-fit index criteria within themselves, a higher value indicates a more successful model. Criteria such as R^2, R^2 adjusted, and KGE approaching 1 signify strong model performance, while AIC, BIC, and AICC consider smaller values as indicative of success. Other goodness-of-fit indices assess model success based on how closely the predicted values match the actual values.

3. Criteria such as MSE, RMSE, and SE solely compare the error between actual and predicted values (SSE). In contrast, R^2-based criteria factor in both error (SSE) values and the deviation between the actual value and the mean of the actual value (SST). Additionally, AIC, BIC, and AICC consider the number of model terms alongside the difference between actual and predicted values as criteria influencing model success. Models with fewer terms and lower error rates are deemed more successful by these criteria.

Overall, while various assessment criteria aid in determining model performance and simplicity, none offer as clear an indication of model usability as the boundedness control criterion. While other criteria provide insight into model simplicity and fit with the data, they do not inherently address the consistency and practicality of the results obtained.

REFERENCES

[1] Doğan I, Doğan N. Model Performans Kriterlerinin Kronolojisine ve Metodolojik Yönlerine Genel Bir Bakış: Bir Gözden Geçirme.

[2] Savran M. A new systematic approach for design, modeling and optimization of the engineering problems based on stochastic multiple-nonlinear neuro-regression analysis and non-traditional search algorithms (PhD thesis). İzmir Katip Çelebi University, 2023.

[3] Hosseinzadeh Y, Jalili S, Khani R. Investigating the effects of flax fibers application on multi-objective optimization of laminated composite plates for simultaneous cost minimization and frequency gap maximization. Journal of Building Engineering 2020; 32: 101477.

[4] Savran M. Development of vibration performances of hybrid laminated composite materials by using stochastic methods (Master thesis). İzmir Katip Çelebi University, 2017. https://tez.yok.gov.tr/UlusalTezMerkezi/tezSorguSonucYeni.jsp

APPENDIX

TABLE A8.1
Proposed Mathematical Models for Modeling Natural Frequency Gap Output (Related to Table 8.5)

(*SON*)

In[•]:=

$SON = -0.0418694\, x1^2 - 0.0379109\, x1\, x2 - 0.0127177\, x1\, x3 - 0.00132491\, x1\, x4 - 0.0109363\, x1\, x5 + 6.15363\, x1 - 0.00695212\, x2^2 - 0.00353891\, x2\, x3 - 0.000719938\, x2\, x4 - 0.0174869\, x2\, x5 + 2.52115\, x2 + 0.000479553\, x3^2 + 0.000742398\, x3\, x4 - 0.00120618\, x3\, x5 + 0.660331\, x3 + 0.00179019\, x4^2 - 0.038216\, x4\, x5 - 0.0378256\, x4 - 10.9735\, x5^2 + 54.1622\, x5 + 317.918;$

(*SONR*)

In[•]:=

$SONR = (179.174\, x1^2 + 32.0567\, x1\, x2 - 1.76239\, x1\, x3 - 1.84316\, x1\, x4 - 171.291\, x1\, x5 - 19543.1\, x1 + 46.6488\, x2^2 - 41.0781\, x2\, x3 - 1.55866\, x2\, x4 - 421.703\, x2\, x5 - 978.882\, x2 - 24.4545\, x3^2 - 43.6841\, x3\, x4 + 1006.96\, x3\, x5 + 4701.15\, x3 + 16.9695\, x4^2 + 1317.61\, x4\, x5 - 2245.31\, x4 - 60762.8\, x5^2 + 261554.x5 + 1.45603 \times 10^6)/(0.634801\, x1^2 + 0.366667\, x1\, x2 + 0.102514\, x1\, x3 + 0.0140228\, x1\, x4 - 0.470294\, x1\, x5 - 83.0299\, x1 + 0.155831\, x2^2 - 0.0349425\, x2\, x3 + 0.0138589\, x2\, x4 - 0.949508\, x2\, x5 - 24.4668\, x2 - 0.0430634\, x3^2 - 0.0725062\, x3\, x4 + 1.86008\, x3\, x5 + 0.999806\, x3 + 0.0246568\, x4^2 + 2.60215\, x4\, x5 - 5.67045\, x4 - 54.1404\, x5^2 + 229.483\, x5 + 4477.05);$

(*TON*)

In[•]:=

$TON = -0.000404502\, x1^3 - 0.000250232\, x1^2\, x2 - 0.0000890122\, x1^2\, x3 + 0.0000831404\, x1^2\, x4 + 0.00653114\, x1^2\, x5 + 0.00648563\, x1^2 + 0.000048935\, x1\, x2^2 - 0.000149523\, x1\, x2\, x3 + 0.000197701\, x1\, x2\, x4 + 0.00957572\, x1\, x2\, x5 - 0.0298132\, x1\, x2 + 0.0000768682\, x1\, x3^2 + 0.0000269734\, x1\, x3\, x4 + 0.006965\, x1\, x3\, x5 - 0.0140711\, x1\, x3 + 0.000272482\, x1\, x4^2 + 0.00663502\, x1\, x4\, x5 - 0.0540915\, x1\, x4 + 0.150768\, x1\, x5^2 - 2.24381\, x1\, x5 + 6.87297\, x1 - 0.0000166565\, x2^3 + 0.00007294\, x2^2\, x3 - 0.000187604\, x2^2\, x4 - 0.00238227\, x2^2\, x5 + 0.00377455\, x2^2 - 0.0000909671\, x2\, x3^2 - 0.000246194\, x2\, x3\, x4 - 0.010339\, x2\, x3\, x5 + 0.0383794\, x2\, x3 + 0.000082483\, x2\, x4^2 - 0.000927458\, x2\, x4\, x5 + 0.00487426\, x2\, x4 + 0.0597587\, x2\, x5^2 - 0.0792058\, x2\, x5 + 1.10944\, x2 + 0.0000300144\, x3^3 - 0.000128413\, x3^2\, x4 - 0.0056586\, x3^2\, x5 + 0.000473396\, x3^2 - 0.0000351938\, x3\, x4^2 + 0.00176217\, x3\, x4\, x5 + 0.0149382\, x3\, x4 + 0.0159276\, x3\, x5^2 + 0.460582\, x3\, x5 - 0.270421\, x3 - 0.000107692\, x4^3 - 0.00621192\, x4^2\, x5 + 0.00170474\, x4^2 + 0.154088\, x4\, x5^2 - 0.483072\, x4\, x5 + 1.88864\, x4 + 5.5211\, x5^3 - 64.5026\, x5^2 + 214.212\, x5 + 242.098;$

(Continued)

TABLE A8.1 *(Continued)*

Proposed Mathematical Models for Modeling Natural Frequency Gap Output (Related to Table 8.5)

(*FOTN*)

FOTN = 149.504 sin $(x1)$ + 91.3208 cos $(x1)$ + 26.9389 sin $(x2)$ + 18.6041 cos $(x2)$ +
2.49409 sin $(x3)$ − 3.4727 cos $(x3)$ + 2.84445 sin $(x4)$ + 9.57027 cos $(x4)$ +
19.003 sin $(x5)$ − 28.7334 cos $(x5)$ + 350.983;

(*FONTR*)

$In[\bullet]:=$

FOTNR = (2744.61 sin $(x1)$ + 1592.82 cos $(x1)$ − 3.95231 sin $(x2)$ + 24.6611 cos $(x2)$ −
17.456 sin $(x3)$ + 2.95951 cos $(x3)$ + 24.7127 sin $(x4)$ + 11.738 cos $(x4)$ + 3.52148 sin $(x5)$ −
9.73771 cos $(x5)$ − 1679.18)/(4.88574 sin $(x1)$ + 2.80258 cos $(x1)$ + 0.0083006 sin $(x2)$ +
0.0270827 cos $(x2)$ − 0.0427886 sin $(x3)$ − 0.0102948 cos $(x3)$ + 0.0435043 sin $(x4)$ +
0.0180178 cos $(x4)$ + 0.00513953 sin $(x5)$ − 0.0191834 cos $(x5)$ − 2.96784);

(*SOTN*)

$In[\bullet]:=$

SOTN = 118.075 sin $(x1)$ sin $(x2)$ − 8.57883 cos $(x1)$ cos $(x2)$ + 146.491 cos $(x1)$ sin $(x2)$ +
149.881 sin $(x1)$ cos $(x2)$ − 10.9097 sin $(x1)$ sin $(x3)$ − 27.0236 cos $(x1)$ cos $(x3)$ +
12.505 cos $(x1)$ sin $(x3)$ + 24.0472 sin $(x1)$ cos $(x3)$ − 10.7671 sin $(x1)$ sin $(x4)$ −
14.5021 cos $(x1)$ cos $(x4)$ − 10.3808 cos $(x1)$ sin $(x4)$ − 3.61454 sin $(x1)$ cos $(x4)$ +
28.3213 sin $(x1)$ sin $(x5)$ + 6.14044 cos $(x1)$ cos $(x5)$ + 19.1797 cos $(x1)$ sin $(x5)$ +
7.31262 sin $(x1)$ cos $(x5)$ + 22.2625 sin^2 $(x1)$ + 82.7692 sin $(x1)$ + 26.4783 cos^2 $(x1)$ +
63.5203 cos $(x1)$ − 39.0814 sin $(x1)$ cos $(x1)$ + 38.4637 sin $(x2)$ sin $(x3)$ +
10.3081 cos $(x2)$ cos $(x3)$ + 38.3244 sin $(x2)$ cos $(x3)$ + 25.3963 cos $(x2)$ sin $(x3)$ +
4.50543 sin $(x2)$ sin $(x4)$ − 2.06009 cos $(x2)$ cos $(x4)$ − 7.56106 sin $(x2)$ cos $(x4)$ −
0.375977 cos $(x2)$ sin $(x4)$ − 0.560657 sin $(x2)$ sin $(x5)$ − 0.740198 cos $(x2)$ cos $(x5)$ −
4.33747 sin $(x2)$ cos $(x5)$ − 1.85291 cos $(x2)$ sin $(x5)$ + 8.20429 sin^2 $(x2)$ + 3.80265 sin $(x2)$ +
33.1046 cos^2 $(x2)$ + 18.2553 cos $(x2)$ − 115.953 sin $(x2)$ cos $(x2)$ + 21.6014 sin $(x3)$ sin $(x4)$ +
3.32196 cos $(x3)$ cos $(x4)$ + 20.3564 sin $(x3)$ cos $(x4)$ − 1.33796 cos $(x3)$ sin $(x4)$ +
6.92061 sin $(x3)$ sin $(x5)$ + 5.35687 cos $(x3)$ cos $(x5)$ − 3.7441 sin $(x3)$ cos $(x5)$ +
0.891637 cos $(x3)$ sin $(x5)$ + 18.2122 sin^2 $(x3)$ + 101.147 sin $(x3)$ + 24.7606 cos^2 $(x3)$ +
99.5018 cos $(x3)$ − 143.278 sin $(x3)$ cos $(x3)$ + 6.40776 sin $(x4)$ sin $(x5)$ −
1.2349 cos $(x4)$ cos $(x5)$ − 4.5544 sin $(x4)$ cos $(x5)$ + 11.1808 cos $(x4)$ sin $(x5)$ +
0.838281 sin^2 $(x4)$ + 31.6113 sin $(x4)$ − 8.68523 cos^2 $(x4)$ + 41.255 cos $(x4)$ −
50.655 sin $(x4)$ cos $(x4)$ + 13.9942 sin^2 $(x5)$ − 8.87075 sin $(x5)$ + 6.93408 cos^2 $(x5)$ −
23.6028 cos $(x5)$ + 13.9655 sin $(x5)$ cos $(x5)$ + 87.7048;

(*BESSELJ*)

$In[\bullet]:=$

BESSELJ = 3150.84J_1 $(x1)J_0$ $(x2)$ + 17000.1J_1 $(x1)J_3$ $(x3)$ + 96.6818J_1 $(x1)J_0$ $(x4)$ +
4051.01J_1 $(x1)J_6$ $(x5)$ − 58842.3J_1 $(x1)^2$ + 4021.84J_1 $(x1)$ − 621.261J_0 $(x2)J_3$ $(x3)$ −
6.53599J_0 $(x2)J_0$ $(x4)$ + 48.1768J_0 $(x2)J_6$ $(x5)$ − 173.658J_0 $(x2)^2$ + 43.812J_0 $(x2)$ +
410.382J_3 $(x3)J_0$ $(x4)$ + 7265.79J_3 $(x3)J_6$ $(x5)$ − 2710.18J_3 $(x3)^2$ − 785.22J_3 $(x3)$ +
673.609J_0 $(x4)J_6$ $(x5)$ − 77.5835J_0 $(x4)^2$ + 95.3576J_0 $(x4)$ − 72128.J_6 $(x5)^2$ +
3497.48J_6 $(x5)$ + 475.981;

(Continued)

TABLE A8.1 *(Continued)*

Proposed Mathematical Models for Modeling Natural Frequency Gap Output (Related to Table 8.5)

(*HYBRID*)

$HM = 0.0484221\ x1^2 - 0.230104\ x1\ x2 + 2.11204\ x1 \sin (x2) + 3.05742\ x2 \sin (x1) +$
157.665 sin (x1) sin (x2) − 3.19845 x1 cos (x2) − 5.26771 x2 cos (x1) + 77.8708 cos (x1) cos (x2) +
98.3701 sin (x1) cos (x2) + 97.8725 cos (x1) sin (x2) − 0.0977908 x1 x3 + 1.6217 x1 sin (x3) +
0.906943 x3 sin (x1) + 3.48318 sin (x1) sin (x3) − 2.19196 x1 cos (x3) − 1.31724 x3 cos (x1) +
19.0801 cos (x1) cos (x3) + 5.3593 sin (x1) cos (x3) + 6.39743 cos (x1) sin (x3) − 0.1848 x1 x4 +
4.03355 x1 sin (x4) + 1.36452 x4 sin (x1) − 4.01124 sin (x1) sin (x4) − 5.67721 x1 cos (x4) −
2.09262 x4 cos (x1) + 15.9605 cos (x1) cos (x4) − 1.91224 sin (x1) cos (x4) +
1.21752 cos (x1) sin (x4) + 0.0511367 x1 x5 − 0.00487597 x1 sin (x5) + 37.5693 x5 sin (x1) +
61.2767 sin (x1) sin (x5) + 0.239913 x1 cos (x5) + 20.0214 x5 cos (x1) + 44.0608 cos (x1) cos (x5) +
61.5316 sin (x1) cos (x5) + 35.4178 cos (x1) sin (x5) + 7.37601 x1 − 0.683867 sin² (x1) +
3.78 x1 sin (x1) − 1.77429 sin (x1) + 4.94709 cos² (x1) + 1.66405 x1 cos (x1) + 17.1523 cos (x1) −
0.375357 sin (x1) cos (x1) − 0.0437802 x2² − 0.238137 x2 x3 + 3.36106 x3 sin (x2) +
3.46743 x2 sin (x3) + 43.665 sin (x2) sin (x3) − 4.76196 x3 cos (x2) − 5.44131 x2 cos (x3) +
31.465 cos (x2) cos (x3) + 37.102 sin (x2) cos (x3) + 28.4824 cos (x2) sin (x3) − 0.113069 x2 x4 +
2.22817 x4 sin (x2) + 1.23909 x2 sin (x4) + 2.08708 sin (x2) sin (x4) − 3.1064 x4 cos (x2) −
1.65768 x2 cos (x4) + 16.7956 cos (x2) cos (x4) + 0.395497 sin (x2) cos (x4) +
2.72461 cos (x2) sin (x4) − 0.0665936 x2 x5 + 3.98629 x5 sin (x2) − 0.0594785 x2 sin (x5) +
2.59475 sin (x2) sin (x5) + 1.98955 x5 cos (x2) + 0.048775 x2 cos (x5) + 12.2809 cos (x2) cos (x5) +
5.38062 sin (x2) cos (x5) − 2.29107 cos (x2) sin (x5) + 10.0031 x2 + 0.676809 sin² (x2) +
8.26809 x2 sin (x2) − 2.31882 sin (x2) + 2.75955 cos² (x2) − 4.08508 x2 cos (x2) +
14.9008 cos (x2) + 0.944184 sin (x2) cos (x2) − 0.0159531 x3² − 0.195846 x3 x4 +
3.13639 x4 sin (x3) + 2.70289 x3 sin (x4) + 19.8786 sin (x3) sin (x4) − 4.90034 x4 cos (x3) −
3.4378 x3 cos (x4) + 24.0983 cos (x3) cos (x4) + 16.2377 sin (x3) cos (x4) +
17.0746 cos (x3) sin (x4) + 0.124594 x3 x5 − 8.101 x5 sin (x3) + 0.244086 x3 sin (x5) −
3.22584 sin (x3) sin (x5) − 4.87099 x5 cos (x3) + 0.104137 x3 cos (x5) − 10.075 cos (x3) cos (x5) −
19.3075 sin (x3) cos (x5) + 3.24714 cos (x3) sin (x5) + 6.92015 x3 + 1.62556 sin² (x3) +
5.42784 x3 sin (x3) − 1.24957 sin (x3) + 3.98734 cos² (x3) − 1.4852 x3 cos (x3) +
16.5756 cos (x3) + 1.4324 sin (x3) cos (x3) + 0.0244998 x4² − 0.042066 x4 x5 +
19.8468 x5 sin (x4) − 0.276828 x4 sin (x5) + 23.0981 sin (x4) sin (x5) + 13.0217 x5 cos (x4) +
0.049634 x4 cos (x5) + 25.1185 cos (x4) cos (x5) + 22.4141 sin (x4) cos (x5) +
9.2063 cos (x4) sin (x5) + 3.78623 x4 + 3.79226 sin² (x4) + 5.7143 x4 sin (x4) + 1.13225 sin (x4) +
3.5538 cos² (x4) − 0.212512 x4 cos (x4) + 15.2192 cos (x4) + 0.251539 sin (x4) cos (x4) +
1.42024 x5² + 1.02293 x5 − 1.969 sin² (x5) + 10.0043 x5 sin (x5) + 0.644124 sin (x5) +
3.516 cos² (x5) − 23.3744 x5 cos (x5) + 6.79139 cos (x5) − 7.38167 sin (x5) cos (x5) + 3.8471;

9 Comparison of the Effects of Experimental Design Methods on Mathematical Modeling

Experimental design, modeling, and optimization are interdependent processes that must be approached holistically rather than in isolation. Decisions made during the experimental design stage, such as selecting design parameters and determining their levels, directly influence the subsequent modeling process. Incorrect selection of design parameters or their level values can lead to erroneous conclusions about the behavior of the output parameter during mathematical modeling.

Ideally, the experimental design method should generate a dataset covering the entire range of defined design parameters and their values. Failure to do so may result in the mathematical model capturing only local trends instead of the global trend of the output parameter. Even if the mathematical model accurately predicts actual values based on a clustered dataset, errors may occur due to the dataset points being concentrated within specific ranges of design parameters.

For instance, in Figure 9.1, where the relationship between input and output parameters exhibits linear behavior, clustering of data within certain marked regions may not significantly impact the modeling phase. However, this scenario may not hold true for more complex relationships between input and output parameters.

A comprehensive understanding of the relationships between experimental design, modeling, and optimization is crucial for achieving reliable and accurate results. It is essential to ensure that experimental design captures the full range of design parameters to obtain a dataset that accurately represents the behavior of the system being studied.

However, making a similar evaluation for Figure 9.2 is impossible. Here, choosing data clustered in specific regions while creating the dataset will cause errors in modeling. For example, intensely selecting data from the region marked in black shows that the relationship between input-output parameters has a linear behavior. In contrast, the region marked in orange indicates that logarithmic behavior is dominant, and the region marked in red indicates that polynomial or trigonometric behavior is dominant. In such a case, only selecting the points in the dataset to represent all input parameter value ranges will lead us to the correct result.

DOI: 10.1201/9781003494843-9

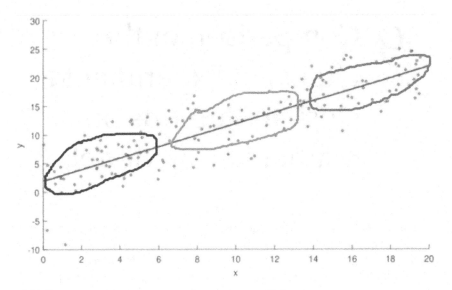

FIGURE 9.1 Expressing the relationship of input and output parameters with a linear model

Suppose the correct strategy is not followed in the experimental design and modeling stages. In that case, the results obtained will be far from reality, even if methods that have proven successful in many different problem types are used in optimization. The main elements of the optimization process are the objective function, constraints, and design parameters. Since the objective function, whose

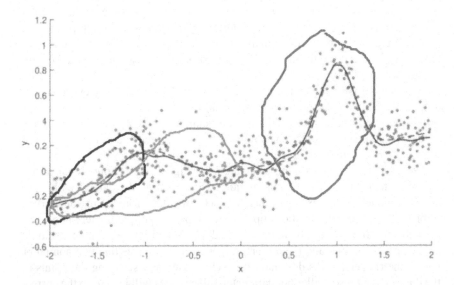

FIGURE 9.2 Expressing the relationship of input and output parameters with a nonlinear model

value is desired to be maximized or minimized, is directly linked to the experimental design and modeling processes, a robust optimization result can only be achieved by having a reliable objective function.

Table 9.1 presents the prevalent experimental design methods found in the literature, along with examples of problems to which these methods are applied and the required number of experiments if Full Factorial design (FFD) were utilized.

TABLE 9.1

Experimental Design Methods and Reference Problems from the Literature

Year	Study	Scope	Design of Experiment Method	Experiment Number	FFD Experiment Number
2021	Kashyzadeh et al. [4]	Composite storage tank design	CCD	24	625
2019	Zhang et al. [5]	Optimization of electric vehicle battery pack enclosure	CCD	79	78,125
2016	Mohamed et al. [6]	Optimization of dynamic-mechanical properties of samples produced by additive-manufacturing	I-Optimal	60	31,104
2016	Pahange and Abolbashari [7]	Airplane wing design	Taguchi	18	486
2016	Arora et al. [8]	Fatigue life optimization for leaf spring	CCD	50	3125
2016	GC et al. [9]	Mathematical modeling the casting process	BBD CCD	27	81
2021	Huang et al. [3]	Laser welding process	Taguchi	25	15,625
2020	Ye et al. [10]	Dynamic performance analysis of Stirling engine	BBD	55	729
2019	Kun et al. [11]	Fiber reinforced composites injection molding process	Taguchi	27	243
2019	Veljković et al. [12]	Modeling the biodiesel production process	FFD BBD CCD	27 14 14	27
2022	Jafari and Akyuz [2]	Vehicle brake disc design	Taguchi	16	1728
2021	Hong et al. [1]	Train brake disc design	Taguchi	32	262,144
2020	Souzangarzadeh et al. [13]	Energy absorbing cylindrical tubes	D-Optimal	34	72
2020	Li et al. [14]	Modeling vehicle stability	D-Optimal	28	729
2021	Kumar et al. [15]	Drilling process in polymer composites	Taguchi	16	64

230   Modeling and Optimization in Engineering and Science

Among the commonly used experimental design methods are Box-Behnken design (BBD), Central Composite design (CCD), and Taguchi, with D-optimal and FFD also frequently employed. Particularly in scenarios necessitating numerous experiments, the Taguchi method stands out for offering a design requiring significantly fewer experiments compared to alternative methods, making it appealing for practitioners in engineering fields where cost is a primary concern. Examination of Table 9.1, inclusive of case study problems, reveals that Taguchi substantially reduces the number of experiments required.

However, the crucial consideration here is the impact of reducing the number of experiments on modeling and optimization outcomes. For instance, in studies by Hong et al. [1], Jafari and Akyuz [2], and Huang et al. [3], which utilized the Taguchi method, a stark contrast is observed when comparing the suggested number of experiments with that of the FFD method. This discrepancy raises questions regarding the efficacy of the Taguchi method. At this juncture, the expectation from experimental design methods is to construct a dataset that most accurately elucidates the relationship between inputs and outputs within the constraints of the design parameters.

In the study conducted by Kumar et al. [15], the focus was on the mathematical modeling and optimization of the drilling process in polymer composites. The Taguchi method was employed as the experimental design approach, resulting in a set of 16 experiments. Conversely, the FFD experimental design method would necessitate 64 experiments. Figure 9.3a illustrates the experiments required for the FFD method, while Figure 9.3b depicts those proposed by the Taguchi method.

In achieving successful mathematical modeling, it is crucial to select experimental data from diverse points rather than concentrating on specific areas. Ideally, the experiments recommended by Taguchi should yield a homogeneous distribution across the cubic element depicted in Figure 9.3b. However, upon scrutiny of the experiments proposed by Taguchi, it becomes apparent that they do not cover all corners of the cubic element. Instead, a total of 16 test points are selected, with four points from the corners and 12 from the surface. Notably, no points located within the interior of the cubic element are included in the experimental set suggested by Taguchi.

Given these observations, a pertinent question arises: Could the utilization of the Taguchi experimental design method instead of FFD lead to errors in the mathematical modeling phase of the problem? This inquiry warrants careful examination and can be extended to various experimental design methods. It is imperative to emphasize that the experimental design process profoundly influences modeling and optimization; therefore, these aspects cannot be considered independently. Choosing the design parameters, their levels, and the experimental design method with a judicious strategy has a profound impact on problem-solving.

However, a critical observation reveals that experimental design, particularly the widespread use of the Taguchi method to minimize the number of experiments, is not accorded the necessary importance in both the reference studies provided in Table 9.1 and the broader literature. Hence, this study delves into the composite shaft problem with an analytical solution to scrutinize the effects of

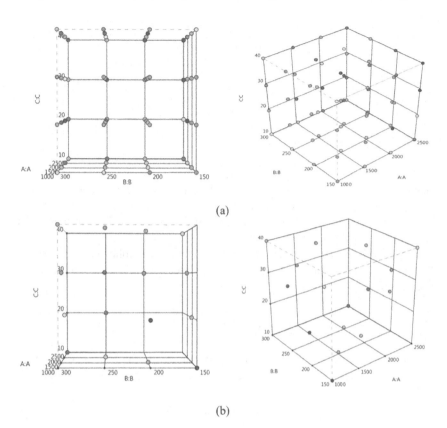

(a)

(b)

FIGURE 9.3 Comparison of experimental sets created by Full Factorial design (a) and Taguchi (b) methods

experimental design methods on modeling and optimization. Consequently, by leveraging analytical formulas, accurate results were obtained, allowing for the comparison of prediction performances of various models and the testing of models with previously unencountered datasets. This approach facilitated a genuine evaluation of the performance of experimental design methods.

CASE STUDY 9.1

The composite shaft examined in this study consists of 16 layers of symmetrical/ balanced glass-epoxy material [16]. Composite fiber angles were chosen as design parameters. The objective was to maximize the critical torsional buckling load, which required at least 550 Nm. Various experimental design methods, including FFD, Taguchi, BBD, CCD, and D-Optimal, were used to examine their effects on modeling and optimization comparatively. To solve the composite shaft problem, Classical Lamination Theory was considered, and the reference study provides detailed information and assumptions regarding this theory (Figure 9.4).

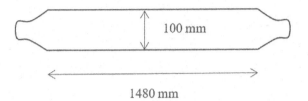

FIGURE 9.4   Composite shaft [16]

To calculate the critical torsional buckling load using the analytical formula based on composite material mechanics, the angle-dependent properties of the composite shaft must be determined. Chapter 6, Equations 6.1 and 6.2 contain the transformed reduced stiffness matrix and the calculated engineering constants needed for this calculation.

The extension stiffness matrix [A] can be calculated as follows, depending on the transformed reduced stiffness matrix calculated for each layer and the thickness of each layer [17].

$$A_{ij} = \sum_{k=1}^{n} [(\overline{Q}_{ij})]_k (h_k - h_{k-1}), \; i, j = 1, 2, 6 \tag{9.1}$$

Effective modules for fiber (Ex) and perpendicular to the fiber (Eh) direction can be expressed as follows depending on the [A] matrix and layer thickness [17].

$$E_x = \frac{1}{t}\left[A_{11} - \frac{A_{12}^2}{A_{22}}\right] \qquad E_h = \frac{1}{t}\left[A_{22} - \frac{A_{12}^2}{A_{11}}\right] \tag{9.2}$$

The critical torsional buckling load (Tcr) selected as the objective function is calculated as follows [17].

$$T_{cr} = (2\pi r_m^2 t)(0.272)(E_x E_h^3)^{1/4}\left(\frac{t}{r_m}\right)^{3/2} \tag{9.3}$$

Here, r_m indicates the average radius of the composite shaft. Calculations were carried out considering the cylindrical composite shaft as the laminated composite shown in Figure 9.5.

Table 9.2 displays the usage fiber angle design parameters and their levels while creating an experimental set. In this context, while all fiber angle levels are utilized in the experiment set creation phase with FFD, D-Optimal, Taguchi, and CCD experimental design methods, BBD can only recommend an experiment set for design parameters consisting of three levels. Therefore, the formed experiment set using BBD will include 0, 45, and 90 levels.

While BBD, CCD, and Taguchi, among the experimental design methods, suggested experiment sets of 25 lines, D-Optimal recommended an experiment set of 118 lines, and FFD suggested an experiment set of 625 lines.

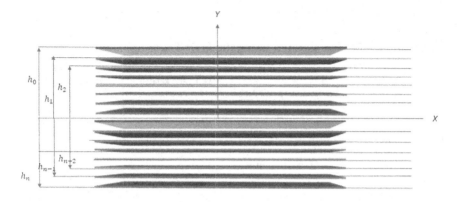

FIGURE 9.5   The 16-layered glass/epoxy composite

TABLE 9.2
Design Parameters and Their Levels Used in Shaft Design

	Level 1	Level 2	Level 3	Level 4	Level 5
x_1	0	22.5	45	67.5	90
x_2	0	22.5	45	67.5	90
x_3	0	22.5	45	67.5	90
x_4	0	22.5	45	67.5	90

The mathematical models obtained using the experimental sets created by the experimental design methods were subjected to optimization with the Differential Evolution (DE) algorithm, and the maximum and minimum critical torsional buckling loads were determined.

Table 9.3 presents mathematical models derived from the dataset generated by the FFD experimental design method, comprising 625 data points, along with their prediction performance, critical torsional buckling loads, and corresponding design parameters. A total of 14 mathematical models, encompassing polynomial, rational, trigonometric, and logarithmic expressions, were proposed to characterize the critical torsional buckling behavior of the glass/epoxy material shaft. Model performance was evaluated using the R-squared (R^2) goodness-of-fit index and the boundedness check criteria, systematically employed throughout the study.

During the training, testing, and validation phases, models demonstrating adequate prediction performance included SON, TON, TONR, SOTN, SOTNR, and SOLN. However, in the boundedness check phase, the TONR and SOLN models were deemed unsuccessful due to unrealistic minimum critical torsional buckling load values from an engineering perspective. Optimization using the specified

TABLE 9.3
Mathematical Models Formed Using the FFD* Experiment Set

Model*	R² Train	R² Adj	R² Test	R² Validation	Max. T$_{cr}$ and Related Design Parameters	Min. T$_{cr}$ and Related Design Parameters
L	0.65	0.65	0.55	0.63	1819.73 [90/90/90/90]	784.04 [0/0/0/0]
LR	0.66	0.66	0.54	0.63	1733.35 [90/90/90/90]	662.36 [0/0/0/0]
SON	0.96	0.96	0.95	0.93	1825.24 [90/0/90/0]	766.01 [0/0/0/0]
SONR	0.09	0.08	0.01	−0.14	4.26E+12 [48.27/0.33/25.48/17.23]	−1.59E+09 [8.90/12.83/6.03/59.18]
TON	0.97	0.97	0.97	0.95	1827 [90/90/90/0]	739.49 [13.15/12.74/12.65/13.39]
TONR	0.98	0.98	0.98	0.97	1805.96 [90/90/90/0]	−8.73 [4.51/7.44/0.06/26.46]
FOTN	0.14	0.14	−0.02	−0.08	1542.37 [90/90/46.62/90]	1026.99 [31.02/18.22/68.61/24.64]
FOTNR	0.11	0.11	0.01	−0.05	8.13E+08 [10.59/12.13/5.17/58.37]	−1.76E+12 [15.49/16.03/89.93/13.79]
SOTN	0.93	0.93	0.94	0.91	2011.55 [30.41/21.09/90/17.85]	797.13 [50.86/28.93/28.93/19.44]
SOTNR	0.98	0.98	0.97	0.97	1937.82 [65.34/71.64/49.61/90]	423.93 [75.80/31.81/56.92/35]
FOLN	0.18	0.17	0.02	0.02	1443.54 [90/90/90/90]	930.61 [0/0/0/0]
FOLNR	0.03	0.02	−0.003	−0.13	1.22E+16 [0.26/0.003/30.42/30.33]	1303.64 [90/90/90/90]
SOLN	0.8	0.8	0.76	0.71	1827.09 [90/90/90/90]	−144.83 [1.37/1.40/1.36/1.41]
SOLNR	0.41	0.4	0.59	0.3	1359.26 [90/90/90/90]	−5.23E+12 [0.001/10.93/1.2730.99]

* Full form of models and FFD experiment set are given in appendix (Tables A9.1 and A9.9).

analytical formula for Tcr yielded maximum and minimum values of 1783.96 Nm and 769.16 Nm, respectively.

In this context, both the SON and TON models exhibited high prediction performance and alignment with actual results. The average goodness-of-fit index for the SON model across the training, adjusted, testing, and validation phases was 0.95, with maximum and minimum Tcr values of 1825.24 Nm and 766.01 Nm, respectively. Conversely, for the TON model, these values were 0.965, 1827 Nm, and 739.49 Nm, respectively. Notably, the optimal design parameter values for maximum Tcr were 0 and 90 for the SON and TON models, respectively. However, for minimum Tcr, these values differed between the models, with the SON model indicating [0/0/0/0] and the TON model suggesting [13.15/12.74/12.65/13.39].

TABLE 9.4
Mathematical Models Formed Using the CCD* Experiment Set

Model*	R² Train	R² Adj	R² Test	R² Validation	Max. T$_{cr}$ and Related Design Parameters	Min. T$_{cr}$ and Related Design Parameters
L	0.62	0.52	−0.3	0.56	1734.55 [90/90/90/90]	562.96 [0/0/0/0]
LR	0.63	0.54	−0.2	0.64	1525.65 [90/90/90/90]	−105.96 [0/0/0/0]
SON	0.77	0.71	0.92	0.78	2502.29 [90/90/90/0]	738.62 [27.87/27.92/27.87/27.17]
SONR	−2	−3.25	−48	−19.82	2.48E+09 [90/0/44.16/82.73]	−1.72E+08 [90/78.47/0/0.02]
TON	0.8	0.75	0.9	0.66	367558 [0/90/50.94/24.43]	−364197 [90/3.27/39.06/65.40]
TONR	−7	−9.87	−149	−39.6	3.14E+13 [85.68/82.97/81.97/25.10]	−7.65E+09 [46.03/4.59/56.77/73.36]
FOTN	0.44	0.3	0.7	0.49	1514.28 [36.84/43.13/74.54/5.38]	764.86 [33.70/71.40/71.40/2.24]
FOTNR	0.99	0.99	0.43	−0.48	4.69E+14 [25.12/15.21/89.28/14.39]	−1.83E+08 [80.92/69.82/10.61/86.71]
SOTN	0.99	0.99	0.99	0.99	2504.55 [55.82/73.88/40.16/24.82]	−130.87 [41.96/82.63/57.62/23.18]
SOTNR	−30	−39.41	−150	−45.25	1.59E+07 [31.93/10.35/82.86/15.67]	−8.97E+11 [34.77/30.23/8.59/68.39]
FOLN	0.45	0.3	0.39	0.84	1.38E+03 [90/90/90/90]	−9.42E+01 [0/0/0/0]
FOLNR	0.49	0.36	0.25	0.8	1.59E+03 [90/90/90/90]	6.98E+02 [0/0/0/0]
SOLN	0.64	0.54	0.9	0.97	2.22E+03 [90/0/90/90]	−2.20E+04 [0/0/0/0]
SOLNR	−0.5	−0.99	0.3	−4.16	1.60E+10 [23.32/49.67/62.83/89.95]	−3.16E+10 [89.97/68.64/90/0]

* Full form of models and CCD experiment set are given in appendix (Tables A9.2 and A9.10).

Table 9.4 presents the mathematical models established using the dataset obtained through the CCD experimental design method, comprising 25 data points, along with their prediction performance, critical torsional buckling loads, and corresponding design parameters. Notably, the SOTN model demonstrates precise prediction performance during the training, testing, and validation stages. However, it yields inconsistent maximum and minimum Tcr values during the boundedness check phase, underscoring the critical role of the boundedness check as an assessment criterion.

In comparison with other models, only the SON model can be regarded as relatively successful concerning its R² prediction performance and boundedness check criterion. The SON model exhibits an average R² prediction performance of 0.8, with maximum and minimum Tcr values of 2502.29 Nm and 738.62 Nm, respectively. It is noteworthy that there is a substantial 29% difference between the maximum Tcr value obtained by the SON model (2502.29 Nm) and the actual value derived from the analytical formula (1783.96 Nm). Despite this discrepancy, the SON model remains preferable compared to other models.

Table 9.5 presents the mathematical models created using a dataset obtained by the BBD experimental design method, which includes 25 lines and their prediction performance. If you wish to create an experiment set using CCD, you can use five different level values of 0, 22.5, 45, 67.5, and 90 degrees for the design parameters. In contrast, the BBD experimental design method uses three levels of 0.45 and 90 degrees to create the experiment set. According to

TABLE 9.5
Mathematical Models Formed Using the BBD* Experiment Set

Model*	R^2 Train	R^2 Adj	R^2 Test	R^2 Validation	Max. T_{cr} and Related Design Parameters	Min. T_{cr} and Related Design Parameters
L	0.75	0.69	0.89	0.61	1691.95 [90/90/90/90]	757.04 [0/0/0/0]
LR	0.77	0.71	0.91	0.57	1580.85 [90/90/90/90]	561.86 [0/0/0/0]
SON	1	1	1	1	1864.27 [90/90/0/90]	744.84 [24.81/24.81/24.81/24.81]
SONR	−19.19	−24.58	−35.97	−142.72	2.40E+10 [87.08/56.88/73.05/53.03]	−6.72E+09 [24.84/44.32/37.92/68.83]
TON	1	1	0.78	0.95	2032.21 [90/90/90/0]	759.19 [0/0/0/0]
TONR	−27.26	−34.79	−47.39	−340.81	3.33E+08 [56.36/38.41/31.70/67.54]	−7.25E+07 [61.69/70.13/40.65/5.17]
FOTN	0.95	0.94	0.92	0.65	4229.06 [72.95/28.97/66.67/60.39]	760.38 [76.09/63.53/57.24/19.55]
FOTNR	1	1	1	1	6.05E+14 [58.83/63.75/63.32/31]	−2.82E+11 [76.38/39.80/68.94/69.09]
SOTN	1	1	1	1	2162.32 [89.98/58.72/18.32/80.59]	−1501.95 [35.64/10.44/41.66/60.44]
SOTNR	−7.48	−9.74	−83.32	−281.71	1.45E+16 [89.96/38.45/89.9/20.55]	−5.09E+09 [78.78/88.75/57.5/87.16]
FOLN	0.53	0.4	0.89	0.74	1463.5 [90/90/90/0]	656.91 [0/0/0/0]
FOLNR	0.62	0.52	0.99	0.35	1357.49 [90/90/90/0]	−1.80E+15 [0.002/0.1/81.53/0.03]
SOLN	1	1	1	1	2004.23 [90/90/90/0]	−1534.65 [1.79/1.79/1.79/1.79]
SOLNR	−9.58	−12.41	−37.78	−166.46	1.16E+16 [42.91/88.69/30.43/54.73]	−3.37E+11 [36.31/34.52/83.20/55.92]

* Full form of models and BBD experiment set are given in appendix (Tables A9.3 and A9.11).

the results in Table 9.5, LR, SON, TON, FOTN, FOTNR, SOTN, and SOLN are successful based on the R^2 goodness-of-fit index. However, among these models, only the SON and TON both accurately predict the actual Tcr values and meet the boundedness check criterion. The maximum and minimum Tcr values obtained from these two models are consistent with the actual values.

Table 9.6 shows the prediction performance of the models obtained when the D-optimal method is used in the mathematical modeling of the Tcr output. The D-optimal recommends a dataset of 118 lines for modeling. Although the number of data suggested by D optimal is not as little as BBD and CCD, it is more advantageous than the FFD method, which consists of 625 lines. When the performance of the models in Table 9.6 is examined, it is seen that the SON, SONR,

TABLE 9.6
Mathematical Models Formed Using the D-Optimal* Experiment Set

Model*	R² Train	R² Adj	R² Test	R² Validation	Max. T_{cr} and Related Design Parameters	Min. T_{cr} and Related Design Parameters
L	0.58	0.56	0.83	0.71	1805.79 [90/90/90/90]	780.91 [0/0/0/0]
LR	0.61	0.59	0.8	0.61	1698.07 [90/90/90/90]	743.84 [0/0/0/0]
SON	0.94	0.94	0.98	0.98	1838.6 [90/90/90/0]	786.36 [3.92/0.68/4.69/1.79]
SONR	0.95	0.95	0.97	0.98	1807.16 [90/90/90/0]	784.63 [14.83/11.64/9.07/11.58]
TON	0.96	0.96	0.99	0.99	1938.86 [90/90/90/0]	757.45 [11.32/12.96/11.55/12.04]
TONR	0.99	0.99	−0.65	0.82	2.29E+18 [72.54/82.77/74.75/1.96]	−2.11E+17 [11.14/25.31/6.97/2.73]
FOTN	0.13	0.1	0.12	0.14	1563.94 [65.58/65.37/27.80/27.63]	1039.58 [12.17/87.36/37.22/37.05]
FOTNR	0.28	0.25	−0.83	0.45	4.40E+09 [87.45/53.67/89.97/5.66]	−1.50E+12 [60.41/70.42/59.96/65.02]
SOTN	0.92	0.91	0.94	0.89	2072.09 [30.42/80.63/65.08/77.61]	686.59 [75.96/38.26/19.42/50.86]
SOTNR	0.85	0.85	0.05	−0.17	1.67E+09 [65.55/71.11/49.60/89.33]	−7.86E+17 [10.36/0.94/86.08/0.05]
FOLN	0.11	0.07	0.37	0.26	1.44E+03 [90/90/90/90]	9.31E+02 [0/0/0/0]
FOLNR	0.29	0.26	0.19	−0.12	1.95E+03 [90/90/90/0]	1.15E+03 [90/0/0/0]
SOLN	0.76	0.75	0.83	0.9	1.80E+03 [90/90/90/90]	−8.73E+01 [1.36/1.43/1.33/1.36]
SOLNR	0.61	0.59	−2.53	−0.84	1.54E+16 [89.95/87.41/90/24.54]	−4.78E+15 [13.78/8.44/51.75/10.57]

* Full form of models and D-Optimal experiment set are given in appendix (Tables A9.4 and A9.12).

TON, and SOTN models successfully meet both the R^2 goodness-of-fit index and the boundedness check criterion. Among these models, although TON showed better prediction performance based on the R^2 criterion, the SONR model produced more consistent results with the actual Tcr found using the analytical formula when the maximum and minimum values were considered. Although some models found the maximum and minimum Tcr load to be realistic in the boundedness check stage, they were considered unsuccessful because their performance was insufficient in estimating actual Tcr. L and LR are models in this category.

Table 9.7 displays the outcomes obtained from employing the Taguchi experimental design method to construct a dataset in the mathematical modeling

TABLE 9.7
Mathematical Models Formed Using the Taguchi* Experiment Set

Model*	R² Train	R² Adj	R² Test	R² Validation	Max. T$_{cr}$ and Related Design Parameters	Min. T$_{cr}$ and Related Design Parameters
L	0.81	0.76	−1.33	0.41	1924.38 [90/90/90/90]	687.74 [0/0/0/0]
LR	0.98	0.97	−14.49	−1.67	Inf –	Inf –
SON	0.96	0.95	0.63	0.88	2140.4 [0/90/0/90]	748.13 [0/0/14.10/0]
SONR	−5.39	−7.1	−348	−72	1.22E+14 [20.88/6.21/47.23/43.20]	−1.06E+14 [55.05/81.65/45.24/67.79]
TON	0.99	0.99	0.94	0.99	3.85E+06 [48.31/0/0/90]	−1.31E+07 [0/0/90/90]
TONR	−5.72	−7.51	−205.44	−73.24	1.46E+14 [80.38/89.92/0.23/89.91]	−3.68E+09 [8.09/25.82/32.48/54.21]
FOTN	0.33	0.16	−55.44	0.53	1786.58 [90/9020.88/90]	657.53 [43.47/5.52/49.15/30.41]
FOTNR	0.99	0.99	−31.76	0.71	5.22E+13 [29.35/30.26/78.27/29.05]	−2.90E+12 [62.32/34.19/18.88/88.57]
SOTN	1	1	1	1	3690.22 [83.53/71.46/45.29/38.79]	−852.43 [21.49/53.54/10.92/47.94]
SOTNR	−9.24	−11.97	−489.36	−76.16	2.01E+12 [59.03/25.26/65.37/34.30]	−3.23E+09 [50.75/68.94/45.07/83.89]
FOLN	0.34	0.17	−7.7	−0.95	1.48E+03 [90/90/90/90]	7.53E+02 [0/0/0/0]
FOLNR	0.55	0.43	−8.16	−0.32	4.10E+15 [90/0.002/58.33/88.78]	−2.30E+14 [0.3/0.007/0.22/90]
SOLN	0.95	0.94	0.88	0.95	4.21E+03 [90/0/0/90]	−4.08E+03 [0/0/90/6.44]
SOLNR	0.97	0.96	0.7	0.99	3.66E+08 [27.15/90/0/90]	−1.66E+10 [0.11/6.93/3.68/0.08]

* Full form of models and Taguchi experiment set are given in appendix (Tables A9.5 and A9.13).

process and utilizing this dataset to discern the most accurate model expressing the behavior of the Tcr output. In the Taguchi experimental design method, a mathematical model was devised using a dataset comprising 25 data points, akin to BBD and CCD. Hence, comparing the results of these three methods with an identical number of data rows will offer valuable insights into which method better captures the Tcr output.

The modeling conducted using the Taguchi experiment set revealed that the SON, TON, SOTN, SOLN, and SOLNR models demonstrated sufficient performance in predicting the actual values, as indicated by the R^2 goodness-of-fit index. However, among these models, only the SON model satisfies the boundedness check criteria.

When the experimental design methods, FFD, CCD, BBD, D-optimal, and Taguchi, whose effects have been examined so far in modeling the Tcr output, are compared with each other obtained results are as follows:

1. SON is the only model recommended by these five different experimental design methods in modeling the Tcr behavior of the composite shaft.
2. Among these five experimental design methods, SONR is found successful only by D-optimal.
3. Although BBD, CCD, and Taguchi methods have produced datasets of 25 lines, the prediction performances of the mathematical models obtained using these datasets are significantly different.

For example, SON models formulated using BBD, CCD, and Taguchi experimental design methods exhibit markedly different performances. While the SON model derived from BBD demonstrated exceptional prediction capabilities, replicating this success with the SON models derived from CCD and Taguchi proved challenging. These findings raise the question of whether employing hybrid experimental design methods could offer advantages. To address this query, experiment sets were generated by combining BBD, CCD, and Taguchi methods. The impact of these experiment sets on mathematical modeling and the Tcr output is presented in Table 9.8, showcasing the models that emerged as the most successful from these trials.

The combination of BBD and CCD methods yielded the most successful models, including SON, TON, and SOTN. However, the performance of SON and TON models in predicting actual values was inferior compared to when the BBD method was used alone. Conversely, the SOTN model exhibited similar prediction performance to the case where only BBD and CCD methods were employed. Nonetheless, the SOTN model was deemed unsuccessful when these methods were used individually or in combination due to unrealistic maximum and minimum values of the Tcr output, as per the boundedness check criterion. Therefore, the hybrid approach of using BBD and CCD methods did not confer any advantage.

When BBD and Taguchi methods were combined as a hybrid (BBD–Taguchi), the L and LR models demonstrated higher success rates compared to when used

TABLE 9.8

Mathematical Models Formed Using the Combined BBD–CCD–Taguchi* Method

	Model	R² Train	R² Adj	R² Test	R² Validation	Max. T$_{cr}$ and Related Design Parameters	Min. T$_{cr}$ and Related Design Parameters
BBD–CCD	SON	0.85	0.83	0.7	0.75	1692.82 [90/90/90/0]	620.62 [0/0/0/0]
	TON	0.86	0.84	0.77	0.74	1857.65 [90/0/90/90]	550.91 [0/0/0/0]
	SOTN	0.99	0.99	0.99	0.99	2754.15 [4.80/70.89/39.47/29.87]	495.62 [56.97/75.82/0.42/6.70]
BBD–Taguchi	L	0.74	0.71	0.95	0.91	1767.22 [0/90/90/90]	731.22 [0/0/0/0]
	LR	0.76	0.74	0.97	0.95	1746.53 [0/90/90/90]	735.87 [0/0/0/0]
	SON	0.94	0.94	0.93	0.93	1792.37 [90/90/90/0]	757.62 [12.99/17.69/11.12/17.55]
	TON	0.98	0.98	0.99	0.96	2506.16 [0/90/90/90]	548.72 [0/0/0/90]
BBD–CCD–Taguchi	SON	0.88	0.87	0.92	0.87	1787.68 [0/90/90/90]	720.65 [0/0/0/0]
	TON	0.91	0.9	0.93	0.96	1959.41 [90/90/0/90]	734.24 [12.71/11.11/14.33/12.64]
	SOTN	0.99	0.99	0.99	0.98	2263.96 [55.57/90/90/86.85]	545.14 [60.45/0.40/19.33/0.42]

* Full form of models and BBD-CCD-Taguchi experiment set are given in appendix (Tables A9.6, A9.7, A9.8, and A9.14).

individually. However, the SON and TON models exhibited greater success with the BBD method alone. Despite the Taguchi method proposing L and LR models, these were eliminated due to failure in the testing and validation stages. Additionally, the L and LR models suggested by the BBD method were eliminated due to insufficient R^2 index and failure to meet the boundedness check criterion, respectively. With the hybrid BBD–Taguchi method, the average R^2 values of the L and LR models were 0.83 and 0.86, respectively, both of which passed the boundedness check tests. Consequently, while the combined BBD–Taguchi method improved the L and LR models, it did not lead to any enhancements in the SON and TON models.

When hybrid combinations of BBD, CCD, and Taguchi methods were assessed, it was observed that combining experimental methods did not enhance the performance of SON and TON models. In contrast, employing the BBD experimental set alone resulted in more successful SON and TON models compared to the hybrid approach. Additionally, the SOTN model generated using the BBD–CCD-Taguchi hybrid method exhibited performance similar to that of the SOTN model produced using non-hybrid methods. However, the hybrid BBD–CCD–Taguchi method yielded more realistic results for the boundedness check of the SOTN model.

Based on the conducted evaluations, BBD emerged as the most successful experimental design method, while CCD was identified as the least successful. The SON model developed using the BBD experimental set accurately predicted actual values, with the maximum and minimum Tcr values obtained from the boundedness test aligning with the maximum and minimum values derived from the analytical formula. However, it is essential to consider that the success of the BBD experimental design method may potentially be misleading. Therefore, it is pertinent to discuss whether utilizing the 25-line BBD experimental set, as opposed to the 625-line FFD set, is adequate and realistic for modeling the critical torsional buckling behavior of the composite shaft.

Despite the initial modeling results yielding favorable outcomes, it is imperative to validate the accuracy of predictions made by models generated using the experimental design methods scrutinized in the study. To facilitate a more rigorous evaluation, it is necessary to test these models using data exclusively included in the FFD dataset but not in other experimental methods.

For instance, the reliability of the BBD method can be deemed satisfactory if the SON model proposed by BBD continues to demonstrate high consistency when its performance in predicting actual TCr values is assessed using 600 lines from the FFD dataset.

Table 9.9 displays the models derived from the datasets of various experimental design methods, which initially demonstrated success in modeling the critical torsional buckling behavior of the composite shaft. The table further presents the performance of these models when tested with an FFD dataset previously unencountered by them. The assessment is conducted based on four criteria: Mean Absolute Percentage Error (MAPE), Mean Absolute Error (MAE), Root Mean Squared Error (RMSE), and R^2.

Analyzing the results in Table 9.9 for each experimental design method reveals the following conclusions:

i. The SON model proposed by CCD exhibited poor performance, failing to adequately predict actual results in the FFD dataset, as indicated in Table 9.4.

ii. Although the SON and TON models generated using the BBD dataset did not achieve the same high performance levels as seen in Table 9.5, they still demonstrated good predictive capability across all four evaluation criteria.

iii. All models derived from the D-Optimal method showcased notable predictive prowess, maintaining similar performance levels to those observed in the initial evaluation results presented in Table 9.6.

iv. While there was a slight decline in the prediction performance of the SON model derived from the Taguchi method, the decrease was not as pronounced.

TABLE 9.9

Test Phase Assessment of Models Produced Using Different Experimental Design Methods

		R^2	MAPE	MAE	RMSE
CCD	SON	−0.66	17.5	238.44	292.95
BBD	SON	0.83	5.99	76.79	94
	TON	0.81	6.15	80.31	100.59
D-Optimal	SON	0.95	3.3	40.88	49.07
	TON	0.96	2.87	35.83	44.47
	SONR	0.96	3.03	38.2	46.83
	SOTN	0.92	4.06	50.38	65.61
Taguchi	SON	0.64	7.9	102.09	136.3
BBD–CCD	SON	0.88	5.08	66.06	78.02
	TON	0.88	5.2	65.7	79.57
	SOTN	−5.45	42.24	522.51	582.28
BBD–Taguchi	L	0.58	9.5	121.68	147.2
	LR	0.57	9.67	122.65	149.91
	SON	0.91	4.19	54.65	69.31
	TON	0.31	10.06	135.49	189.61
BBD–CCD–Taguchi	SON	0.95	3.36	42.44	52.65
	TON	0.89	4.54	59.16	76.82
	SOTN	0.75	6.88	88.09	115.04

v. Comparatively, the hybrid approaches of BBD–Taguchi and BBD–CCD–Tagufchi resulted in a significant enhancement in the performance of the SON model, surpassing that of using non-hybrid methods such as BBD, CCD, and Taguchi alone. Additionally, the SOTN model, which failed to meet assessment criteria when employed individually in the three non-hybrid methods, demonstrated success when integrated into the hybrid experimental design method.

It is noteworthy that all four models proposed by D-Optimal consistently exhibited success during the testing phase using the FFD dataset. This suggests that the models derived from the D-Optimal method are more robust compared to those from other experimental design methods. However, it's important to mention that the FFD models have not undergone testing yet. To address this, all experimental design methods, including FFD, were tested using a common dataset comprising 25 lines. The design parameters, namely fiber angles, were selected as 0, 15, 30, 45, 60, 75, and 90 degrees, representing commonly preferred angles in manufacturing.

Table 9.10 presents the prediction performances obtained by testing the models proposed by the experimental design methods using a 25-row experiment set.

TABLE 9.10

Test Phase Performance Assessment of Models Produced Using Different Experimental Design Methods

Design of Experiment Method	Model	R^2	MAPE	MAE	RMSE
FFD	SON	0.93	3.68	45.24	52.79
	TON	0.95	3.09	37.63	47.21
	SOTN	−0.23	15.68	188.92	228.33
	SOTNR	−0.72	19.23	231.7	269.78
CCD	SON	−0.24	14.23	188.25	229.1
BBD	SON	0.77	6.61	86.3	99
	TON	0.74	6.83	87.28	104.45
D-Optimal	SON	0.93	3.63	45.24	53.18
	TON	0.95	3.2	38.33	47.38
	SONR	0.95	3.08	39.37	40.01
	SOTN	−0.14	14.73	177.01	219.4
Taguchi	SON	0.61	7.83	97.93	127.55
BBD–CCD	SON	0.82	6.03	76.3	88.22
	TON	0.82	5.9	73.52	87.78
	SOTN	−5.34	37.46	450.19	517.54
BBD–Taguchi	L	0.49	9.25	114.48	146.56
	LR	0.41	9.85	121.26	157.87
	SON	0.86	4.97	64.01	76.35
	TON	0.42	9.7	128.29	156.58
BBD–CCD–Taguchi	SON	0.9	4.38	54.54	64.16
	TON	0.82	5.51	68.76	86.79
	SOTN	−0.82	18.95	232.8	277.42

The results indicate that the TON and SONR models suggested by D-Optimal, along with the TON proposed by FFD, demonstrate the highest levels of success. However, models like SOTN and SOTNR, which showed promise in the initial stages of FFD (as seen in Table 9.3), failed during the testing phase (as shown in Table 9.10). Similarly, the SOTN model proposed by D-Optimal, which exhibited high success rates in Table 9.9, failed according to the test results in Table 9.10. Additionally, the hybrid utilization of experimental design methods such as BBD–CCD, BBD–Taguchi, or BBD–CCD–Taguchi improved the prediction performance of the created models compared to their individual usage.

After evaluating the results obtained throughout the entire study, it is evident that selecting the appropriate experimental design method in the modeling process is a crucial step that must be carefully considered. Both the FFD and D-Optimal experimental design methods have proven to be the most successful in modeling the critical torsional buckling behavior of the composite shaft. Unfortunately, the CCD experimental design method did not yield a model that met the necessary success criteria. However, the BBD and Taguchi methods have produced models that pass the tests, albeit with lower success rates compared to FFD and D-Optimal.

In the problem addressed in this study, the experimental sets created by BBD, CCD, and Taguchi are all of the same size, allowing for a meaningful comparison of their contributions to successful model creation. In such a comparison, BBD appears to be more successful than Taguchi and CCD.

Interestingly, upon examining the level values of the design parameters during the creation of the experimental set, it becomes apparent that the experimental sets generated by CCD and Taguchi are more comprehensive than the BBD set. This is because while the BBD method is used when each design parameter consists of three levels, the CCD and Taguchi methods account for scenarios where the design parameters consist of five levels when creating the experiment set.

One possible explanation for the CCD and Taguchi methods not being as successful as BBD, despite offering more comprehensive experimental sets, is that although the design parameters can take on more values, the number of experiments does not increase at the same rate. In other words, increasing the number of level values that design parameters can take may not always lead to a more accurate problem definition.

The notable disparities in modeling among three datasets of the same size underscore the need for a detailed examination of this issue across different problem domains. An intriguing question arises: does the combined utilization of BBD, CCD, and Taguchi methods positively impact modeling? Particularly evident in the test phases conducted using the FFD dataset and a common dataset of 25 lines, as shown in Tables 9.9 and 9.10, it becomes apparent that the combined application of these experimental design methods enhances success rates compared to individual use. This observation lends support to the notion that CCD and Taguchi methods are less effective than BBD alone. Moreover, specific to this study, it can be inferred that the success rate of models increases with the size of the experimental set, albeit not linearly. Thus, a more comprehensive understanding can be gained by evaluating the success of models created using five experimental sets varying in size.

Figure 9.6 provides a comparative analysis of the outcomes derived from five distinct experimental design methods employed in modeling the critical torsional buckling behavior of a composite shaft. The average R^2 index, illustrating the prediction performance of actual Tcr as depicted in Tables 9.9 and 9.10, focuses on the SON models proposed by these methods. Additionally, the optimization results attained through the utilization of the SON models suggested by the experimental design methods are depicted. The evaluation was solely conducted for the SON model, recognized as successful by all experimental design methods except CCD.

The findings indicate that augmenting the dataset size enhances models' capacity to depict the output more accurately. However, there was no significant discrepancy in prediction performance between the hybrid experimental design method comprising 72 rows and FFD comprising 625 rows. Consequently, it becomes meaningful to prioritize D-Optimal or hybrid methods incorporating

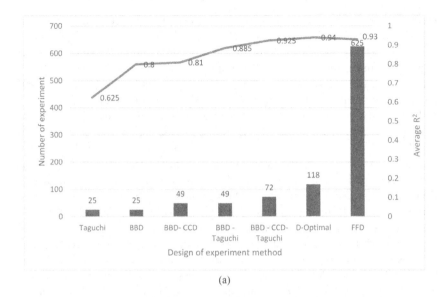

(a)

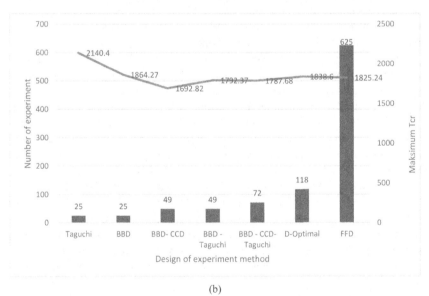

(b)

FIGURE 9.6 Comparison of experimental design methods: (a) prediction performance of mathematical models and (b) optimization results

BBD, CCD, and Taguchi methods over FFD. D-Optimal stands out among the experimental design methods, having demonstrated high prediction performance throughout all tests conducted in the study.

Additionally, while other experimental design methods failed to propose a successful rational type model, the SONR model suggested by the D-Optimal method met all criteria and test stages effectively. However, it's crucial to emphasize that although the SON model identified by D-Optimal in Figure 9.6 exhibits a slightly higher success rate than FFD, this doesn't inherently signify its superiority. As highlighted in earlier sections of this study, success rates alone do not provide a definitive measure of a model's performance. It's imperative to consider how well the model performs during boundedness checks and optimization stages.

Despite D-Optimal appearing more successful based on these assessments, the key insight is that a model offering the best description of critical torsional buckling behavior using the FFD dataset remains elusive. It is mathematically implausible for FFD, a method encompassing datasets generated by D-Optimal and other experimental design methods, to propose a less successful model. Consistent conclusions were drawn throughout the study. For instance, when comparing the SON models generated by FFD and BBD in Tables 9.3 and 9.5, the BBD method emerges as more successful than FFD.

Indeed, while the SON model proposed by BBD demonstrated accurate predictions of actual values throughout all training, testing, and validation phases, it later exhibited a lack of similar success in the subsequent test phases outlined in Tables 9.9 and 9.10. From a mathematical standpoint, this outcome is to be expected. It's essential to remember that the FFD design consistently offers the most accurate representation of the problem, and efforts are made to approximate this design with other experimental methods.

Furthermore, the optimization results obtained using the SON models proposed by each experimental design method align with their respective modeling achievements. For instance, the model suggested by Taguchi estimated the Tcr value as 2140.4 Nm, with an approximate 17% error compared to the actual value obtained using the analytical formula. This underscores the importance of thorough evaluation and consideration of various factors when assessing the performance of different experimental design methods in modeling complex phenomena.

Indeed, in comparison, D-Optimal and FFD estimated the Tcr value as 1838 Nm and 1825 Nm, respectively, with an error of approximately 3%. On the other hand, the models proposed by the BBD–Taguchi and BBD–CCD–Taguchi methods, utilizing hybrid experimental design approaches, yielded the closest results, estimating the Tcr value as 1792 Nm and 1787 Nm, respectively, with an error of approximately 0.5%, aligning closely with the optimization result obtained using the analytical formula.

It's noteworthy that in this specific problem, the success of D-Optimal, BBD, and hybrid methods, presented as alternatives to FFD, is remarkable. As a suggestion for future studies, given that the D-Optimal method is an optimization-based experimental design approach, further exploration into its application across various problems, comparison with other experimental design methods, and evaluation of its advantages and disadvantages would be beneficial. Such investigations could provide valuable insights into the efficacy and versatility of different experimental design methodologies.

REFERENCES

[1] Hong H, Kim G, Lee H, Kim J, Lee D, Kim M, Lee J ve diğ. Optimal location of brake pad for reduction of temperature deviation on brake disc during high-energy braking. Journal of Mechanical Science and Technology 2021; 35: 1109–1120.

[2] Jafari R, Akyüz R. Optimization and thermal analysis of radial ventilated brake disc to enhance the cooling performance. Case Studies in Thermal Engineering 2022; 3: 101731.

[3] Huang Y, Gao X, Ma B, Liu G, Zhang N, Zhang Y, You D. Optimization of weld strength for laser welding of steel to PMMA using Taguchi design method. Optics & Laser Technology 2021; 136: 106726.

[4] Kashyzadeh KR, Rahimian Koloor SS, Omidi Bidgoli M, Petrů M, Amiri Asfarjani A. An optimum fatigue design of polymer composite compressed natural gas tank using hybrid finite element-response surface methods. Polymers 2021; 13(4): 483.

[5] Zhang Y, Chen S, Shahin ME, Niu X, Gao L, Chin CMM, Goyal A, et al. Multi-objective optimization of lithium-ion battery pack casing for electric vehicles: Key role of materials design and their influence. International Journal of Energy Research 2020; 44(12): 9414–9437.

[6] Mohamed OA, Masood SH, Bhowmik JL. Analytical modelling and optimization of the temperature-dependent dynamic mechanical properties of fused deposition fabricated parts made of PC-ABS. Materials 2016; 9(11): 895.

[7] Pahange H, Abolbashari MH. Mass and performance optimization of an airplane wing leading edge structure against bird strike using Taguchi-based grey relational analysis. Chinese Journal of Aeronautics 2016; 29(4): 934–944.

[8] Arora VK, Bhushan G, Aggarwal ML. Enhancement of fatigue life of multi-leaf spring by parameter optimization using RSM. Journal of the Brazilian Society of Mechanical Sciences and Engineering 2017; 39: 1333–1349.

[9] GC MP, Krishna P, Parappagoudar MB. Squeeze casting process modeling by a conventional statistical regression analysis approach. Applied Mathematical Modelling 2016; 40(15–16): 6869–6888.

[10] Ye W, Wang X, Liu Y, Chen J. Analysis and prediction of the performance of free-piston Stirling engine using response surface methodology and artificial neural network. Applied Thermal Engineering 2021; 188: 116557.

[11] Kun L, Shilin Y, Yucheng Z, Wenfeng P, Gang Z. Multi-objective optimization of the fiber-reinforced composite injection molding process using Taguchi method, RSM, and NSGA-II. Simulation Modelling Practice and Theory 2019; 91: 69–82.

[12] Veljković VB, Veličković AV, Avramović JM, Stamenković OS. Modeling of biodiesel production: Performance comparison of Box–Behnken, face central composite and full factorial design. Chinese Journal of Chemical Engineering 2019; 27(7): 1690–1698.

[13] Souzangarzadeh H, Jahan A, Rezvani MJ, Milani AS. Multi-objective optimization of cylindrical segmented tubes as energy absorbers under oblique crushes: D-optimal design and integration of MULTIMOORA with combinative weighting. Structural and Multidisciplinary Optimization 2020; 62: 249–268.

[14] Li B, Ge W, Liu D, Tan C, Sun B. Optimization method of vehicle handling stability based on response surface model with D-optimal test design. Journal of Mechanical Science and Technology 2020; 34: 2267–2276.

[15] Kumar J, Verma RK, Mondal AK. Taguchi-grey theory based harmony search algorithm (GR-HSA) for predictive modeling and multi-objective optimization in drilling of polymer composites. Experimental Techniques 2021; 45: 531–548.

[16] Savran M, Ayakdaş O, Aydin L, Dizlek ME. Design and optimization of glass reinforced composite driveshafts for automotive industry. Designing Engineering Structures Using Stochastic Optimization Methods 2020; 44–58.

[17] Kaw AK. Mechanics of Composite Materials. London: CRC, 2006.

APPENDIX

TABLE A9.1

FFD Experiment Set (The Experiment Set Used to Form the Models in Table 9.3)

Trial	x_1	x_2	x_3	x_4	T_{cr}
1	90	67.5	0	45	1448.41
2	0	67.5	45	22.5	1142.76
3	90	0	45	67.5	1448.41
4	22.5	0	90	67.5	1451.31
5	67.5	22.5	90	0	1451.31
6	45	45	90	0	1283.42
7	22.5	67.5	45	0	1142.76
8	67.5	22.5	0	0	1129.89
9	45	22.5	0	22.5	876.791
10	67.5	45	22.5	67.5	1259.5
11	45	67.5	0	90	1448.41
12	90	22.5	45	90	1496.78
13	45	45	22.5	45	886.052
14	22.5	0	45	45	927.346
15	22.5	0	45	22.5	876.791
16	90	0	22.5	22.5	1238.2
17	90	22.5	90	0	1562.14
18	22.5	22.5	22.5	0	802.671
19	90	45	0	22.5	1273.73
20	22.5	67.5	90	67.5	1521.02
21	45	0	67.5	0	1179.65
22	67.5	67.5	22.5	45	1259.5
23	90	22.5	0	90	1562.14
24	45	67.5	0	0	1179.65
25	90	22.5	45	0	1273.73
26	67.5	45	45	90	1313.57
27	90	45	22.5	0	1273.73
28	67.5	45	67.5	90	1431.08
29	22.5	90	90	67.5	1624.48
30	67.5	0	22.5	22.5	1105.63
31	22.5	67.5	0	45	1142.76
32	0	90	90	0	1595.7

(Continued)

TABLE A9.1 *(Continued)*
FFD Experiment Set (The Experiment Set Used to Form the Models in Table 9.3)

Trial	x_1	x_2	x_3	x_4	T_{cr}
33	45	67.5	67.5	45	1187.43
34	67.5	0	45	90	1448.41
35	0	0	0	0	822.617
36	90	22.5	22.5	22.5	1208.48
37	90	45	45	67.5	1313.57
38	67.5	67.5	22.5	67.5	1409.37
39	0	22.5	67.5	22.5	1105.63
40	67.5	90	22.5	90	1624.48
41	0	45	67.5	22.5	1142.76
42	45	22.5	90	67.5	1382.41
43	45	22.5	22.5	22.5	845.169
44	45	22.5	90	45	1220.97
45	45	90	22.5	45	1220.97
46	67.5	45	22.5	90	1382.41
47	45	90	0	67.5	1448.41
48	67.5	45	67.5	0	1329.82
49	0	45	22.5	67.5	1142.76
50	90	90	90	67.5	1701.07
51	22.5	0	90	90	1562.14
52	0	22.5	0	90	1260.15
53	45	45	0	22.5	927.346
54	0	67.5	67.5	22.5	1332.75
55	45	45	45	0	950.755
56	22.5	22.5	22.5	22.5	782.853
57	90	67.5	0	67.5	1591.73
58	22.5	22.5	90	0	1238.2
59	90	90	67.5	0	1691.55
60	0	45	45	90	1283.42
61	67.5	0	22.5	67.5	1332.75
62	45	0	90	0	1307.69
63	22.5	90	0	90	1562.14
64	45	0	67.5	90	1448.41
65	45	67.5	0	67.5	1329.82
66	90	90	45	90	1634.29
67	67.5	22.5	45	22.5	1095.44
68	45	22.5	22.5	45	881.309
69	0	0	45	22.5	900.073
70	67.5	67.5	0	90	1591.73
71	45	45	90	90	1430.54
72	22.5	0	67.5	67.5	1332.75

(Continued)

TABLE A9.1 *(Continued)*
FFD Experiment Set (The Experiment Set Used to Form the Models in Table 9.3)

Trial	x_1	x_2	x_3	x_4	T_{cr}
73	22.5	90	45	67.5	1382.41
74	45	90	22.5	22.5	1229.94
75	45	45	67.5	90	1313.57
76	90	45	45	0	1283.42
77	0	45	90	22.5	1273.73
78	90	22.5	90	22.5	1519
79	90	90	0	45	1558.93
80	67.5	67.5	22.5	22.5	1283.42
81	22.5	22.5	0	0	815.435
82	67.5	90	45	22.5	1382.41
83	67.5	45	0	45	1152.75
84	45	90	90	90	1634.29
85	22.5	22.5	67.5	45	1095.44
86	0	90	0	67.5	1487.22
87	90	22.5	45	22.5	1229.94
88	45	45	0	67.5	1152.75
89	22.5	0	67.5	0	1129.89
90	0	0	90	90	1595.7
91	22.5	67.5	90	22.5	1405.28
92	22.5	22.5	90	22.5	1208.48
93	90	90	0	90	1783.96
94	45	90	45	0	1283.42
95	0	0	22.5	67.5	1129.89
96	67.5	90	0	45	1448.41
97	45	0	67.5	22.5	1142.76
98	90	45	22.5	45	1220.97
99	22.5	0	90	45	1273.73
100	90	22.5	67.5	67.5	1521.02
101	0	90	22.5	90	1562.14
102	45	45	67.5	22.5	1085.77
103	22.5	45	67.5	90	1382.41
104	67.5	0	22.5	90	1451.31
105	67.5	22.5	67.5	22.5	1283.42
106	22.5	45	90	0	1273.73
107	22.5	0	0	90	1260.15
108	0	45	45	0	963.085
109	67.5	0	0	22.5	1129.89
110	90	22.5	0	0	1260.15
111	45	0	0	22.5	900.073
112	67.5	90	0	0	1487.22

(Continued)

TABLE A9.1 *(Continued)*
FFD Experiment Set (The Experiment Set Used to Form the Models in Table 9.3)

Trial	x_1	x_2	x_3	x_4	T_{cr}
113	45	45	0	45	950.755
114	22.5	67.5	22.5	67.5	1283.42
115	90	67.5	22.5	22.5	1405.28
116	22.5	45	0	67.5	1142.76
117	45	45	90	22.5	1220.97
118	90	22.5	22.5	67.5	1405.28
119	90	45	67.5	90	1537.11
120	0	22.5	45	45	927.346
121	22.5	67.5	0	0	1129.89
122	0	67.5	90	0	1487.22
123	67.5	90	67.5	67.5	1520.7
124	67.5	67.5	67.5	22.5	1409.37
125	45	0	0	67.5	1179.65
126	90	67.5	90	90	1701.07
127	90	67.5	45	0	1448.41
128	67.5	90	67.5	90	1615.54
129	22.5	0	67.5	90	1451.31
130	45	22.5	0	67.5	1142.76
131	90	90	67.5	67.5	1615.54
132	67.5	22.5	45	0	1142.76
133	0	45	0	90	1307.69
134	90	90	90	45	1634.29
135	90	22.5	67.5	90	1624.48
136	22.5	45	0	0	900.073
137	0	22.5	90	0	1260.15
138	22.5	45	67.5	22.5	1095.44
139	0	67.5	90	22.5	1451.31
140	67.5	22.5	45	67.5	1259.5
141	45	22.5	0	0	900.073
142	90	22.5	90	67.5	1624.48
143	45	67.5	67.5	90	1431.08
144	67.5	22.5	90	90	1624.48
145	22.5	45	22.5	22.5	845.169
146	90	22.5	67.5	22.5	1405.28
147	45	90	90	22.5	1496.78
148	67.5	22.5	22.5	0	1105.63
149	67.5	45	45	0	1152.75
150	22.5	67.5	90	0	1451.31
151	0	0	22.5	45	900.073
152	90	67.5	22.5	0	1451.31

(Continued)

TABLE A9.1 *(Continued)*

FFD Experiment Set (The Experiment Set Used to Form the Models in Table 9.3)

Trial	x_1	x_2	x_3	x_4	T_{cr}
153	67.5	67.5	90	45	1431.08
154	45	90	45	67.5	1313.57
155	0	0	0	22.5	821.884
156	22.5	45	45	0	927.346
157	90	0	45	0	1307.69
158	90	0	0	67.5	1487.22
159	90	45	45	22.5	1220.97
160	67.5	90	22.5	45	1382.41
161	45	90	45	22.5	1220.97
162	0	0	90	22.5	1260.15
163	0	67.5	45	90	1448.41
164	67.5	45	0	0	1179.65
165	67.5	90	45	0	1448.41
166	22.5	22.5	67.5	22.5	1073.1
167	0	67.5	67.5	67.5	1484.23
168	90	90	90	22.5	1720.03
169	0	45	22.5	0	900.073
170	0	22.5	45	0	900.073
171	90	90	0	67.5	1691.55
172	45	90	45	45	1170.26
173	0	0	0	45	916.046
174	0	67.5	0	90	1487.22
175	67.5	90	90	67.5	1615.54
176	67.5	45	67.5	67.5	1316.02
177	0	67.5	67.5	0	1371.35
178	90	22.5	67.5	45	1382.41
179	22.5	45	0	90	1273.73
180	22.5	45	90	67.5	1382.41
181	0	22.5	67.5	67.5	1332.75
182	45	0	90	67.5	1448.41
183	0	0	45	67.5	1179.65
184	45	22.5	67.5	0	1142.76
185	22.5	22.5	0	45	876.791
186	67.5	0	90	22.5	1451.31
187	67.5	90	0	90	1691.55
188	90	45	67.5	0	1448.41
189	0	22.5	90	67.5	1451.31
190	22.5	45	0	22.5	876.791
191	22.5	90	22.5	45	1229.94
192	0	67.5	45	0	1179.65

(Continued)

TABLE A9.1 *(Continued)*
FFD Experiment Set (The Experiment Set Used to Form the Models in Table 9.3)

Trial	x_1	x_2	x_3	x_4	T_{cr}
193	67.5	22.5	0	67.5	1332.75
194	67.5	0	90	45	1448.41
195	67.5	0	67.5	0	1371.35
196	22.5	0	45	0	900.073
197	90	0	0	22.5	1260.15
198	22.5	0	22.5	67.5	1105.63
199	45	45	90	67.5	1313.57
200	45	22.5	22.5	0	876.791
201	67.5	45	22.5	22.5	1095.44
202	0	90	45	22.5	1273.73
203	67.5	90	90	0	1691.55
204	22.5	67.5	0	22.5	1105.63
205	0	22.5	0	67.5	1129.89
206	0	45	22.5	45	927.346
207	67.5	22.5	22.5	45	1095.44
208	90	22.5	22.5	45	1229.94
209	90	67.5	0	22.5	1451.31
210	0	22.5	45	67.5	1142.76
211	90	45	90	45	1430.54
212	90	90	22.5	90	1720.03
213	0	45	0	67.5	1179.65
214	22.5	90	22.5	67.5	1405.28
215	0	90	67.5	45	1448.41
216	90	67.5	90	0	1691.55
217	67.5	67.5	67.5	67.5	1416.54
218	0	0	67.5	90	1487.22
219	0	67.5	90	45	1448.41
220	45	22.5	0	45	927.346
221	67.5	0	22.5	0	1129.89
222	90	0	67.5	67.5	1591.73
223	22.5	0	90	0	1260.15
224	67.5	67.5	67.5	45	1316.02
225	67.5	90	22.5	22.5	1405.28
226	90	67.5	67.5	45	1431.08
227	90	0	67.5	45	1448.41
228	22.5	67.5	22.5	22.5	1073.1
229	67.5	45	67.5	45	1187.43
230	22.5	0	0	67.5	1129.89
231	90	67.5	45	45	1313.57
232	22.5	22.5	90	90	1519

(Continued)

TABLE A9.1 *(Continued)*
FFD Experiment Set (The Experiment Set Used to Form the Models in Table 9.3)

Trial	x_1	x_2	x_3	x_4	T_{cr}
233	0	90	45	67.5	1448.41
234	67.5	67.5	67.5	0	1484.23
235	0	90	0	0	1275.35
236	67.5	67.5	0	67.5	1484.23
237	0	90	90	90	1783.96
238	22.5	0	22.5	45	876.791
239	67.5	0	67.5	45	1329.82
240	67.5	90	0	67.5	1591.73
241	0	45	0	0	916.046
242	22.5	67.5	22.5	45	1095.44
243	0	0	45	45	963.085
244	90	90	22.5	0	1562.14
245	22.5	67.5	45	22.5	1095.44
246	90	22.5	90	90	1720.03
247	0	22.5	67.5	90	1451.31
248	45	0	67.5	45	1152.75
249	45	22.5	45	45	886.052
250	67.5	67.5	22.5	0	1332.75
251	45	67.5	45	67.5	1187.43
252	0	22.5	22.5	67.5	1105.63
253	22.5	90	45	45	1220.97
254	45	22.5	45	22.5	881.309
255	0	45	90	67.5	1448.41
256	22.5	67.5	0	67.5	1332.75
257	90	22.5	0	45	1273.73
258	22.5	90	0	22.5	1238.2
259	90	0	22.5	67.5	1451.31
260	67.5	45	22.5	45	1085.77
261	22.5	67.5	67.5	67.5	1409.37
262	0	0	67.5	67.5	1371.35
263	0	67.5	0	45	1179.65
264	0	90	67.5	90	1691.55
265	90	45	22.5	90	1496.78
266	22.5	90	90	0	1562.14
267	0	67.5	22.5	0	1129.89
268	67.5	0	90	67.5	1591.73
269	90	67.5	67.5	90	1615.54
270	45	90	22.5	90	1496.78
271	0	22.5	22.5	0	815.435
272	22.5	90	67.5	45	1382.41

(Continued)

TABLE A9.1 *(Continued)*
FFD Experiment Set (The Experiment Set Used to Form the Models in Table 9.3)

Trial	x_1	x_2	x_3	x_4	T_{cr}
273	22.5	45	90	90	1496.78
274	0	0	45	90	1307.69
275	67.5	90	45	90	1537.11
276	67.5	67.5	0	0	1371.35
277	0	45	90	45	1283.42
278	67.5	45	67.5	22.5	1259.5
279	67.5	22.5	0	22.5	1105.63
280	45	45	45	67.5	1032.22
281	67.5	67.5	45	22.5	1259.5
282	45	0	0	0	916.046
283	0	67.5	22.5	90	1451.31
284	0	0	22.5	22.5	815.435
285	90	0	67.5	0	1487.22
286	22.5	0	0	0	821.884
287	22.5	0	0	45	900.073
288	45	67.5	67.5	0	1329.82
289	0	90	90	45	1558.93
290	0	90	0	90	1595.7
291	45	90	90	67.5	1537.11
292	90	67.5	90	45	1537.11
293	90	0	22.5	0	1260.15
294	22.5	22.5	45	45	881.309
295	45	90	0	22.5	1273.73
296	90	90	0	0	1595.7
297	0	45	90	0	1307.69
298	45	0	45	90	1283.42
299	67.5	22.5	22.5	22.5	1073.1
300	67.5	45	90	45	1313.57
301	22.5	22.5	45	22.5	845.169
302	22.5	90	0	45	1273.73
303	45	67.5	45	45	1032.22
304	67.5	0	0	45	1179.65
305	0	22.5	90	90	1562.14
306	90	45	90	67.5	1537.11
307	22.5	0	67.5	45	1142.76
308	67.5	90	22.5	0	1451.31
309	22.5	0	45	67.5	1142.76
310	67.5	90	45	45	1313.57
311	22.5	0	22.5	22.5	802.671
312	67.5	45	90	90	1537.11

(Continued)

TABLE A9.1 *(Continued)*

FFD Experiment Set (The Experiment Set Used to Form the Models in Table 9.3)

Trial	x_1	x_2	x_3	x_4	T_{cr}
313	90	0	90	22.5	1562.14
314	90	67.5	90	22.5	1624.48
315	0	0	67.5	0	1146.99
316	67.5	67.5	45	45	1187.43
317	45	22.5	90	90	1496.78
318	45	67.5	90	0	1448.41
319	0	90	67.5	22.5	1451.31
320	45	22.5	45	67.5	1085.77
321	0	90	22.5	67.5	1451.31
322	45	22.5	67.5	90	1382.41
323	90	22.5	0	22.5	1238.2
324	67.5	45	45	22.5	1085.77
325	22.5	67.5	67.5	22.5	1283.42
326	45	0	22.5	90	1273.73
327	22.5	22.5	45	67.5	1095.44
328	90	45	0	90	1558.93
329	22.5	90	67.5	22.5	1405.28
330	67.5	67.5	45	67.5	1316.02
331	0	90	45	45	1283.42
332	45	22.5	22.5	90	1229.94
333	22.5	0	67.5	22.5	1105.63
334	0	0	45	0	916.046
335	45	67.5	45	90	1313.57
336	0	45	0	22.5	900.073
337	67.5	67.5	45	90	1431.08
338	0	22.5	0	22.5	815.435
339	45	22.5	90	0	1273.73
340	90	90	22.5	22.5	1519
341	45	67.5	90	45	1313.57
342	67.5	67.5	0	45	1329.82
343	67.5	22.5	67.5	67.5	1409.37
344	67.5	22.5	45	45	1085.77
345	90	45	0	45	1283.42
346	67.5	22.5	67.5	45	1259.5
347	22.5	90	67.5	67.5	1521.02
348	45	22.5	67.5	45	1085.77
349	90	0	67.5	22.5	1451.31
350	90	67.5	45	22.5	1382.41
351	0	90	22.5	45	1273.73
352	90	45	67.5	45	1313.57

(Continued)

TABLE A9.1 *(Continued)*
FFD Experiment Set (The Experiment Set Used to Form the Models in Table 9.3)

Trial	x_1	x_2	x_3	x_4	T_{cr}
353	67.5	45	90	67.5	1431.08
354	67.5	0	90	90	1691.55
355	90	45	45	90	1430.54
356	45	90	0	90	1558.93
357	0	22.5	0	0	821.884
358	45	0	22.5	0	900.073
359	45	90	67.5	90	1537.11
360	0	0	22.5	0	821.884
361	90	45	67.5	67.5	1431.08
362	0	90	90	22.5	1562.14
363	0	0	90	45	1307.69
364	67.5	0	67.5	22.5	1332.75
365	0	22.5	90	45	1273.73
366	67.5	22.5	22.5	67.5	1283.42
367	45	0	90	90	1558.93
368	22.5	0	45	90	1273.73
369	22.5	0	90	22.5	1238.2
370	45	45	67.5	45	1032.22
371	22.5	90	22.5	90	1519
372	90	90	0	22.5	1562.14
373	45	22.5	90	22.5	1229.94
374	90	90	67.5	90	1701.07
375	67.5	90	22.5	67.5	1521.02
376	90	90	67.5	45	1537.11
377	0	90	90	67.5	1691.55
378	90	0	0	90	1595.7
379	45	0	45	45	950.755
380	90	0	45	22.5	1273.73
381	0	45	67.5	67.5	1329.82
382	22.5	67.5	45	45	1085.77
383	90	22.5	67.5	0	1451.31
384	45	90	45	90	1430.54
385	90	22.5	45	67.5	1382.41
386	0	22.5	90	22.5	1238.2
387	67.5	22.5	90	67.5	1521.02
388	45	90	90	45	1430.54
389	90	67.5	22.5	45	1382.41
390	67.5	0	0	0	1146.99
391	90	0	90	67.5	1691.55
392	45	22.5	67.5	22.5	1095.44

(Continued)

TABLE A9.1 *(Continued)*
FFD Experiment Set (The Experiment Set Used to Form the Models in Table 9.3)

Trial	x_1	x_2	x_3	x_4	T_{cr}
393	90	22.5	22.5	0	1238.2
394	0	0	90	67.5	1487.22
395	90	45	0	67.5	1448.41
396	90	67.5	67.5	0	1591.73
397	67.5	67.5	90	90	1615.54
398	90	67.5	45	67.5	1431.08
399	90	67.5	22.5	67.5	1521.02
400	0	90	45	90	1558.93
401	90	0	22.5	45	1273.73
402	67.5	22.5	45	90	1382.41
403	45	0	0	90	1307.69
404	67.5	0	45	0	1179.65
405	22.5	45	45	67.5	1085.77
406	22.5	67.5	22.5	0	1105.63
407	0	67.5	45	67.5	1329.82
408	67.5	22.5	0	45	1142.76
409	67.5	67.5	90	67.5	1520.7
410	0	22.5	22.5	45	876.791
411	0	67.5	22.5	22.5	1105.63
412	45	45	0	0	963.085
413	22.5	45	45	45	886.052
414	22.5	45	90	45	1220.97
415	90	67.5	22.5	90	1624.48
416	67.5	90	90	45	1537.11
417	22.5	90	0	0	1260.15
418	0	90	0	45	1307.69
419	45	45	22.5	22.5	881.309
420	45	45	22.5	67.5	1085.77
421	45	90	67.5	67.5	1431.08
422	22.5	45	22.5	45	881.309
423	0	0	67.5	45	1179.65
424	0	22.5	45	22.5	876.791
425	45	67.5	45	22.5	1085.77
426	67.5	67.5	67.5	90	1520.7
427	22.5	45	22.5	67.5	1095.44
428	22.5	67.5	90	90	1624.48
429	90	0	90	0	1595.7
430	22.5	67.5	67.5	90	1521.02
431	22.5	22.5	22.5	67.5	1073.1
432	45	0	67.5	67.5	1329.82

(Continued)

TABLE A9.1 *(Continued)*
FFD Experiment Set (The Experiment Set Used to Form the Models in Table 9.3)

Trial	x_1	x_2	x_3	x_4	T_{cr}
433	67.5	22.5	90	45	1382.41
434	90	90	45	67.5	1537.11
435	67.5	0	45	22.5	1142.76
436	22.5	67.5	45	67.5	1259.5
437	22.5	67.5	67.5	0	1332.75
438	67.5	22.5	90	22.5	1405.28
439	45	67.5	0	45	1152.75
440	67.5	90	90	22.5	1624.48
441	67.5	0	67.5	67.5	1484.23
442	90	0	90	90	1783.96
443	22.5	22.5	90	67.5	1405.28
444	22.5	90	45	0	1273.73
445	0	67.5	45	45	1152.75
446	22.5	90	90	22.5	1519
447	0	45	67.5	90	1448.41
448	45	90	0	45	1283.42
449	22.5	90	67.5	0	1451.31
450	22.5	90	22.5	0	1238.2
451	90	67.5	90	67.5	1615.54
452	22.5	22.5	45	0	876.791
453	22.5	90	90	90	1720.03
454	22.5	90	45	90	1496.78
455	45	45	90	45	1170.26
456	90	67.5	0	90	1691.55
457	0	67.5	67.5	90	1591.73
458	0	22.5	0	45	900.073
459	22.5	45	45	22.5	881.309
460	45	90	22.5	67.5	1382.41
461	90	90	45	22.5	1496.78
462	67.5	45	0	67.5	1329.82
463	90	45	22.5	22.5	1229.94
464	67.5	45	0	90	1448.41
465	22.5	45	22.5	0	876.791
466	0	90	67.5	67.5	1591.73
467	22.5	90	45	22.5	1229.94
468	0	0	90	0	1275.35
469	67.5	22.5	67.5	90	1521.02
470	22.5	22.5	22.5	90	1208.48
471	45	90	67.5	45	1313.57
472	45	0	22.5	67.5	1142.76

(Continued)

TABLE A9.1 *(Continued)*
FFD Experiment Set (The Experiment Set Used to Form the Models in Table 9.3)

Trial	x_1	x_2	x_3	x_4	T_{cr}
473	45	0	45	67.5	1152.75
474	22.5	45	0	45	927.346
475	45	67.5	67.5	22.5	1259.5
476	90	90	22.5	67.5	1624.48
477	0	90	0	22.5	1260.15
478	67.5	0	0	90	1487.22
479	0	45	67.5	45	1152.75
480	22.5	22.5	22.5	45	845.169
481	22.5	67.5	67.5	45	1259.5
482	67.5	0	22.5	45	1142.76
483	22.5	90	22.5	22.5	1208.48
484	90	0	45	45	1283.42
485	67.5	90	67.5	22.5	1521.02
486	67.5	45	22.5	0	1142.76
487	0	45	22.5	90	1273.73
488	67.5	90	67.5	0	1591.73
489	67.5	0	0	67.5	1371.35
490	45	67.5	67.5	67.5	1316.02
491	45	22.5	45	0	927.346
492	45	67.5	22.5	67.5	1259.5
493	22.5	22.5	67.5	67.5	1283.42
494	90	0	45	90	1558.93
495	0	67.5	0	67.5	1371.35
496	0	67.5	67.5	45	1329.82
497	45	67.5	22.5	45	1085.77
498	90	67.5	45	90	1537.11
499	0	22.5	67.5	0	1129.89
500	45	0	45	0	963.085
501	45	45	45	90	1170.26
502	0	22.5	22.5	90	1238.2
503	0	22.5	22.5	22.5	802.671
504	45	45	45	22.5	886.052
505	45	45	22.5	0	927.346
506	90	0	0	0	1275.35
507	22.5	22.5	45	90	1229.94
508	45	0	90	45	1283.42
509	90	45	67.5	22.5	1382.41
510	22.5	22.5	0	67.5	1105.63
511	0	67.5	90	90	1691.55
512	0	22.5	67.5	45	1142.76

(Continued)

TABLE A9.1 *(Continued)*
FFD Experiment Set (The Experiment Set Used to Form the Models in Table 9.3)

Trial	x_1	x_2	x_3	x_4	T_{cr}
513	67.5	0	90	0	1487.22
514	67.5	22.5	0	90	1451.31
515	67.5	0	45	45	1152.75
516	0	67.5	0	0	1146.99
517	67.5	45	90	0	1448.41
518	67.5	45	90	22.5	1382.41
519	22.5	0	0	22.5	815.435
520	0	45	0	45	963.085
521	90	45	90	90	1634.29
522	90	22.5	22.5	90	1519
523	22.5	22.5	67.5	0	1105.63
524	45	67.5	22.5	0	1142.76
525	45	67.5	45	0	1152.75
526	67.5	90	45	67.5	1431.08
527	0	90	45	0	1307.69
528	90	90	90	90	1777.21
529	67.5	0	67.5	90	1591.73
530	0	0	0	90	1275.35
531	22.5	45	67.5	67.5	1259.5
532	45	67.5	90	67.5	1431.08
533	90	67.5	67.5	67.5	1520.7
534	0	67.5	22.5	67.5	1332.75
535	90	0	22.5	90	1562.14
536	90	90	67.5	22.5	1624.48
537	0	22.5	45	90	1273.73
538	90	90	45	0	1558.93
539	90	90	45	45	1430.54
540	45	0	22.5	45	927.346
541	22.5	0	22.5	90	1238.2
542	67.5	45	45	45	1032.22
543	67.5	90	67.5	45	1431.08
544	90	0	90	45	1558.93
545	45	90	22.5	0	1273.73
546	90	90	90	0	1783.96
547	67.5	90	0	22.5	1451.31
548	67.5	67.5	22.5	90	1521.02
549	90	22.5	45	45	1220.97
550	0	90	22.5	22.5	1238.2
551	0	45	22.5	22.5	876.791
552	45	90	67.5	0	1448.41

(Continued)

TABLE A9.1 *(Continued)*

FFD Experiment Set (The Experiment Set Used to Form the Models in Table 9.3)

Trial	x_1	x_2	x_3	x_4	T_{cr}
553	45	0	0	45	963.085
554	90	45	22.5	67.5	1382.41
555	22.5	22.5	0	90	1238.2
556	67.5	67.5	45	0	1329.82
557	90	45	0	0	1307.69
558	90	67.5	67.5	22.5	1521.02
559	0	45	67.5	0	1179.65
560	45	0	45	22.5	927.346
561	45	45	67.5	67.5	1187.43
562	22.5	0	22.5	0	815.435
563	67.5	22.5	22.5	90	1405.28
564	45	67.5	0	22.5	1142.76
565	22.5	45	67.5	0	1142.76
566	90	0	0	45	1307.69
567	0	67.5	90	67.5	1591.73
568	45	67.5	22.5	22.5	1095.44
569	67.5	67.5	90	22.5	1521.02
570	22.5	45	22.5	90	1229.94
571	0	90	22.5	0	1260.15
572	90	90	22.5	45	1496.78
573	0	45	45	22.5	927.346
574	90	45	90	22.5	1496.78
575	22.5	67.5	90	45	1382.41
576	22.5	22.5	90	45	1229.94
577	67.5	67.5	0	22.5	1332.75
578	45	0	90	22.5	1273.73
579	22.5	90	0	67.5	1451.31
580	45	0	22.5	22.5	876.791
581	67.5	22.5	67.5	0	1332.75
582	22.5	22.5	67.5	90	1405.28
583	22.5	67.5	0	90	1451.31
584	45	22.5	45	90	1220.97
585	45	45	45	45	849.685
586	0	90	67.5	0	1487.22
587	45	45	67.5	0	1152.75
588	0	0	22.5	90	1260.15
589	0	67.5	0	22.5	1129.89
590	45	45	0	90	1283.42
591	0	0	0	67.5	1146.99
592	45	90	0	0	1307.69

(Continued)

TABLE A9.1 *(Continued)*
FFD Experiment Set (The Experiment Set Used to Form the Models in Table 9.3)

Trial	x_1	x_2	x_3	x_4	T_{cr}
593	90	22.5	90	45	1496.78
594	22.5	45	67.5	45	1085.77
595	0	45	45	45	950.755
596	0	45	45	67.5	1152.75
597	22.5	45	90	22.5	1229.94
598	90	22.5	0	67.5	1451.31
599	45	67.5	22.5	90	1382.41
600	67.5	0	45	67.5	1329.82
601	45	90	67.5	22.5	1382.41
602	45	67.5	90	90	1537.11
603	67.5	45	0	22.5	1142.76
604	45	22.5	22.5	67.5	1095.44
605	22.5	45	45	90	1220.97
606	45	90	90	0	1558.93
607	0	67.5	22.5	45	1142.76
608	90	45	90	0	1558.93
609	45	67.5	90	22.5	1382.41
610	0	0	67.5	22.5	1129.89
611	22.5	90	67.5	90	1624.48
612	22.5	90	90	45	1496.78
613	67.5	67.5	90	0	1591.73
614	45	22.5	0	90	1273.73
615	0	45	90	90	1558.93
616	45	22.5	67.5	67.5	1259.5
617	90	67.5	0	0	1487.22
618	90	0	67.5	90	1691.55
619	22.5	67.5	45	90	1382.41
620	67.5	45	45	67.5	1187.43
621	45	45	22.5	90	1220.97
622	90	45	45	45	1170.26
623	22.5	67.5	22.5	90	1405.28
624	67.5	90	90	90	1701.07
625	22.5	22.5	0	22.5	802.671

TABLE A9.2
CCD Experiment Set (The Experiment Set Used to Form the Models in Table 9.4)

Trial	x_1	x_2	x_3	x_4	T_{cr}
1	22.5	22.5	67.5	22.5	1073.1
2	67.5	67.5	22.5	67.5	1409.37
3	22.5	67.5	67.5	67.5	1409.37
4	22.5	22.5	22.5	67.5	1073.1
5	67.5	22.5	22.5	67.5	1283.42
6	22.5	22.5	22.5	22.5	782.853
7	90	45	45	45	1170.26
8	22.5	22.5	67.5	67.5	1283.42
9	45	45	45	90	1170.26
10	0	45	45	45	950.755
11	45	0	45	45	950.755
12	45	45	0	45	950.755
13	22.5	67.5	67.5	22.5	1283.42
14	45	90	45	45	1170.26
15	22.5	67.5	22.5	22.5	1073.1
16	67.5	67.5	67.5	67.5	1416.54
17	67.5	22.5	22.5	22.5	1073.1
18	67.5	22.5	67.5	67.5	1409.37
19	22.5	67.5	22.5	67.5	1283.42
20	45	45	90	45	1170.26
21	67.5	67.5	22.5	22.5	1283.42
22	67.5	22.5	67.5	22.5	1283.42
23	67.5	67.5	67.5	22.5	1409.37
24	45	45	45	45	849.685
25	45	45	45	0	950.755

TABLE A9.3
BBD Experiment Set (The Experiment Set Used to Form the Models in Table 9.5)

Trial	x_1	x_2	x_3	x_4	T_{cr}
1	45	90	45	0	1283.42
2	0	45	0	45	963.085
3	45	0	45	90	1283.42
4	45	0	45	0	963.085
5	45	0	0	45	963.085
6	45	45	0	0	963.085

(Continued)

TABLE A9.3 *(Continued)*
BBD Experiment Set (The Experiment Set Used to Form the Models in Table 9.5)

Trial	x_1	x_2	x_3	x_4	T_{cr}
7	45	0	90	45	1283.42
8	45	45	90	0	1283.42
9	90	45	0	45	1283.42
10	0	90	45	45	1283.42
11	0	45	45	90	1283.42
12	45	45	0	90	1283.42
13	90	45	45	90	1430.54
14	90	0	45	45	1283.42
15	0	45	45	0	963.085
16	90	90	45	45	1430.54
17	90	45	90	45	1430.54
18	90	45	45	0	1283.42
19	45	45	45	45	849.685
20	45	45	90	90	1430.54
21	45	90	90	45	1430.54
22	0	0	45	45	963.085
23	45	90	0	45	1283.42
24	0	45	90	45	1283.42
25	45	90	45	90	1430.54

TABLE A9.4
D-Optimal Experiment Set (The Experiment Set Used to Form the Models in Table 9.6)

Trial	x_1	x_2	x_3	x_4	T_{cr}
1	67.5	90	67.5	67.5	1520.7
2	90	22.5	45	45	1220.97
3	90	67.5	22.5	0	1451.31
4	67.5	45	67.5	45	1187.43
5	0	22.5	0	22.5	815.435
6	45	90	22.5	67.5	1382.41
7	22.5	67.5	90	90	1624.48
8	45	0	67.5	0	1179.65
9	90	67.5	67.5	90	1615.54
10	22.5	90	45	22.5	1229.94
11	67.5	90	45	90	1537.11
12	0	45	45	22.5	927.346

(Continued)

TABLE A9.4 *(Continued)*
D-Optimal Experiment Set (The Experiment Set Used to Form the Models in Table 9.6)

Trial	x_1	x_2	x_3	x_4	T_{cr}
13	22.5	22.5	67.5	45	1095.44
14	90	45	90	45	1430.54
15	90	90	45	67.5	1537.11
16	67.5	67.5	22.5	67.5	1409.37
17	90	45	45	90	1430.54
18	45	45	22.5	0	927.346
19	67.5	22.5	22.5	45	1095.44
20	0	45	22.5	90	1273.73
21	67.5	67.5	0	90	1591.73
22	0	67.5	67.5	45	1329.82
23	90	0	22.5	90	1562.14
24	22.5	67.5	45	67.5	1259.5
25	22.5	0	0	45	900.073
26	0	0	67.5	22.5	1129.89
27	90	90	90	90	1777.21
28	0	22.5	45	67.5	1142.76
29	45	0	0	90	1307.69
30	45	67.5	45	90	1313.57
31	22.5	22.5	90	22.5	1208.48
32	45	0	22.5	45	927.346
33	90	67.5	0	45	1448.41
34	22.5	90	67.5	90	1624.48
35	0	90	22.5	22.5	1238.2
36	90	22.5	0	90	1562.14
37	0	45	0	45	963.085
38	45	22.5	67.5	22.5	1095.44
39	45	22.5	90	0	1273.73
40	0	90	90	67.5	1691.55
41	90	45	22.5	67.5	1382.41
42	22.5	90	0	67.5	1451.31
43	67.5	90	0	45	1448.41
44	22.5	22.5	0	0	815.435
45	22.5	0	67.5	67.5	1332.75
46	45	0	90	67.5	1448.41
47	0	90	0	90	1595.7
48	22.5	67.5	22.5	45	1095.44
49	67.5	0	67.5	90	1591.73
50	67.5	67.5	90	0	1591.73
51	67.5	67.5	45	45	1187.43
52	90	90	67.5	22.5	1624.48

(Continued)

TABLE A9.4 *(Continued)*
D-Optimal Experiment Set (The Experiment Set Used to Form the Models
in Table 9.6)

Trial	x_1	x_2	x_3	x_4	T_{cr}
53	0	67.5	0	67.5	1371.35
54	90	67.5	45	22.5	1382.41
55	0	45	90	0	1307.69
56	22.5	45	90	67.5	1382.41
57	90	22.5	67.5	67.5	1521.02
58	22.5	22.5	22.5	67.5	1073.1
59	90	90	0	0	1595.7
60	22.5	67.5	0	22.5	1105.63
61	90	90	22.5	45	1496.78
62	67.5	0	90	45	1448.41
63	0	22.5	90	90	1562.14
64	90	0	90	22.5	1562.14
65	67.5	45	22.5	22.5	1095.44
66	67.5	67.5	67.5	22.5	1409.37
67	0	45	67.5	67.5	1329.82
68	67.5	22.5	90	67.5	1521.02
69	90	45	0	22.5	1273.73
70	67.5	90	90	22.5	1624.48
71	0	90	45	45	1283.42
72	22.5	67.5	67.5	0	1332.75
73	90	45	67.5	0	1448.41
74	22.5	45	0	90	1273.73
75	45	67.5	67.5	67.5	1316.02
76	0	0	22.5	67.5	1129.89
77	67.5	22.5	45	22.5	1095.44
78	45	67.5	0	0	1179.65
79	0	90	67.5	0	1487.22
80	45	0	45	22.5	927.346
81	67.5	0	45	0	1179.65
82	22.5	22.5	45	90	1229.94
83	67.5	22.5	0	67.5	1332.75
84	45	45	45	45	849.685
85	45	45	90	22.5	1220.97
86	67.5	45	90	90	1537.11
87	90	22.5	45	0	1273.73
88	67.5	45	0	0	1179.65
89	45	90	90	90	1634.29
90	0	0	45	90	1307.69
91	0	0	90	45	1307.69
92	22.5	90	90	45	1496.78

(Continued)

TABLE A9.4 *(Continued)*
D-Optimal Experiment Set (The Experiment Set Used to Form the Models in Table 9.6)

Trial	x_1	x_2	x_3	x_4	T_{cr}
93	67.5	22.5	67.5	90	1521.02
94	0	22.5	22.5	0	815.435
95	22.5	0	90	0	1260.15
96	22.5	45	45	0	927.346
97	67.5	45	45	67.5	1187.43
98	45	67.5	22.5	22.5	1095.44
99	22.5	90	22.5	90	1519
100	22.5	0	22.5	22.5	802.671
101	0	67.5	90	22.5	1451.31
102	67.5	90	22.5	0	1451.31
103	45	90	45	0	1283.42
104	45	45	0	67.5	1152.75
105	45	45	67.5	90	1313.57
106	45	90	0	22.5	1273.73
107	0	0	0	0	822.617
108	45	67.5	90	45	1313.57
109	22.5	45	67.5	22.5	1095.44
110	45	22.5	0	45	927.346
111	45	22.5	22.5	90	1229.94
112	90	22.5	22.5	22.5	1208.48
113	90	0	67.5	45	1448.41
114	90	67.5	90	67.5	1615.54
115	67.5	0	0	22.5	1129.89
116	0	67.5	45	0	1179.65
117	90	0	0	67.5	1487.22
118	45	90	67.5	45	1313.57

TABLE A9.5
Taguchi Experiment Set (The Experiment Set Used to Form the Models in Table 9.7)

Trial	x_1	x_2	x_3	x_4	T_{cr}
1	0	0	0	0	822.617
2	0	22.5	22.5	22.5	802.671
3	0	45	45	45	950.755
4	0	67.5	67.5	67.5	1484.23
5	0	90	90	90	1783.96

(Continued)

TABLE A9.5 *(Continued)*
Taguchi Experiment Set (The Experiment Set Used to Form the Models in Table 9.7)

Trial	x_1	x_2	x_3	x_4	T_{cr}
6	22.5	0	22.5	45	876.791
7	22.5	22.5	45	67.5	1095.44
8	22.5	45	67.5	90	1382.41
9	22.5	67.5	90	0	1451.31
10	22.5	90	0	22.5	1238.2
11	45	0	45	90	1283.42
12	45	22.5	67.5	0	1142.76
13	45	45	90	22.5	1220.97
14	45	67.5	0	45	1152.75
15	45	90	22.5	67.5	1382.41
16	67.5	0	67.5	22.5	1332.75
17	67.5	22.5	90	45	1382.41
18	67.5	45	0	67.5	1329.82
19	67.5	67.5	22.5	90	1521.02
20	67.5	90	45	0	1448.41
21	9	0	90	67.5	1691.55
22	90	22.5	0	90	1562.14
23	90	45	22.5	0	1273.73
24	90	67.5	45	22.5	1382.41
25	90	90	67.5	45	1537.11

TABLE A9.6
BBD–CCD Hybrid Experiment Set (The Experiment Set Used to Form the Models in Table 9.8)

Trial	x_1	x_2	x_3	x_4	T_{cr}
1	45	90	45	0	1283.42
2	0	45	0	45	963.085
3	45	0	45	90	1283.42
4	45	0	45	0	963.085
5	45	0	0	45	963.085
6	45	45	0	0	963.085
7	45	0	90	45	1283.42
8	45	45	90	0	1283.42
9	90	45	0	45	1283.42
10	0	90	45	45	1283.42
11	0	45	45	90	1283.42

(Continued)

TABLE A9.6 *(Continued)*
BBD–CCD Hybrid Experiment Set (The Experiment Set Used to Form the Models in Table 9.8)

Trial	x_1	x_2	x_3	x_4	T_{cr}
12	45	45	0	90	1283.42
13	90	45	45	90	1430.54
14	90	0	45	45	1283.42
15	0	45	45	0	963.085
16	90	90	45	45	1430.54
17	90	45	90	45	1430.54
18	90	45	45	0	1283.42
19	45	45	45	45	849.685
20	45	45	90	90	1430.54
21	45	90	90	45	1430.54
22	0	0	45	45	963.085
23	45	90	0	45	1283.42
24	0	45	90	45	1283.42
25	45	90	45	90	1430.54
26	22.5	22.5	67.5	22.5	1073.1
27	67.5	67.5	22.5	67.5	1409.37
28	22.5	67.5	67.5	67.5	1409.37
29	22.5	22.5	22.5	67.5	1073.1
30	67.5	22.5	22.5	67.5	1283.42
31	22.5	22.5	22.5	22.5	782.853
32	90	45	45	45	1170.26
33	22.5	22.5	67.5	67.5	1283.42
34	45	45	45	90	1170.26
35	0	45	45	45	950.755
36	45	0	45	45	950.755
37	45	45	0	45	950.755
38	22.5	67.5	67.5	22.5	1283.42
39	45	90	45	45	1170.26
40	22.5	67.5	22.5	22.5	1073.1
41	67.5	67.5	67.5	67.5	1416.54
42	67.5	22.5	22.5	22.5	1073.1
43	67.5	22.5	67.5	67.5	1409.37
44	22.5	67.5	22.5	67.5	1283.42
45	45	45	90	45	1170.26
46	67.5	67.5	22.5	22.5	1283.42
47	67.5	22.5	67.5	22.5	1283.42
48	67.5	67.5	67.5	22.5	1409.37
49	45	45	45	0	950.755

TABLE A9.7
BBD–Taguchi Hybrid Experiment Set (The Experiment Set Used to Form the Models in Table 9.8)

Trial	x_1	x_2	x_3	x_4	T_{cr}
1	45	90	45	0	1283.42
2	0	45	0	45	963.085
3	45	0	45	0	963.085
4	45	0	0	45	963.085
5	45	45	0	0	963.085
6	45	0	90	45	1283.42
7	45	45	90	0	1283.42
8	90	45	0	45	1283.42
9	0	90	45	45	1283.42
10	0	45	45	90	1283.42
11	45	45	0	90	1283.42
12	90	45	45	90	1430.54
13	90	0	45	45	1283.42
14	0	45	45	0	963.085
15	90	90	45	45	1430.54
16	90	45	90	45	1430.54
17	90	45	45	0	1283.42
18	45	45	45	45	849.685
19	45	45	90	90	1430.54
20	45	90	90	45	1430.54
21	0	0	45	45	963.085
22	45	90	0	45	1283.42
23	0	45	90	45	1283.42
24	45	90	45	90	1430.54
25	0	0	0	0	822.617
26	0	22.5	22.5	22.5	802.671
27	0	45	45	45	950.755
28	0	67.5	67.5	67.5	1484.23
29	0	90	90	90	1783.96
30	22.5	0	22.5	45	876.791
31	22.5	22.5	45	67.5	1095.44
32	22.5	45	67.5	90	1382.41
33	22.5	67.5	90	0	1451.31
34	22.5	90	0	22.5	1238.2
35	45	0	45	90	1283.42
36	45	22.5	67.5	0	1142.76
37	45	45	90	22.5	1220.97
38	45	67.5	0	45	1152.75
39	45	90	22.5	67.5	1382.41
40	67.5	0	67.5	22.5	1332.75

(Continued)

TABLE A9.7 *(Continued)*
BBD–Taguchi Hybrid Experiment Set (The Experiment Set Used to Form the Models in Table 9.8)

Trial	x_1	x_2	x_3	x_4	T_{cr}
41	67.5	22.5	90	45	1382.41
42	67.5	45	0	67.5	1329.82
43	67.5	67.5	22.5	90	1521.02
44	67.5	90	45	0	1448.41
45	90	0	90	67.5	1691.55
46	90	22.5	0	90	1562.14
47	90	45	22.5	0	1273.73
48	90	67.5	45	22.5	1382.41
49	90	90	67.5	45	1537.11

TABLE A9.8
BBD–CCD–Taguchi Hybrid Experiment Set (The Experiment Set Used to Form the Models in Table 9.8)

Trial	x_1	x_2	x_3	x_4	T_{cr}
1	45	90	45	0	1283.42
2	0	45	0	45	963.085
3	45	0	45	90	1283.42
4	45	0	45	0	963.085
5	45	0	0	45	963.085
6	45	45	0	0	963.085
7	45	0	90	45	1283.42
8	45	45	90	0	1283.42
9	90	45	0	45	1283.42
10	0	90	45	45	1283.42
11	0	45	45	90	1283.42
12	45	45	0	90	1283.42
13	90	45	45	90	1430.54
14	90	0	45	45	1283.42
15	0	45	45	0	963.085
16	90	90	45	45	1430.54
17	90	45	90	45	1430.54
18	90	45	45	0	1283.42
19	45	45	45	45	849.685
20	45	45	90	90	1430.54
21	45	90	90	45	1430.54
22	0	0	45	45	963.085

(Continued)

TABLE A9.8 *(Continued)*
BBD–CCD–Taguchi Hybrid Experiment Set (The Experiment Set Used to Form the Models in Table 9.8)

Trial	x_1	x_2	x_3	x_4	T_{cr}
23	45	90	0	45	1283.42
24	0	45	90	45	1283.42
25	45	90	45	90	1430.54
26	22.5	22.5	67.5	22.5	1073.1
27	67.5	67.5	22.5	67.5	1409.37
28	22.5	67.5	67.5	67.5	1409.37
29	22.5	22.5	22.5	67.5	1073.1
30	67.5	22.5	22.5	67.5	1283.42
31	22.5	22.5	22.5	22.5	782.853
32	90	45	45	45	1170.26
33	22.5	22.5	67.5	67.5	1283.42
34	45	45	45	90	1170.26
35	45	0	45	45	950.755
36	45	45	0	45	950.755
37	22.5	67.5	67.5	22.5	1283.42
38	45	90	45	45	1170.26
39	22.5	67.5	22.5	22.5	1073.1
40	67.5	67.5	67.5	67.5	1416.54
41	67.5	22.5	22.5	22.5	1073.1
42	67.5	22.5	67.5	67.5	1409.37
43	22.5	67.5	22.5	67.5	1283.42
44	45	45	90	45	1170.26
45	67.5	67.5	22.5	22.5	1283.42
46	67.5	22.5	67.5	22.5	1283.42
47	67.5	67.5	67.5	22.5	1409.37
48	45	45	45	0	950.755
49	0	0	0	0	822.617
50	0	22.5	22.5	22.5	802.671
51	0	45	45	45	950.755
52	0	67.5	67.5	67.5	1484.23
53	0	90	90	90	1783.96
54	22.5	0	22.5	45	876.791
55	22.5	22.5	45	67.5	1095.44
56	22.5	45	67.5	90	1382.41
57	22.5	67.5	90	0	1451.31
58	22.5	90	0	22.5	1238.2
59	45	22.5	67.5	0	1142.76
60	45	45	90	22.5	1220.97
61	45	67.5	0	45	1152.75
62	45	90	22.5	67.5	1382.41

(Continued)

TABLE A9.8 *(Continued)*

BBD–CCD–Taguchi Hybrid Experiment Set (The Experiment Set Used to Form the Models in Table 9.8)

Trial	x_1	x_2	x_3	x_4	T_{cr}
63	67.5	0	67.5	22.5	1332.75
64	67.5	22.5	90	45	1382.41
65	67.5	45	0	67.5	1329.82
66	67.5	67.5	22.5	90	1521.02
67	67.5	90	45	0	1448.41
68	90	0	90	67.5	1691.55
69	90	22.5	0	90	1562.14
70	90	45	22.5	0	1273.73
71	90	67.5	45	22.5	1382.41
72	90	90	67.5	45	1537.11

TABLE A9.9

Mathematical Models Created Using the FFD Experiment Set (Related to Table 9.3)

(*L*)

$$L = 2.8496\,x1 + 2.91635\,x2 + 2.81214\,x3 + 2.92952\,x4 + 784.041$$

(*LR*)

$$LR = \frac{537.832\,x1 + 394.419\,x2 + 490.701\,x3 + 451.788\,x4 + 54193}{0.188599\,x1 + 0.0664453\,x2 + 0.153146\,x3 + 0.111684\,x4 + 81.8179}$$

(*SON*)

$$SON = 0.0659605\,x1^2 - 0.0238031\,x1\,x2 - 0.0233561\,x1\,x3 - 0.0214101\,x1\,x4 + 0.0475433\,x1 + 0.0677409\,x2^2 - 0.0240302\,x2\,x3 - 0.023447\,x2\,x4 - 0.0229136\,x2 + 0.0669032\,x3^2 - 0.0237751\,x3\,x4 + 0.0209458\,x3 + 0.0670618\,x4^2 - 0.0976322\,x4 + 770.22$$

(*SONR*)

$$SONR = (473.043\,x1^2 + 665.455\,x1\,x2 + 690.867\,x1\,x3 + 932.628\,x1\,x4 - 40693.6\,x1 + 229.283\,x2^2 + 547.991\,x2\,x3 + 652.623\,x2\,x4 - 20884.x2 + 306.979\,x3^2 + 633.446\,x3\,x4 - 25235.3\,x3 + 563.349\,x4^2 - 44634.8\,x4 - 2223.35)/(0.352134\,x1^2 + 0.498436\,x1\,x2 + 0.516191\,x1\,x3 + 0.706769\,x1\,x4 - 30.1423\,x1 + 0.176812\,x2^2 + 0.426458\,x2\,x3 + 0.497592\,x2\,x4 - 16.1052\,x2 + 0.237085\,x3^2 + 0.470722\,x3\,x4 - 19.3107\,x3 + 0.436463\,x4^2 - 34.4024\,x4 - 2.86592)$$

(*TON*)

$$TON = -(0.000667089\,x1^3) + 0.0000372361\,x1^2\,x2 + 0.0000478081\,x1^2\,x3 + 0.0000243818\,x1^2\,x4 + 0.15091\,x1^2 + 0.0000630569\,x1\,x2^2 - 0.0000850818\,x1\,x2\,x3 - 0.0000786581\,x1\,x2\,x4 - 0.0249615\,x1\,x2 + 0.0000365121\,x1\,x3^2 - 0.0000698035\,x1\,x3\,x4 - 0.0240607\,x1\,x3 +$$

(Continued)

TABLE A9.9 *(Continued)*
Mathematical Models Created Using the FFD Experiment Set (Related to Table 9.3)

$0.0000287276\ x1\ x4^2 - 0.020364\ x1\ x4 - 2.74678\ x1 - 0.000649226\ x2^3 +$
$0.0000381678\ x2^2\ x3 + 0.0000683315\ x2^2\ x4 + 0.14726\ x2^2 + 0.0000327375\ x2\ x3^2 -$
$0.0000631168\ x2\ x3\ x4 - 0.0234297\ x2\ x3 + 0.0000452957\ x2\ x4^2 - 0.0270435\ x2\ x4 -$
$2.4869\ x2 - 0.000669827\ x3^3 + 0.0000727497\ x3^2\ x4 + .150044\ x3^2 + 0.0000271207\ x3\ x4^2 -$
$0.0269203\ x3\ x4 - 2.52924\ x3 - 0.00067689\ x4^3 + 0.153684\ x4^2 - 2.82566\ x4 + 805.635$

(*TONR*)

TONR = $(-7569.53 + 278.917\ x1 - 6.63071\ x1^2 + 0.0842038\ x1^3 + 289.119\ x2 - 0.114051\ x1\ x2 +$
$0.00371097\ x1^2\ x2 - 6.74546\ x2^2 - 0.0062751\ x1\ x2^2 + 0.0898434\ x2^3 + 279.555\ x3 +$
$0.151883\ x1\ x3 - 0.00250055\ x1^2\ x3 - 0.084848\ x2\ x3 - 0.00818415\ x1\ x2\ x3 +$
$0.00116624\ x2^2\ x3 - 6.58921\ x3^2 - 0.00455424\ x1\ x3^2 + 0.000706366\ x2\ x3^2 + 0.0880142$
$x3^3 + 307.457\ x4 - 0.0966578\ x1\ x4 + 0.00483134\ x1^2\ x4 - 0.340043\ x2\ x4 -$
$0.0104066\ x1\ x2\ x4 + 0.00172169\ x2^2\ x4 + 0.0613931\ x3\ x4 - 0.0063888\ x1\ x3\ x4 -$
$0.0162241\ x2\ x3\ x4 + 0.000922097\ x3^2\ x4 - 7.1509\ x4^2 - 0.00267128\ x1\ x4^2 - 0.00298757$
$x2\ x4^2 - 0.00584524\ x3\ x4^2 + 0.0919878\ x4^3\ )/(- 9.57655 + 0.318789\ x1 -$
$0.0056927\ x1^2 + 0.0000649325\ x1^3 + 0.329745\ x2 - 0.00032029\ x1\ x2 - 5.33739\ *^\wedge -$
$6\ x1^2\ x2 - 0.00566113\ x2^2 - 0.0000126079\ x1\ x2^2 + 0.0000676779\ x2^3 + 0.320447\ x3 -$
$0.000125669\ x1\ x3 - 0.0000101961\ x1^2\ x3 - 0.00029349\ x2\ x3 + 9.83623\ *^\wedge - 6\ x1\ x2\ x3 -$
$7.46596\ *^\wedge - 6\ x2^2\ x3 - 0.00552633\ x3^2 - 0.0000105025\ x1\ x3^2 - 7.89197\ *^\wedge - 6\ x2\ x3^2 +$
$0.0000661927\ x3^3 + 0.354409\ x4 - 0.000249231\ x1\ x4 - 5.03427\ *^\wedge - 6\ x1^2\ x4 -$
$0.000636817\ x2\ x4 + 8.494849999999999\ *^\wedge - 6\ x1\ x2\ x4 - 6.67156\ *^\wedge - 6\ x2^2\ x4 -$
$0.00023257\ x3\ x4 + 0.0000104701\ x1\ x3\ x4 + 5.014789999999999\ *^\wedge - 6\ x2\ x3\ x4 -$
$7.41974\ *^\wedge - 6\ x3^2\ x4 - 0.0062506\ x4^2 - 9.940609999999998\ *^\wedge - 6\ x1\ x4^2 - 9.0828\ *^\wedge -$
$6\ x2\ x4^2 - 0.0000119214\ x3\ x4^2 + 0.0000709892\ x4^3)$

(*FOTN*)

FOTN = $31.5631\ \text{Sin}\ [x1] - 74.6143\ \text{Cos}\ [x1] + 39.0651\ \text{Sin}\ [x2] - 53.2335\ \text{Cos}\ [x2] +$
$28.7416\ \text{Sin}\ [x3] - 51.7785\ \text{Cos}\ [x3] + 33.6597\ \text{Sin}\ [x4] - 62.7918\ \text{Cos}\ [x4] + 1304.5$

(*FOTNR*)

FOTNR = $(1378.58\ \text{Sin}\ [x1] + 1546.53\ \text{Cos}\ [x1] + 1359.9\ \text{Sin}\ [x2] + 1552.7\ \text{Cos}\ [x2] +$
$1376.98\ \text{Sin}\ [x3] + 1548.75\ \text{Cos}\ [x3] + 1047.52\ \text{Sin}\ [x4] + 1647.77\ \text{Cos}\ [x4] +$
$920.833)/(1.04153\ \text{Sin}\ [x1] + 1.19184\ \text{Cos}\ [x1] + 1.02861\ \text{Sin}\ [x2] + 1.19515\ \text{Cos}\ [x2] +$
$1.04029\ \text{Sin}\ [x3] + 1.19237\ \text{Cos}\ [x3] + 0.793088\ \text{Sin}\ [x4] + 1.27206\ \text{Cos}\ [x4] + 0.716988)$

(*SOTN*)

SOTN = $1.09343\ \text{Sin}\ [x1]\ \text{Sin}\ [x2] - 7.37557\ \text{Cos}\ [x1]\ \text{Cos}\ [x2] + 4.67355\ \text{Cos}\ [x1]\ \text{Sin}\ [x2] +$
$5.24043\ \text{Sin}\ [x1]\ \text{Cos}\ [x2] + 1.82499\ \text{Sin}\ [x1]\ \text{Sin}\ [x3] - 6.85035\ \text{Cos}\ [x1]\ \text{Cos}\ [x3] +$
$4.74853\ \text{Cos}\ [x1]\ \text{Sin}\ [x3] + 3.47443\ \text{Sin}\ [x1]\ \text{Cos}\ [x3] - 4.08693\ \text{Sin}\ [x1]\ \text{Sin}\ [x4] -$
$5.90723\ \text{Cos}\ [x1]\ \text{Cos}\ [x4] + 9.52502\ \text{Cos}\ [x1]\ \text{Sin}\ [x4] + 6.44875\ \text{Sin}\ [x1]\ \text{Cos}\ [x4] +$
$342.659\ \text{Sin}\ [x1]^2 - 1.48947\ \text{Sin}\ [x1] + 236.799\ \text{Cos}\ [x1]^2 - 32.2678\ \text{Cos}\ [x1] -$
$280.771\ \text{Sin}\ [x1]\ \text{Cos}\ [x1] + 0.365213\ \text{Sin}\ [x2]\ \text{Sin}\ [x3] - 4.16157\ \text{Cos}\ [x2]\ \text{Cos}\ [x3] +$
$4.00382\ \text{Sin}\ [x2]\ \text{Cos}\ [x3] - 0.691984\ \text{Cos}\ [x2]\ \text{Sin}\ [x3] - 4.8343\ \text{Sin}\ [x2]\ \text{Sin}\ [x4] -$
$11.5662\ \text{Cos}\ [x2]\ \text{Cos}\ [x4] + 3.69343\ \text{Sin}\ [x2]\ \text{Cos}\ [x4] + 6.67829\ \text{Cos}\ [x2]\ \text{Sin}\ [x4] +$
$343.259\ \text{Sin}\ [x2]^2 - 1.40853\ \text{Sin}\ [x2] + 240.401\ \text{Cos}\ [x2]^2 - 31.7125\ \text{Cos}\ [x2] -$

(Continued)

TABLE A9.9 *(Continued)*
Mathematical Models Created Using the FFD Experiment Set
(Related to Table 9.3)

288.609 Sin [$x2$] Cos [$x2$] − 2.6375 Sin [$x3$] Sin [$x4$] − 10.0159 Cos [$x3$] Cos [$x4$] +
3.44834 Sin [$x3$] Cos [$x4$] + 6.40372 Cos [$x3$] Sin [$x4$] + 341.842 Sin [$x3$]^2 −
4.16604 Sin [$x3$] + 237.765 Cos [$x3$]^2 − 29.5008 Cos [$x3$] − 286.581 Sin [$x3$] Cos [$x3$] +
344.804 Sin [$x4$]^2 − 2.89699 Sin [$x4$] + 238.996 Cos [$x4$]^2 − 29.172 Cos [$x4$] −
284.473 Sin [$x4$] Cos [$x4$] + 233.32

(*SOTNR*)

SOTNR = (−(4053.81 Sin [$x1$] Sin [$x2$]) − 39.451 Cos [$x1$] Cos [$x2$] + 1585.92 Cos [$x1$] Sin [$x2$] +
2093.74 Sin [$x1$] Cos [$x2$] − 3310.64 Sin [$x1$] Sin [$x3$] − 407.821 Cos [$x1$] Cos [$x3$] +
1881.78 Cos [$x1$] Sin [$x3$] + 1701.06 Sin [$x1$] Cos [$x3$] − 2527.12 Sin [$x1$] Sin [$x4$] −
777.725 Cos [$x1$] Cos [$x4$] + 593.268 Cos [$x1$] Sin [$x4$] + 1109.73 Sin [$x1$] Cos [$x4$] −
130 056. Sin [$x1$]^2 + 2566.79 Sin [$x1$] − 168884 Cos [$x1$]^2 − 5430.86 Cos [$x1$] −
38938.2 Sin [$x1$] Cos [$x1$] − 3078.53 Sin [$x2$] Sin [$x3$] − 461.67 Cos [$x2$] Cos [$x3$] +
1253.43 Sin [$x2$] Cos [$x3$] + 2484.07 Cos [$x2$] Sin [$x3$] − 27.8341 Sin [$x2$] Sin [$x4$] +
455.826 Cos [$x2$] Cos [$x4$] + 1308.55 Sin [$x2$] Cos [$x4$] + 96.9721 Cos [$x2$] Sin [$x4$] +
332386.Sin [$x2$]^2 + 2191.74 Sin [$x2$] + 291767 Cos [$x2$]^2 − 4388.98 Cos [$x2$] −
34635.8 Sin [$x2$] Cos [$x2$] − 2416.37 Sin [$x3$] Sin [$x4$] + 603.53 Cos [$x3$] Cos [$x4$] +
1224.4 Sin [$x3$] Cos [$x4$] + 1628.06 Cos [$x3$] Sin [$x4$] + 120774.Sin [$x3$]^2 +
2409.87 Sin [$x3$] + 80614.8 Cos [$x3$]^2 − 4414.11 Cos [$x3$] − 34903.2 Sin [$x3$] Cos [$x3$] +
27725.9 Sin [$x4$]^2 + 2222.84 Sin [$x4$] − 12198.4 Cos [$x4$]^2 − 4530.37 Cos [$x4$] −
33799.2 Sin [$x4$] Cos [$x4$] − 134515)/(−(2.84955 Sin [$x1$] Sin [$x2$]) +
0.358408 Cos [$x1$] Cos [$x2$] + 1.13942 Cos [$x1$] Sin [$x2$] + 1.36465 Sin [$x1$] Cos [$x2$] −
2.34509 Sin [$x1$] Sin [$x3$] + 0.00515941 Cos [$x1$] Cos [$x3$] + 1.37415 Sin [$x1$] Sin [$x3$] +
1.22731 Sin [$x1$] Cos [$x3$] − 1.6927 Sin [$x1$] Sin [$x4$] − 0.642884 Cos [$x1$] Cos [$x4$] +
0.121594 Cos [$x1$] Sin [$x4$] + 0.725779 Sin [$x1$] Cos [$x4$] − 90.4148 Sin [$x1$]^2 +
2.07223 Sin [$x1$] − 113.257 Cos [$x1$]^2 − 1.52869 Cos [$x1$] − 8.00859 Sin [$x1$] Cos [$x1$] −
2.20618 Sin [$x2$] Sin [$x3$] − 0.109362 Cos [$x2$] Cos [$x3$] + 0.857622 Sin [$x2$] Cos [$x3$] +
1.93069 Cos [$x2$] Sin [$x3$] + 0.0282477 Sin [$x2$] Sin [$x4$] + 0.602507 Cos [$x2$] Cos [$x4$] +
0.968859 Sin [$x2$] Cos [$x4$] − 0.00495416 Cos [$x2$] Sin [$x4$] + 55.6575 Sin [$x2$]^2 +
1.99896 Sin [$x2$] + 31.1051 Cos [$x2$]^2 − 0.681228 Cos [$x2$] − 4.41083 Sin [$x2$] Cos [$x2$] −
1.76428 Sin [$x3$] Sin [$x4$] + 1.02414 Cos [$x3$] Cos [$x4$] + 0.984694 Sin [$x3$] Cos [$x4$] +
1.1594 Cos [$x3$] Sin [$x4$] + 57.9921 Sin [$x3$]^2 + 2.2849 Sin [$x3$] + 33.7302 Cos [$x3$]^2 −
0.830282 Cos [$x3$] − 4.57768 Sin [$x3$] Cos [$x3$] − 6.18878 Sin [$x4$]^2 + 2.02932 Sin [$x4$] −
30.0755 Cos [$x4$]^2 − 0.948982 Cos [$x4$] − 4.0714 Sin [$x4$] Cos [$x4$] + 125.356)

(*FOLN*)

FOLN = 20.7344 Log [$x1 + 0.1$] + 18.3479 Log [$x2 + 0.1$] + 17.9236 Log [$x3 + 0.1$] +
18.3864 Log [$x4 + 0.1$] + 1104.21

(*FOLNR*)

FOLNR = (658.525 Log [$x1 + 0.1$] + 610.473 Log [$x2 + 0.1$] + 658.223 Log [$x3 + 0.1$] +
657.603 Log [$x4 + 0.1$] − 2421.94)/(0.505621 Log [$x1 + 0.1$] + 0.468919 Log [$x2 + 0.1$] +
0.505522 Log [$x3 + 0.1$] + 0.505059 Log [$x4 + 0.1$] − 1.86841)

(Continued)

TABLE A9.9 *(Continued)*

Mathematical Models Created Using the FFD Experiment Set (Related to Table 9.3)

(*SOLN*)

$\text{SOLN} = -(2.35115 \text{ Log } [x1 + 0.1] \text{ Log } [x2 + 0.1]) - 2.51003 \text{ Log } [x1 + 0.1] \text{ Log } [x3 + 0.1] -$
$\quad 2.28903 \text{ Log } [x1 + 0.1] \text{ Log } [x4 + 0.1] + 31.6074 \text{ Log } [x1 + 0.1]^2 - 21.4188 \text{ Log } [x1 + 0.1] -$
$\quad 2.31421 \text{ Log } [x2 + 0.1] \text{ Log } [x3 + 0.1] - 2.50788 \text{ Log } [x2 + 0.1] \text{ Log } [x4 + 0.1] +$
$\quad 33.4833 \text{ Log } [x2 + 0.1]^2 - 24.2985 \text{ Log } [x2 + 0.1] - 2.60857 \text{ Log } [x3 + 0.1] \text{ Log } [x4 + 0.1] +$
$\quad 32.7419 \text{ Log } [x3 + 0.1]^2 - 21.8903 \text{ Log } [x3 + 0.1] + 33.7117 \text{ Log } [x4 + 0.1]^2 -$
$\quad 24.778 \text{ Log } [x4 + 0.1] - 126.566$

(*SOLNR*)

$\text{SOLNR} = (1206.4 \text{ Log } [x1 + 0.1] \text{ Log } [x2 + 0.1] + 1260.55 \text{ Log } [x1 + 0.1] \text{ Log } [x3 + 0.1] +$
$\quad 1294.73 \text{ Log } [x1 + 0.1] \text{ Log } [x4 + 0.1] - 745.018 \text{ Log } [x1 + 0.1]^2 - 79.3788 \text{ Log } [x1 + 0.1] +$
$\quad 1290.25 \text{ Log } [x2 + 0.1] \text{ Log } [x3 + 0.1] + 1255.17 \text{ Log } [x2 + 0.1] \text{ Log } [x4 + 0.1] -$
$\quad 754.173 \text{ Log } [x2 + 0.1]^2 + 19.824 \text{ Log } [x2 + 0.1] + 1206.27 \text{ Log } [x3 + 0.1] \text{ Log } [x4 + 0.1] -$
$\quad 772.321 \text{ Log } [x3 + 0.1]^2 + 126.303 \text{ Log } [x3 + 0.1] - 719.13 \text{ Log } [x4 + 0.1]^2 -$
$\quad 53.8599 \text{ Log } [x4 + 0.1] + 20003.9)/(0.878754 \text{ Log } [x1 + 0.1] \text{ Log } [x2 + 0.1] +$
$\quad 0.915035 \text{ Log } [x1 + 0.1] \text{ Log } [x3 + 0.1] + 0.955078 \text{ Log } [x1 + 0.1] \text{ Log } [x4 + 0.1] -$
$\quad 0.705227 \text{ Log } [x1 + 0.1]^2 + 0.554506 \text{ Log } [x1 + 0.1] + 0.933765 \text{ Log } [x2 + 0.1] \text{ Log } [x3 + 0.1] +$
$\quad 0.9131 \text{ Log } [x2 + 0.1] \text{ Log } [x4 + 0.1] - 0.722659 \text{ Log } [x2 + 0.1]^2 + 0.709573 \text{ Log } [x2 + 0.1] +$
$\quad 0.87151 \text{ Log } [x3 + 0.1] \text{ Log } [x4 + 0.1] - 0.735581 \text{ Log } [x3 + 0.1]^2 + 0.821515 \text{ Log } [x3 + 0.1] -$
$\quad 0.695857 \text{ Log } [x4 + 0.1]^2 + 0.612184 \text{ Log } [x4 + 0.1] + 17.1913)$

TABLE A9.10

Mathematical Models Created Using the CCD Experiment Set (Related to Table 9.4)

(*L*)

$L = 562.96 + 3.49214 \, x1 + 3.06171 \, x2 + 3.08422 \, x3 + 3.37963 \, x4$

(*LR*)

$$LR = \frac{319505 \, x1 + 612370 \, x2 + 608157 \, x3 + 177771 \, x4 - 2.93654 \times 10^6}{88.7947 \, x1 + 369.238 \, x2 + 361.42 \, x3 - 22.8274 \, x4 + 27714.2}$$

(*SON*)

$\text{SON} = 0.163695 \, x1^2 + 0.0250804 \, x1 \, x2 - 0.168816 \, x1 \, x3 + 0.0780883 \, x1 \, x4 - 8.68439 \, x1 +$
$\quad 0.141182 \, x2^2 + 0.0401847 \, x2 \, x3 - 0.130912 \, x2 \, x4 - 6.15161 \, x2 + 0.141182 \, x3^2 +$
$\quad 0.062984 \, x3 \, x4 - 6.0939 \, x3 + 0.141182 \, x4^2 - 10.3581 \, x4 + 1257.6$

(*SONR*)

$\text{SONR} = (1885.05 \, x1^2 + 3885.77 \, x1 \, x2 + 3189.35 \, x1 \, x3 + 1564.62 \, x1 \, x4 + 13.2286 \, x1 + 1221.24 \, x2^2 +$
$\quad 1643.79 \, x2 \, x3 - 1086.9 \, x2 \, x4 - 14.5205 \, x2 - 2340.44 \, x3^2 - 3149.11 \, x3 \, x4 - 72.1223 \, x3 -$
$\quad 1290.75 \, x4^2 - 102.246 \, x4 + 0.331437)/(-(4.08236 \, x1^2) - 150.847 \, x1 \, x2 + 88.195 \, x1 \, x3 +$
$\quad 85.2069 \, x1 \, x4 + 47.3128 \, x1 + 90.0918 \, x2^2 + 77.5747 \, x2 \, x3 - 84.3812 \, x2 \, x4 + 3474.53 \, x2 -$
$\quad 119.609 \, x3^2 + 33.9966 \, x3 \, x4 - 2120.09 \, x3 - 3.67361 \, x4^2 - 1536.85 \, x4 + 45.3441)$

(Continued)

TABLE A9.10 *(Continued)*
Mathematical Models Created Using the CCD Experiment Set
(Related to Table 9.4)

(*TON*)

$TON = -(0.00310742\ x1^3) + 0.00198676\ x1^2\ x2 + 0.00128569\ x1^2\ x3 + 0.00148489\ x1^2\ x4 +$
$0.461596\ x1^2 + 0.00299216\ x1\ x2^2 + 0.00151795\ x1\ x2\ x3 - 0.00194432\ x1\ x2\ x4 -$
$0.477979\ x1\ x2 + 0.0022293\ x1\ x3^2 - 0.0035362\ x1\ x3\ x4 - 0.563331\ x1\ x3 +$
$0.00464561\ x1\ x4^2 - 0.0580455\ x1\ x4 - 2.53019\ x1 - 0.00114605\ x2^3 + 0.00189146\ x2^2\ x3 -$
$0.00151895\ x2^2\ x4 + 0.107417\ x2^2 + 0.00168015\ x2\ x3^2 - 0.00599812\ x2\ x3\ x4 -$
$0.0776898\ x2\ x3 + 0.00102944\ x2\ x4^2 + 0.268592\ x2\ x4 + 4.84553\ x2 - 0.00132382\ x3^3 +$
$0.000199055\ x3^2\ x4 + 0.0979428\ x3^2 + 0.00257778\ x3\ x4^2 + 0.334995\ x3\ x4 + 6.05818\ x3 +$
$0.00173075\ x4^3 - 0.500919\ x4^2 - 1.56062\ x4 + 759.775$

(*TONR*)

$TONR = (-(436.197\ x1^3) - 788.758\ x1^2\ x2 + 420.89\ x1^2\ x3 + 2052.45\ x1^2\ x4 - 66.158\ x1^2 -$
$763.508\ x1\ x2^2 + 3605.65\ x1\ x2\ x3 + 3146.81\ x1\ x2\ x4 - 93.8154\ x1\ x2 + 2356.83\ x1\ x3^2 +$
$7216.33\ x1\ x3\ x4 - 42.3721\ x1\ x3 + 2982.49\ x1\ x4^2 + 101.838\ x1\ x4 + 1.50611\ x1 -$
$302.589\ x2^3 - 135.001\ x2^2\ x3 - 831.8\ x2^2\ x4 - 35.4704\ x2^2 + 1775.69\ x2\ x3^2 +$
$2563.4\ x2\ x3\ x4 - 28.3011\ x2\ x3 + 72.9887\ x2\ x4^2 + 12.4261\ x2\ x4 + 2.06252\ x2 +$
$361.318\ x3^3 + 1628.26\ x3^2\ x4 + 18.9532\ x3^2 + 622.363\ x3\ x4^2 + 34.5239\ x3\ x4 + 2.18257\ x3 +$
$193.294\ x4^3 + 27.6622\ x4^2 + 2.92601\ x4 + 0.995333)/(111.733\ x1^3 - 86.7755\ x1^2\ x2 -$
$221.274\ x1^2\ x3 - 202.875\ x1^2\ x4 - 1064.46\ x1^2 + 35.307\ x1\ x2^2 + 16.0502\ x1\ x2\ x3 -$
$9.90743\ x1\ x2\ x4 + 4467.99\ x1\ x2 + 78.8567\ x1\ x3^2 + 368.491\ x1\ x3\ x4 - 1409.58\ x1\ x3 +$
$207.761\ x1\ x4^2 - 683.898\ x1\ x4 + 31.6351\ x1 + 73.1522\ x2^3 - 146.509\ x2^2\ x3 - 161.295\ x2^2\ x4 -$
$1686.71\ x2^2 + 31.5389\ x2\ x3^2 + 146.034\ x2\ x3\ x4 - 225.183\ x2\ x3 + 127.258\ x2\ x4^2 +$
$499.021\ x2\ x4 + 31.2827\ x2 + 85.042\ x3^3 - 116.945\ x3^2\ x4 + 1234.06\ x3^2 - 6.44063\ x3\ x4^2 -$
$5378.51\ x3\ x4 + 31.0718\ x3 - 142.348\ x4^3 + 768.395\ x4^2 + 29.8975\ x4 + 1.28682)$

(*FOTN*)

$FOTN = -(72.7587\ \text{Sin}\ [x1]) + 62.6321\ \text{Cos}\ [x1] - 73.6938\ \text{Sin}\ [x2] + 64.1467\ \text{Cos}\ [x2] -$
$72.7587\ \text{Sin}\ [x3] + 62.6321\ \text{Cos}\ [x3] - 66.6573\ \text{Sin}\ [x4] + 52.7492\ \text{Cos}\ [x4] + 1139.57$

(*FOTNR*)

$FOTNR = (886.043\ \text{Sin}\ [x1] + 506.156\ \text{Cos}\ [x1] - 163.425\ \text{Sin}\ [x2] - 132.539\ \text{Cos}\ [x2] -$
$160.006\ \text{Sin}\ [x3] - 141.355\ \text{Cos}\ [x3] - 370.881\ \text{Sin}\ [x4] - 356.418\ \text{Cos}\ [x4] - 57.5392)/$
$(0.754108\ \text{Sin}\ [x1] + 0.445757\ \text{Cos}\ [x1] - 0.144012\ \text{Sin}\ [x2] - 0.100074\ \text{Cos}\ [x2] -$
$0.139404\ \text{Sin}\ [x3] - 0.107532\ \text{Cos}\ [x3] - 0.337987\ \text{Sin}\ [x4] - 0.292121\ \text{Cos}\ [x4] -$
$0.037041)$

(*SOTN*)

$SOTN = 80.9809\ \text{Sin}\ [x1]\ \text{Sin}\ [x2] - 120.733\ \text{Cos}\ [x1]\ \text{Cos}\ [x2] - 16.3735\ \text{Cos}\ [x1]\ \text{Sin}\ [x2] +$
$87.28\ \text{Sin}\ [x1]\ \text{Cos}\ [x2] + 61.7962\ \text{Sin}\ [x1]\ \text{Sin}\ [x3] - 44.6598\ \text{Cos}\ [x1]\ \text{Cos}\ [x3] +$
$91.1827\ \text{Cos}\ [x1]\ \text{Sin}\ [x3] + 91.1006\ \text{Sin}\ [x1]\ \text{Cos}\ [x3] - 63.3258\ \text{Sin}\ [x1]\ \text{Sin}\ [x4] -$
$156.385\ \text{Cos}\ [x1]\ \text{Cos}\ [x4] - 11.893\ \text{Cos}\ [x1]\ \text{Sin}\ [x4] - 64.0402\ \text{Sin}\ [x1]\ \text{Cos}\ [x4] +$
$105.141\ \text{Sin}\ [x1]^2 - 3.67287\ \text{Sin}\ [x1] + 262.557\ \text{Cos}\ [x1]^2 + 34.6483\ \text{Cos}\ [x1] -$
$293.032\ \text{Sin}\ [x1]\ \text{Cos}\ [x1] - 149.08\ \text{Sin}\ [x2]\ \text{Sin}\ [x3] - 36.3376\ \text{Cos}\ [x2]\ \text{Cos}\ [x3] -$
$12.718\ \text{Sin}\ [x2]\ \text{Cos}\ [x3] + 78.293\ \text{Cos}\ [x2]\ \text{Sin}\ [x3] + 119.496\ \text{Sin}\ [x2]\ \text{Sin}\ [x4] -$
$137.077\ \text{Cos}\ [x2]\ \text{Cos}\ [x4] + 12.1298\ \text{Sin}\ [x2]\ \text{Cos}\ [x4] + 56.0811\ \text{Cos}\ [x2]\ \text{Sin}\ [x4] +$

(Continued)

TABLE A9.10 *(Continued)*
Mathematical Models Created Using the CCD Experiment Set (Related to Table 9.4)

436.909 Sin $[x2]^2$ − 88.2864 Sin $[x2]$ + 344.712 Cos $[x2]^2$ − 78.9145 Cos $[x2]$ − 321.28 Sin $[x2]$ Cos $[x2]$ + 180.136 Sin $[x3]$ Sin $[x4]$ − 112.075 Cos $[x3]$ Cos $[x4]$ + 97.2345 Sin $[x3]$ Cos $[x4]$ + 48.9103 Cos $[x3]$ Sin $[x4]$ + 96.7295 Sin $[x3]^2$ − 10.2973 Sin $[x3]$ + 223.409 Cos $[x3]^2$ + 11.0165 Cos $[x3]$ − 399.411 Sin $[x3]$ Cos $[x3]$ − 44.0008 Sin $[x4]^2$ + 5.92138 Sin $[x4]$ + 281.369 Cos $[x4]^2$ + 80.43 Cos $[x4]$ − 286.08 Sin $[x4]$ Cos $[x4]$ + 547.046

(*SOTNR*)

SOTNR = (9.81676 × 10^7 Cos $[x1]^2$ − 7.2599 × 10^8 Cos $[x1]$ + 3.27021 × 10^8 Cos $[x2]$ Cos $[x1]$ + 8.17952 × 10^7 Cos $[x3]$ Cos $[x1]$ + 1.39964 × 10^8 Cos $[x4]$ Cos $[x1]$ + 1.43613 × 10^8 Sin $[x1]$ Cos $[x1]$ + 2.2274 × 10^8 Sin $[x2]$ Cos $[x1]$ + 3.73867 × 10^8 Sin $[x3]$ Cos $[x1]$ + 3.37391 × 10^8 Sin $[x4]$ Cos $[x1]$ + 9.14068 × 10^7 + 1.39251 × 10^6 Cos $[x2]^2$ − 4.32263 × 10^7 Cos $[x3]^2$ − 3.08643 × 10^7 Cos $[x4]^2$ − 6.76082 × 10^6 Sin $[x1]^2$ + 9.00143 × 10^7 Sin $[x2]^2$ + 1.34633 × 10^8 Sin $[x3]^2$ + 1.22271 × 10^8 Sin $[x4]^2$ − 3.78653 × 10^8 Cos $[x2]$ − 1.76505 × 10^8 Cos $[x3]$ − 2.84398 × 10^8 Cos $[x2]$ Cos $[x3]$ − 2.28983 × 10^8 Cos $[x4]$ − 2.41488 × 10^8 Cos $[x2]$ Cos $[x4]$ − 2.14483 × 10^8 Cos $[x3]$ Cos $[x4]$ − 7.19569 × 10^8 Sin $[x1]$ − 2.10126 × 10^8 Cos $[x2]$ Sin $[x1]$ − 2.56342 × 10^8 Cos $[x3]$ Sin $[x1]$ − 2.37782 × 10^8 Cos $[x4]$ Sin $[x1]$ − 8.97526 × 10^8 Sin $[x2]$ + 2.5566 × 10^6 Cos $[x2]$ Sin $[x2]$ − 1.1103 × 10^7 Cos $[x3]$ Sin $[x2]$ + 1.68781 × 10^7 Cos $[x4]$ Sin $[x2]$ + 2.21524 × 10^8 Sin $[x1]$ Sin $[x2]$ − 1.02284 × 10^9 Sin $[x3]$ + 1.86241 × 10^8 Cos $[x2]$ Sin $[x3]$ − 8.76351 × 10^7 Cos $[x3]$ Sin $[x3]$ − 61584.3 Cos $[x4]$ Sin $[x3]$ + 2.49623 × 10^8 Sin $[x1]$ Sin $[x3]$ + 3.96498 × 10^8 Sin $[x2]$ Sin $[x3]$ − 9.9151 × 10^8 Sin $[x4]$ + 1.59186 × 10^8 Cos $[x2]$ Sin $[x4]$ − 5.50976 × 10^7 Cos $[x3]$ Sin $[x4]$ − 6.441 × 10^7 Cos $[x4]$ Sin $[x4]$ + 2.37251 × 10^8 Sin $[x1]$ Sin $[x4]$ + 3.78309 × 10^8 Sin $[x2]$ Sin $[x4]$ + 5.10168 × 10^8 Sin $[x3]$ Sin $[x4]$)/ (2.16087 × 10^7 Cos $[x1]^2$ − 6.43454 × 10^8 Cos $[x1]$ + 6.02081 × 10^8 Cos $[x2]$ Cos $[x1]$ + 7.89858 × 10^7 Cos $[x3]$ Cos $[x1]$ − 3.37558 × 10^8 Cos $[x4]$ Cos $[x1]$ − 1.19903 × 10^8 Sin $[x1]$ Cos $[x1]$ − 3.52351 × 10^8 Sin $[x2]$ Cos $[x1]$ − 3.04825 × 10^7 Sin $[x3]$ Cos $[x1]$ + 2.28972 × 10^8 Sin $[x4]$ Cos $[x1]$ − 1.95709 × 10^8 + 1.2049 × 10^7 Cos $[x2]^2$ + 4.61843 × 10^6 Cos $[x3]^2$ − 2.71754 × 10^7 Cos $[x4]^2$ − 2.17318 × 10^8 Sin $[x1]^2$ − 2.07758 × 10^8 Sin $[x2]^2$ − 2.00327 × 10^8 Sin $[x3]^2$ − 1.68534 × 10^8 Sin $[x4]^2$ − 7.3943 × 10^8 Cos $[x2]$ − 6.89755 × 10^8 Cos $[x3]$ − 1.67807 × 10^8 Cos $[x2]$ Cos $[x3]$ − 5.28341 × 10^8 Cos $[x4]$ − 3.25949 × 10^8 Cos $[x2]$ Cos $[x4]$ + 3.97196 × 10^8 Cos $[x3]$ Cos $[x4]$ − 4.8711 × 10^8 Sin $[x1]$ − 3.99898 × 10^8 Cos $[x2]$ Sin $[x1]$ − 1.06721 × 10^8 Cos $[x3]$ Sin $[x1]$ + 2.15459 × 10^7 Cos $[x4]$ Sin $[x1]$ − 3.67526 × 10^8 Sin $[x2]$ − 5.14785 × 10^7 Cos $[x2]$ Sin $[x2]$ + 7.73351 × 10^7 Cos $[x3]$ Sin $[x2]$ + 4.60724 × 10^7 Cos $[x4]$ Sin $[x2]$ − 6.47473 × 10^8 Sin $[x1]$ Sin $[x2]$ − 4.0024 × 10^8 Sin $[x3]$ + 1.06026 × 10^8 Cos $[x2]$ Sin $[x3]$ − 6.93459 × 10^7 Cos $[x3]$ Sin $[x3]$ − 4.0145 × 10^8 Cos $[x4]$ Sin $[x3]$ − 8.30213 × 10^8 Sin $[x1]$ Sin $[x3]$ − 8.63153 × 10^8 Sin $[x2]$ Sin $[x3]$ − 4.95528 × 10^8 Sin $[x4]$ + 2.05951 × 10^8 Cos $[x2]$ Sin $[x4]$ − 2.70263 × 10^8 Cos $[x3]$ Sin $[x4]$ − 1.26751 × 10^8 Cos $[x4]$ Sin $[x4]$ − 9.05687 × 10^8 Sin $[x1]$ Sin $[x4]$ − 8.40138 × 10^8 Sin $[x2]$ Sin $[x4]$ − 5.47879 × 10^8 Sin $[x3]$ Sin $[x4]$)

(*FOLN*)

FOLN = 54.3148 Log $[x1 + 0.1]$ + 54.3148 Log $[x2 + 0.1]$ + 54.3148 Log $[x3 + 0.1]$ + 54.3148 Log $[x4 + 0.1]$ + 406.022

(Continued)

TABLE A9.10 *(Continued)*

Mathematical Models Created Using the CCD Experiment Set (Related to Table 9.4)

(*FOLNR*)

FOLNR = (−(1.13286 Log [$x1$ + 0.1]) − 1.35086 Log [$x2$ + 0.1] − 1.26777 Log [$x3$ + 0.1] −
1.35134 Log [$x4$ + 0.1] + 43.7918)/(−(0.00229618 Log [$x1$ + 0.1]) −
0.00251852 Log [$x2$ + 0.1] − 0.00243396 Log [$x3$ + 0.1] − 0.00251901 Log [$x4$ + 0.1] +
0.0570514)

(*SOLN*)

SOLN = −(98.9188 Log [$x1$ + 0.1] Log [$x2$ + 0.1]) − 98.9188 Log [$x1$ + 0.1] Log [$x3$ + 0.1] −
98.9188 Log [$x1$ + 0.1] Log [$x4$ + 0.1] + 4.25497 Log [$x1$ + 0.1]^2 + 1169.37 Log [$x1$ + 0.1] −
98.9188 Log [$x2$ + 0.1] Log [$x3$ + 0.1] − 98.9188 Log [$x2$ + 0.1] Log [$x4$ + 0.1] +
4.25497 Log [$x2$ + 0.1]^2 + 1169.37 Log [$x2$ + 0.1] − 98.9188 Log [$x3$ + 0.1] Log [$x4$ + 0.1] +
4.25497 Log [$x3$ + 0.1]^2 + 1169.37 Log [$x3$ + 0.1] + 4.25497 Log [$x4$ + 0.1]^2 +
1169.37 Log [$x4$ + 0.1] − 8219.27

(*SOLNR*)

SOLNR = (14.2638 Log [$x1$ + 0.1] Log [$x2$ + 0.1] + 2.1717 Log [$x1$ + 0.1] Log [$x3$ + 0.1] +
21.7519 Log [$x1$ + 0.1] Log [$x4$ + 0.1] + 3.49434 Log [$x1$ + 0.1]^2 + 0.642613 Log [$x1$ + 0.1] −
0.0340597 Log [$x2$ + 0.1] Log [$x3$ + 0.1] − 0.2381 Log [$x2$ + 0.1] Log [$x4$ + 0.1] −
2.53694 Log [$x2$ + 0.1]^2 − 1.59087 Log [$x2$ + 0.1] − 20.4531 Log [$x3$ + 0.1] Log [$x4$ + 0.1] −
6.79204 Log [$x3$ + 0.1]^2 − 3.49871 Log [$x3$ + 0.1] − 0.312002 Log [$x4$ + 0.1]^2 −
0.525086 Log [$x4$ + 0.1] − 0.902372)/(−(0.00509613 Log [$x1$ + 0.1] Log [$x2$ + 0.1]) −
0.0112121 Log [$x1$ + 0.1] Log [$x3$ + 0.1] − 0.00196712 Log [$x1$ + 0.1] Log [$x4$ + 0.1] −
0.0351031 Log [$x1$ + 0.1]^2 + 0.453019 Log [$x1$ + 0.1] + 0.000609488 Log [$x2$ +
0.1] Log [$x3$ + 0.1] − 0.00557515 Log [$x2$ + 0.1] Log [$x4$ + 0.1] − 0.0681879 Log [$x2$ +
0.1]^2 + 0.555066 Log [$x2$ + 0.1] + 0.025246 Log [$x3$ + 0.1] Log [$x4$ + 0.1] −
0.00534022 Log [$x3$ + 0.1]^2 − 0.14296 Log [$x3$ + 0.1] − 0.00467694 Log [$x4$ + 0.1]^2 −
0.057961 Log [$x4$ + 0.1] − 1.31326)

TABLE A9.11

Mathematical Models Created Using the BBD Experiment Set (Related to Table 9.5)

(*L*)

L = 2.59697 $x1$ + 2.59697 $x2$ + 2.59697 $x3$ + 2.59697 $x4$ + 757.044

(*LR*)

$$LR = \frac{13364.8\ x1 + 12424.2\ x2 + 13465.3\ x3 + 10845.2\ x4 + 1.00047 \times 10^6}{5.49668\ x1 + 4.8431\ x2 + 5.8447\ x3 + 2.75385\ x4 + 1780.64}$$

(*SON*)

SON = 0.0964028 $x1^2$ − 0.0213846 $x1$ $x2$ − 0.0213846 $x1$ $x3$ − 0.0213846 $x1$ $x4$ − 3.19236 $x1$ +
0.0964028 $x2^2$ − 0.0213846 $x2$ $x3$ − 0.0213846 $x2$ $x4$ − 3.19236 $x2$ + 0.0964028 $x3^2$ −
0.0213846 $x3$ $x4$ − 3.19236 $x3$ + 0.0964028 $x4^2$ − 3.19236 $x4$ + 903.27

(Continued)

TABLE A9.11 *(Continued)*
Mathematical Models Created Using the BBD Experiment Set
(Related to Table 9.5)

(*SONR*)

SONR = (300194 $x1^2$ + 106875.$x1$ $x2$ + 551033 $x1$ $x3$ + 301189 $x1$ $x4$ + 13208.5 $x1$ + 230590 $x2^2$ −
549472 $x2$ $x3$ − 365843 $x2$ $x4$ − 1871.5 $x2$ + 148188 $x3^2$ − 617537 $x3$ $x4$ + 18206 $x3$ −
3181.98 $x4^2$ + 23499.8 $x4$ + 160.562)/(1833.11 $x1^2$ − 6977.1 $x1$ $x2$ − 8361.18 $x1$ $x3$ +
6537.42 $x1$ $x4$ − 87968.$x1$ + 3798.96 $x2^2$ + 9480.02 $x2$ $x3$ − 7408.19 $x2$ $x4$ + 188644.$x2$ +
3780.13 $x3^2$ − 8474.51 $x3$ $x4$ + 343234.$x3$ + 1342.41 $x4^2$ − 93211.2 $x4$ + 1753.02)

(*TON*)

TON = 0.000436468 $x1^3$ + 0.000186534 $x1^2$ $x2$ + 0.000139301 $x1^2$ $x3$ + 0.0000412858 $x1^2$ $x4$ +
0.0258201 $x1^2$ + 0.000107365 $x1$ $x2^2$ − 0.0001773 $x1$ $x2$ $x3$ − 0.0000909765 $x1$ $x2$ $x4$ −
0.0359532 $x1$ $x2$ + 0.000111591 $x1$ $x3^2$ − 0.000103801 $x1$ $x3$ $x4$ − 0.0313153 $x1$ $x3$ +
0.000369425 $x1$ $x4^2$ − 0.0611861 $x1$ $x4$ + 0.367408 $x1$ + 0.000248364 $x2^3$ + 0.000351113 $x2^2$ $x3$ +
0.000345142 $x2^2$ $x4$ + 0.0220401 $x2^2$ + 0.00039412 $x2$ $x3^2$ − 0.000450763 $x2$ $x3$ $x4$ −
0.0697242 $x2$ $x3$ + 0.00022833 $x2$ $x4^2$ − 0.0465477 $x2$ $x4$ + 1.64578 $x2$ + 0.00026195 $x3^3$ +
0.000257324 $x3^2$ $x4$ + 0.0218421 $x3^2$ + 0.000181096 $x3$ $x4^2$ − 0.0377678 $x3$ $x4$ + 1.50556 $x3$ +
0.000342064 $x4^3$ + 0.0219174 $x4^2$ + 0.644812 $x4$ + 759.193

(*TONR*)

TONR = (0.231212 $x1^3$ + 0.238605 $x1^2$ $x2$ − 0.836673 $x1^2$ $x3$ − 2.64817 $x1^2$ $x4$ + 1.94565 $x1^2$ −
0.295754 $x1$ $x2^2$ − 6.0289 $x1$ $x2$ $x3$ − 8.15929 $x1$ $x2$ $x4$ − 7.4442 $x1$ $x2$ + 0.214973 $x1$ $x3^2$ −
7.69545 $x1$ $x3$ $x4$ − 3.94153 $x1$ $x3$ − 2.67154 $x1$ $x4^2$ + 5.5323 $x1$ $x4$ + 2.59531 $x1$ −
0.37288 $x2^3$ − 0.496619 $x2^2$ $x3$ − 0.801099 $x2^2$ $x4$ − 1.62752 $x2^2$ − 5.99277 $x2$ $x3^2$ −
5.08864 $x2$ $x3$ $x4$ + 5.07901 $x2$ $x3$ + 4.07797 $x2$ $x4^2$ + 5.49804 $x2$ $x4$ − 1.60081 $x2$ −
0.796928 $x3^3$ − 0.539158 $x3^2$ $x4$ + 6.22205 $x3^2$ − 1.43502 $x3$ $x4^2$ − 5.43743 $x3$ $x4$ −
1.03754 $x3$ − 2.3245 $x4^3$ − 3.10472 $x4^2$ − 1.39118 $x4$ + 0.281533)/(3.28414 $x1^3$ −
5.71307 $x1^2$ $x2$ + 1.82433 $x1^2$ $x3$ + 1.51692 $x1^2$ $x4$ − 1.0149 $x1^2$ − 1.98195 $x1$ $x2^2$ −
21.1072 $x1$ $x2$ $x3$ − 7.04063 $x1$ $x2$ $x4$ + 6.92799 $x1$ $x2$ + 9.3479 $x1$ $x3^2$ + 2.51474 $x1$ $x3$ $x4$ −
7.87201 $x1$ $x3$ − 3.25954 $x1$ $x4^2$ − 0.558072 $x1$ $x4$ − 5.01454 $x1$ + 4.23335 $x2^3$ +
5.85734 $x2^2$ $x3$ + 6.36665 $x2^2$ $x4$ + 2.20641 $x2^2$ + 7.54257 $x2$ $x3^2$ + 2.3179 $x2$ $x3$ $x4$ +
1.95103 $x2$ $x3$ + 5.82205 $x2$ $x4^2$ − 4.21564 $x2$ $x4$ − 1.59128 $x2$ + 2.56199 $x3^3$ +
3.58537 $x3^2$ $x4$ − 1.31882 $x3^2$ − 1.87876 $x3$ $x4^2$ + 5.90054 $x3$ $x4$ − 2.03958 $x3$ − 1.42625 $x4^3$ −
3.84362 $x4^2$ − 0.0138111 $x4$ − 0.837879)

(*FOTN*)

FOTN = −(277.794 Sin [$x1$]) − 332.907 Cos [$x1$] − 277.794 Sin [$x2$] − 332.907 Cos [$x2$] −
277.794 Sin [$x3$] − 332.907 Cos [$x3$] − 277.794 Sin [$x4$] − 332.907 Cos [$x4$] + 2494.72

(*FOTNR*)

FOTNR = (2917.82 Sin [$x1$] − 326.911 Cos [$x1$] + 2917.82 Sin [$x2$] − 326.911 Cos [$x2$] +
2917.82 Sin [$x3$] − 326.911 Cos [$x3$] + 2917.82 Sin [$x4$] − 326.911 Cos [$x4$] − 1209.44)/
(4.9171 Sin [$x1$] + 1.87112 Cos [$x1$] + 4.9171 Sin [$x2$] + 1.87112 Cos [$x2$] + 4.9171 Sin [$x3$] +
1.87112 Cos [$x3$] + 4.9171 Sin [$x4$] + 1.87112 Cos [$x4$] − 11.2115)

(*SOTN*)

SOTN = −(164.476 Sin [$x1$] Sin [$x2$]) − 26.8819 Cos [$x1$] Cos [$x2$] + 31.1773 Cos [$x1$] Sin [$x2$] −
42.4618 Sin [$x1$] Cos [$x2$] − 103.566 Sin [$x1$] Sin [$x3$] − 49.601 Cos [$x1$] Cos [$x3$] +
68.7815 Cos [$x1$] Sin [$x3$] − 79.2616 Sin [$x1$] Cos [$x3$] − 161.736 Sin [$x1$] Sin [$x4$] −

(Continued)

TABLE A9.11 *(Continued)*
Mathematical Models Created Using the BBD Experiment Set
(Related to Table 9.5)

56.3317 Cos [x1] Cos [x4] + 32.8689 Cos [x1] Sin [x4] − 90.1639 Sin [x1] Cos [x4] +
267.31 Sin [x1]² + 412.426 Sin [x1] + 76.6955 Cos [x1]² + 29.2005 Cos [x1] −
416.032 Sin [x1] Cos [x1] − 35.7199 Sin [x2] Sin [x3] − 69.2043 Cos [x2] Cos [x3] −
37.3754 Sin [x2] Cos [x3] + 37.0285 Cos [x2] Sin [x3] − 93.8902 Sin [x2] Sin [x4] −
75.9351 Cos [x2] Cos [x4] − 48.2777 Sin [x2] Cos [x4] + 1.11596 Cos [x2] Sin [x4] +
94.0199 Sin [x2]² + 426.022 Sin [x2] + 120.224 Cos [x2]² + 134.957 Cos [x2] −
410.55 Sin [x2] Cos [x2] − 32.98 Sin [x3] Sin [x4] − 98.6541 Cos [x3] Cos [x4] −
10.6736 Sin [x3] Cos [x4] − 35.6838 Cos [x3] Sin [x4] + 24.4999 Sin [x3]² + 152.318 Sin [x3] +
127.179 Cos [x3]² + 94.2157 Cos [x3] − 228.455 Sin [x3] Cos [x3] + 84.8145 Sin [x4]² +
136.981 Sin [x4] + 25.56 Cos [x4]² + 100.263 Cos [x4] − 229.779 Sin [x4] Cos [x4] + 448.86

(*SOTNR*)

SOTNR = (1734.26 Sin [x1] Sin [x2] + 92.1602 Cos [x1] Cos [x2] − 565.411 Cos [x1] Sin [x2] −
299.118 Sin [x1] Cos [x2] − 450.402 Sin [x1] Sin [x3] − 2331.93 Cos [x1] Cos [x3] −
249.323 Cos [x1] Sin [x3] + 980.653 Sin [x1] Cos [x3] + 3280.99 Sin [x1] Sin [x4] −
2747.38 Cos [x1] Cos [x4] + 1255.9 Cos [x1] Sin [x4] + 619.055 Sin [x1] Cos [x4] +
464.448 Sin [x1]² + 2134.43 Sin [x1] − 95.4568 Cos [x1]² − 688.158 Cos [x1] +
234.407 Sin [x1] Cos [x1] − 56.6885 Sin [x2] Sin [x3] − 1461.59 Cos [x2] Cos [x3] +
1352.37 Sin [x2] Cos [x3] − 60.6445 Cos [x2] Sin [x3] + 3388.91 Sin [x2] Sin [x4] −
960.687 Cos [x2] Cos [x4] − 1429.46 Sin [x2] Cos [x4] − 1775.41 Cos [x2] Sin [x4] +
377.976 Sin [x2]² + 1695.2 Sin [x2] − 8.9843 Cos [x2]² − 1129.51 Cos [x2] −
625.485 Sin [x2] Cos [x2] + 3120.25 Sin [x3] Sin [x4] − 162.833 Cos [x3] Cos [x4] −
4452.96 Sin [x3] Cos [x4] + 804.614 Cos [x3] Sin [x4] − 61.1233 Sin [x3]² −
366.329 Sin [x3] + 430.115 Cos [x3]² + 75.6616 Cos [x3] − 1657.9 Sin [x3] Cos [x3] +
850.665 Sin [x4]² + 3924.08 Sin [x4] − 481.673 Cos [x4]² − 2229.77 Cos [x4] +
676.586 Sin [x4] Cos [x4] + 367.991)/(19.9282 Sin [x1] Sin [x2] + 93.1191 Cos [x1] Cos [x2] +
41.6555 Cos [x1] Sin [x2] − 23.6683 Sin [x1] Cos [x2] − 22.8234 Sin [x1] Sin [x3] −
33.5515 Cos [x1] Cos [x3] − 60.8992 Cos [x1] Sin [x3] − 3.36762 Sin [x1] Cos [x3] +
35.9423 Sin [x1] Sin [x4] − 32.5057 Cos [x1] Cos [x4] − 13.5877 Cos [x1] Sin [x4] −
2.66814 Sin [x1] Cos [x4] + 5.51563 Sin [x1]² + 22.6515 Sin [x1] + 3.13524 Cos [x1]² +
5.00572 Cos [x1] + 1.7782 Sin [x1] Cos [x1] − 10.5415 Sin [x2] Sin [x3] +
72.8763 Cos [x2] Cos [x3] + 20.6909 Sin [x2] Cos [x3] − 76.6225 Cos [x2] Sin [x3] −
56.5176 Sin [x2] Sin [x4] + 35.5252 Cos [x2] Cos [x4] − 3.74343 Sin [x2] Cos [x4] −
71.6818 Cos [x2] Sin [x4] + 2.38535 Sin [x2]² + 6.7972 Sin [x2] + 6.26552 Cos [x2]² −
10.0367 Cos [x2] − 28.4458 Sin [x2] Cos [x2] + 114.448 Sin [x3] Sin [x4] +
32.0043 Cos [x3] Cos [x4] − 34.5157 Sin [x3] Cos [x4] − 69.3037 Cos [x3] Sin [x4] +
1.59109 Sin [x3]² + 3.66516 Sin [x3] + 7.05979 Cos [x3]² + 4.32715 Cos [x3] −
18.5263 Sin [x3] Cos [x3] + 3.8063 Sin [x4]² + 14.2558 Sin [x4] + 4.84458 Cos [x4]² +
2.13419 Cos [x4] − 9.55749 Sin [x4] Cos [x4] + 7.65087)

(*FOLN*)

FOLN = 28.3155 Log [x1 + 0.1] + 31.1611 Log [x2 + 0.1] + 30.9739 Log [x3 + 0.1] +
28.1049 Log [x4 + 0.1] + 929.894

(Continued)

TABLE A9.11 *(Continued)*
Mathematical Models Created Using the BBD Experiment Set
(Related to Table 9.5)

(*FOLNR*)

FOLNR = (3768.89 Log [$x1$ + 0.1] + 3855.62 Log [$x2$ + 0.1] + 3823.35 Log [$x3$ + 0.1] +
  3688.42 Log [$x4$ + 0.1] − 434.613)/(2.5514 Log [$x1$ + 0.1] + 2.64179 Log [$x2$ +
  0.1] + 2.6089 Log [$x3$ + 0.1] + 2.46754 Log [$x4$ + 0.1] + 3.64338)

(*SOLN*)

SOLN = −(3.74214 Log [$x1$ + 0.1] Log [$x2$ + 0.1]) − 3.74214 Log [$x1$ + 0.1] Log [$x3$ + 0.1] −
  3.74214 Log [$x1$ + 0.1] Log [$x4$ + 0.1] + 64.9191 Log [$x1$ + 0.1]^2 − 75.74 Log [$x1$ + 0.1] −
  3.74214 Log [$x2$ + 0.1] Log [$x3$ + 0.1] − 3.74214 Log [$x2$ + 0.1] Log [$x4$ + 0.1] +
  64.9191 Log [$x2$ + 0.1]^2 − 75.74 Log [$x2$ + 0.1] − 3.74214 Log [$x3$ + 0.1] Log [$x4$ + 0.1] +
  64.9191 Log [$x3$ + 0.1]^2 − 75.74 Log [$x3$ + 0.1] + 64.9191 Log [$x4$ + 0.1]^2 −
  75.74 Log [$x4$ + 0.1] − 1437.92

(*SOLNR*)

SOLNR = (663.62 Log [$x1$ + 0.1] Log [$x2$ + 0.1] + 454.599 Log [$x1$ + 0.1] Log [$x3$ + 0.1] −
  295.508 Log [$x1$ + 0.1] Log [$x4$ + 0.1] − 17.8307 Log [$x1$ + 0.1]^2 + 28.9579 Log [$x1$ +
  0.1] + 196.939 Log [$x2$ + 0.1] Log [$x3$ + 0.1] + 437.42 Log [$x2$ + 0.1] Log [$x4$ + 0.1] −
  25.3299 Log [$x2$ + 0.1]^2 + 40.8876 Log [$x2$ + 0.1] + 134.425 Log [$x3$ + 0.1] Log [$x4$ +
  0.1] − 96.8317 Log [$x3$ + 0.1]^2 − 82.5126 Log [$x3$ + 0.1] − 104.812 Log [$x4$ + 0.1]^2 −
  156.574 Log [$x4$ + 0.1] − 3.86353)/(7.89354 Log [$x1$ + 0.1] Log [$x2$ + 0.1] +
  7.76931 Log [$x1$ + 0.1] Log [$x3$ + 0.1] − 66.2613 Log [$x1$ + 0.1] Log [$x4$ + 0.1] −
  4.83686 Log [$x1$ + 0.1]^2 + 251.112 Log [$x1$ + 0.1] + 68.5184 Log [$x2$ + 0.1] Log [$x3$ + 0.1] +
  7.68086 Log [$x2$ + 0.1] Log [$x4$ + 0.1] − 4.11405 Log [$x2$ + 0.1]^2 − 312.367 Log [$x2$ +
  0.1] + 7.45293 Log [$x3$ + 0.1] Log [$x4$ + 0.1] − 4.61397 Log [$x3$ + 0.1]^2 −
  311.972 Log [$x3$ + 0.1] − 4.89245 Log [$x4$ + 0.1]^2 + 250.853 Log [$x4$ + 0.1] + 276.499)

TABLE A9.12
Mathematical Models Created Using the D-Optimal Experiment Set
(Related to Table 9.6)

(*L*)

L = 2.66945 $x1$ + 2.93261 $x2$ + 2.76502 $x3$ + 3.02049 $x4$ + 780.911

(*LR*)

$$LR = \frac{0.0165999\ x1 - 0.0129258\ x2 + 0.0730232\ x3 + 0.0538571\ x4 + 5.40547}{-(3.92078 * {}^\wedge - 6x1) - 0.000032041\ x2 + 0.0000418118\ x3 + 0.000025659\ x4 + 0.00726703}$$

(*SON*)

SON = 0.068041 $x1^2$ − 0.0231708 $x1$ $x2$ − 0.0186847 $x1$ $x3$ − 0.0237037 $x1$ $x4$ − 0.388343 $x1$ +
  0.0653761 $x2^2$ − 0.022254 $x2$ $x3$ − 0.0253877 $x2$ $x4$ + 0.151987 $x2$ + 0.0687399 $x3^2$ −
  0.0213256 $x3$ $x4$ − 0.518098 $x3$ + 0.0662825 $x4^2$ − 0.0266569 $x4$ + 788.312

(Continued)

TABLE A9.12 *(Continued)*

Mathematical Models Created Using the D-Optimal Experiment Set (Related to Table 9.6)

(*SONR*)

SONR = $(0.0854675\ x1^2 - 0.0114555\ x1\ x2 - 0.0478801\ x1\ x3 - 0.0364522\ x1\ x4 - 3.17797\ x1 +$
$0.0636386\ x2^2 - 0.0433277\ x2\ x3 - 0.0124285\ x2\ x4 - 2.6025\ x2 + 0.0623331\ x3^2 -$
$0.0136053\ x3\ x4 - 1.16656\ x3 + 0.052252\ x4^2 - 2.16715\ x4 + 501.288)/(0.0000434724\ x1^2 +$
$1.02777\ *^\wedge - 6\ x1\ x2 - 0.000023776\ x1\ x3 - 0.0000149024\ x1\ x4 - 0.00299359\ x1 +$
$0.0000294761\ x2^2 - 0.0000191463\ x2\ x3 + 3.1341999999999996\ *^\wedge - 6\ x2\ x4 -$
$0.00289311\ x2 + 0.0000277873\ x3^2 +)1.24492\ *^\wedge - 6\ x3\ x4 - 0.00173781\ x3 +$
$0.0000208915\ x4^2 - 0.00256103\ x4 + 0.628557)$

(*TON*)

TON = $-0.000430873\ x1^3 + 4.39935\ *^\wedge - 6\ x1^2\ x2 + 0.000131179\ x1^2\ x3 - 0.0000367849\ x1^2\ x4 +$
$0.119618\ x1^2 + 0.000104948\ x1\ x2^2 + 0.0000168618\ x1\ x2\ x3 - 0.000177397\ x1\ x2\ x4 -$
$0.0249442\ x1\ x2 + 0.000138113\ x1\ x3^2 - 0.00012221\ x1\ x3\ x4 - 0.0434977\ x1\ x3 +$
$0.000141769\ x1\ x4^2 - 0.0197997\ x1\ x4 - 1.51915\ x1 - 0.0006408\ x2^3 + 0.000189146\ x2^2\ x3 +$
$0.0000151845\ x2^2\ x4 + 0.141168\ x2^2 + 0.000139468\ x2\ x3^2 - 0.000352051\ x2\ x3\ x4 -$
$0.03622\ x2\ x3 + 0.0000597464\ x2\ x4^2 - 0.00743366\ x2\ x4 - 2.59457\ x2 - 0.000578686\ x3^3 +$
$0.0000277338\ x3^2\ x4 + 0.130871\ x3^2 + 0.000248375\ x3\ x4^2 - 0.0254045\ x3\ x4 - 1.62383\ x3 -$
$0.000713585\ x4^3 + 0.14199\ x4^2 - 2.52836\ x4 + 805.815$

(*TONR*)

TONR = $(96.0484\ x1^3 - 86.0517\ x1^2\ x2 + 30.0618\ x1^2\ x3 - 81.2474\ x1^2\ x4 - 3010.83\ x1^2 -$
$71.0582\ x1\ x2^2 + 51.4571\ x1\ x2\ x3 + 64.4966\ x1\ x2\ x4 + 9219.99\ x1\ x2 - 30.5523\ x1\ x3^2 +$
$14.6237\ x1\ x3\ x4 - 4722.97\ x1\ x3 - 51.1365\ x1\ x4^2 + 5477.3\ x1\ x4 - 281483\ x1 +$
$100.378\ x2^3 - 23.0242\ x2^2\ x3 + 75.8766\ x2^2\ x4 - 5402.9\ x2^2 - 98.967\ x2\ x3^2 -$
$20.3102\ x2\ x3\ x4 - 2278.16\ x2\ x3 - 85.7494\ x2\ x4^2 + 1580.36\ x2\ x4 - 42570.2\ x2 +$
$192.723\ x3^3 - 74.8162\ x3^2\ x4 - 11302.9\ x3^2 + 78.5064\ x3\ x4^2 - 8377.06\ x3\ x4 + 864424.x3 +$
$90.2271\ x4^3 - 5840.22\ x4^2 + 523812.x4 - 2.90354 \times 10^6)/(0.0929542\ x1^3 - 0.0746584\ x1^2\ x2 +$
$0.0137635\ x1^2\ x3 - 0.0770622\ x1^2\ x4 - 4.1552\ x1^2 - 0.055147\ x1\ x2^2 + 0.0495151\ x1\ x2\ x3 +$
$0.0610914\ x1\ x2\ x4 + 7.49744\ x1\ x2 - 0.0167458\ x1\ x3^2 + 0.0281148\ x1\ x3\ x4 -$
$4.11644\ x1\ x3 - 0.0291262\ x1\ x4^2 + 3.48152\ x1\ x4 - 189.626\ x1 + 0.0783728\ x2^3 -$
$0.0154865\ x2^2\ x3 + 0.0255622\ x2^2\ x4 - 3.99426\ x2^2 - 0.0723911\ x2\ x3^2 + 0.021166\ x2\ x3\ x4 -$
$3.7975\ x2\ x3 - 0.0669619\ x2\ x4^2 + 1.94054\ x2\ x4 - 27.0634\ x2 + 0.154929\ x3^3 -$
$0.0647604\ x3^2\ x4 - 10.9904\ x3^2 + 0.0659972\ x3\ x4^2 - 8.984\ x3\ x4 + 882.108\ x3 +$
$0.0754942\ x4^3 - 7.4999\ x4^2 + 666.634\ x4 - 3527.15)$

(*FOTN*)

FOTN = $25.8834\ \mathrm{Sin}\ [x1] - 61.9506\ \mathrm{Cos}\ [x1] + 39.0573\ \mathrm{Sin}\ [x2] - 56.3602\ \mathrm{Cos}\ [x2] +$
$29.8056\ \mathrm{Sin}\ [x3] - 57.8959\ \mathrm{Cos}\ [x3] + 36.8953\ \mathrm{Sin}\ [x4] - 49.0167\ \mathrm{Cos}\ [x4] + 1301.76$

(*FOTNR*)

FOTNR = $(1500.08\ \mathrm{Sin}\ [x1] + 1408.1\ \mathrm{Cos}\ [x1] + 1202.26\ \mathrm{Sin}\ [x2] + 2706.87\ \mathrm{Cos}\ [x2] +$
$664.99\ \mathrm{Sin}\ [x3] + 889.987\ \mathrm{Cos}\ [x3] + 508.035\ \mathrm{Sin}\ [x4] + 1165.63\ \mathrm{Cos}\ [x4] + 1545.31)/$
$(1.12085\ \mathrm{Sin}\ [x1] + 1.12413\ \mathrm{Cos}\ [x1] + 0.917771\ \mathrm{Sin}\ [x2] + 2.08648\ \mathrm{Cos}\ [x2] +$
$0.479894\ \mathrm{Sin}\ [x3] + 0.725769\ \mathrm{Cos}\ [x3] + 0.380481\ \mathrm{Sin}\ [x4] + 0.901048\ \mathrm{Cos}\ [x4] + 1.18939)$

(Continued)

TABLE A9.12 *(Continued)*
Mathematical Models Created Using the D-Optimal Experiment Set (Related to Table 9.6)

(*SOTN*)

SOTN = 1.48447 Sin [$x1$] Sin [$x2$] + 4.78741 Cos [$x1$] Cos [$x2$] + 1.26459 Cos [$x1$] Sin [$x2$] +
9.18037 Sin [$x1$] Cos [$x2$] − 18.2085 Sin [$x1$] Sin [$x3$] − 14.0036 Cos [$x1$] Cos [$x3$] +
23.4543 Cos [$x1$] Sin [$x3$] + 19.6764 Sin [$x1$] Cos [$x3$] − 9.86498 Sin [$x1$] Sin [$x4$] −
3.54911 Cos [$x1$] Cos [$x4$] + 9.74927 Cos [$x1$] Sin [$x4$] + 9.25914 Sin [$x1$] Cos [$x4$] +
355.892 Sin [$x1$]^2 − 4.06437 Sin [$x1$] + 246.447 Cos [$x1$]^2 − 31.0033 Cos [$x1$] −
289.054 Sin [$x1$] Cos [$x1$] − 17.4738 Sin [$x2$] Sin [$x3$] − 1.45756 Cos [$x2$] Cos [$x3$] +
10.0571 Sin [$x2$] Cos [$x3$] − 4.13204 Cos [$x2$] Sin [$x3$] − 9.85889 Sin [$x2$] Sin [$x4$] −
3.72955 Cos [$x2$] Cos [$x4$] + 15.2321 Sin [$x2$] Cos [$x4$] − 8.26753 Cos [$x2$] Sin [$x4$] +
236.243 Sin [$x2$]^2 + 1.07657 Sin [$x2$] + 123.064 Cos [$x2$]^2 − 25.8363 Cos [$x2$] −
288.506 Sin [$x2$] Cos [$x2$] − 14.1083 Sin [$x3$] Sin [$x4$] − 8.28909 Cos [$x3$] Cos [$x4$] +
10.0061 Sin [$x3$] Cos [$x4$] − 4.22614 Cos [$x3$] Sin [$x4$] + 140.089 Sin [$x3$]^2 + 1.59627 Sin [$x3$] +
43.3783 Cos [$x3$]^2 − 38.3675 Cos [$x3$] − 273.06 Sin [$x3$] Cos [$x3$] + 475.291 Sin [$x4$]^2 −
1.27548 Sin [$x4$] + 361.505 Cos [$x4$]^2 − 29.749 Cos [$x4$] − 275.061 Sin [$x4$] Cos [$x4$] + 401.328

(*SOTNR*)

SOTNR = (−(50381.5 Sin [$x1$] Sin [$x2$]) + 3720.74 Cos [$x1$] Cos [$x2$] − 6494.87 Cos [$x1$] Sin [$x2$] +
21147.8 Sin [$x1$] Cos [$x2$] + 57536.7 Sin [$x1$] Sin [$x3$] − 2877.99 Cos [$x1$] Cos [$x3$] +
5270.54 Cos [$x1$] Sin [$x3$] + 27210.9 Sin [$x1$] Cos [$x3$] + 68721 Sin [$x1$] Sin [$x4$] −
5573.92 Cos [$x1$] Cos [$x4$] + 1577.17 Cos [$x1$] Sin [$x4$] − 11325.8 Sin [$x1$] Cos [$x4$] −
62616.3 Sin [$x1$]^2 − 1693.75 Sin [$x1$] − 3658.53 Cos [$x1$]^2 + 2711.84 Cos [$x1$] −
8231.45 Sin [$x1$] Cos [$x1$] − 3777.28 Sin [$x2$] Sin [$x3$] + 310.712 Cos [$x2$] Cos [$x3$] +
32183.4 Sin [$x2$] Cos [$x3$] − 6952.79 Cos [$x2$] Sin [$x3$] − 25992.1 Sin [$x2$] Sin [$x4$] −
1573.71 Cos [$x2$] Cos [$x4$] + 33547.3 Sin [$x2$] Cos [$x4$] + 1226.33 Cos [$x2$] Sin [$x4$] −
31240.7 Sin [$x2$]^2 + 14030.9 Sin [$x2$] − 3693.79 Cos [$x2$]^2 − 4362.08 Cos [$x2$] −
45608.6 Sin [$x2$] Cos [$x2$] + 6158.56 Sin [$x3$] Sin [$x4$] − 2438.96 Cos [$x3$] Cos [$x4$] −
2692.97 Sin [$x3$] Cos [$x4$] − 1863.37 Cos [$x3$] Sin [$x4$] − 6616.96 Sin [$x3$]^2 +
4650.06 Sin [$x3$] − 8884.07 Cos [$x3$]^2 + 4267.28 Cos [$x3$] + 7532.71 Sin [$x3$] Cos [$x3$] −
6111.86 Sin [$x4$]^2 + 5111.12 Sin [$x4$] − 5843.81 Cos [$x4$]^2 + 3206.2 Cos [$x4$] +
3762.94 Sin [$x4$] Cos [$x4$] − 4142.91)/(−(37.0336 Sin [$x1$] Sin [$x2$]) +
4.11897 Cos [$x1$] Cos [$x2$] − 5.24489 Cos [$x1$] Sin [$x2$] + 16.1091 Sin [$x1$] Cos [$x2$] +
41.5249 Sin [$x1$] Sin [$x3$] − 3.20451 Cos [$x1$] Cos [$x3$] + 3.52993 Cos [$x1$] Sin [$x3$] +
21.4915 Sin [$x1$] Cos [$x3$] + 49.6295 Sin [$x1$] Sin [$x4$] − 4.7189 Cos [$x1$] Cos [$x4$] +
2.70344 Cos [$x1$] Sin [$x4$] − 6.33043 Sin [$x1$] Cos [$x4$] − 37.9896 Sin [$x1$]^2 −
0.320027 Sin [$x1$] + 0.311111 Cos [$x1$]^2 + 1.62388 Cos [$x1$] − 11.1013 Sin [$x1$] Cos [$x1$] −
3.90932 Sin [$x2$] Sin [$x3$] + 2.16238 Cos [$x2$] Cos [$x3$] + 23.9466 Sin [$x2$] Cos [$x3$] −
4.4242 Cos [$x2$] Sin [$x3$] − 18.7453 Sin [$x2$] Sin [$x4$] + 0.0873923 Cos [$x2$] Cos [$x4$] +
25.1867 Sin [$x2$] Cos [$x4$] − 0.419094 Cos [$x2$] Sin [$x4$] − 26.1829 Sin [$x2$]^2 + 10.3315 Sin [$x2$] −
8.24371 Cos [$x2$]^2 − 4.34238 Cos [$x2$] − 42.8916 Sin [$x2$] Cos [$x2$] + 5.8569 Sin [$x3$] Sin [$x4$] −
4.0056 Cos [$x3$] Cos [$x4$] − 0.502995 Sin [$x3$] Cos [$x4$] + 2.37668 Cos [$x3$] Sin [$x4$] −
4.17548 Sin [$x3$]^2 + 4.09363 Sin [$x3$] − 14.4979 Cos [$x3$]^2 + 3.65808 Cos [$x3$] +
0.089098 Sin [$x3$] Cos [$x3$] − 3.9175 Sin [$x4$]^2 + 2.56239 Sin [$x4$] − 8.15188 Cos [$x4$]^2 +
1.87976 Cos [$x4$] + 0.193029 Sin [$x4$] Cos [$x4$] − 0.440286)

(Continued)

TABLE A9.12 *(Continued)*

Mathematical Models Created Using the D-Optimal Experiment Set (Related to Table 9.6)

(*FOLN*)

FOLN = 17.0629 Log [$x1$ + 0.1] + 18.8783 Log [$x2$ + 0.1] + 17.4205 Log [$x3$ + 0.1] +
    20.8854 Log [$x4$ + 0.1] + 1101.55

(*FOLNR*)

FOLNR = (−(295.489 Log [$x1$ + 0.1]) − 296.67 Log [$x2$ + 0.1] + 10.5809 Log [$x3$ + 0.1] +
    23.9707 Log [$x4$ + 0.1] + 2766.33)/(−(0.243748 Log [$x1$ + 0.1]) − 0.258781 Log [$x2$ + 0.1] +
    0.00507703 Log [$x3$ + 0.1] + 0.0153947 Log [$x4$ + 0.1] + 2.32245)

(*SOLN*)

SOLN = −(1.96837 Log [$x1$ + 0.1] Log [$x2$ + 0.1]) − 2.92119 Log [$x1$ + 0.1] Log [$x3$ + 0.1] −
    2.2404 Log [$x1$ + 0.1] Log [$x4$ + 0.1] + 30.8853 Log [$x1$ + 0.1]^2 − 20.6516 Log [$x1$ + 0.1] −
    2.38878 Log [$x2$ + 0.1] Log [$x3$ + 0.1] − 2.10625 Log [$x2$ + 0.1] Log [$x4$ + 0.1] +
    32.1865 Log [$x2$ + 0.1]^2 − 24.7907 Log [$x2$ + 0.1] − 2.38691 Log [$x3$ + 0.1] Log [$x4$ + 0.1] +
    30.9389 Log [$x3$ + 0.1]^2 − 19.2188 Log [$x3$ + 0.1] + 31.623 Log [$x4$ + 0.1]^2 −
    21.2749 Log [$x4$ + 0.1] − 70.6913

(*SOLNR*)

SOLNR = (−(667.178 Log [$x1$ + 0.1] Log [$x2$ + 0.1]) + 266.049 Log [$x1$ + 0.1] Log [$x3$ + 0.1] +
    953.933 Log [$x1$ + 0.1] Log [$x4$ + 0.1] − 2750.77 Log [$x1$ + 0.1]^2 + 19820.1 Log [$x1$ + 0.1] +
    768.903 Log [$x2$ + 0.1] Log [$x3$ + 0.1] + 599.106 Log [$x2$ + 0.1] Log [$x4$ + 0.1] −
    291.353 Log [$x2$ + 0.1]^2 − 1502.06 Log [$x2$ + 0.1] + 1313.51 Log [$x3$ + 0.1] Log [$x4$ + 0.1] −
    895.132 Log [$x3$ + 0.1]^2 − 2302.68 Log [$x3$ + 0.1] − 447.125 Log [$x4$ + 0.1]^2 −
    7491.46 Log [$x4$ + 0.1] − 12326.1)/(−(0.533369 Log [$x1$ + 0.1] Log [$x2$ + 0.1]) −
    0.00533381 Log [$x1$ + 0.1] Log [$x3$ + 0.1] + 0.612442 Log [$x1$ + 0.1] Log [$x4$ + 0.1] −
    2.13836 Log [$x1$ + 0.1]^2 + 16.6938 Log [$x1$ + 0.1] + 0.571267 Log [$x2$ + 0.1] Log [$x3$ + 0.1] +
    0.44612 Log [$x2$ + 0.1] Log [$x4$ + 0.1] − 0.234628 Log [$x2$ + 0.1]^2 − 0.986269 Log [$x2$ + 0.1] +
    0.997535 Log [$x3$ + 0.1] Log [$x4$ + 0.1] − 0.719107 Log [$x3$ + 0.1]^2 −
    0.838528 Log [$x3$ + 0.1] − 0.374204 Log [$x4$ + 0.1]^2 − 5.16576 Log [$x4$ + 0.1] − 14.2478)

TABLE A9.13

Mathematical Models Created Using the Taguchi Experiment Set (Related to Table 9.7)

(*L*)

L = 3.53153 $x1$ + 3.35242 $x2$ + 3.47331 $x3$ + 3.38313 $x4$ + 687.743

(*LR*)

$$LR = \frac{27.2484\ x1 + 18.6917\ x2 - 17.6767\ x3 - 14.0325\ x4 + 1613.31}{0.014199\ x1 + 0.0106099\ x2 - 0.0151251\ x3 - 0.014881\ x4 + 1.99435}$$

(*SON*)

SON = 0.0723376 $x1^2$ − 0.0781774 $x1$ $x2$ + 0.0505781 $x1$ $x3$ − 0.0561948 $x1$ $x4$ + 0.172382 $x1$ +
    0.0477587 $x2^2$ + 0.0049273 $x2$ $x3$ + 0.00908891 $x2$ $x4$ + 1.89024 $x2$ + 0.0400973 $x3^2$ −
    0.0651175 $x3$ $x4$ − 1.1304 $x3$ + 0.0790627 $x4^2$ + 1.25899 $x4$ + 756.093

(Continued)

TABLE A9.13 *(Continued)*
Mathematical Models Created Using the Taguchi Experiment Set (Related to Table 9.7)

(*SONR*)

SONR = $(1014.11\,x1^2 + 555.562\,x1\,x2 + 420.441\,x1\,x3 + 3959.7\,x1\,x4 + 67.7767\,x1 + 240.566\,x2^2 + 1385.66\,x2\,x3 + 407.302\,x2\,x4 + 81.7553\,x2 + 680.892\,x3^2 - 1707.86\,x3\,x4 + 101.015\,x3 + 998.458\,x4^2 + 103.653\,x4 + 967.936)/(-(10.7295\,x1^2) + 22.553\,x1\,x2 - 41.8346\,x1\,x3 + 22.0155\,x1\,x4 - 3393.28\,x1 + 149.471\,x2^2 - 89.6913\,x2\,x3 - 207.946\,x2\,x4 + 3417.86\,x2 + 59.0878\,x3^2 - 27.6154\,x3\,x4 + 12.878\,x3 + 117.317\,x4^2 - 3387.58\,x4 + 410.009)$

(*TON*)

TON = $-(0.214779\,x1^3) + 0.0163257\,x1^2\,x2 + 10.1227\,x1^2\,x3 - 7.95128\,x1^2\,x4 - 325.401\,x1^2 + 12.0995\,x1\,x2^2 - 12.6669\,x1\,x2\,x3 - 2.66026\,x1\,x2\,x4 - 1199.34\,x1\,x2 + 0.401412\,x1\,x3^2 + 1.00741\,x1\,x3\,x4 + 419.672\,x1\,x3 + 1.81739\,x1\,x4^2 + 533.604\,x1\,x4 + 39331.7\,x1 + 7.58854\,x2^3 - 16.3126\,x2^2\,x3 - 0.232668\,x2^2\,x4 - 1359.57\,x2^2 + 7.84641\,x2\,x3^2 + 6.86905\,x2\,x3\,x4 + 1963.56\,x2\,x3 + 1.09866\,x2\,x4^2 - 237.416\,x2\,x4 + 56918.2\,x2 - 4.34341\,x3^3 - 1.16645\,x3^2\,x4 - 0.0436212\,x3^2 - 1.87264\,x3\,x4^2 - 364.01\,x3\,x4 - 68149.3\,x3 + 0.521804\,x4^3 - 1.92883\,x4^2 + 11215.6\,x4 + 836.109$

(*TONR*)

TONR = $(5348.57\,x1^3 + 26034.5\,x1^2\,x2 + 6069.52\,x1^2\,x3 + 26016.8\,x1^2\,x4 + 559.299\,x1^2 + 20741.6\,x1\,x2^2 + 49751.7\,x1\,x2\,x3 + 46571.7\,x1\,x2\,x4 + 1474.02\,x1\,x2 + 6245.62\,x1\,x3^2 - 5559.39\,x1\,x3\,x4 + 493.297\,x1\,x3 + 25599.5\,x1\,x4^2 + 1194.75\,x1\,x4 + 9.93545\,x1 + 3411.99\,x2^3 + 15838.4\,x2^2\,x3 + 11884.2\,x2^2\,x4 + 597.912\,x2^2 + 16930.6\,x2\,x3^2 + 50372.7\,x2\,x3\,x4 + 1770.41\,x2\,x3 + 20900.5\,x2\,x4^2 + 1806.84\,x2\,x4 + 24.8289\,x2 + 1791.57\,x3^3 + 11343.1\,x3^2\,x4 + 418.821\,x3^2 + 11753.8\,x3\,x4^2 + 1213.64\,x3\,x4 + 21.3183\,x3 + 4371.99\,x4^3 + 609.071\,x4^2 + 22.7102\,x4 - 3449.36)/(2503.68\,x1^3 + 5223.54\,x1^2\,x2 + 1865.56\,x1^2\,x3 + 70.3134\,x1^2\,x4 + 293.072\,x1^2 + 1816.32\,x1\,x2^2 + 6017.69\,x1\,x2\,x3 - 5929.28\,x1\,x2\,x4 - 820.235\,x1\,x2 - 4577.02\,x1\,x3^2 + 1746.09\,x1\,x3\,x4 - 494.346\,x1\,x3 - 1030.87\,x1\,x4^2 - 32.0357\,x1\,x4 + 8.57338\,x1 - 1797.08\,x2^3 - 11834.1\,x2^2\,x3 + 7290.66\,x2^2\,x4 + 784.147\,x2^2 + 1875.31\,x2\,x3^2 + 12444.4\,x2\,x3\,x4 + 1763.73\,x2\,x3 - 7352.34\,x2\,x4^2 + 955.561\,x2\,x4 - 6081.48\,x2 + 14.9701\,x3^3 - 856.411\,x3^2\,x4 + 330.034\,x3^2 - 152.832\,x3\,x4^2 + 1044.15\,x3\,x4 - 6079.08\,x3 + 398.632\,x4^3 + 199.375\,x4^2 - 6074.07\,x4 - 204.459)$

(*FOTN*)

FOTN = $82.7389\,\mathrm{Sin}\,[x1] - 146.4\,\mathrm{Cos}\,[x1] + 108.747\,\mathrm{Sin}\,[x2] - 113.237\,\mathrm{Cos}\,[x2] + 131.71\,\mathrm{Sin}\,[x3] - 64.4169\,\mathrm{Cos}\,[x3] + 94.6297\,\mathrm{Sin}\,[x4] - 59.5955\,\mathrm{Cos}\,[x4] + 1241.14$

(*FOTNR*)

FOTNR = $(-(369.988\,\mathrm{Sin}\,[x1]) - 537.996\,\mathrm{Cos}\,[x1] + 1242.11\,\mathrm{Sin}\,[x2] + 399.549\,\mathrm{Cos}\,[x2] - 1166.6\,\mathrm{Sin}\,[x3] + 1270.64\,\mathrm{Cos}\,[x3] - 52.6396\,\mathrm{Sin}\,[x4] - 1655.68\,\mathrm{Cos}\,[x4] + 531.316)/(-(0.325155\,\mathrm{Sin}\,[x1]) - 0.403205\,\mathrm{Cos}\,[x1] + 0.944297\,\mathrm{Sin}\,[x2] + 0.36702\,\mathrm{Cos}\,[x2] - 0.841984\,\mathrm{Sin}\,[x3] + 0.922245\,\mathrm{Cos}\,[x3] - 0.0776861\,\mathrm{Sin}\,[x4] - 1.27031\,\mathrm{Cos}\,[x4] + 0.393761)$

(*SOTN*)

SOTN = $-(32.9282\,\mathrm{Sin}\,[x1]\,\mathrm{Sin}\,[x2]) - 72.2327\,\mathrm{Cos}\,[x1]\,\mathrm{Cos}\,[x2] + 25.6623\,\mathrm{Cos}\,[x1]\,\mathrm{Sin}\,[x2] + 28.2206\,\mathrm{Sin}\,[x1]\,\mathrm{Cos}\,[x2] + 3.06705\,\mathrm{Sin}\,[x1]\,\mathrm{Sin}\,[x3] - 122.224\,\mathrm{Cos}\,[x1]\,\mathrm{Cos}\,[x3] + 64.3269\,\mathrm{Cos}\,[x1]\,\mathrm{Sin}\,[x3] - 20.7846\,\mathrm{Sin}\,[x1]\,\mathrm{Cos}\,[x3] - 50.4616\,\mathrm{Sin}\,[x1]\,\mathrm{Sin}\,[x4] +$

(Continued)

TABLE A9.13 *(Continued)*
Mathematical Models Created Using the Taguchi Experiment Set (Related to Table 9.7)

36.3195 Cos [$x1$] Cos [$x4$] + 69.9264 Cos [$x1$] Sin [$x4$] + 7.97754 Sin [$x1$] Cos [$x4$] + 318.164 Sin [$x1$]^2 + 43.7149 Sin [$x1$] + 197.474 Cos [$x1$]^2 − 29.314 Cos [$x1$] − 43.6856 Sin [$x1$] Cos [$x1$] − 12.7418 Sin [$x2$] Sin [$x3$] − 18.8956 Cos [$x2$] Cos [$x3$] − 16.0274 Sin [$x2$] Cos [$x3$] + 38.3065 Cos [$x2$] Sin [$x3$] − 20.9751 Sin [$x2$] Sin [$x4$] − 12.515 Cos [$x2$] Cos [$x4$] + 66.2476 Sin [$x2$] Cos [$x4$] − 176.978 Cos [$x2$] Sin [$x4$] + 286.352 Sin [$x2$]^2 + 86.4909 Sin [$x2$] + 254.013 Cos [$x2$]^2 − 39.4859 Cos [$x2$] − 204.182 Sin [$x2$] Cos [$x2$] − 95.749 Sin [$x3$] Sin [$x4$] + 17.189 Cos [$x3$] Cos [$x4$] − 63.8891 Sin [$x3$] Cos [$x4$] − 37.9622 Cos [$x3$] Sin [$x4$] + 281.256 Sin [$x3$]^2 + 33.2738 Sin [$x3$] + 250.906 Cos [$x3$]^2 + 3.56065 Cos [$x3$] + 59.0065 Sin [$x3$] Cos [$x3$] + 311.337 Sin [$x4$]^2 + 51.7521 Sin [$x4$] + 186.551 Cos [$x4$]^2 − 31.483 Cos [$x4$] − 365.281 Sin [$x4$] Cos [$x4$] + 202.754

(*SOTNR*)

SOTNR = (7612.65 Sin [$x1$] Sin [$x2$] − 136.129 Cos [$x1$] Cos [$x2$] + 8512.25 Cos [$x1$] Sin [$x2$] + 18881.9 Sin [$x1$] Cos [$x2$] + 4896.87 Sin [$x1$] Sin [$x3$] + 2202.25 Cos [$x1$] Cos [$x3$] + 8436.19 Cos [$x1$] Sin [$x3$] + 19726.3 Sin [$x1$] Cos [$x3$] − 11344.9 Sin [$x1$] Sin [$x4$] − 9461.42 Cos [$x1$] Cos [$x4$] + 17840.6 Cos [$x1$] Sin [$x4$] − 5275.12 Sin [$x1$] Cos [$x4$] − 4016.49 Sin [$x1$]^2 − 41933 Sin [$x1$] + 95.2353 Cos [$x1$]^2 + 45286.4 Cos [$x1$] + 170.355 Sin [$x1$] Cos [$x1$] + 37431.4 Sin [$x2$] Sin [$x3$] + 29415.7 Cos [$x2$] Cos [$x3$] + 27596.3 Sin [$x2$] Cos [$x3$] − 1510.55 Cos [$x2$] Sin [$x3$] + 3133.53 Sin [$x2$] Sin [$x4$] + 3275.67 Cos [$x2$] Cos [$x4$] + 13090.8 Sin [$x2$] Cos [$x4$] + 8890.66 Cos [$x2$] Sin [$x4$] − 484.911 Sin [$x2$]^2 − 17073.7 Sin [$x2$] − 3436.34 Cos [$x2$]^2 − 3861.57 Cos [$x2$] − 13099.5 Sin [$x2$] Cos [$x2$] + 30002.1 Sin [$x3$] Sin [$x4$] + 34123.5 Cos [$x3$] Cos [$x4$] + 26 068.9 Sin [$x3$] Cos [$x4$] − 8046.45 Cos [$x3$] Sin [$x4$] − 1900.33 Sin [$x3$]^2 + 17859.8 Sin [$x3$] − 2020.92 Cos [$x3$]^2 − 14841.2 Cos [$x3$] + 5848.66 Sin [$x3$] Cos [$x3$] − 3436.83 Sin [$x4$]^2 + 31475.6 Sin [$x4$] − 484.421 Cos [$x4$]^2 − 4591.06 Cos [$x4$] − 13937.4 Sin [$x4$] Cos [$x4$] − 3922.25)/(317.893 Sin [$x1$] Sin [$x2$] − 4720.94 Cos [$x1$] Cos [$x2$] − 2997.76 Cos [$x1$] Sin [$x2$] + 1149.08 Sin [$x1$] Cos [$x2$] − 10282.2 Sin [$x1$] Sin [$x3$] − 1227.96 Cos [$x1$] Cos [$x3$] − 3158.73 Cos [$x1$] Sin [$x3$] + 4632.92 Sin [$x1$] Cos [$x3$] − 4131.57 Sin [$x1$] Sin [$x4$] + 2653.67 Cos [$x1$] Cos [$x4$] + 5231 Cos [$x1$] Sin [$x4$] + 7061.62 Sin [$x1$] Cos [$x4$] + 747.799 Sin [$x1$]^2 + 11522.1 Sin [$x1$] − 3089.52 Cos [$x1$]^2 + 4258.91 Cos [$x1$] − 2377.96 Sin [$x1$] Cos [$x1$] + 5780.07 Sin [$x2$] Sin [$x3$] − 1006.58 Cos [$x2$] Cos [$x3$] + 4751.04 Sin [$x2$] Cos [$x3$] + 16280.6 Cos [$x2$] Sin [$x3$] + 6900.9 Sin [$x2$] Sin [$x4$] − 19993.3 Cos [$x2$] Cos [$x4$] + 3768.12 Sin [$x2$] Cos [$x4$] − 3899.41 Cos [$x2$] Sin [$x4$] − 173.642 Sin [$x2$]^2 − 6098.61 Sin [$x2$] − 2168.08 Cos [$x2$]^2 − 13221.8 Cos [$x2$] − 5546.04 Sin [$x2$] Cos [$x2$] + 1689.64 Sin [$x3$] Sin [$x4$] − 995.288 Cos [$x3$] Cos [$x4$] − 4263.58 Sin [$x3$] Cos [$x4$] + 6602.48 Cos [$x3$] Sin [$x4$] + 2497.47 Sin [$x3$]^2 − 5130.23 Sin [$x3$] − 4839.19 Cos [$x3$]^2 − 5773.5 Cos [$x3$] − 83.1008 Sin [$x3$] Cos [$x3$] − 2420.63 Sin [$x4$]^2 + 9165.71 Sin [$x4$] + 78.9129 Cos [$x4$]^2 + 9867.61 Cos [$x4$] + 443.93 Sin [$x4$] Cos [$x4$] − 2342.72)

(*FOLN*)

FOLN = 42.6875 Log [$x1$ + 0.1] + 25.6833 Log [$x2$ + 0.1] + 11.5252 Log [$x3$ + 0.1] + 27.2784 Log [$x4$ + 0.1] + 1000.14

(Continued)

TABLE A9.13 *(Continued)*
**Mathematical Models Created Using the Taguchi Experiment Set
(Related to Table 9.7)**

(*FOLNR*)

FOLNR = (−(180.617 Log [$x1$ + 0.1]) + 491.519 Log [$x2$ + 0.1] + 371.833 Log [$x3$ + 0.1] −
    578.569 Log [$x4$ + 0.1] + 3118.76)/(−(0.240152 Log [$x1$ + 0.1]) + 0.342412 Log [$x2$ + 0.1] +
    0.251026 Log [$x3$ + 0.1] − 0.517566 Log [$x4$ + 0.1] + 3.16325)

(*SOLN*)

SOLN = −(123.223 Log [$x1$ + 0.1] Log [$x2$ + 0.1]) + 45.8291 Log [$x1$ + 0.1] Log [$x3$ + 0.1] +
    15.3297 Log [$x1$ + 0.1] Log [$x4$ + 0.1] + 19.8663 Log [$x1$ + 0.1]^2 + 234.297 Log [$x1$ + 0.1] +
    65.7898 Log [$x2$ + 0.1] Log [$x3$ + 0.1] − 6.55856 Log [$x2$ + 0.1] Log [$x4$ + 0.1] +
    22.5385 Log [$x2$ + 0.1]^2 + 230.329 Log [$x2$ + 0.1] − 0.867123 Log [$x3$ + 0.1] Log [$x4$ + 0.1] +
    9.22541 Log [$x3$ + 0.1]^2 − 430.531 Log [$x3$ + 0.1] + 21.9395 Log [$x4$ + 0.1]^2 −
    58.2935 Log [$x4$ + 0.1] + 400.769

(*SOLNR*)

SOLNR = (−(22.366 Log [$x1$ + 0.1] Log [$x2$ + 0.1]) + 20.7046 Log [$x1$ + 0.1] Log [$x3$ + 0.1] +
    27.4164 Log [$x1$ + 0.1] Log [$x4$ + 0.1] + 22.407 Log [$x1$ + 0.1]^2 − 1.44664 Log [$x1$ + 0.1] −
    4.726 Log [$x2$ + 0.1] Log [$x3$ + 0.1] − 9.72794 Log [$x2$ + 0.1] Log [$x4$ + 0.1] +
    9.00736 Log [$x2$ + 0.1]^2 + 11.0767 Log [$x2$ + 0.1] + 30.3794 Log [$x3$ + 0.1] Log [$x4$ + 0.1] +
    15.031 Log [$x3$ + 0.1]^2 + 28.5337 Log [$x3$ + 0.1] + 5.86192 Log [$x4$ + 0.1]^2 +
    25.6735 Log [$x4$ + 0.1] + 4.92629)/(0.0610545 Log [$x1$ + 0.1] Log [$x2$ + 0.1] −
    0.0204488 Log [$x1$ + 0.1] Log [$x3$ + 0.1] − 0.00861943 Log [$x1$ + 0.1] Log [$x4$ + 0.1] −
    0.016232 Log [$x1$ + 0.1]^2 − 0.0148 Log [$x1$ + 0.1] − 0.00365476 Log [$x2$ + 0.1] Log [$x3$ + 0.1] −
    0.0128307 Log [$x2$ + 0.1] Log [$x4$ + 0.1] − 0.0038294 Log [$x2$ + 0.1]^2 −
    0.275811 Log [$x2$ + 0.1] + 0.0125046 Log [$x3$ + 0.1] Log [$x4$ + 0.1] −
    0.00288017 Log [$x3$ + 0.1]^2 + 0.216578 Log [$x3$ + 0.1] − 0.0277066 Log [$x4$ + 0.1]^2 +
    0.24296 Log [$x4$ + 0.1] + 0.941456)

TABLE A9.14
**Mathematical Models Created Using the BBD–CCD/BBD–Taguchi/
BBD–CCD–Taguchi Hybrid Experiment Set (Related to Table 9.8)**

(*Mathematical models created using the BBD–CCD hybrid experiment set*)
(*SON*)

SON = 0.0490454 $x1^2$ − 0.0242195 $x1$ $x2$ − 0.0291227 $x1$ $x3$ − 0.0291227 $x1$ $x4$ + 2.18623 $x1$ +
    0.0415222 × 2^2 − 0.0275632 $x2$ $x3$ − 0.0257144 $x2$ $x4$ + 2.52583 $x2$ + 0.0493345 $x3^2$ −
    0.0232382 $x3$ $x4$ + 1.89154 $x3$ + 0.0453226 $x4^2$ + 2.0229 $x4$ + 620.62

(*TON*)

TON = −(0.000912653 $x1^3$) + 0.0000584186 $x1^2$ $x2$ + 0.0000780384 $x1^2$ $x3$ + 0.000431245 $x1^2$ $x4$ +
    0.14572 $x1^2$ + 0.0000976582 $x1$ $x2^2$ − 0.000267824 $x1$ $x2$ $x3$ − 0.000267824 $x1$ $x2$ $x4$ −
    0.015171 $x1$ $x2$ + 0.0000780384 $x1$ $x3^2$ + 0.000428949 $x1$ $x3$ $x4$ − 0.0503676 $x1$ $x3$ +

(Continued)

TABLE A9.14 *(Continued)*
Mathematical Models Created Using the BBD–CCD/BBD–Taguchi/ BBD–CCD–Taguchi Hybrid Experiment Set (Related to Table 9.8)

$0.0000780384\ x1\ x4^2 - 0.0821562\ x1\ x4 + 0.765923\ x1 - 0.000736499\ x2^3 + 0.0000780384\ x2^2\ x3 + 0.00041145\ x2^2\ x4 + 0.113003\ x2^2 + 0.0000780384\ x2\ x3^2 - 0.000158546\ x2\ x3\ x4 - 0.0212865\ x2\ x3 + 0.0000582429\ x2\ x4^2 - 0.0487993\ x2\ x4 + 0.704424\ x2 - 0.000491103\ x3^3 + 0.000431245\ x3^2\ x4 + 0.0882814\ x3^2 + 0.0000780384\ x3\ x4^2 - 0.0812945\ x3\ x4 + 2.49628\ x3 - 0.000615437\ x4^3 + 0.119029\ x4^2 + 2.09443\ x4 + 550.906$

(*SOTN*)

SOTN = $-(135.606$ Sin $[x1]$ Sin $[x2]) - 73.8341$ Cos $[x1]$ Cos $[x2] - 22.3308$ Cos $[x1]$ Sin $[x2] - 22.3308$ Sin $[x1]$ Cos $[x2] - 135.606$ Sin $[x1]$ Sin $[x3] - 73.8342$ Cos $[x1]$ Cos $[x3] - 22.3308$ Cos$[x1]$ Sin$[x3] - 22.3308$ Sin $[x1]$ Cos $[x3] - 135.606$ Sin $[x1]$ Sin $[x4] - 73.8341$ Cos $[x1]$ Cos $[x4] - 22.3307$ Cos $[x1]$ Sin $[x4] - 22.3308$ Sin $[x1]$ Cos $[x4] + 413.481$ Sin $[x1]^2 - 30.6974$ Sin $[x1] + 152.44$ Cos $[x1]^2 - 58.0285$ Cos $[x1] - 110.274$ Sin $[x1]$ Cos $[x1] - 135.606$ Sin $[x2]$ Sin $[x3] - 73.8343$ Cos $[x2]$ Cos $[x3] - 22.331$ Sin $[x2]$ Cos $[x3] - 22.331$ Cos $[x2]$ Sin $[x3] - 135.606$ Sin $[x2]$ Sin $[x4] - 73.8342$ Cos $[x2]$ Cos $[x4] - 22.3309$ Sin $[x2]$ Cos $[x4] - 22.3308$ Cos $[x2]$ Sin $[x4] + 373.435$ Sin $[x2]^2 - 5.72443$ Sin $[x2] + 121.014$ Cos $[x2]^2 - 42.611$ Cos $[x2] - 127.475$ Sin $[x2]$ Cos $[x2] - 135.606$ Sin $[x3]$ Sin $[x4] - 73.8343$ Cos $[x3]$ Cos $[x4] - 22.331$ Sin $[x3]$ Cos $[x4] - 22.331$ Cos $[x3]$ Sin $[x4] + 298.773$ Sin $[x3]^2 + 77.2865$ Sin $[x3] + 75.0102$ Cos $[x3]^2 + 8.63746$ Cos $[x3] - 184.652$ Sin $[x3]$ Cos $[x3] + 611.632$ Sin $[x4]^2 - 57.5048$ Sin $[x4] + 341.336$ Cos $[x4]^2 - 74.5786$ Cos $[x4] - 91.8082$ Sin $[x4]$ Cos $[x4] + 593.06$

(*Mathematical models created using the BBD–Taguchi hybrid experiment set*)

(*L*)

L = $3.08948\ x1 + 2.88517\ x2 + 3.09786\ x3 + 2.4386\ x4 + 731.217$

(*LR*)

$$LR = \frac{8.7849\ x1 + 15.2515\ x2 - 1.78311\ x3 - 2.27986\ x4 + 1162.49}{0.00307678\ x1 + 0.00870131\ x2 - 0.00527048\ x3 - 0.00522889\ x4 + 1.57976}$$

(*SON*)

SON = $0.0839686\ x1^2 - 0.0236045\ x1\ x2 - 0.0420583\ x1\ x3 - 0.0258407\ x1\ x4 - 0.84227\ x1 + 0.0857084\ x2^2 - 0.0291635\ x2\ x3 - 0.0100829\ x2\ x4 - 2.2249\ x2 + 0.0855286\ x3^2 - 0.0250752\ x3\ x4 - 0.39968\ x3 + 0.0886218\ x4^2 - 2.31712\ x4 + 805.315$

(*TON*)

TON = $0.000585059\ x1^3 + 0.000572779\ x1^2\ x2 + 0.0021811\ x1^2\ x3 - 0.0000250394\ x1^2\ x4 - 0.123465\ x1^2 + 0.000246091\ x1\ x2^2 - 0.00143697\ x1\ x2\ x3 - 0.0000207062\ x1\ x2\ x4 - 0.0268451\ x1\ x2 + 0.0002215\ x1\ x3^2 - 0.00379363\ x1\ x3\ x4 - 0.0509555\ x1\ x3 + 0.000127239\ x1\ x4^2 + 0.138634\ x1\ x4 + 4.52138\ x1 + 0.000481942\ x2^3 + 0.000867954\ x2^2\ x3 - 0.0000654923\ x2^2\ x4 - 0.0236721\ x2^2 + 0.00054952\ x2\ x3^2 - 0.000551786\ x2\ x3\ x4 - 0.0737556\ x2\ x3 + 0.000528596\ x2\ x4^2 - 0.0386481\ x2\ x4 + 2.60959\ x2 - 0.0000166844\ x3^3 - 0.000436762\ x3^2\ x4 + 0.0616339\ x3^2 + 0.000656012\ x3\ x4^2 + 0.128914\ x3\ x4 - 2.07048\ x3 + 0.000368965\ x4^3 - 0.0320964\ x4^2 - 3.19113\ x4 + 826.925$

(Continued)

TABLE A9.14 *(Continued)*

Mathematical Models Created Using the BBD–CCD/BBD–Taguchi/ BBD–CCD–Taguchi Hybrid Experiment Set (Related to Table 9.8)

(*Mathematical models created using the BBD–CCD–Taguchi hybrid experiment set*)

(*SON*)

$SON = 0.0592866\,x1^2 - 0.028899\,x1\,x2 - 0.0255154\,x1\,x3 - 0.019975\,x1\,x4 + 0.772061\,x1 + 0.0582182\,x2^2 - 0.0249413\,x2\,x3 - 0.0176238\,x2\,x4 + 0.75727\,x2 + 0.0560612\,x3^2 - 0.0236733\,x3\,x4 + 1.14626\,x3 + 0.0570566\,x4^2 + 0.493633\,x4 + 720.647$

(*TON*)

$TON = -(0.000489444\,x1^3) + 0.000236416\,x1^2\,x2 + 0.0000315189\,x1^2\,x3 - 0.000206312\,x1^2\,x4 + 0.116034\,x1^2 + 0.000323644\,x1\,x2^2 - 0.000828321\,x1\,x2\,x3 - 0.0000658763\,x1\,x2\,x4 - 0.040268\,x1\,x2 + 0.000225894\,x1\,x3^2 - 0.000990549\,x1\,x3\,x4 + 0.0311779\,x1\,x3 + 0.000225095\,x1\,x4^2 + 0.0225478\,x1\,x4 - 2.81135\,x1 - 0.000712716\,x2^3 + \frac{4.39398\,x2^2\,x3}{10^6} - 0.000272519\,x2^2\,x4 + 0.146699\,x2^2 + 0.000213982\,x2\,x3^2 - 0.00029154\,x2\,x3\,x4 - 0.0000336478\,x2\,x3 + 0.000150756\,x2\,x4^2 + 0.00186569\,x2\,x4 - 2.4154\,x2 - 0.000739029\,x3^3 - 0.000273693\,x3^2 \times 4 + 0.14164\,x3^2 + 0.0000145958\,x3\,x4^2 + 0.051092\,x3\,x4 - 4.38809\,x3 - 0.00081854\,x4^3 + 0.143654\,x4^2 - 4.03885\,x4 + 817.9$

(*SOTN*)

$SOTN = -(63.2027\ Sin\ [x1]\ Sin\ [x2]) - 4.55782\ Cos\ [x1]\ Cos\ [x2] + 37.1812\ Cos\ [x1]\ Sin\ [x2] + 34.4701\ Sin\ [x1]\ Cos\ [x2] - 33.9498\ Sin\ [x1]\ Sin\ [x3] - 4.15805\ Cos\ [x1]\ Cos\ [x3] + 30.6007\ Cos\ [x1]\ Sin\ [x3] + 44.6083\ Sin\ [x1]\ Cos\ [x3] - 14.1306\ Sin\ [x1]\ Sin\ [x4] - 37.6575\ Cos\ [x1]\ Cos\ [x4] - 13.4393\ Cos\ [x1]\ Sin\ [x4] + 75.8923\ Sin\ [x1]\ Cos\ [x4] + 396.973\ Sin\ [x1]^2 + 10.0446\ Sin\ [x1] + 273.374\ Cos\ [x1]^2 - 22.4726\ Cos\ [x1] - 316.751\ Sin\ [x1]\ Cos\ [x1] - 8.98342\ Sin\ [x2]\ Sin\ [x3] - 57.8857\ Cos\ [x2]\ Cos\ [x3] + 32.2562\ Sin\ [x2]\ Cos\ [x3] - 13.9116\ Cos\ [x2]\ Sin\ [x3] - 41.8961\ Sin\ [x2]\ Sin\ [x4] - 58.1672\ Cos\ [x2]\ Cos\ [x4] + 15.4343\ Sin\ [x2]\ Cos\ [x4] + 49.2925\ Cos\ [x2]\ Sin\ [x4] + 348.793\ Sin\ [x2]^2 + 2.32012\ Sin\ [x2] + 203.512\ Cos\ [x2]^2 - 25.551\ Cos\ [x2] - 279.823\ Sin\ [x2]\ Cos\ [x2] + 17.4741\ Sin\ [x3]\ Sin\ [x4] - 42.4754\ Cos\ [x3]\ Cos\ [x4] + 26.2096\ Sin\ [x3]\ Cos\ [x4] + 38.1508\ Cos\ [x3]\ Sin\ [x4] + 428.249\ Sin\ [x3]^2 + 16.9252\ Sin\ [x3] + 293.756\ Cos\ [x3]^2 - 22.5371\ Cos\ [x3] - 337.689\ Sin\ [x3]\ Cos\ [x3] + 262.444\ Sin\ [x4]^2 - 38.0174\ Sin\ [x4] + 133.818\ Cos\ [x4]^2 - 47.1705\ Cos\ [x4] - 308.28\ Sin\ [x4]\ Cos\ [x4] + 238.368$

10 Special Functions in Mathematical Modeling

This chapter delves into the exploration of special functions as potential alternative model types in mathematical modeling. It examines various special function types including BesselJ, ChebyshevT, Erf, ExpIntegralE, Fresnel, Hermit, HyperGeometric, and Legendre, evaluating their suitability for modeling purposes and comparing their performance with the basic mathematical functions utilized in previous chapters of this book.

Now, let's examine the basic properties of the special functions we will use in this section.

10.1 BESSEL FUNCTION OF THE FIRST KIND-$J_n(x)$

Bessel functions play a crucial role in addressing various problems related to wave propagation and static potentials. When dealing with problem scenarios in cylindrical coordinate systems, Bessel functions of integer order are commonly encountered. Conversely, in spherical problems, half-integer order Bessel functions are typically obtained [1].

The Bessel functions of the first kind $J_n(x)$ are defined as the solutions to the Bessel differential equation

$$x^2 \frac{d^2 y}{dx^2} + x \frac{dy}{dx} + \left(x^2 - n^2\right) y = 0$$

Figure 10.1 shows $J_n(x)$ for n=0, 1, 2, ..., 5. The Bessel functions of the first kind $J_n(x)$ are defined from their power series representation:

$$J_n(x) := \sum_{k=0}^{\infty} \frac{(-1)^k}{\Gamma(k+1)\,\Gamma(k+n+1)} \left(\frac{x}{2}\right)^{2k+n},$$

Where x is a complex variable and n is a parameter that can take arbitrary real or complex values. When n is an integer, it turns out as an entire function; in this case:

$$J_{-n}(x) = (-1)^n J_n(x),\ n = 1, 2 \ldots$$

In fact,

$$J_n(z) = \sum_{k=0}^{\infty} \frac{(-1)^k}{k!(k+n)!} \left(\frac{z}{2}\right)^{2k+n},$$

DOI: 10.1201/9781003494843-10

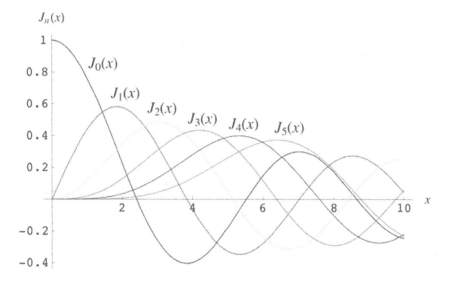

FIGURE 10.1   The Bessel functions of the first kind

$$J_{-n}(x) = \sum_{k=n}^{\infty} \frac{(-1)^k}{k!(k-n)!}\left(\frac{x}{2}\right)^{2k-n} = \sum_{s=0}^{\infty} \frac{(-1)^{n+s}}{(n+s)!s!}\left(\frac{x}{2}\right)^{2s+n}.$$

The Bessel functions of order $\pm 1/2$ are therefore defined as:

$$J_{-1/2}(x) \equiv \sqrt{\frac{2}{\pi\, x}}\, \cos x$$

$$J_{1/2}(x) \equiv \sqrt{\frac{2}{\pi\, x}}\, \sin x.$$

10.2   CHEBYSHEV FUNCTION-$T_n(x)$

The Chebyshev polynomials comprise two sequences of polynomials, denoted as $T_n(x)$ and $U_n(x)$, which are associated with the cosine and sine functions, respectively. These polynomials can be defined in various equivalent ways, with one common approach beginning with trigonometric functions [2].

The Chebyshev polynomials of the first kind T_n are defined by:

$$T_n(\cos\theta) = \cos(n\theta)$$

Similarly, the Chebyshev polynomials of the second kind U_n are defined by:

$$U_n(\cos\theta)\sin\,\theta = \sin((n+1)\theta).$$

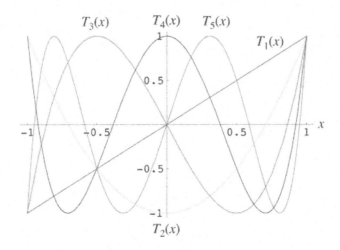

FIGURE 10.2 The first five Chebyshev polynomials of the first kind

Figure 10.2 represents the first five Chebyshev polynomials of the first kind. The Chebyshev polynomials of the first kind are a set of orthogonal polynomials defined as the solutions to the Chebyshev differential equation

$$\left(1-x^2\right)\frac{d^2y}{dx^2}-x\frac{dy}{dx}+\alpha^2y=0$$

The Chebyshev polynomials of the first kind can be obtained as in the following recurrence relation:

$$T_0\left(x\right)=1$$

$$T_1\left(x\right)=x$$

$$T_{n+1}\left(x\right)=2\;x\;T_n\left(x\right)-T_{n-1}\left(x\right).$$

10.3   ERROR FUNCTION-*erf(x)*

The **error function** *erf* (*x*) is a special function that gets its name for its importance in the study of errors and is also used in probability theory, mathematical physics, and a wide variety of other theoretical and practical applications. It occurs frequently in engineering problems; e.g., in heat conduction problems. The error function is a monotonically increasing odd function of *x*. It has an integral representation, as in the following form [2]:

$$\mathrm{erf}\left(x\right)=\frac{2}{\sqrt{\pi}}\int_{0}^{x}e^{-t^2}\,dt.$$

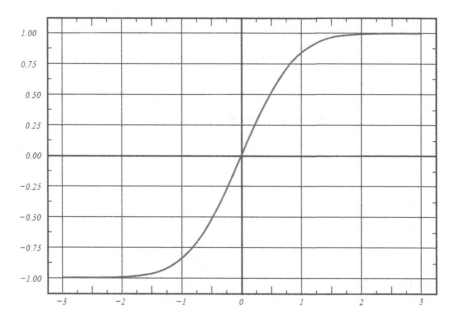

FIGURE 10.3 The representation of the Error function

Its *Maclaurin series* is given by:

$$\text{erf}(x) = \frac{2}{\sqrt{\pi}} \sum_{n=0}^{\infty} \frac{(-1)^n x^{2n+1}}{n!(2n+1)}$$

The graphical representation of erf(x) in the interval [−3,3] is as follows in Figure 10.3.

10.4 THE EXPONENTIAL INTEGRAL FUNCTION-$E_i(x)$

The exponential integral function - $E_i(x)$ is a special function on the complex plane and defined as one particular definite integral of the ratio between an exponential function and its argument [2].

$$E_i(x) = -\int_{-x}^{\infty} \frac{e^{-t} dt}{t}.$$

For real non-zero values of x, the exponential integral $E_i(x)$ is defined as:

$$E_i(x) = -\int_{-x}^{\infty} \frac{e^{-t}}{t} dt = \int_{-\infty}^{x} \frac{e^t dt}{t}.$$

The exponential integral may also be generalized to

$$E_n(x) = -\int\limits_1^\infty \frac{e^{-xt}dt}{t^n},$$

which can be written as a special case of upper incomplete gamma function [3]:

$$E_n(x) = x^{n-1}\Gamma(1-n,x).$$

Series expansion around the branch point at the origin can be obtained by the Mathematica command:

Series[*ExpIntegralEi*[*x*], {*x*,0,5}, *Assumptions* → *x* > 0]

$$\left(EulerGamma + Log[x]\right) + x + \frac{x^2}{4} + \frac{x^3}{18} + \frac{x^4}{96} + \frac{x^5}{600} + O[x]^6$$

Series[*ExpIntegralEi*[*x*], {*x*,0,5}, *Assumptions* → *x* < 0]

$$\left(EulerGamma + Log[-x]\right) + x + \frac{x^2}{4} + \frac{x^3}{18} + \frac{x^4}{96} + \frac{x^5}{600} + O[x]^6$$

Series for $E_n(x)$ generic and logarithmic cases:

Series[*ExpIntegralE*[*n*, *x*], {*x*,0,4}]
$x \wedge n \ (Gamma[1-n]/x + O[x] \wedge 5) + (1/(-1+n) + x/(2-n) + x \wedge 2/(2(-3+n))$
$-x \wedge 3/(6 \ (-4+n)) + x \wedge 4/(24 \ (-5+n)) + O[x] \wedge 5)$
Series[*ExpIntegralE*[1, *x*], {*x*,0,4}]
$(-EulerGamma - Log[x]) + x - x \wedge 2/4 + x \wedge 3/18 - x \wedge 4/96 + O[x] \wedge 5$

Series expansion of $E_n(x)$ at infinity:

Series[*ExpIntegralE*[*n*, *x*], {*x*,*Infinity*,4}]
$E \wedge -x + O[1/x] \wedge 5(1/x - n/x \wedge 2 + (n \ (1+n))/x \wedge 3 - (n \ (1+n)$
$(2+n))/x \wedge 4 + O[1/x] \wedge 5)$

The graphical representation of erf(x) in the interval [0,3] is as follows in Figure 10.4.

Plot over a subset of the reals for integer values of the parameter n:

```
Plot[Table[ExpIntegralE[n,x],{n,6}]//Evaluate,{x,0,3},Frame->
True, GridLines->Automatic,PlotLegends->"Expressions"]
```

10.5   FRESNEL INTEGRALS-S(z) AND C(z)

The Fresnel integrals appeared in the works by A. J. Fresnel (1798, 1818, 1826) who investigated an optical problem. Later, K. W. Knochenhauer (1839) found series representations of these integrals. N. Nielsen (1906) studied various properties of these integrals [3].

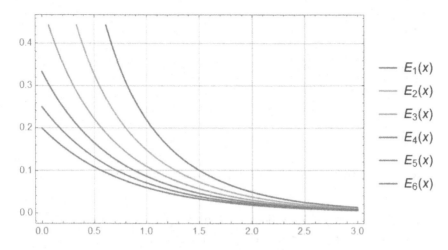

FIGURE 10.4 The representation of the Exponential Integral function

The Fresnel integrals $S(z)$ and $C(z)$ are defined as values of the following definite integrals:

$$S(z) = \int_0^z \sin\left(\pi t^2/2\right) dt$$

$$C(z) = \int_0^z \cos\left(\pi t^2/2\right) dt$$

Graphics for the Fresnel integrals along the real axis are given in Figures 10.5 and 10.6.

```
Plot[{FresnelS[z],FresnelC[z]},{z,-10,10}]
```

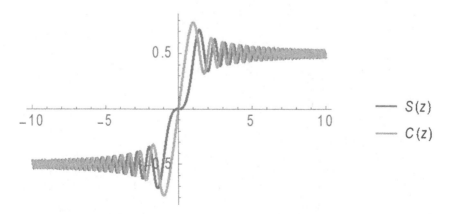

FIGURE 10.5 The representation of Fresnel integrals $S(z)$ and $C(z)$

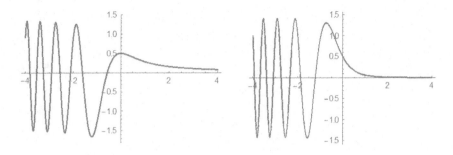

FIGURE 10.6 The representation of the FresnelF and FresnelG functions

FresnelF and FresnelG functions are defined depending on $S(z)$ and $C(z)$.
$$F(z) = (1/2 - S(z))\cos(\pi z^2/2) - (1/2 - C(z))\sin(\pi z^2/2)$$
$$G(z) = (1/2 - C(z))\,\cos(\pi\,z2/2) + (1/2 - S(z))\,\sin(\pi\,z2/2)$$

```
Plot[FresnelF[z],{z,-4,4}]    Plot[FresnelG[z],{z,-4,4}]
```

Series expansions of Fresnel functions are:

Series[FresnelS[z],{z,0,20}]

$$\frac{\pi x^3}{6} - \frac{\pi^3 x^7}{336} + \frac{\pi^5 x^{11}}{42240} - \frac{\pi^7 x^{15}}{9676800} + \frac{\pi^9 x^{19}}{3530096640} + O[x]^{21}$$

Series[FresnelC[z],{z,0,20}]

$$z - \frac{\pi^2 z^5}{40} + \frac{\pi^4 z^9}{3456} - \frac{\pi^6 z^{13}}{599040} + \frac{\pi^8 z^{17}}{175472640} + O[z]^{21}$$

Series[FresnelF[z],{z,0,20}]

$$\frac{1}{2} - \frac{\pi z^2}{4} + \frac{\pi z^3}{3} - \frac{\pi^2 z^4}{16} + \frac{\pi^3 z^6}{96} - \frac{\pi^3 z^7}{105} + \frac{\pi^4 z^8}{768} - \frac{\pi^5 z^{10}}{7680} + \frac{\pi^5 z^{11}}{10395} - \frac{\pi^6 z^{12}}{92160}$$

$$+ \frac{\pi^7 z^{14}}{1290240} - \frac{\pi^7 z^{15}}{2027025} + \frac{\pi^8 z^{16}}{20643840} - \frac{\pi^9 z^{18}}{371589120} + \frac{\pi^9 z^{19}}{654729075}$$

$$- \frac{\pi^{10} z^{20}}{7431782400} + O[z]^{21}$$

Series[FresnelG[z], {z,0,20}]

$$\frac{1}{2} - z + \frac{\pi z^2}{4} - \frac{\pi^2 z^4}{16} + \frac{\pi^2 z^5}{15} - \frac{\pi^3 z^6}{96} + \frac{\pi^4 z^8}{768} - \frac{\pi^4 z^9}{945} + \frac{\pi^5 z^{10}}{7680} - \frac{\pi^6 z^{12}}{92160} + \frac{\pi^6 z^{13}}{135135}$$

$$- \frac{\pi^7 z^{14}}{1290240} + \frac{\pi^8 z^{16}}{20643840} - \frac{\pi^8 z^{17}}{34459425} + \frac{\pi^9 z^{18}}{371589120} - \frac{\pi^{10} z^{20}}{7431782400} + O[z]^{21}$$

10.6 HERMITE POLYNOMIAL-$H_n(x)$

Similar to other classical orthogonal polynomials, the Hermite polynomials can be derived from various initial conditions. It is noteworthy that there exist two commonly employed standardizations. One convenient approach is outlined as follows:
The **"probabilist's Hermite polynomials"** are given by

$$He_n(x) = (-1)^n e^{\frac{x^2}{2}} \frac{d^n}{dx^n} e^{-\frac{x^2}{2}},$$

the **"physicist's Hermite polynomials"** are given by

$$H_n(x) = (-1)^n e^{x^2} \frac{d^n}{dx^n} e^{-x^2}.$$

The first five probabilist's Hermite polynomials are:

$He_0(x) = 1,$
$He_1(x) = x,$
$He_2(x) = x^2 - 1,$
$He_3(x) = x^3 - 3x,$
$He_4(x) = x^4 - 6x^2 + 3,$
$He_5(x) = x^5 - 10x^3 + 15x.$

The first five physicist's Hermite polynomials are:

$H_0(x) = 1,$
$H_1(x) = 2x,$
$H_2(x) = 4x^2 - 2,$
$H_3(x) = 8x^3 - 12x,$
$H_4(x) = 16x^4 - 48x^2 + 12,$
$H_5(x) = 32x^5 - 160x^3 + 120x.$

10.7 HYPERGEOMETRIC FUNCTION

HyperGeometric functions are probably the most useful, but least understood, class of functions.

A generalized HyperGeometric function ${}_pF_q(a_1,\ldots,a_p; b_1,\ldots,b_q; z)$ is a function that can be defined in the form of a HyperGeometric Series, i.e., a series for which the ratio of successive terms can be written

$$\frac{c_{k+1}}{c_k} = \frac{P(k)}{Q(k)} = \frac{(k+a_1)(k+a_2)\ldots(k+a_p)}{(k+b_1)(k+b_2)\ldots(k+b_q)(k+1)} x.$$

$$1 + \frac{a_1 \ldots a_p}{b_1 \ldots b_p} \frac{z}{1!} + \frac{a_1(a_1+1)\ldots a_p(a_p+1)}{b_1(b_1+1)\ldots b_q(b_q+1)} \frac{z^2}{2!} + \ldots$$

This has the form of an exponential generating function. This series is usually denoted by

$$_pF_q\left(a_1,\ldots,a_p;b_1,\ldots,b_q;z\right)$$

Or

$$_pF_q\left[\begin{array}{cccc} a_1 & a_2 & \cdots & a_p \\ b_1 & b_2 & \cdots & b_q \end{array} \; ; \; z \right].$$

Using the rising factorial or Pochhammer symbol

$$(a)_0 = 1$$

$$(a)_n = a(a+1)(a+2)\ldots(a+n-1), \; n \geq 1$$

this can be written

$$_pF_q\left(a_1,\ldots,a_p;b_1,\ldots,b_q;z\right) = \sum_{n=0}^{\infty} \frac{(a_1)_n \ldots (a_p)_n}{(b_1)_n \ldots (b_q)_n} \frac{z^n}{n!}.$$

The HyperGeometric function is defined for $|z|<1$ by the power series

$$_2F_1\left(a,b;c;z\right) = \sum_{n=0}^{\infty} \frac{(a)_n (b)_n}{c_n} \frac{z^n}{n!} = 1 + \frac{ab}{c}\frac{z}{1!} + \frac{a(a+1)b(b+1)}{c(c+1)}\frac{z^2}{2!} + \cdots.$$

The HyperGeometric functions are solutions to the HyperGeometric differential equation, which has a regular singular point at the origin. To derive the HyperGeometric function from the HyperGeometric differential equation [2].

The HyperGeometric function is a solution of Euler's HyperGeometric differential equation

$$z(1-z)\frac{d^2w}{dz^2} + \left[c-(a+b+1)z\right]\frac{dw}{dz} - abw = 0.$$

An infinite family of rational values for well-poised HyperGeometric functions with rational arguments is given by

$$_kF_{k-1}\left(\frac{1}{k+1},\ldots,\frac{k}{k+1};\frac{2}{k},\frac{3}{k},\ldots,\frac{k-1}{k},\frac{k+1}{k};\left(\frac{x\left(1-x^k\right)}{f_k}\right)^k\right) = \frac{1}{1-x^k}$$

For $k=2, 3, \ldots, 0 \leq x \leq (k+1)^{-1/k}$, and

$$f_k \equiv \frac{k}{(1+k)^{1+1/k}}$$

Since the space has dimension 2, any of these $p+q+2$ functions are linearly independent. These dependencies can be written out to generate a large number of identities involving $_pF_q$.

For example, in the simplest non-trivial case,

$$_0F_1(;a;z) = (1)\ _0F_1(;a;z),$$

$$_0F_1(;a-1;z) = \left(\frac{\vartheta}{a-1}+1\right)_0F_1(;a;z),$$

$$z\ _0F_1(;a+1;z) = (a\vartheta)\ _0F_1(;a;z),$$

So

$$_0F_1(;a-1;z) - _0F_1(;a;z) = \frac{z}{a(a-1)}\ _0F_1(;a+1;z).$$

10.8 LEGENDRE FUNCTION

The following second-order linear differential equation with variable coefficients is known as Legendre's differential equation [2]:

$$\left(1-x^2\right)y'' - 2xy' + n(n+1)y = 0$$

where n is a non-negative integer. Legendre's differential equation occurs in many physical and engineering problems involving spherical geometry and gravitation. The graphical representation of Legendre's function is given in Figure 10.7.

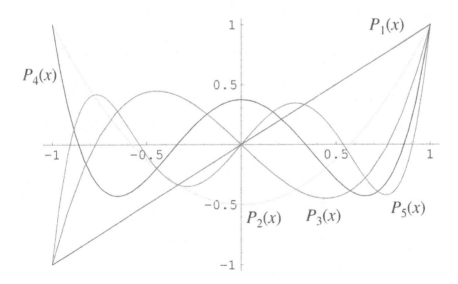

FIGURE 10.7 The representation of Legendre's function of the first kind

Legendre's function of the first kind or Legendre's polynomial of degree n is denoted by $P_n(x)$ and is defined by:

$$P_n(x) = \frac{1.3.5\ldots(2n-1)}{n}\left[x^n - \frac{n(n-1)}{2(2n-1)}x^{n-2} + \frac{n(n-1)(n-2)(n-3)}{2.4(2n-1)(2n-3)}x^{n-4} - \ldots\right].$$

We can also write Pn(x) in a compact from as:

$$P_n(x) = \sum_{r=0}^{[n/2]}(-1)^r \frac{(2n-2r)!}{2^n r!(n-2r)!(n-r)}x^{n-2r}$$

Where $[n/2] = \begin{cases} n/2, & \text{if } n \text{ is even} \\ (n-1)/2, & \text{if } n \text{ is odd.} \end{cases}$

Legendre's function of the second kind is denoted by $Q_n(x)$ and is defined by:

$$Q_n(x) = \frac{n!}{1.3.5\ldots(2n+1)}$$

$$\left[x^{-n-1} + \frac{(n+1)(n+2)}{2(2n+3)}x^{-n-3} + \frac{(n+1)(n+2)(n+3)(n+4)}{2.4(2n+3)(2n+5)}x^{-n-5} - \ldots\right].$$

Using the definition (1) or (2), the first few Legendre's polynomials are given by

$$P_0(x) = 1, \ P_1(x) = x, \ P_2(x) = \frac{1}{2}(3x^2 - 1), \ P_3(x) = \frac{1}{2}(5x^3 - 3x)$$

$$P_4(x) = \frac{1}{8}(35x^4 - 30x^2 + 3), \ P_5(x) = \frac{1}{8}(63x^5 - 70x^3 + 15x), \ etc.$$

10.9 RAMANUJANTAUTHETA FUNCTION

The RamanujanTauTheta function is an analytic function of t except for branch cuts on the imaginary axis running from $\pm 6\,i$ to $\pm\,i\,\infty$ called the Ramanujan Tau functions. This function is defined for all positive integers n by:

$$q\prod_{m=1}^{\infty}(1-q^m)^{24} = \sum_{n=1}^{\infty}\tau(n)q^n, \ q \in \mathbb{C}, |q| < 1,$$

where $q = \exp(2\pi i z)$ *with* $Im\ z > 0,$

$$\theta(t) = \frac{-i}{2}\log\frac{\Gamma(6+i\,t)}{\Gamma(6-i\,t)} - t\log(2\ \pi) \text{ for real } t.$$

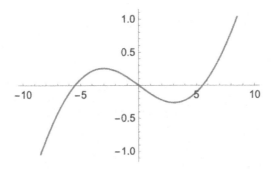

FIGURE 10.8   The representation of the RamanujanTauTheta function

Figure 10.8 shows the representation of the RamanujanTauTheta special function in the interval [−10,10].

```
Plot[RamanujanTauTheta [t],{t,-10,10}]
```

10.10   RIEMANN-SIEGEL THETA FUNCTION

In mathematics, the Riemann-Siegel Theta function is defined in terms of the *gamma function* as

$$\theta(t) = \arg\left(\Gamma\left(\frac{1}{4} + \frac{it}{2} \right) \right) - \frac{\log \pi}{2} t$$

for real values of t. Here, the argument is chosen in such a way that a continuous function is obtained and $\theta(t) = 0$ holds, i.e., in the same way that the principal branch of the log-gamma functions is defined [1].

It has an asymptotic expansion

$$\theta(t) \sim \frac{t}{2} \log \frac{t}{2\pi} - \frac{t}{2} - \frac{\pi}{8} + \frac{1}{48t} + \frac{7}{5760t^3} + \cdots$$

which is not convergent, but whose first few terms give a good approximation for *t*, »1.

Its Taylor-series at 0, which converges for $|t| < 1/2$, is

$$\vartheta(t) = -\frac{t}{2}\log\pi + \sum_{k=0}^{\infty} \frac{(-1)^k \, \psi^{2k} \frac{1}{4}}{(2k+1)!} \left(\frac{t}{2} \right)^{2k+1}$$

Where ψ^{2k} denotes the polygamma function of order *2k*. The Riemann-Siegel Theta function is of interest in studying the Riemann Zeta function, since it can rotate the Riemann Zeta function such that it becomes the totally real valued Z function on the critical line $s = 1/2 + it$.

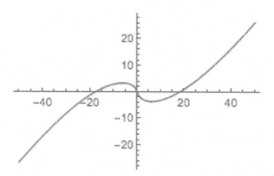

FIGURE 10.9   The representation of Riemann-Siegel Theta function

For real t, the Riemann-Siegel Theta function $\vartheta\,(t)$ is defined as:

$$\vartheta(t) = I\left[ln\Gamma\left(\frac{1}{4} + \frac{1}{2}\,i\,t \right) \right] - \frac{1}{2}t\,ln\pi$$

$$= arg\left[\Gamma\left(\frac{1}{4} + \frac{1}{2}\,i\,t \right) \right] - \frac{1}{2}t\,ln\pi.$$

The series expansion of $\vartheta\,(t)$ about 0 is given by

$$\vartheta(t) = -\frac{1}{2}\,t\,ln\pi + \sum_{k=0}^{\infty} \frac{(-1)^k}{(4k+2)!!}\,\Psi_{2k}\left(\frac{1}{4} \right) t^{2\,k+1}$$

$$= \frac{1}{2}\left[-ln\pi + \Psi\left(\frac{1}{4} \right) \right] t - \frac{1}{48}\Psi_2\left(\frac{1}{4} \right) t^3 + \frac{1}{3840}\Psi_4\left(\frac{1}{4} \right) t^5 - \frac{1}{645120}\Psi_6\left(\frac{1}{4} \right) t^7$$

$$= -\frac{1}{4}\left[2\gamma + \pi + 2\ln(8\pi) \right] t + \frac{1}{24}\left[\pi^3 + 28\zeta(3) \right] t^3 - \left[\frac{\pi^5}{96} + \frac{31\xi(5)}{10} \right] + \dots$$

Figure 10.9 shows the graphical presentation of the Riemann-Siegel Theta function.

```
Plot[RiemannSiegelTheta[t],{t,-50,50}]
```

CASE STUDY 10.1

This study examines a laminated composite problem to demonstrate the use of special functions as an alternative model type for mathematical modeling [4]. The material being used is a 16-layered symmetric balanced flax/epoxy material, and the material's properties can be found in Chapter 8, Table 8.2. The hybrid composite is illustrated in Figure 10.10.

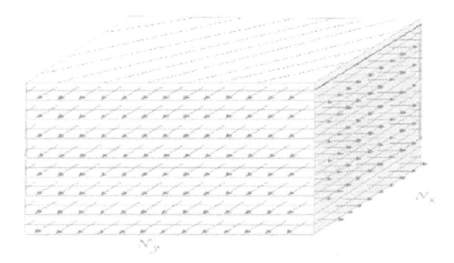

FIGURE 10.10 The 16-layered symmetric balanced flax/epoxy composite

The design of laminated composites involves determining the optimum values of specific parameters such as composite fiber angles, plate aspect ratio (a/b), and the plate in-plane load applied in the y-axis direction (Ny). The aim is to maximize the critical buckling load based on these design parameters.

To solve the critical buckling load problem for a flax/epoxy composite, the Classical Lamination Theory is taken into account. For a more detailed understanding of this theory and its assumptions, please refer to Reference [5].

To calculate the critical buckling load using an analytical formula based on composite material mechanics, it is necessary to determine the angle-dependent properties of the laminated composite. Transformed reduced stiffness matrix elements $[\bar{Q}_{ij}]$ for angular laminated composite, the expressions of these elements based on engineering constants (E_{ij}, G_{ij}, v_{ij}), and the bending stiffness matrix [D] are given previously in Equations 6.1, 6.2, and 6.3 in Chapter 6.

The critical buckling load selected as the objective function is calculated as follows for a composite plate with four edges simply supported [4]:

$$\lambda_b(m,n) = \frac{\pi^2 \left[D_{11}\left(\dfrac{m}{a}\right)^4 + 2(D_{12}+2D_{66})\left(\dfrac{m}{a}\right)^2\left(\dfrac{n}{b}\right)^2 + D_{22}\left(\dfrac{n}{b}\right)^4 \right]}{N_x\left(\dfrac{m}{a}\right)^2 + N_y\left(\dfrac{n}{b}\right)^2 + N_{xy}\left(\dfrac{m}{a}\right)\left(\dfrac{n}{b}\right)} \qquad (10.1)$$

Here, λ_b indicates the critical buckling load factor, and N_x and N_y indicate the applied loads per unit length. The critical buckling loads N_{xb} and N_{yb} are defined as:

$$N_{xb} = N_x\,\lambda_b, \; N_{yb} = N_y\,\lambda_b$$

TABLE 10.1

Design Parameters and Their Levels Affecting Buckling Behavior

	Level 1	Level 2	Level 3
B (x_1)	0.254	0.508	0.762
N_y (x_2)	0.5	1	1.5
Q_1 (x_3)	0	45	90
Q_2 (x_4)	0	45	90
Q_3 (x_5)	0	45	90
Q_4 (x_6)	0	45	90

The smallest value of the buckling load (λ_b) is determined as the critical buckling load factor. Depending on the change of m and n modes in the formula used in buckling load factor calculation, among the different buckling loads ($\{\lambda_b (1,1), \lambda_b (1,2), \lambda_b (2,1), \lambda_b (2,2)\}$), the smallest one is selected and multiplied by the unit applied load, the critical buckling load is obtained.

The design parameters and their corresponding level values, utilized to construct an experimental set for mathematical modeling, are outlined in Table 10.1. The investigation encompassed six design parameters, each spanning three levels. Among these parameters, four pertain to the angular orientation of the laminated composite. Specifically, the composite plate comprises 16 layers featuring four distinct fiber angles to ensure symmetry and balance. Design parameters Q1, Q2, Q3, and Q4 denote the fiber angles. Notably, the width of the plate (a) remains constant at 0.508, while the length (b) varies. The final design parameter, Ny, pertains to the load applied across the plate width. The three feasible levels for this design parameter are also delineated in Table 10.1, mirroring the setup of other design parameters for the specified buckling problem.

Table 10.2 illustrates an experimental set generated through the Box-Behnken design (BBD) method. While the BBD method advocates for 49 lines for modeling, utilizing the Full Factorial design (FFD) method would necessitate 729 analyses. By employing the BBD experimental design method, the most crucial 49 lines are identified and recommended from the 729-line experiment set. The output parameter, critical buckling factor results, was derived using an analytical formula. The BBD method proposes designs yielding maximum and minimum critical buckling factors of 3129.39 and 373.90, respectively. The design parameters corresponding to the maximum critical buckling factor are x1=0.254, x2=0.5, x3=45, x4=90, x5=45, x6=45. Conversely, the design parameters for the minimum critical buckling factor are x1=0.762, x2=1, x3=90, x4=45, x5= 45, x6=0. These results are ascertained by the analytical formula utilizing the designs recommended by the BBD experimental set.

Table 10.3 presents various mathematical models proposed to predict the critical buckling load of a 16-layer flax/epoxy composite plate using a 49-row test set.

TABLE 10.2
Box-Behnken Design Dataset Created to Model Buckling Behavior

		B (x_1)	Ny (x_2)	Q1 (x_3)	Q2 (x_4)	Q3 (x_5)	Q4 (x_6)	Critical Buckling Load Factor
Training	1	0.254	1	0	45	45	0	1153.97
	2	0.254	1	90	45	45	90	2057.79
	3	0.762	1.5	45	90	45	45	389.152
	4	0.508	1	90	90	45	0	562.044
	5	0.508	1.5	45	45	90	90	584.629
	6	0.508	1	90	0	45	90	562.044
	7	0.508	0.5	45	45	0	0	974.381
	8	0.508	0.5	90	45	0	45	809.083
	9	0.762	1	45	0	0	45	535.866
	10	0.508	1.5	45	45	0	0	584.629
	11	0.762	1	90	45	45	90	373.904
	12	0.762	1	45	0	90	45	505.034
	13	0.508	0.5	90	45	90	45	809.083
	14	0.762	1	0	45	45	90	536.873
	15	0.762	1.5	45	0	45	45	461.681
	16	0.508	1	45	45	45	45	758.335
	17	0.762	0.5	45	90	45	45	530.662
	18	0.254	1.5	45	90	45	45	1341.17
	19	0.254	0.5	45	90	45	45	3129.39
	20	0.508	1	90	90	45	90	562.044
	21	0.254	1	0	45	45	90	1177.75
	22	0.508	1	90	0	45	0	562.044
	23	0.254	1	45	90	90	45	1944.01
	24	0.762	1	0	45	45	0	541.278
	25	0.508	1.5	45	45	0	90	584.629
	26	0.508	0.5	45	45	90	0	974.381
	27	0.508	1	0	90	45	0	562.044
	28	0.254	1	90	45	45	0	2034
	29	0.254	0.5	45	0	45	45	2376.21
	30	0.508	0.5	0	45	90	45	809.083
	31	0.508	1.5	0	45	0	45	485.45
	32	0.254	1	45	0	0	45	1325.6
	33	0.508	1	0	90	45	90	562.044
	34	0.254	1.5	45	0	45	45	1018.37
	35	0.508	0.5	45	45	90	90	974.381
	36	0.508	0.5	0	45	0	45	809.083
	37	0.762	1	45	90	90	45	421.347
	38	0.508	1.5	90	45	90	45	485.45
	39	0.508	1.5	45	45	90	0	584.629

(Continued)

TABLE 10.2 *(Continued)*
Box-Behnken Design Dataset Created to Model Buckling Behavior

		B (x_1)	Ny (x_2)	Q1 (x_3)	Q2 (x_4)	Q3 (x_5)	Q4 (x_6)	Critical Buckling Load Factor
Testing	40	0.508	1	0	0	45	0	562.044
	41	0.508	1.5	90	45	0	45	485.45
	42	0.762	1	90	45	45	0	378.309
	43	0.762	0.5	45	0	45	45	629.565
	44	0.762	1	45	90	0	45	452.179
	45	0.508	1	0	0	45	90	562.044
	46	0.508	0.5	45	45	0	90	974.381
Validation	47	0.254	1	45	0	90	45	1492.1
	48	0.508	1.5	0	45	90	45	485.45
	49	0.254	1	45	90	0	45	1777.51

The models were evaluated based on their R-squared (R^2) index. Successful models, which demonstrated high prediction performance of 0.8 or higher, included SON, TON, FOTN, SOTN, SOLN, LR, and FOTNR. However, it was observed that other models were inadequate in modeling the composite plate's critical buckling load behavior during the training phase. Among the successful models,

TABLE 10.3
Proposed Mathematical Models for Modeling Critical Buckling Load and Assessment of Their Prediction Performance

Model*	R^2 Train	R^2 Test	R^2 Valid	Maximum	Minimum
L	0.66	0.63	0.6	1953.98	−244.49
SON	0.97	0.19	0.93	2941.73	−292.92
TON	1	−1.45	0.69	4161.62	−1567.36
FOTN	0.88	0.43	0.88	2374.5	−1087.22
SOTN	1	−8.22	0.66	6764.28	−4198.63
FOLN	0.78	0.72	0.8	2079.34	−285.53
SOLN	0.99	0.59	0.99	3647.93	162.96
LR	0.98	0.61	0.79	Inf	Inf
SONR	−0.42	−9	−4.77	7.80E+16	−7.4E+11
TONR	0.47	−8.55	−5.18	3.83E+14	−3.39E+14
FOTNR	0.99	0.95	−1.02	1.02E+14	−6.14E+08
SOTNR	−0.39	−12.55	−5.29	2.89E+07	−5.23E+10
FOLNR	0.39	−1.71	−0.4	2.28E+10	−2.50E+15
SOLNR	−0.95	−10.67	−5.1	4.03E+12	−1.80E+12

* Full form of models are given in appendix (Table A10.1).

only the FOTNR model exhibited high prediction performance (0.95) in the testing phase. Nevertheless, the prediction performance of the same model was unsatisfactory in the validation phase, unlike in the training and testing phases. Consequently, no model was found to successfully meet the training, testing, and validation criteria.

Upon evaluating the critical buckling factor of the composite plate generated by the model, it was determined that only the SOLN model yielded appropriate results according to the boundedness control criterion. All other models produced results that were unattainable and, therefore, unusable.

Using the SOLN model, the highest critical buckling factor value obtained was 3647.93, while the lowest value was 162.96. These findings align with the maximum and minimum values obtained using the analytical formula, which are 3129.39 and 373.904, respectively. Based on these assessments, it can be concluded that the SOLN model is the most effective among all the proposed models. However, it exhibited a low prediction performance of 0.59 R^2 during the testing phase.

The results from Table 10.3 indicate that conventional mathematical models are inadequate for modeling the critical buckling behavior of laminated composite plates. Consequently, there is a necessity to develop alternative mathematical models. Special functions can be beneficial in modeling data exhibiting nonlinear and unstable behavior that cannot be expressed in basic functions. Therefore, special functions were preferred for modeling the buckling behavior of the examined composite plates.

Table 10.4 presents evaluations of various types of special functions that can be utilized to model composite plate buckling behavior. During the training, testing, and validation phases, all proposed special functions demonstrated high

---

TABLE 10.4

Performance Assessment of the Special Functions in Modeling Composite Plate Buckling Behavior

Model*	R² Train	R² Test	R² Valid	Max.	Min.
BesselJ	0.97	0.98	0.97	3160.01	−87767.5
ChebyShevT	0.99	0.99	0.99	3480.46	12.707
Erf	0.98	0.98	0.98	2914.08	253.748
ExpIntegralE	0.98	0.98	0.99	2909.36	341.248
FresnelG	0.94	0.85	0.95	2886.16	56.81
FresnelG	0.91	0.95	0.93	2838.27	117.864
HermiteH	0.9	0.77	0.78	2537.53	100.513
HyperGeometric0F1	0.96	0.79	0.94	3096.61	48.92
LegendreP	0.9	0.77	0.78	2537.37	101.086
RamanujanTauTheta	0.95	0.93	0.99	3228.72	45.7072
RiemannSiegelTheta	0.97	0.92	0.99	3448.3	84.55

* Full form of models are given in appendix (Table A10.2).

performance in predicting actual buckling values. The lowest R^2 index recorded was 0.77 in the testing phase. All models successfully met the R^2-based assessment criteria across the training, testing, and validation stages.

When examining the maximum and minimum critical buckling factor produced by the models within the boundedness check criterion, it becomes clear that the Erf and ExpIntegralE special functions produce consistent results with the actual buckling load found using the analytical formula. While the maximum critical buckling factor found by other models, except for LegendreP and HermiteH, is close to the actual value found by the analytical formula, the same cannot be said for the minimum critical buckling factor.

Therefore, it can be concluded that the Erf and ExpIntegralE special functions are more effective in modeling the critical buckling factor due to their consistency in both maximum and minimum critical buckling factor values and their high prediction performance in the training, testing, and validation stages.

In Table 10.5, the modeling process was reiterated ten times using the special functions Erf and ExpIntegralE. The efficacy of these models was assessed

TABLE 10.5

Comparison of Prediction Performance of Erf and ExpIntegralE Special Functions

		R^2 Train	R^2 Test	R^2 Valid	Max.	Min.
Erf Model	1	0.97	0.99	0.96	2909.8	295.5
	2	0.98	0.99	0.89	2905.54	258
	3	0.96	0.99	0.97	3021.5	227.85
	4	0.97	0.99	0.91	2921.91	218.47
	5	0.97	0.98	0.99	2930.79	242.44
	6	0.98	0.96	0.92	2797.58	297.4
	7	0.98	0.88	0.8	2880.16	259.42
	8	0.98	0.99	0.97	2907.96	274.12
	9	0.98	0.95	0.85	2923.21	291.88
	10	0.95	0.99	0.94	2993.94	199.13
	Mean	0.97	0.97	0.92	2919.24	256.42
ExpIntegralE Model	1	0.98	0.99	0.98	2909.83	308.68
	2	0.98	0.99	0.97	2922.05	291.05
	3	0.98	0.95	0.96	2944.51	319.48
	4	0.98	0.96	0.96	2932.05	313.29
	5	0.98	0.97	0.63	2911.58	336.38
	6	0.98	0.98	0.94	2964.82	314.62
	7	0.97	0.99	0.94	2943.2	319.81
	8	0.98	0.93	0.54	2902.59	318.04
	9	0.98	0.94	0.87	2829.72	288.27
	10	0.98	0.95	0.99	2982.66	315.73
	Mean	0.98	0.97	0.88	2924.30	312.54

based on the R^2 criterion across the training, testing, and validation stages. Each iteration of the modeling process involved the random selection of data, enabling an assessment of the effectiveness of Erf and ExpIntegralE special functions in modeling the critical buckling factor using different datasets for training, testing, and validation.

According to the R^2 criterion, Erf exhibited a prediction performance of 0.95 in the training phase, 0.88 in the testing phase, and 0.8 in the validation phase during the repeated running process. An R^2 value nearing 1 signifies high prediction performance, indicating that the Erf model successfully met the R^2 criterion in repeated iterations. Furthermore, the boundedness control analysis indicates that the obtained maximum and minimum critical buckling factors align with the actual values derived from the analytical formula.

When evaluating the ExpIntegralE special function, the R^2 index for this model was 0.97 in the training phase, 0.93 in the testing phase, and 0.54 in the validation phase. Although the prediction performance in the validation phase was lower compared to the Erf model, the ExpIntegralE model still demonstrated a high average success rate across all evaluation stages regarding R^2, considering the results from the ten iterations.

The study explores the applicability of special functions as an alternative model type for mathematical modeling. It was observed that while basic mathematical functions failed to effectively model the critical buckling factor parameter of the laminated composite plate, the proposed special functions exhibited high prediction performance. This suggests that special functions may be preferable over basic mathematical functions in cases where the latter are insufficient for modeling purposes. Since special functions encompass a wide and diverse range in mathematics, a more comprehensive assessment can be made by comparing the results obtained when special functions are utilized as mathematical models in problems across various fields.

REFERENCES

[1] Andrews GE, Askey R, Roy R. Special Functions (Vol. 71, pp. xvi+-664). Cambridge: Cambridge University Press, 1999.

[2] Bell WW. Special Functions for Scientists and Engineers. Courier Corporation, 2004.

[3] Wolfram Mathematica software 12.

[4] Savran M, Aydin L. Natural frequency and buckling optimization considering weight saving for hybrid graphite/epoxy-sitka spruce and graphite-flax/epoxy laminated composite plates using stochastic methods. Mechanics of Advanced Materials and Structures 2022;30(13):2637–2650.

[5] Kaw AK. Mechanics of Composite Materials. London: CRC, 2006.

APPENDIX

TABLE A10.1
Mathematical Models Proposed for Modeling Critical Buckling Load Behavior (Related to Table 10.3)

(*L*)

$$L = -(2620.52\ x1) - 482.753\ x2 - 0.865144\ x3 + 2.0262\ x4 + 1.10895\ x5 - 0.271943\ x6 + 2578.81$$

(*LR*)

$$LR = \frac{4956.68\ x1 + 684.165\ x2 - 168.315\ x3 - 35.3051\ x4 - 65.7556\ x5 - 14.691\ x6 + 39646.9}{98.3453\ x1 + 17.2821\ x2 - 0.201182\ x3 - 0.04697\ x4 - 0.103662\ x5 - 0.00846491\ x6 - 6.7137}$$

(*SON*)

$$
\begin{aligned}
SON =\ & 7418.83\ x1^2 + 3724.92\ x1\ x2 - 7.37003\ x1\ x3 - 13.0484\ x1\ x4 + 4.05459\ x1\ x5 - \\
& 0.192695\ x1\ x6 - 13517.4\ x1 + 231.99\ x2^2 + 0.244885\ x2\ x3 - 2.4869\ x2\ x4 - \\
& 0.122443\ x2\ x5 + \frac{1.68166\ x2\ x6}{10^{13}} - 2700.27\ x2 - 0.0383005\ x3^2 + \frac{4.72987\ x3\ x4}{10^{16}} + \\
& 0.00272095\ x3\ x5 + \frac{5.05264\ x3\ x6}{10^{16}} + 6.88491\ x3 - 0.0883568\ x4^2 + 0.00687159\ x4\ x5 - \\
& \frac{4.48872\ x4\ x6}{10^{16}} + 18.2151\ x4 - 0.0479098\ x5^2 + \frac{3.60097\ x5\ x6}{10^{16}} + 2.86207\ x5 + \\
& 0.0297234\ x6^2 - 2.57722\ x6 + 5903.13
\end{aligned}
$$

(*SONR*)

$$
\begin{aligned}
SONR =\ & (955555.x1^2 + 7.50563 \times 10^6\ x1\ x2 + 2.24704 \times 10^8\ x1\ x3 + 2.02603 \times 10^8\ x1\ x4 + 2.53529 \times \\
& 10^8\ x1\ x5 + 3.03621 \times 10^8\ x1\ x6 + 7.31679 \times 10^6\ x1 + 4.7614 \times 10^6\ x2^2 + 6.61791 \times 10^8\ x2\ x3 + \\
& 5.36601 \times 10^8\ x2\ x4 + 5.82654 \times 10^8\ x2\ x5 + 6.31645 \times 10^8\ x2\ x6 + 1.72435 \times 10^7\ x2 + \\
& 9.57368 \times 10^9\ x3^2 + 1.70885 \times 10^{10}\ x3\ x4 + 2.7227 \times 10^{10}\ x3\ x5 + 2.80911 \times 10^{10}\ x3\ x6 + \\
& 6.43821 \times 10^8\ x3 + 8.29713 \times 10^9\ x4^2 + 2.59841 \times 10^{10}\ x4\ x5 + 2.13471 \times 10^{10}\ x4\ x6 + \\
& 5.34105 \times 10^8\ x4 + 9.42402 \times 10^9\ x5^2 + 2.16394 \times 10^{10}\ x5\ x6 + 6.45748 \times 10^8\ x5 + 8.93041 \times \\
& 10^9\ x6^2 + 6.8181 \times 10^8\ x6 + 4.35474 \times 10^6)(-(6.07425 \times 10^7\ x1^2) - 8.70174 \times 10^7\ x1\ x2 + \\
& 1.07754 \times 10^{10}\ x1\ x3 + 1.52592 \times 10^{10}\ x1\ x4 + 1.2302 \times 10^{10}\ x1\ x5 - 6.24158 \times 10^9\ x1\ x6 - \\
& 3.30035 \times 10^8\ x1 + 1.5783 \times 10^8\ x2^2 + 6.03272 \times 10^9\ x2\ x3 + 7.67241 \times 10^9\ x2\ x4 - 8.8872 \times \\
& 10^8\ x2\ x5 - 6.81011 \times 10^9\ x2\ x6 + 1.62153 \times 10^8\ x2 - 2.1581 \times 10^9\ x3^2 + 1.72138 \times 10^9\ x3\ x4 + \\
& 3.08629 \times 10^9\ x3\ x5 - 2.17759 \times 10^8\ x3\ x6 - 1.03805 \times 10^{10}\ x3 + 2.73895 \times 10^8\ x4^2 - \\
& 5.52165 \times 10^9\ x4\ x5 + 3.11232 \times 10^9\ x4\ x6 - 1.32991 \times 10^{10}\ x4 + 5.41831 \times 10^9\ x5^2 - \\
& 8.10284 \times 10^9\ x5\ x6 + 5.00198 \times 10^9\ x5 + 2.52936 \times 10^9\ x6^2 + 4.40846 \times 10^9\ x6 - 7.95964 \times 10^7)
\end{aligned}
$$

(*TON*)

$$
\begin{aligned}
TON =\ & 1713.49\ x1^3 + 779.944\ x1^2\ x2 + 4.89363\ x1^2\ x3 + 9.69314\ x1^2\ x4 + 1.21838\ x1^2\ x5 + \\
& 6.48799\ x1^2\ x6 - 1336.86\ x1^2 + 641.472\ x1\ x2^2 + 1.29335\ x1\ x2\ x3 + 8.94281\ x1\ x2\ x4 + \\
& 5.26732\ x1\ x2\ x5 + 3.55674\ x1\ x2\ x6 - 751.24\ x1\ x2 - 0.00702183\ x1\ x3^2 + \\
& 0.00536006\ x1\ x3\ x4 - 0.145111\ x1\ x3\ x5 - 0.813165\ x1\ x3\ x6 - 10.424\ x1\ x3 - \\
& 0.191665\ x1\ x4^2 + 0.0311873\ x1\ x4\ x5 - 0.0124094\ x1\ x4\ x6 - 15.7455\ x1\ x4 + \\
& 0.127495\ x1\ x5^2 - 0.111963\ x1\ x5\ x6 - 7.64037\ x1\ x5 + 0.0413324\ x1\ x6^2 - 13.864\ x1\ x6 - \\
& 662.446\ x1 + 232.119\ x2^3 - 0.00132166\ x2^2\ x3 + 6.17512\ x2^2\ x4 - 0.917799\ x2^2\ x5 + \\
& 0.791585\ x2^2\ x6 - 433.502\ x2^2 + 0.0543096\ x2\ x3^2 + 0.0450418\ x2\ x3\ x4 -
\end{aligned}
$$

(Continued)

TABLE A10.1 *(Continued)*
Mathematical Models Proposed for Modeling Critical Buckling Load Behavior (Related to Table 10.3)

$$\frac{2.96488\ x2\ x3\ x5}{10^{16}} + 0.0255897\ x2\ x3\ x6 - 8.72066\ x2\ x3 - 0.205166\ x2\ x4^2 +$$

$0.00472452\ x2\ x4\ x5 + 0.0591139\ x2\ x4\ x6 - 5.83835\ x2\ x4 + 0.108687\ x2\ x5^2 +$

$\dfrac{2.12645\ x2\ x5\ x6}{10^{15}} - 10.8346\ x2\ x5 + 0.0216582\ x2\ x6^2 - 9.1509\ x2\ x6 - 668.725\ x2 -$

$0.00119744\ x3^3 - 0.000276688\ x3^2\ x4 - 0.000409589\ x3^2\ x5 + 0.0000402164\ x3^2\ x6 +$

$0.0240028\ x3^2 - 0.000339602\ x3\ x4^2 - 0.000440241\ x3\ x4\ x5 - \dfrac{7.48555\ x3\ x4\ x6}{10^{18}} +$

$0.0275122\ x3\ x4 - 0.000196899\ x3\ x5^2 + 0.0017822\ x3\ x5\ x6 + 0.0679123\ x3\ x5 +$

$0.00014254\ x3\ x6^2 + 0.290851\ x3\ x6 + 17.3884\ x3 + 0.00206407\ x4^3 + 0.000373749\ x4^2\ x5 -$

$0.000470578\ x4^2\ x6 + 0.113249\ x4^2 + 0.000290043\ x4\ x5^2 - 0.000209464\ x4\ x5\ x6 -$

$0.0431506\ x4\ x5 + 0.0000413722\ x4\ x6^2 - 0.00475551\ x4\ x6 - 0.150731\ x4 -$

$0.00133191\ x5^3 - 0.000528088\ x5^2\ x6 - 0.0192204\ x5^2 - 0.000409589\ x5\ x6^2 +$

$0.0704952\ x5\ x6 + 11.6446\ x5 - 0.000822072\ x6^3 + 0.027801\ x6^2 + 14.3542\ x6 + 1597.52$

(*TONR*)

TONR = $(1.00015\ x1^3 + 3.00056\ x2\ x1^2 + 2.83522\ x3\ x1^2 + 3.28101\ x4\ x1^2 + 3.16863\ x5\ x1^2 +$

$2.8546\ x6\ x1^2 + 3.0008\ x1^2 + 2.99762\ x2^2\ x1 - 3.25436\ x3^2\ x1 + 16.8462\ x4^2\ x1 +$

$4.7681\ x5^2\ x1 - 12.8309\ x6^2\ x1 + 5.99275\ x2\ x1 + 4.64501\ x2\ x3\ x1 + 4.58902\ x3\ x1 +$

$7.26952\ x2\ x4\ x1 + 16.0809\ x3\ x4\ x1 + 7.15802\ x4\ x1 + 6.62406\ x2\ x5\ x1 - 5.59534\ x3\ x5\ x1 +$

$50.7854\ x4\ x5\ x1 + 6.49642\ x5\ x1 + 4.67919\ x2\ x6\ x1 - 19.4655\ x3\ x6\ x1 + 38.0495\ x4\ x6\ x1 +$

$0.405579\ x5\ x6\ x1 + 4.80338\ x6\ x1 + 2.99807x1 + 0.998348\ x2^3 - 25.3647\ x3^3 - 41.9651\ x4^3 -$

$58.3227\ x5^3 - 400.633\ x6^3 + 2.98828\ x2^2 - 23.9273\ x2\ x3^2 - 27.2266\ x3^2 + 12.9826\ x2\ x4^2 -$

$497.015\ x3\ x4^2 + 4.63711\ x4^2 + 4.98445\ x2\ x5^2 - 275.485\ x3\ x5^2 + 426.225\ x4\ x5^2 -$

$4.65681\ x5^2 - 50.688\ x2\ x6^2 - 1713.49\ x3\ x6^2 - 146.635\ x4\ x6^2 - 1421.83\ x5\ x6^2 -$

$50.5806\ x6^2 + 2.98928\ x2 + 2.14188\ x2^2\ x3 + 2.55214\ x2\ x3 + 2.05103\ x3 + 3.31861\ x2^2\ x4 +$

$358.521\ x3^2\ x4 + 7.04947\ x2\ x4 - 3.51859\ x2\ x3\ x4 - 26.895\ x3\ x4 + 3.15005\ x4 +$

$3.00053\ x2^2\ x5 + 415.904\ x3^2\ x5 + 293.821\ x4^2\ x5 + 5.99544\ x2\ x5 - 30.0398\ x2\ x3\ x5 -$

$60.1791\ x3\ x5 + 70.0859\ x2\ x4\ x5 + 93.1743\ x3\ x4\ x5 + 40.3629\ x4\ x5 + 2.8766\ x5 +$

$2.06459\ x2^2\ x6 - 799.09\ x3^2\ x6 + 563.767\ x4^2\ x6 + 36.5326\ x5^2\ x6 + 2.41929\ x2\ x6 -$

$109.919\ x2\ x3\ x6 - 112.776\ x3\ x6 + 14.76\ x2\ x4\ x6 + 1055.71\ x3\ x4\ x6 + 7.34523\ x4\ x6 -$

$55.9785\ x2\ x5\ x6 + 283.221\ x3\ x5\ x6 + 1311.07\ x4\ x5\ x6 - 57.3713\ x5\ x6 + 2.1065\ x6 +$

$0.998637)/(0.991965\ x1^3 + 2.90672\ x2\ x1^2 + 58.3207\ x3\ x1^2 - 55.1498\ x4\ x1^2 - 52.6211\ x5\ x1^2 +$

$57.1537\ x6\ x1^2 + 2.89387\ x1^2 + 2.94848\ x2^2\ x1 - 270.679\ x3^2\ x1 + 227.63\ x4^2\ x1 +$

$75.5125\ x5^2\ x1 - 358.612\ x6^2\ x1 + 5.47895\ x2\ x1 + 220.05\ x2\ x3\ x1 + 215.204\ x3\ x1 -$

$241.072\ x2\ x4\ x1 + 206.533\ x3\ x4\ x1 - 227.702\ x4\ x1 - 221.01\ x2\ x5\ x1 + 655.217\ x3\ x5\ x1 -$

$213.137\ x4\ x5\ x1 - 220.67\ x5\ x1 + 221.451\ x2\ x6\ x1 - 926.651\ x3\ x6\ x1 - 29.2518\ x4\ x6\ x1 +$

$493.648\ x5\ x6\ x1 + 210.668\ x6\ x1 + 2.81794\ x1 + 1.02315\ x2^3 + 32.1867\ x3^3 - 72.3699\ x4^3 +$

$74.7638\ x5^3 + 266.989\ x6^3 + 3.01776\ x2^2 - 154.946\ x2\ x3^2 - 356.672\ x3^2 + 336.548\ x2\ x4^2 -$

$567.264\ x3\ x4^2 - 182.442\ x4^2 - 435.723\ x2\ x5^2 - 53.1979\ x3\ x5^2 - 168.435\ x4\ x5^2 -$

$293.088\ x5^2 + 84.3711\ x2\ x6^2 - 161.27\ x3\ x6^2 - 742.593\ x4\ x6^2 + 50.1595\ x5\ x6^2 -$

$347.099\ x6^2 + 2.83332\ x2 + 5.85279\ x2^2\ x3 - 8.24771\ x2\ x3 - 9.05837\ x3 + 3.32954\ x2^2\ x4 +$

$228.803\ x3^2\ x4 - 3.7246\ x2\ x4 - 60.6588\ x2\ x3\ x4 - 309.741\ x3\ x4 - 3.01836\ x4 +$

$2.50827\ x2^2\ x5 - 89.5253\ x3^2\ x5 + 213.78\ x4^2\ x5 - 10.3563\ x2\ x5 - 436.458\ x2\ x3\ x5 -$

(Continued)

TABLE A10.1 *(Continued)*
Mathematical Models Proposed for Modeling Critical Buckling Load Behavior (Related to Table 10.3)

772.097 $x3\,x5$ − 888.314 $x2\,x4\,x5$ + 269.404 $x3\,x4\,x5$ − 678.036 $x4\,x5$ − 7.03351 $x5$ + 11.3303 $x2^2\,x6$ − 89.0839 $x3^2\,x6$ + 641.492 $x4^2\,x6$ + 97.428 $x5^2\,x6$ + 2.62888 $x2\,x6$ + 265.129 $x2\,x3\,x6$ − 1393.55 $x3\,x6$ + 366.469 $x2\,x4\,x6$ + 570.352 $x3\,x4\,x6$ − 408.544 $x4\,x6$ + 564.747 $x2\,x5\,x6$ − 40.6366 $x3\,x5\,x6$ − 406.872 $x4\,x5\,x6$ − 724.807 $x5\,x6$ − 9.26106 $x6$ + 0.95506)

(*FOTN*)

FOTN = − (9352.27 Sin [$x1$]) − 11861.3 Cos [$x1$] − 1496.64 Sin [$x2$] − 194.352 Cos [$x2$] + 265.686 Sin [$x3$] + 168.04 Cos [$x3$] + 144.542 Sin [$x4$] − 36.6956 Cos [$x4$] + 271.891 Sin [$x5$] + 153.159 Cos [$x5$] − 0.591071 Sin [$x6$] − 7.68234 Cos [$x6$] + 16310

(*FOTNR*)

FOTNR = (− (11.4539 Sin [$x1$]) − 20.5784 Cos [$x1$] − 197.183 Sin [$x2$] − 126.609 Cos [$x2$] − 101.49 Sin [$x3$] − 62.657 Cos [$x3$] − 151.558 Sin [$x4$] − 93.5672 Cos [$x4$] − 51.4225 Sin [$x5$] − 31.7467 Cos [$x5$] − 101.49 Sin [$x6$] − 62.6569 Cos [$x6$] + 620.988)/(0.24491 Sin [$x1$] + 0.439894 Cos [$x1$] − 0.230547 Sin [$x2$] − 0.148032 Cos [$x2$] − 0.149206 Sin [$x3$] − 0.092115 Cos [$x3$] − 0.238287 Sin [$x4$] − 0.147111 Cos [$x4$] − 0.0601237 Sin [$x5$] − 0.0371185 Cos [$x5$] − 0.149205 Sin [$x6$] − 0.0921147 Cos [$x6$] + 0.321589)

(*SOTN*)

SOTN = 1337.28 Sin [$x1$] Sin [$x2$] + 679.798 Cos [$x1$] Cos [$x2$] − 378.04 Cos [$x1$] Sin [$x2$] − 3275.79 Sin [$x1$] Cos [$x2$] − 1206.02 Sin [$x1$] Sin [$x3$] + 21.0682 Cos [$x1$] Cos [$x3$] + 220.247 Cos [$x1$] Sin [$x3$] + 831.831 Sin [$x1$] Cos [$x3$] + 225.662 Sin [$x1$] Sin [$x4$] + 44.8764 Cos [$x1$] Cos [$x4$] − 83.6904 Cos [$x1$] Sin [$x4$] + 814.812 Sin [$x1$] Cos [$x4$] − 1888.62 Sin [$x1$] Sin [$x5$] + 94.1782 Cos [$x1$] Cos [$x5$] + 389.767 Cos [$x1$] Sin [$x5$] − 716.558 Sin [$x1$] Cos [$x5$] − 674.859 Sin [$x1$] Sin [$x6$] + 78.3864 Cos [$x1$] Cos [$x6$] + 179.542 Cos [$x1$] Sin [$x6$] − 390.376 Sin [$x1$] Cos [$x6$] + 1037.04 Sin [$x1$]^2 − 70.9201 Sin [$x1$] + 43.8692 Cos [$x1$]^2 + 53.2294 Cos [$x1$] − 676.929 Sin [$x1$] Cos [$x1$] − 92.7729 Sin [$x2$] Sin [$x3$] + 255.86 Cos [$x2$] Cos [$x3$] − 23.113 Sin [$x2$] Cos [$x3$] + 378.905 Cos [$x2$] Sin [$x3$] − 47.0029 Sin $x2$] Sin [$x4$] + 63.3913 Cos [$x2$] Cos [$x4$] + 516.422 Sin [$x2$] Cos [$x4$] + 132.055 Cos [$x2$] Sin [$x4$] − 80.4845 Sin [$x2$] Sin [$x5$] + 185.191 Cos [$x2$] Cos [$x5$] − 131.126 Sin [$x2$] Cos [$x5$] + 384.666 Cos [$x2$] Sin [$x5$] − 55.1138 Sin [$x2$] Sin [$x6$] + 150.016 Cos [$x2$] Cos [$x6$] − 46.7119 Sin [$x2$] Cos [$x6$] + 256.187 Cos [$x2$] Sin [$x6$] − 27.1612 Sin [$x2$]^2 + 16.8104 Sin [$x2$] + 644.86 Cos [$x2$]^2 + 209.704 Cos [$x2$] − 253.255 Sin [$x2$] Cos [$x2$] + 112.259 Sin [$x3$] Sin [$x4$] − 639.546 Cos [$x3$] Cos [$x4$] − 572.964 Sin [$x3$] Cos [$x4$] − 1137.4 Cos [$x3$] Sin [$x4$] + 126.346 Sin [$x3$] Sin [$x5$] − 64.9239 Cos [$x3$] Cos [$x5$] − 5.057 Sin [$x3$] Cos [$x5$] − 22.1029 Sin [$x3$] Sin [$x5$] + 166.562 Sin [$x3$] Sin [$x6$] + 265.464 Cos [$x3$] Cos [$x6$] + 115.682 Sin [$x3$] Cos [$x6$] + 417.136 Cos [$x3$] Sin [$x6$] + 129.277 Sin [$x3$]^2 + 113.227 Sin [$x3$] + 196.409 Cos [$x3$]^2 + 152.032 Cos [$x3$] + 174.003 Sin [$x3$] Cos [$x3$] + 394.828 Sin [$x4$] Sin [$x5$] + 95.391 Cos [$x4$] Cos [$x5$] + 341.505 Sin [$x4$] Cos [$x5$] − 12.6603 Cos [$x4$] Sin [$x5$] − 388.419 Sin [$x4$] Sin [$x6$] − 438.701 Cos [$x4$] Cos [$x6$] − 252.693 Sin [$x4$] Cos [$x6$] − 697.704 Cos [$x4$] Sin [$x6$] + 81.7808 Sin [$x4$]^2 + 71.5765 Sin [$x4$] + 290.382 Cos [$x4$]^2 + 199.791 Cos [$x4$] + 111.859 Sin [$x4$] Cos [$x4$] + 79.9747 Sin [$x5$] Sin [$x6$] − 18.7786 Cos [$x5$] Cos [$x6$] + 52.642 Sin [$x5$] Cos [$x6$] − 33.6852 Cos [$x5$] Sin [$x6$] +

(Continued)

TABLE A10.1 *(Continued)*
Mathematical Models Proposed for Modeling Critical Buckling Load Behavior (Related to Table 10.3)

162.871 Sin $[x5]^2 + 140.853$ Sin $[x5] + 149.348$ Cos $[x5]^2 + 75.5998$ Cos $[x5] +$
88.9264 Sin $[x5]$ Cos $[x5] + 148.33$ Sin $[x6]^2 + 128.18$ Sin $[x6] + 160.531$ Cos $[x6]^2 +$
85.2981 Cos $[x6] + 50.426$ Sin $[x6]$ Cos $[x6] + 110.954$

(*SOTNR*)

SOTNR = $(195.025$ Cos $[x1]^2 + 613.012$ Cos $[x2]$ Cos $[x1] + 414.422$ Cos $[x3]$ Cos $[x1] +$
264.401 Cos $[x4]$ Cos $[x1] + 401.836$ Cos $[x5]$ Cos $[x1] + 970.229$ Cos $[x6]$ Cos $[x1] +$
39.3235 Sin $[x1]$ Cos $[x1] + 380.815$ Sin $[x2]$ Cos $[x1] + 568.061$ Sin $[x3]$ Cos $[x1] +$
708.239 Sin $[x4]$ Cos $[x1] + 638.579$ Sin $[x5]$ Cos $[x1] + 187.914$ Sin $[x6]$ Cos $[x1] +$
738.027 Cos $[x1] + 132.764$ Cos $[x2]^2 + 62.6737$ Cos $[x3]^2 + 30.0644$ Cos $[x4]^2 +$
47.2017 Cos $[x5]^2 + 147.993$ Cos $[x6]^2 - 22.7444$ Sin $[x1]^2 + 39.5174$ Sin $[x2]^2 +$
109.607 Sin $[x3]^2 + 142.217$ Sin $[x4]^2 + 125.079$ Sin $[x5]^2 + 24.2876$ Sin $[x6]^2 +$
592.932 Cos $[x2] + 350.175$ Cos $[x2]$ Cos $[x3] + 432.141$ Cos $[x3] +$
69.8715 Cos $[x2]$ Cos $[x4] - 249.436$ Cos $[x3]$ Cos $[x4] + 328.436$ Cos $[x4] +$
336.083 Cos $[x2]$ Cos $[x5] + 247.649$ Cos $[x3]$ Cos $[x5] + 148.823$ Cos $[x4]$ Cos $[x5] +$
414.22 Cos $[x5] + 656.943$ Cos $[x2]$ Cos $[x6] + 37.4828$ Cos $[x3]$ Cos $[x6] +$
42.2769 Cos $[x4]$ Cos $[x6] + 529.573$ Cos $[x5]$ Cos $[x6] + 986.917$ Cos $[x6] +$
59.8842 Cos $[x2]$ Sin $[x1] + 129.71$ Cos $[x3]$ Sin $[x1] + 202.142$ Cos $[x4]$ Sin $[x1] +$
81.4819 Cos $[x5]$ Sin $[x1] + 233.5$ Cos $[x6]$ Sin $[x1] - 2.74422$ Sin $[x1] +$
275.434 Cos $[x2]$ Sin $[x2] + 231.861$ Cos $[x3]$ Sin $[x2] + 392.454$ Cos $[x4]$ Sin $[x2] +$
226.758 Cos $[x5]$ Sin $[x2] + 696.56$ Cos $[x6]$ Sin $[x2] - 68.9097$ Sin $[x1]$ Sin $[x2] +$
324.705 Sin $[x2] + 467.263$ Cos $[x2]$ Sin $[x3] + 302.93$ Cos $[x3]$ Sin $[x3] +$
174.671 Cos $[x4]$ Sin $[x3] + 312.772$ Cos $[x5]$ Sin $[x3] + 772.012$ Cos $[x6]$ Sin $[x3] -$
23.303 Sin $[x1]$ Sin $[x3] + 220.867$ Sin $[x2]$ Sin $[x3] + 514.208$ Sin $[x3] +$
566.251 Cos $[x2]$ Sin $[x4] + 427.14$ Cos $[x3]$ Sin $[x4] + 365.859$ Cos $[x4]$ Sin $[x4] +$
377.862 Cos $[x5]$ Sin $[x4] + 905.036$ Cos $[x6]$ Sin $[x4] + 22.1173$ Sin $[x1]$ Sin $[x4] +$
329.933 Sin $[x2]$ Sin $[x4] + 513.548$ Sin $[x3]$ Sin $[x4] + 666.609$ Sin $[x4] +$
511.321 Cos $[x2]$ Sin $[x5] + 371.593$ Cos $[x3]$ Sin $[x5] + 294.879$ Cos $[x4]$ Sin $[x5] +$
363.053 Cos $[x5]$ Sin $[x5] + 858.653$ Cos $[x6]$ Sin $[x5] - 8.22658$ Sin $[x1]$ Sin $[x5] +$
281.438 Sin $[x2]$ Sin $[x5] + 447.004$ Sin $[x3]$ Sin $[x5] + 557.132$ Sin $[x4]$ Sin $[x5] +$
588.05 Sin $[x5] + 258.417$ Cos $[x2]$ Sin $[x6] + 416.386$ Cos $[x3]$ Sin $[x6] -$
27.6755 Cos $[x4]$ Sin $[x6] + 158.744$ Cos $[x5]$ Sin $[x6] + 412.485$ Cos $[x6]$ Sin $[x6] -$
107.984 Sin $[x1]$ Sin $[x6] - 97.7069$ Sin $[x2]$ Sin $[x6] + 109.979$ Sin $[x3]$ Sin $[x6] +$
182.467 Sin $[x4]$ Sin $[x6] + 107.796$ Sin $[x5]$ Sin $[x6] + 129.348$ Sin $[x6] +$
$171.281)/(55.6127$ Cos $[x1]^2 - 11.8359$ Cos $[x2]$ Cos $[x1] - 101.401$ Cos $[x3]$ Cos $[x1] +$
74.2328 Cos $[x4]$ Cos $[x1] + 54.58$ Cos $[x5]$ Cos $[x1] + 98.5831$ Cos $[x6]$ Cos $[x1] +$
199.3 Sin $[x1]$ Cos $[x1] - 201.76$ Sin $[x2]$ Cos $[x1] + 170.825$ Sin $[x3]$ Cos $[x1] +$
304.091 Sin $[x4]$ Cos $[x1] + 258.053$ Sin $[x5]$ Cos $[x1] - 108.221$ Sin $[x6]$ Cos $[x1] +$
210.848 Cos $[x1] + 102.797$ Cos $[x2]^2 + 11.646$ Cos $[x3]^2 - 12.6727$ Cos $[x4]^2 +$
4.61374 Cos $[x5]^2 + 54.7306$ Cos $[x6]^2 - 20.757$ Sin $[x1]^2 - 67.9417$ Sin $[x2]^2 +$
23.2097 Sin $[x3]^2 + 47.5284$ Sin $[x4]^2 + 30.242$ Sin $[x5]^2 - 19.8749$ Sin $[x6]^2 -$
147.097 Cos $[x2] - 181.155$ Cos $[x2]$ Cos $[x3] + 15.2459$ Cos $[x3] -$
206.383 Cos $[x2]$ Cos $[x4] + 126.32$ Cos $[x3]$ Cos $[x4] - 3.78795$ Cos $[x4] +$
111.7 Cos $[x2]$ Cos $[x5] + 15.5695$ Cos $[x3]$ Cos $[x5] + 446.598$ Cos $[x4]$ Cos $[x5] +$

(Continued)

TABLE A10.1 *(Continued)*
Mathematical Models Proposed for Modeling Critical Buckling Load Behavior (Related to Table 10.3)

    39.3394 Cos [$x5$] − 209.543 Cos [$x2$] Cos [$x6$] + 235.065 Cos [$x3$] Cos [$x6$] −
    8.39491 Cos [$x4$] Cos [$x6$] − 207.962 Cos [$x5$] Cos [$x6$] + 60.2719 Cos [$x6$] −
    170.112 Cos [$x2$] Sin [$x1$] + 264.282 Cos [$x3$] Sin [$x1$] − 59.0511 Cos [$x4$] Sin [$x1$] −
    45.794 Cos [$x5$] Sin [$x1$] + 103.727 Cos [$x6$] Sin [$x1$] + 87.6494 Sin [$x1$] −
    691.265 Cos [$x2$] Sin [$x2$] − 123.912 Cos [$x3$] Sin [$x2$] − 284.133 Cos [$x4$] Sin [$x2$] −
    10.8965 Cos [$x5$] Sin [$x2$] − 163.857 Cos [$x6$] Sin [$x2$] − 36.0337 Sin [$x1$] Sin [$x2$] −
    276.613 Sin [$x2$] − 51.5397 Cos [$x2$] Sin [$x3$] − 1.13532 Cos [$x3$] Sin [$x3$] +
    101.874 Cos [$x4$] Sin [$x3$] + 132.021 Cos [$x5$] Sin [$x3$] + 262.758 Cos [$x6$] Sin [$x3$] +
    32.6446 Sin [$x1$] Sin [$x3$] − 224.634 Sin [$x2$] Sin [$x3$] + 103.469 Sin [$x3$] +
    43.3538 Cos [$x2$] Sin [$x4$] + 358.533 Cos [$x3$] Sin [$x4$] + 102.201 Cos [$x4$] Sin [$x4$] −
    1298.59 Cos [$x5$] Sin [$x4$] + 0.514268 Cos [$x6$] Sin [$x4$] + 43.6886 Sin [$x1$] Sin [$x4$] −
    27.6095 Sin [$x2$] Sin [$x4$] + 160.662 Sin [$x3$] Sin [$x4$] + 219.98 Sin [$x4$] −
    170.877 Cos [$x2$] Sin [$x5$] + 86.4831 Cos [$x3$] Sin [$x5$] + 73.4326 Cos [$x4$] Sin [$x5$] +
    57.3798 Cos [$x5$] Sin [$x5$] − 150.848 Cos [$x6$] Sin [$x5$] − 20.7853 Sin [$x1$] Sin [$x5$] −
    166.69 Sin [$x2$] Sin [$x5$] + 313.191 Sin [$x3$] Sin [$x5$] + 119.635 Sin [$x4$] Sin [$x5$] +
    138.611 Sin [$x5$] − 128.601 Cos [$x2$] Sin [$x6$] − 286.831 Cos [$x3$] Sin [$x6$] +
    91.4992 Cos [$x4$] Sin [$x6$] + 1059.55 Cos [$x5$] Sin [$x6$] − 173.277 Cos [$x6$] Sin [$x6$] +
    45.1486 Sin [$x1$] Sin [$x6$] − 324.949 Sin [$x2$] Sin [$x6$] − 491.378 Sin [$x3$] Sin [$x6$] +
    230.173 Sin [$x4$] Sin [$x6$] − 137.651 Sin [$x5$] Sin [$x6$] − 102.397 Sin [$x6$] + 33.8557)

(*FOLN*)

FOLN = − (1260.46 Log [$x1$ + 0.00001]) − 481.86 Log [0.00001 $x2$] +
    6.67871 Log [$x3$ + 0.00001] + 7.0544 Log [$x4$ + 0.00001] +
    10.0976 Log [$x5$ + 0.00001] + 4.32009 Log [$x6$ + 0.00001] − 5656.26

(*FOLNR*)

FOLNR = (15791.7 Log [$x1$ + 0.00001] + 10229.6 Log [$x2$ + 0.00001] + 1418.4 Log [$x3$ + 0.00001] −
    804.651 Log [$x4$ + 0.00001] + 460.59 Log [$x5$ + 0.00001] + 2753.27 Log [$x6$ + 0.00001] −
    2977.68)/(14.6558 Log [$x1$ + 0.00001] + 11.9585 Log [$x2$ + 0.00001] +
    2.12835 Log [$x3$ + 0.00001] − 0.921107Log [$x4$ + 0.00001] +
    0.634241 Log [$x5$ + 0.00001] + 3.42312 Log [$x6$ + 0.00001] − 7.79439)

(*SOLN*)

SOLN = 1632.5 Log [$x1$ + 0.00001] Log [0.00001 $x2$] − 31.9367 Log [$x1$ + 0.00001]
    Log [$x3$ + 0.00001] − 31.7349 Log [$x1$ + 0.00001] Log [$x4$ + 0.00001] − 20.7024
    Log [$x1$ + 0.00001] Log [$x5$ + 0.00001] − 5.35896 Log [$x1$ + 0.00001] Log [$x6$ + 0.00001] +
    1007.75 Log [$x1$ + 0.00001]^2 + 19304. Log [$x1$ + 0.00001] − 2.22906 Log [0.00001 $x2$]
    Log [$x3$ + 0.00001] − 12.648 Log [0.00001 $x2$] Log [$x4$ + 0.00001] − 2.99455
    Log [0.00001 $x2$] Log [$x5$ + 0.00001] − 0.173265 Log [0.00001 $x2$] Log [$x6$ + 0.00001] −
    193.148 Log [0.00001 $x2$]^2 − 3710.26 Log [0.00001 $x2$] + 0.315321 Log [$x3$ + 0.00001]
    Log [$x4$ + 0.00001] + 0.155954 Log [$x3$ + 0.00001] Log [$x5$ + 0.00001] − 0.0627364
    Log [$x3$ + 0.00001] Log [$x6$ + 0.00001] − 9.815 Log [$x3$ + 0.00001]^2 − 115.871
    Log [$x3$ + 0.00001] + 0.0123185 Log [$x4$ + 0.00001] Log [$x5$ + 0.00001] + 0.260976
    Log [$x4$ + 0.00001] Log [$x6$ + 0.00001] − 7.56226 Log [$x4$ + 0.00001]^2 − 215.369

(Continued)

TABLE A10.1 *(Continued)*
Mathematical Models Proposed for Modeling Critical Buckling Load Behavior (Related to Table 10.3)

Log [$x4 + 0.00001$] + 0.06132 Log [$x5 + 0.00001$] Log [$x6 + 0.00001$] – 0.41301
Log [$x5 + 0.00001$]^2 – 52.2252 Log [$x5 + 0.00001$] + 3.02865 Log [$x6 + 0.00001$]^2 +
15.5647 Log [$x6 + 0.00001$] – 15892.3

(*SOLNR*)

SOLNR = (2.15064 Log [$x1 + 0.00001$] Log [$x2 + 0.00001$] + 2.70724 Log [$x1 + 0.00001$]
Log [$x3 + 0.00001$] – 4.89623 Log [$x1 + 0.00001$] Log [$x4 + 0.00001$] + 4.36105
Log [$x1 + 0.00001$] Log [$x5 + 0.00001$] + 5.07604 Log [$x1 + 0.00001$]
Log [$x6 + 0.00001$] + 0.768048 Log [$x1 + 0.00001$]^2 + 2.72884 Log [$x1 + 0.00001$] +
1.38529 Log [$x2 + 0.00001$] Log [$x3 + 0.00001$] + 4.47229 Log [$x2 + 0.00001$]
Log [$x4 + 0.00001$] + 1.40938 Log [$x2 + 0.00001$] Log [$x5 + 0.00001$] + 0.807802
Log [$x2 + 0.00001$] Log [$x6 + 0.00001$] + 0.963566 Log [$x2 + 0.00001$]^2 + 1.94927
Log [$x2 + 0.00001$] + 30.2368 Log [$x3 + 0.00001$] Log [$x4 + 0.00001$] + 6.37977
Log [$x3 + 0.00001$] Log [$x5 + 0.00001$] + 4.11463 Log [$x3 + 0.00001$]
Log [$x6 + 0.00001$] – 6.5281 Log [$x3 + 0.00001$]^2 + 2.2698 Log [$x3 + 0.00001$] +
22.4783 Log [$x4 + 0.00001$] Log [$x5 + 0.00001$] + 19.656 Log [$x4 + 0.00001$]
Log [$x6 + 0.00001$] – 14.9419 Log [$x4 + 0.00001$]^2 + 6.69526 Log [$x4 + 0.00001$] –
4.45532 Log [$x5 + 0.00001$] Log [$x6 + 0.00001$] – 2.6889 Log [$x5 + 0.00001$]^2 +
0.167342 Log [$x5 + 0.00001$] – 1.60856 Log [$x6 + 0.00001$]^2 – 0.477046 Log [$x6 +
0.00001$] + 0.839632)/(2.22212 Log [$x1 + 0.00001$] Log [$x2 + 0.00001$] + 0.468517
Log [$x1 + 0.00001$] Log [$x3 + 0.00001$] + 6.13195 Log [$x1 + 0.00001$] Log [$x4 + 0.00001$] –
2.10411 Log [$x1 + 0.00001$] Log [$x5 + 0.00001$] – 0.373887 Log [$x1 + 0.00001$]
Log [$x6 + 0.00001$] + 1.42561 Log [$x1 + 0.00001$]^2 + 0.831693 Log [$x1 + 0.00001$] +
2.41877 Log [$x2 + 0.00001$] Log [$x3 + 0.00001$] – 0.930476 Log [$x2 + 0.00001$]
Log [$x4 + 0.00001$] + 1.89378 Log [$x2 + 0.00001$] Log [$x5 + 0.00001$] – 2.78164
Log [$x2 + 0.00001$] Log [$x6 + 0.00001$] + 1.04253 Log [$x2 + 0.00001$]^2 + 2.05606
Log [$x2 + 0.00001$] – 16.7708 Log [$x3 + 0.00001$] Log [$x4 + 0.00001$] + 10.2361
Log [$x3 + 0.00001$] Log [$x5 + 0.00001$] + 9.72358 Log [$x3 + 0.00001$] Log [$x6 + 0.00001$] +
1.1859 Log [$x3 + 0.00001$]^2 + 4.46715 Log [$x3 + 0.00001$] – 16. 2082 Log [$x4 + 0.00001$]
Log [$x5 + 0.00001$] – 14.8629 Log [$x4 + 0.00001$] Log [$x6 + 0.00001$] + 16.0119
Log [$x4 + 0.00001$]^2 – 2.45819 Log [$x4 + 0.00001$] + 8.21868 Log [$x5 + 0.00001$]
Log [$x6 + 0.00001$] + 1.90059 Log [$x5 + 0.00001$]^2 + 3.3168 Log [$x5 + 0.00001$] +
2.41692 Log [$x6 + 0.00001$]^2 + 3.5592 Log [$x6 + 0.00001$] + 1.10832)

TABLE A10.2
Mathematical Models Proposed for Modeling Critical Buckling Load Behavior (Related to Table 10.4)

(*BesselJ*)

BESSELJ = 20307.1 BesselJ [$1,x1$] BesselJ [$1, x2$] – 57637.1 BesselJ [$1, x1$] BesselJ [$1, x3$] –
34781.5 BesselJ [$1, x1$] BesselJ [$1, x4$] – 11502.2 BesselJ [$1, x1$] BesselJ [$1, x5$] +
3990.91 BesselJ [$1, x1$] BesselJ [$1, x6$] + 28190.7 BesselJ [$1,x1$]^2 – 24225.BesselJ [$1,x1$] +

(Continued)

TABLE A10.2 *(Continued)*
Mathematical Models Proposed for Modeling Critical Buckling Load Behavior (Related to Table 10.4)

6744.91 BesselJ [1, $x2$] BesselJ [1, $x3$] − 13 511.6 BesselJ [1, $x2$] BesselJ [1, $x4$] + 2887.45 BesselJ [1,$x2$] BesselJ [1,$x5$] − 6556.07 BesselJ [1,$x2$] − 19 564.9 BesselJ [1, $x3$] BesselJ [1, $x4$] − 6477.48 BesselJ [1, $x3$] BesselJ [1, $x5$] − 126 469.BesselJ [1, $x3$]^2 + 23892.9 BesselJ [1,$x3$] + 7021.39 BesselJ [1,$x4$] BesselJ [1,$x5$] − 7997. 28 BesselJ [1, $x4$] BesselJ [1, $x6$] − 52251.1 BesselJ [1,$x4$]^2 + 20633.8 BesselJ [1, $x4$] − 1195.46 BesselJ [1, $x5$] BesselJ [1, $x6$] − 58753.5 BesselJ [1, $x5$]^2 + 6500.18 BesselJ [1,$x5$] − 20135.BesselJ [1,$x6$]^2 + 1005.57 BesselJ [1,$x6$] + 5259.89

(*ChebyShebT*)

ChebyShevT = 18.3778 $x1^6$ − 190.335 $x1^4$ + 687.966 $x1^3$ $x2^3$ − 7394.14 $x1^3$ $x2$ + 40.5167 $x1^3$ $x3$ + 20.8059 $x1^3$ $x4$ − 24.177 $x1^3$ $x5$ + 1.09504 $x1^3$ $x6$ − 25.9708 $x1^3$ + 14512.1 $x1^2$ − 1394. $x1$ $x2^3$ + 11648.8 $x1$ $x2$ − 56.7946 $x1$ $x3$ − 30.1283 $x1$ $x4$ + 15.9614 $x1$ $x5$ − 1.535 $x1$ $x6$ − 19684.2 $x1$ + 45.4059 $x2^6$ − 12.8457 $x2^4$ − $\dfrac{5.90195\, x2^3\, x3}{10^{15}}$ + 0.6226 $x2^3$ $x4$ − 4.17447 $x2^3$ $x5$ − $\dfrac{2.15871\, x2^3\, x6}{10^{15}}$ − 6.85499 $x2^3$ + 1957.59 $x2^2$ + $\dfrac{2.22938\, x2\, x3}{10^{14}}$ − 4.26793 $x2$ $x4$ + 13.567 $x2$ $x5$ + $\dfrac{7.95804\, x2\, x6}{10^{15}}$ − 7540.71 $x2$ − 0.0446289 $x3^2$ + $\dfrac{1.79157\, x3\, x5}{10^{17}}$ + $\dfrac{3.08642\, x3\, x6}{10^7}$ + 27.5566 $x3$ + $\dfrac{1.04854\, x4\, x6}{10^{17}}$ + 16.2229 $x4$ + $\dfrac{7.34742\, x5\, x6}{10^{18}}$ − 11.2006 $x5$ + 0.636208 $x6$ + 7930.36

(*Erf*)

Erf = 7157.98 Erf [$x1$] Erf [$x2$] − 2214.19 Erf [$x1$] Erf [$x3$] − 1453.22 Erf [$x1$] Erf [$x4$] − 379.388 Erf [$x1$] Erf [$x5$] + 171.515 Erf [$x1$] Erf [$x6$] + 8621.02 Erf [$x1$]^2 − 14289.2 Erf [$x1$] − 328.612 Erf [$x2$] Erf [$x3$] + 188.287 Erf [$x2$] Erf [$x4$] − 348.12 Erf [$x2$] Erf [$x5$] − 220.38 Erf [$x2$] Erf [$x6$] − 298.928 Erf [$x2$]^2 − 3912.27 Erf [$x2$] + 206.675 Erf [$x3$] Erf [$x4$] + 132.527 Erf [$x3$] Erf [$x5$] − 48.2014 Erf [$x3$] Erf [$x6$] + 160.592 Erf [$x3$]^2 + 1155.51 Erf [$x3$] − 5.48177 Erf [$x4$] Erf [$x5$] + 45.0942 Erf [$x4$] Erf [$x6$] + 76.7228 Erf [$x4$]^2 + 430.684 Erf [$x4$] + 92.4267 Erf [$x5$] Erf [$x6$] − 57.6913 Erf [$x5$]^2 + 369.02 Erf [$x5$] + 6.52558 Erf [$x6$]^2 + 0.00285268 Erf [$x6$] + 6208.31

(*ExpIntergralE*)

ExpIntegralIE = 184866.ExpIntegralE [5, $x1$] ExpIntegralE [5, $x2$] − 47 023.8 ExpIntegralE [5, $x1$] ExpIntegralE [5, $x3$] − 29940. 2 ExpIntegralE [5, $x1$] ExpIntegralE [5, $x4$] − 10760.6 ExpIntegralE [5,$x1$] ExpIntegralE [5,$x5$] + 69.7381 ExpIntegralE [5, $x1$] ExpIntegralE [5, $x6$] + 213810.ExpIntegralE [5, $x1$]^2 − 52708.3 ExpIntegralE [5, $x1$] − 4679.13 ExpIntegralE [5, $x2$] ExpIntegralE [5, $x3$] + 4060.8 ExpIntegralE [5,$x2$] ExpIntegralE [5,$x4$] − 6348.89 ExpIntegralE [5, $x2$] ExpIntegralE [5, $x5$] + 21595.6 ExpIntegralE [5, $x2$]^2 21830.2 ExpIntegralE [5, $x2$] + 2918.1 ExpIntegralE [5, $x3$] ExpIntegralE [5, $x4$] + 227.359 ExpIntegralE [5, $x3$]^2 + 5783.33 ExpIntegralE [5, $x3$] + 402.007 ExpIntegralE [5, $x4$] 2 + 2989.12ExpIntegralE [5,$x4$] − 252.456 ExpIntegralE [5,$x5$]^2 + 1867.84ExpIntegralE [5,$x5$] + 3633.49

(Continued)

TABLE A10.2 *(Continued)*
Mathematical Models Proposed for Modeling Critical Buckling Load
Behavior (Related to Table 10.4)

(*FresnelG*)

FresnelG2 = − 13482.5 FresnelG [$x1$] × FresnelG [$x2$] × FresnelG [$x3$] + 49776.7 FresnelG [$x1$] ×
FresnelG [$x2$] − 9090.46 FresnelG [$x1$] × FresnelG [$x3$] − 3468.09 FresnelG [$x1$] ×
FresnelG [$x4$] × FresnelG [$x6$] − 7283.29 FresnelG [$x1$] × FresnelG [$x4$] −
1137.FresnelG [$x1$] × FresnelG [$x5$] + 1016.31 FresnelG [$x1$] × FresnelG [$x6$] +
5128.61 FresnelG [$x1$] − 704.6 FresnelG [$x2$] × FresnelG [$x3$] × FresnelG [$x4$] −
5516.83 FresnelG [$x2$] − 112.322 FresnelG [$x3$] × FresnelG [$x4$] + 1423.84 FresnelG [$x3$] +
1273.84 FresnelG [$x4$] + 22.7821 FresnelG [$x5$] − 225.262 FresnelG [$x6$] − 251.799

(*FresnelG*)

FresnelG3 = − 250.041FresnelG [$x1$] × FresnelG [$x2$] × FresnelG [$x3$] + 42287.FresnelG [$x1$] ×
FresnelG [$x2$] − 9130.14 FresnelG [$x1$] × FresnelG [$x3$] − 2215.85 FresnelG [$x1$] ×
FresnelG [$x4$] × FresnelG [$x6$] − 5503.48 FresnelG [$x1$] × FresnelG [$x4$] −
1545.2 FresnelG [$x1$] × FresnelG [$x5$] + 2829.92 FresnelG [$x1$] × FresnelG [$x6$] +
4951.34 FresnelG [$x1$] − 13.6938 FresnelG [$x2$] × FresnelG [$x3$] × FresnelG [$x4$] −
3374.1 FresnelG [$x2$] × FresnelG [$x3$] + 2418.98 FresnelG [$x2$] × FresnelG [$x4$] −
4139.32 FresnelG [$x2$] − 66.6854 FresnelG [$x3$] × FresnelG [$x4$] + 1491.68 FresnelG [$x3$] +
671.988 FresnelG [$x4$] + 82.2361 FresnelG [$x5$] − 605.345 FresnelG [$x6$] − 180.561,

(*HermiteH*)

HermiteH = 9209.49 $x1^4$ + 1042.43 $x1^2$ $x2^2$ + 0.000277079 $x1^2$ $x3^2$ − 0.132289 $x1^2$ $x4^2$ −
0.0213077 $x1^2$ $x5^2$ − 8819.13 $x1^2$ + 259.991 $x2^4$ − 1201.36 $x2^2$ + 0.0616147 $x4^2$ +
0.00858444 $x5^2$ − 0.0123756 $x6^2$ + 2846.96

(*HyperGeometric0F1*)

HyperGeometric0F1 = 53675.4 Hypergeometric0F1 [5, $x1$] Hypergeometric0F1 [5, $x2$] −
0.11998 Hypergeometric0F1 [5, $x1$] Hypergeometric0F1 [5, $x3$] −
0.179875 Hypergeometric0F1 [5, $x1$] Hypergeometric0F1 [5, $x4$] +
0.0264249 Hypergeometric0F1 [5, $x1$] Hypergeometric0F1 [5, $x5$] −
0.00395301 Hypergeometric0F1 [5, $x1$] Hypergeometric0F1 [5, $x6$] +
162 051.Hypergeometric0F1 [5, $x1$]^2 − 433536.Hypergeometric0F1 [5, $x1$] +
0.0188362 Hypergeometric0F1 [5, $x2$] Hypergeometric0F1 [5, $x3$] −
0.074917 Hypergeometric0F1 [5, $x2$] Hypergeometric0F1 [5, $x4$] +
13517.6 Hypergeometric0F1 [5, $x2$]^2 − 94352.6 Hypergeometric0F1 [5, $x2$] +
0.109254 Hypergeometric0F1 [5, $x3$] + 0.293779 Hypergeometric0F1 [5, $x4$] −
0.0287439 Hypergeometric0F1 [5, $x5$] + 0.00564038 Hypergeometric0F1
[5, $x6$] + 304346.

(*LegendreP*)

LegendreP = 9208.7 $x1^4$ + 1042.45 $x1^2$ $x2^2$ − 0.132349 $x1^2$ $x4^2$ − 0.021357 $x1^2$ $x5^2$ − 8817.37 $x1^2$ +
259.863 $x2^4$ − 1201.03 $x2^2$ + 0.0616231 $x4^2$ + 0.00859723 $x5^2$ − 0.0123645 $x6^2$ +
2846.65

(Continued)

TABLE A10.2 *(Continued)*
**Mathematical Models Proposed for Modeling Critical Buckling Load
Behavior (Related to Table 10.4)**

(*RamanujanTauTheta*)

RamanujanTauTheta = 196 806.RamanujanTauTheta [$x1$] × RamanujanTauTheta [$x2$] +
95.0779 RamanujanTauTheta [$x1$] × RamanujanTauTheta [$x3$] +
60.906 RamanujanTauTheta [$x1$] × RamanujanTauTheta [$x4$] +
13.9361 RamanujanTauTheta [$x1$] × RamanujanTauTheta [$x5$] −
5.12448 RamanujanTauTheta [$x1$] × RamanujanTauTheta [$x6$] +
460237.RamanujanTauTheta [$x1$]^2 + 93 913.RamanujanTauTheta [$x1$] −
6.9991 RamanujanTauTheta [$x2$] × RamanujanTauTheta [$x3$] +
21.9628 RamanujanTauTheta [$x2$] × RamanujanTauTheta [$x4$] −
5. 23877 RamanujanTauTheta [$x2$] × RamanujanTauTheta [$x5$] +
46 678.9 RamanujanTauTheta [$x2$]^2 + 28474.6RamanujanTauTheta [$x2$] +
6.00238 RamanujanTauTheta [$x3$] + 7.62129 RamanujanTauTheta [$x4$] +
0.401973 RamanujanTauTheta [$x5$] − 0.265166 RamanujanTauTheta [$x6$] +
5860.21

(*RiemannSiegelTheta*)

RiemannSiegelTheta = 1704.07 θ [$x1$] × θ [$x2$] + 14.4887 θ [$x1$] × θ [$x3$] + 9.70896 θ [$x1$] × θ [$x4$] +
2.30886 θ [$x1$] × θ [$x5$] + 2124.67 θ [$x1$]^2 + 8239.16 θ[$x1$] + 5.29345 θ [$x2$] ×
θ [$x4$] + 391.658 θ [$x2$]^2 + 3546.79 θ [$x2$] + 17.0458 θ [$x3$] + 21.6166 θ [$x4$] +
2.89045 θ [$x5$] + 8753.88

11 Conclusion

The book delves into the intricate relationship between experimental design, mathematical modeling, and optimization, which are crucial components in solving engineering problems. Through the examination of 11 distinct problems, drawn from both literature and original sources, the impact of these processes on problem-solving is thoroughly investigated. Throughout the study, various experimental design methods such as Full Factorial design (FFD), Box-Behnken design (BBD), Central Composite design (CCD), Taguchi, and D-Optimal, alongside modeling techniques like Neuro Regression (NRM) and Stochastic Neuro Regression (SNRM), and optimization algorithms including Differential Evolution (DE), Nelder Mead (NM), Random Search (RS), and Simulated Annealing (SA) are employed.

Key issues addressed in the book include:

1. **Constraints and limitations of Artificial Neural Networks (ANN)**: The study scrutinizes the constraints and limitations of using ANN as a mathematical modeling method, a widely preferred approach in engineering problem-solving.
2. **Evaluation of R-squared (R^2) as a standalone criterion**: The significance and limitations of using R^2 as a sole assessment criterion for model performance are critically examined, considering its widespread use in measuring model accuracy.
3. **Choice of mathematical functions in modeling**: The adequacy of polynomial expressions in regression and ANN modeling is evaluated, exploring whether alternative mathematical functions are necessary to accurately represent the relationship between input and output parameters.
4. **Impact of Taguchi experimental design on modeling**: The influence of Taguchi experimental design method on mathematical modeling is investigated, shedding light on its advantages and limitations in various scenarios.
5. **Effects of data separation methods on modeling**: Different methods for separating datasets into training and testing subsets are compared to assess their impact on mathematical modeling and model performance.
6. **Comparison of assessment criteria for model prediction performance**: The performance of different assessment criteria for measuring model prediction accuracy is compared, aiming to identify the most appropriate criteria for different scenarios.
7. **Effect of experimental design methods on modeling and optimization**: The study analyzes how the choice of experimental design method affects both mathematical modeling and optimization processes, aiming to determine which experimental design method offers the most advantages.

DOI: 10.1201/9781003494843-11

8. **Usability of special functions in mathematical modeling**: The feasibility of using special functions as an alternative model type in mathematical modeling is explored, considering their potential advantages over basic mathematical functions.
9. **Comparison of original modeling methods with commonly used approaches**: The book compares original modeling methods like NRM and SNRM with commonly used methods such as ANN and regression, evaluating their applicability and effectiveness in diverse problem-solving scenarios.

By addressing these key issues and providing solutions through real-life applications, the book aims to offer a methodological approach to efficiently utilizing experimental design, mathematical modeling, and optimization processes in engineering problem-solving endeavors. It emphasizes the interconnectedness of these processes and advocates for a holistic approach in defining, interpreting, and solving engineering problems.

Chapter 3 of the book delves into the utilization of ANN in mathematical modeling, focusing on two distinct problem scenarios:

1. **Modeling Energy Consumption in Electric Arc Furnaces**: The aim was to develop a model estimating energy consumption in electric arc furnaces used in the iron and steel industry. Despite achieving a high R^2 index of 0.88, 0.95, and 0.95 in the training, testing, and validation phases, respectively, using ANN, the model failed the stability test. It was found that the model's high accuracy did not translate into stability during testing.
2. **Modeling Stress Behavior in Flax/Epoxy Laminated Composites**: Another study aimed to model the stress behavior of flax/epoxy laminated composites using an ANN-based model. The model exhibited exceptionally high R^2 values of 0.95, 0.99, and 0.99 in the training, testing, and validation phases, respectively. However, upon further examination using additional test data and comparison with an analytical formula, it was revealed that the model's performance was significantly lower than expected. The R^2 criterion alone, which initially indicated high accuracy, proved misleading, as the model's actual predictive capability was substantially lower.

These findings underscore the limitations of ANN-based models, despite their widespread use and often high predictive performance.

Chapter 4 expands on this by discussing the drawbacks of relying solely on R^2 as a measure of model performance. Through case studies involving the free-piston Stirling engine and additive manufacturing, it was concluded that evaluating models based solely on R^2 may lead to erroneous assessments. The chapter emphasizes the importance of considering testing, validation, boundedness check, and stability criteria to ensure robustness in model selection and performance evaluation, particularly in engineering problems where accurate predictions are crucial.

It has been revealed that creating a model using the entire dataset may mislead us despite the high prediction performance observed in the model's success. To obtain more robust models, it is recommended to divide the dataset into sections such as training, testing, and validation. Also, the stability and boundedness check proposed within this book are original criteria independent of R^2. They give an idea about the model's performance and whether it is usable. For some models, even though they produced successful results in the training, testing, and validation phases, they were considered unsuccessful because they could not meet the stability and boundedness check criteria.

Chapter 5 of the book discusses the limitations of polynomial models in mathematical modeling and explores alternative mathematical functions that may provide more accurate representations of the relationship between input and output variables. Through modeling studies on two engineering problems, it was discovered that trigonometric, rational, logarithmic, and hybrid models can serve as effective alternatives to polynomial models. Specifically, the first-order logarithmic rational model was found to be successful in modeling stresses in a multi-layer armored system subjected to shock loading in an underwater explosion, while the second-order trigonometric model effectively captured the natural frequency behavior of an electric vehicle battery pack enclosure. These findings underscore the necessity of considering various mathematical functions beyond polynomials in modeling studies to achieve more accurate results.

In Chapter 6, the Taguchi method, a popular experimental design, modeling, and optimization approach, is examined in detail. This method is favored for its ability to reduce the number of experiments required, thus saving time and resources. However, the chapter highlights some critical evaluations regarding the Taguchi method:

1. **Limited Effectiveness with Two-Level Design Parameters**: Both the Taguchi and FFD methods failed to produce successful models when design parameters consisted of two-level values. This limitation arises due to the difficulty in accurately determining the intended behavior with only two levels, given the flexibility in intermediate values that may affect the outcomes.
2. **Importance of Proper Definition of Design Parameters**: The success of experimental designs depends not only on the number of experiments but also on the proper definition of design parameters and levels. Despite having a higher number of experiments, the L16 array in the Taguchi method yielded the most successful model compared to other arrays.
3. **Caution in Model Selection**: While the Taguchi method suggested models with high prediction performance during the training phase, these models often failed during the testing phase. Conversely, the FFD method produced models with desired prediction performance in both training and testing phases. This highlights the importance of careful analysis and selection of the appropriate experimental design method to ensure accurate and reliable results.

In conclusion, Chapter 6 emphasizes the need for cautious consideration when utilizing the Taguchi method, particularly in selecting the appropriate experimental design method based on the problem at hand. Proper analysis and selection are crucial to achieving reliable results and avoiding potential pitfalls associated with experimental design and modeling processes.

Chapter 7 of the book serves as a comparative analysis between two original modeling methods: NRM (Normal Regression Model) and SNRM (Stochastic Normal Regression Model). The comparison is based on a reference problem, and particular attention is given to the impact of dataset separation methods on modeling results, a crucial aspect in ensuring the reliability of mathematical models.

The chapter explores three different dataset separation techniques: hold-out, k-fold cross-validation, and bootstrap. These methods are applied to evaluate the modeling performances of NRM and SNRM. The results indicate several key findings:

1. **Advantage of SNRM**: SNRM, which integrates ANN, regression, and optimization techniques, demonstrates superior modeling performance compared to NRM. Utilizing stochastic optimization methods for determining model coefficients provides SNRM with a significant advantage, enabling it to propose multiple alternative models with high prediction accuracy.

2. **Effectiveness of Dataset Separation Methods**: While both bootstrap and cross-validation techniques show more effectiveness in mathematical modeling compared to the other methods, it's important to note that hold-out isn't always insufficient. The choice of dataset separation method significantly impacts the modeling results, emphasizing the need for careful consideration in model development.

3. **Importance of Testing and Validation**: Despite achieving high R^2 values close to 1 during the training phase, both NRM and SNRM models may not consistently demonstrate similar success in the testing and validation stages. This underscores the necessity of rigorously testing and validating proposed models with unseen data to ensure their reliability and generalization capability.

Overall, Chapter 7 underscores the critical role of dataset separation methods and thorough testing and validation procedures in the development of reliable mathematical models. It highlights the advantages of employing optimization techniques, such as those utilized in SNRM, and emphasizes the need for careful consideration of dataset separation methods to enhance modeling effectiveness and accuracy.

Chapter 8 of the book focuses on the evaluation of model performance using 22 different assessment criteria in the context of modeling the natural frequency gap behavior of a hybrid composite material. Several conclusions are drawn from this comprehensive analysis:

1. **Boundedness Control Criterion**: Boundedness control is highlighted as a crucial model assessment criterion that evaluates the compatibility of model results with reality, rather than just the model's success. This

criterion provides direct information about the usability of the model. Unlike other assessment criteria, boundedness control revealed that the maximum and minimum natural frequency gap produced by the hybrid model were impossible to achieve, rendering the model unusable. Only the maximum and minimum natural frequency range values produced by specific models were consistent with reality.

2. **Classification of Model Assessment Criteria**: The assessment criteria are classified into three groups:
   - The first group includes criteria such as R^2, R^2 adjusted, and KGE, which assess the model's success based on how close the values are to 1.
   - The second group includes criteria such as AIC, BIC, and AICC, which consider smaller values indicative of successful models.
   - The third group aims to minimize errors in estimating actual values, including criteria such as MSE, RMSE), and SE.

3. **Effectiveness of Different Criteria**: While various criteria provide insights into the simplicity and fit of models, none offer as clear information about model usability as boundedness control. Although criteria like R^2-based measures and AIC/BIC consider different aspects of model performance, they lack direct information on the consistency and applicability of results.

4. **Importance of Diverse Criteria**: It is emphasized that employing diverse criteria is significant for evaluating model success comprehensively. However, the boundedness control criterion stands out as the only one providing direct information on model usability, underscoring its importance in model assessment.

In summary, Chapter 8 highlights the importance of employing a range of assessment criteria to comprehensively evaluate model performance. While various criteria offer insights into different aspects of model success, boundedness control emerges as a crucial criterion for assessing model usability directly, offering valuable insights into the reliability and applicability of mathematical models.

In Chapter 9, the effect of using different experimental design methods on mathematical modeling and optimization processes was investigated. The study focused on a composite shaft comprising a 16 layered symmetric/balanced glass-epoxy material with composite fiber angles chosen as design parameters. The objective was to maximize the critical torsional buckling load to be at least greater than 550 Nm. Five experimental design methods were compared: FFD, Taguchi, BBD, CCD, and D-Optimal.

The results revealed that CCD was unable to produce a successful model, while the Taguchi method resulted in a prediction error of approximately 17% compared to the actual critical torsional buckling load obtained using the analytical formula. Conversely, D-Optimal and FFD predicted the critical torsional buckling load with an error of roughly 3%. Interestingly, hybrid methods combining BBD with Taguchi or CCD yielded the closest results with an error of approximately 0.5% compared to the optimization result obtained using the analytical formula.

In this specific problem, D-Optimal, BBD, and hybrid methods emerged as successful alternatives to FFD. As a future study, it was suggested to further explore the D-Optimal method, particularly its performance in various problems compared to other experimental design methods, and to evaluate its advantages and disadvantages.

In Chapter 10, the usability of special functions as an alternative model type in mathematical modeling was investigated. Special function types such as BesselJ, ChebyShevT, Erf, ExpIntegralIE, Fresnel, Hermit, HyperGeometric, and Legendre were used for modeling purposes and compared with basic mathematical functions. The objective was to maximize the critical buckling load of a 16-layered symmetric balanced flax/epoxy composite plate based on selected design parameters.

The results indicated that basic function types were unsuccessful in producing a viable model, whereas special functions Erf and ExpIntegralE met the necessary assessment criteria. Therefore, it was suggested that further research could be conducted on using special functions in mathematical modeling.

Overall, the studies in the book highlighted the interconnected nature of experimental design, mathematical modeling, and optimization processes. The success of one process can significantly impact the others, emphasizing the importance of considering these stages as a whole in problem-solving endeavors.

Index

A

Abrasive water Jet machining, 5
Accuracygoal, 44, 45
Adaptive Neural Fuzzy Inference System
(ANFIS), 3, 8, 22, 142
Additive manufacturing, 1, 2, 5
Akaike information criterion (AIC), 214,
218, 219
Albedo, 4
Analysis of Variance (ANOVA), 4, 61, 62, 68
Ant Colony Optimization, 5
Artificial BEE COLONY, 5
Artificial intelligence, 8, 143
Artificial Neural Networks, 8, 22, 48, 57,
60, 142
Aspect ratio, 54, 305

B

Bacterial foraging optimization, 5
Balanced, 54, 232, 304, 325
Battery housing package, 5
Battery pack enclosure, 77, 78, 147, 229, 323
Bayesian information criterion (BIC), 212, 214,
218, 325
Bessel, 9, 218, 292, 293
BesselJ, 218, 292, 326
Blocking, 12, 13
Bootstrap, 9, 145, 147, 152,
Boundedness, 9, 219, 223, 233, 239, 310, 322, 324
Box-behnken design (BBD), 54, 60, 230, 306,
Build orientation, 68
Bukin function, 28, 30, 33, 38,

C

Central composite design (CCD), 1, 17, 74, 77,
143, 229, 231, 239, 321
Chebyshev, 293, 294
Classical lamination theory, 97, 212, 231,
Coefficient of load damping, 61
Cold-end temperature, 61
Composite shafts, 230, 232, 233, 239
Confidence interval, 76, 84
Conjugategradient, 44
Critical buckling load, 231, 233, 243, 305
Critical torsional buckling load, 231, 325
Crossover, 38
Crossprobability, 38, 40

Cross-validation, 150, 152, 200
Cuckoo search algorithm, 5

D

Decision Trees Method (DTM), 3, 22
Deflection, 75, 78, 94
Design Expert, 3
Design of experiment (DOE), 11, 12, 13
Design parameters, 3, 18, 54, 234, 244, 306
Design variables, 11, 12
Desirability approach, 2, 3, 6
Differential Evolution, 38, 40
Displacement, 1, 5, 74
Displacer amplitudes, 60, 61
D-Optimal, 1, 21, 229, 230, 239, 242, 286, 321,
Drilling process, 94, 229,
Driveshaft, 248

E

Eggholder test function, 28, 30, 33, 38
Electric arc furnace, 4, 48, 322
Electric vehicle battery modules, 77
Electrical discharge machining, 94
Energy storage systems, 95
Erf, 292, 294
Exact Global Optimization, 27
Experimental design, 93, 96
ExpIntegralE, 292, 295

F

Face central, 17
Factor, 17, 21
Fatigue life, 229
Feed rate, 2, 66, 68
Fiber angle, 54, 97, 212, 232,
FindMaximum, 27
FindMinimum, 27
Frequency gap, 96, 101, 107
Fresnel, 292, 296
Full Factorial Design (FFD), 1, 15, 321, 323
Fused deposition modeling, 66
Fuzzy Logic (FL), 3, 22, 142

G

Genetic Algorithm, 5, 6
Goodness of fit index, 213, 223, 233
Grey relational analysis, 9

H

Heat exchanger, 95
Hermite, 299
Holdout, 9
Hot-end temperature, 61
Hybrid composite, 212, 324
Hyperbolic tangent, 4, 48
HyperGeometric, 292, 299

I

Initialization, 38
InitialPoints, 30, 38, 40
Injection molding, 94, 229
Input, 2, 321, 323

K

Kling Gupta Efficiency (KGE), 4, 213

L

Laminated composite, 215, 232
Laser welding, 95
Latin Hypercube Sampling (LHS), 1, 77
Layer thickness, 66, 232
Legendre, 292, 301
LegendreP, 9, 309
Level, 15, 17
LevenbergMarquardt, 44
Linear optimization, 27

M

Mathematica, 24, 27
Mathematical modeling, 7, 11, 22, 292, 321
Matlab, 50, 54
MaxIterations, 44
Mean absolute error (MAE), 213, 218
Mean Absolute Percentage Error (MAPE), 213, 218, 220
Mean Squared Error (MSE), 213, 218, 220, 221
Mean Square Relative Error (MSRE), 213, 218, 220, 221
Model assessment criteria, 221, 324
Multi-layer armored system, 74, 323
Multiple nonlinear regression, 25
Mutation, 38

N

Natural frequency, 78, 96
Nelder Mead, 27, 41

Neuro Regression Method (NRM), 23
NMaximize, 27, 28
NMinimize, 27, 28
Non-Dominated Sorting Genetic
    Algorithm, 5
Normal distribution, 76, 77, 84
Numerical global optimization, 27
Numerical local optimization, 27

O

Objective function, 94, 98, 228
One Variable at a Time Design
    (OVAT), 15
Operating frequency, 60, 61
Orthogonal array, 1, 18, 93, 94
Output, 143, 152
Output power, 60, 61, 63
Optimization, 5, 7, 24, 27

P

Particle Swarm Optimization, 5
PenaltyFunction, 30, 38
Piston amplitudes, 60, 61
PLA, 66, 68
Polynomial model, 4, 62, 66
PostProcess, 30, 33, 38
PrecisionGoal, 44
PrincipalAxis, 44
P-value, 62

Q

Quasi-Newton, 30

R

R^2 adjusted, 213, 223, 325
R^2 test, 223, 325
R^2 train, 223, 325
R^2 validation, 223, 325
RamanujanTauTheta, 302
Random Search, 27, 30
Randomness, 12, 13
RandomSeed, 30, 33, 35, 38, 40, 43
Reflect Ratio, 43, 44
Regression, 22, 23
Replication, 12
Response Surface Method (RSM), 5, 8
Rieman Siegel, 303
Root Mean Square Error (RMSE), 213
Root Mean Square Percentage Error
    (RMSPE), 213

Rotatable, 17
R-squared, 213, 218, 223

S

SAS, 3
ScalingFactor, 38
Schwefel test function, 45
SearchPoints, 30, 33, 38
Second order polynomial model, 4, 25
Selection, 38
Shock loading, 5, 74
Shrink Ratio, 43
Simply supported, 305
Simulated Annealing, 27, 33
Special functions, 292
Spherical, 17
Spring stiffnesses of the displacer, 61
Spring stiffnesses of the piston, 61
SPSS, 5
Square error (SE), 27
Stability, 66, 69, 80, 322, 323
Standart error, 77
Statistical experimental design, 12, 13
Stochastic algorithms, 24
Stochastic methods, 24
Stochastic Neuro Regression Method (SNRM), 23

Stress, 54, 74
Support Vector Machines (SVM), 3

T

Taguchi, 18, 20, 93, 99
Taguchi Design, 18, 20, 93, 99
TIG welding, 4
Tolerance, 38, 41
Topsis, 6
Train brake disc, 94, 229
Transformed reduced stiffness matrix, 232

U

U2theil, 213

V

Version of Akaike Information Criterion (AIC), 213

W

Weight, 4, 5, 77
WEKA, 3
Wind turbine, 95
Wolfram Mathematica, 24

Printed in the United States
by Baker & Taylor Publisher Services